Design Decisions under Uncertainty with Limited Information

Structures and Infrastructures Series

ISSN 1747-7735

Book Series Editor:

Dan M. Frangopol

Professor of Civil Engineering and
Fazlur R. Khan Endowed Chair of Structural Engineering and Architecture
Department of Civil and Environmental Engineering
Center for Advanced Technology for Large Structural Systems (ATLSS Center)
Lehigh University
Bethlehem, PA, USA

Volume 7

Design Decisions under Uncertainty with Limited Information

Efstratios Nikolaidis,
Zissimos P. Mourelatos and
Vijitashwa Pandey

CRC Press
Taylor & Francis Group
Boca Raton London New York Leiden

CRC Press is an imprint of the
Taylor & Francis Group, an **informa** business

A BALKEMA BOOK

Colophon

Book Series Editor:
Dan M. Frangopol

Volume Authors:
Efstratios Nikolaidis, Zissimos P. Mourelatos and Vijitashwa Pandey

First issued in paperback 2017

Taylor & Francis is an imprint of the Taylor & Francis Group,
an informa business

© 2011 Taylor & Francis Group, London, UK

Typeset by MPS Ltd, a Macmillan Company, Chennai, India

British Library Cataloguing in Publication Data
A catalogue record for this book is available from the British Library

Library of Congress Cataloging-in-Publication Data

Nikolaidis, Efstratios.
  Design decisions under uncertainty with limited information /
Efstratios Nikolaidis, Zissimos P. Mourelatos, and Vijitashwa Pandey.
    p. cm. – (Structures and infrastructures series, ISSN 1747-7735; v. 7)

  Includes bibliographical references and glossary.
  ISBN 978-0-415-49247-8 (hbk) – ISBN 978-0-203-83498-5 (ebook)
  1. Engineering design. 2. Decision making. 3. Reliability
(Engineering) I. Mourelatos, Zissimos P.   II. Pandey, Vijitashwa.
III. Title. IV. Series.

  TA174.N55   2010
  620'.0042—dc22

                                                    2010052754

Published by: CRC Press/Balkema
                P.O. Box 447, 2300 AK Leiden, The Netherlands
                e-mail: Pub.NL@taylorandfrancis.com
                www.crcpress.com – www.taylorandfrancis.co.uk – www.balkema.nl

ISBN 13: 978-1-138-11509-5 (pbk)
ISBN 13: 978-0-415-49247-8 (hbk)
Structures and Infrastructures Series: ISSN 1747-7735
Volume 7

Table of Contents

Editorial   VII
About the Book Series Editor   IX
Preface   XI
About the Authors   XIII

1. **Design Decision under Uncertainty**   1

    1.1   Decision under Uncertainty   1
    1.2   The Role of Decision Analysis in Engineering Design   6
    1.3   Conclusion   12
    Questions   13

2. **Overview of Theories of Uncertainty and Tools for Modeling
    Uncertainty**   15

    2.1   Introduction: Management of Uncertainty in Design   15
    2.2   Theories of Uncertainty   16
    2.3   Conclusion   35
    Questions and Exercises   36

3. **Objective Probability**   41

    3.1   Probability and Random Variables for Modeling Uncertainty   42
    3.2   Common Probabilistic Models   82
    3.3   Probability Calculations   105
    3.4   Concluding Remarks   123
    Questions and Exercises   124

4. **Statistical Inference – Constructing Probabilistic Models from
    Observations**   133

    4.1   Introduction   133
    4.2   Estimating Mean Values of Random Variables and Probabilities 136
          of Events
    4.3   Statistical Hypothesis Testing   169
    4.4   Selecting Input Probability Distributions   177
    4.5   Modeling Dependent Variables   195

4.6   Conclusion of this Chapter   207
Questions and Exercises   208

5.   **Probabilistic Analysis of Dynamic Systems**   215

5.1   Introduction and Objectives   215
5.2   Modeling Random Processes in Time Domain   217
5.3   Fourier Analysis   227
5.4   Spectral Analysis   241
5.5   Calculation of the Response   264
5.6   Random Process Characterization using Simulation   279
5.7   Failure Analysis   292
5.8   Evaluation of the Stochastic Response of Nonlinear Systems   304
5.9   Concluding Remarks   304
Questions and Exercises   305

6.   **Subjective (Bayesian) Probability**   311

6.1   Introduction   311
6.2   Definition of Subjective Probability   313
6.3   Eliciting Expert's Judgments in Order to Construct Models of   326
      Uncertainty
6.4   Bayesian Analysis   365
6.5   Heuristics and Biases in Probability Judgments   388
6.6   Concluding Remarks   390
Questions and Exercises   392

7.   **Decision Analysis**   401

7.1   Introduction   401
7.2   Framing and Structuring Decisions   410
7.3   Solving Decision Problems   424
7.4   Performing Sensitivity Analysis   437
7.5   Modeling Preferences   451
7.6   Conclusion   478
Questions and Exercises   479

8.   **Multiattribute Considerations in Design**   485

8.1   Tradeoff Between Attributes   485
8.2   Different Multiattribute Formulations   490
8.3   Solving Decision Problems under Uncertainty using Multiattribute   499
      Utility Analysis
8.4   Conclusions   501
Questions and Exercises   501

Glossary   505
References   513
Structures and Infrastructures Series   523

Editorial

Welcome to the Book Series *Structures and Infrastructures*.

Our knowledge to model, analyze, design, maintain, manage and predict the life-cycle performance of structures and infrastructures is continually growing. However, the complexity of these systems continues to increase and an integrated approach is necessary to understand the effect of technological, environmental, economical, social and political interactions on the life-cycle performance of engineering structures and infrastructures. In order to accomplish this, methods have to be developed to systematically analyze structure and infrastructure systems, and models have to be formulated for evaluating and comparing the risks and benefits associated with various alternatives. We must maximize the life-cycle benefits of these systems to serve the needs of our society by selecting the best balance of the safety, economy and sustainability requirements despite imperfect information and knowledge.

In recognition of the need for such methods and models, the aim of this Book Series is to present research, developments, and applications written by experts on the most advanced technologies for analyzing, predicting and optimizing the performance of structures and infrastructures such as buildings, bridges, dams, underground construction, offshore platforms, pipelines, naval vessels, ocean structures, nuclear power plants, and also airplanes, aerospace and automotive structures.

The scope of this Book Series covers the entire spectrum of structures and infrastructures. Thus it includes, but is not restricted to, mathematical modeling, computer and experimental methods, practical applications in the areas of assessment and evaluation, construction and design for durability, decision making, deterioration modeling and aging, failure analysis, field testing, structural health monitoring, financial planning, inspection and diagnostics, life-cycle analysis and prediction, loads, maintenance strategies, management systems, nondestructive testing, optimization of maintenance and management, specifications and codes, structural safety and reliability, system analysis, time-dependent performance, rehabilitation, repair, replacement, reliability and risk management, service life prediction, strengthening and whole life costing.

This Book Series is intended for an audience of researchers, practitioners, and students world-wide with a background in civil, aerospace, mechanical, marine and automotive engineering, as well as people working in infrastructure maintenance, monitoring, management and cost analysis of structures and infrastructures. Some volumes are monographs defining the current state of the art and/or practice in the field, and some are textbooks to be used in undergraduate (mostly seniors), graduate and postgraduate courses. This Book Series is affiliated to *Structure and Infrastructure Engineering* (http://www.informaworld.com/sie), an international peer-reviewed journal which is included in the Science Citation Index.

It is now up to you, authors, editors, and readers, to make *Structures and Infrastructures* a success.

Dan M. Frangopol
Book Series Editor

About the Book Series Editor

Dr. Dan M. Frangopol is the first holder of the Fazlur R. Khan Endowed Chair of Structural Engineering and Architecture at Lehigh University, Bethlehem, Pennsylvania, USA, and a Professor in the Department of Civil and Environmental Engineering at Lehigh University. He is also an Emeritus Professor of Civil Engineering at the University of Colorado at Boulder, USA, where he taught for more than two decades (1983–2006). Before joining the University of Colorado, he worked for four years (1979–1983) in structural design with A. Lipski Consulting Engineers in Brussels, Belgium. In 1976, he received his doctorate in Applied Sciences from the University of Liège, Belgium, and holds two honorary doctorates (Doctor Honoris Causa) from the Technical University of Civil Engineering in Bucharest, Romania, and the University of Liège, Belgium. He is an Honorary Professor at Tongji University and a Visiting Chair Professor at the National Taiwan University of Science and Technology.

Dan Frangopol is a Distinguished Member of the American Society of Civil Engineers (ASCE), a Fellow of the American Concrete Institute (ACI), the International Association for Bridge and Structural Engineering (IABSE), and the International Society for Health Monitoring of Intelligent Infrastructures (ISHMII). He is also an Honorary Member of both the Romanian Academy of Technical Sciences and the Portuguese Association for Bridge Maintenance and Safety. He is the initiator and organizer of the Fazlur R. Khan Distinguished Lecture Series (www.lehigh.edu/frkseries) at Lehigh University.

Dan Frangopol is an experienced researcher and consultant to industry and government agencies, both nationally and abroad. His main areas of expertise are structural reliability, structural optimization, bridge engineering, and life-cycle analysis, design, maintenance, monitoring, and management of structures and infrastructures. His work has been funded by NSF, FHWA, NASA, ONR, WES, AFOSR and by numerous other agencies. He is the Founding President of the International Association for Bridge Maintenance and Safety (IABMAS, www.iabmas.org) and of the International Association for Life-Cycle Civil Engineering (IALCCE, www.ialcce.org), and Past Director of the Consortium on Advanced Life-Cycle Engineering for Sustainable Civil Environments (COALESCE). He is also the Chair of the Executive Board of the International Association for Structural Safety and Reliability (IASSAR, www.columbia.edu/cu/civileng/iassar), the Vice-President of the International Society for Health Monitoring of Intelligent Infrastructures (ISHMII, www.ishmii.org), and the founder and current chair of the ASCE Technical Council on Life-Cycle Performance, Safety, Reliability and Risk of Structural Systems (content.seinstitute.org/committees/strucsafety.html).

Dan Frangopol is the recipient of several prestigious awards including the 2008 IALCCE Senior Award, the 2007 ASCE Ernest Howard Award, the 2006 IABSE OPAC Award, the 2006 Elsevier Munro Prize, the 2006 T. Y. Lin Medal, the 2005 ASCE Nathan M. Newmark Medal, the 2004 Kajima Research Award, the 2003 ASCE Moisseiff Award, the 2002 JSPS Fellowship Award for Research in Japan, the 2001 ASCE J. James R. Croes Medal, the 2001 IASSAR Research Prize, the 1998 and 2004 ASCE State-of-the-Art of Civil Engineering Award, and the 1996 Distinguished Probabilistic Methods Educator Award of the Society of Automotive Engineers (SAE). He has given plenary keynote lectures in numerous major conferences held in Asia, Australia, Europe and North America.

Dan Frangopol is the Founding Editor-in-Chief of *Structure and Infrastructure Engineering* (Taylor & Francis, www.informaworld.com/sie) an international peer-reviewed journal, which is included in the Science Citation Index. This journal is dedicated to recent advances in maintenance, management, and life-cycle performance of a wide range of structures and infrastructures. He is the author or co-author of more than 270 books, book chapters, and refereed journal articles, and over 500 papers in conference proceedings. He is also the editor or co-editor of more than 30 books published by ASCE, Balkema, CIMNE, CRC Press, Elsevier, McGraw-Hill, Taylor & Francis, and Thomas Telford and an editorial board member of several international journals. Additionally, he has chaired and organized several national and international structural engineering conferences and workshops.

Dan Frangopol has supervised 34 Ph.D. and 50 M.Sc. students. Many of his former students are professors at major universities in the United States, Asia, Europe, and South America, and several are prominent in professional practice and research laboratories.

For additional information on Dan M. Frangopol's activities, please visit www.lehigh.edu/~dmf206/

Preface

We all make decisions that greatly affect both our personal and professional lives;

- A homeowner considers buying flood insurance.
- A patient evaluates alternative medical treatments in terms of their effectiveness, side effects and cost.
- A design-and-release-engineer in a car company tests three nominally identical doors of a new car model in crash and obtains satisfactory results. However, crash performance varies from door-to-door. The engineer must decide whether to delay launching in order to test more doors or skip the test and signoff the new design.

The quality of our decisions depends greatly on how we define and frame them, quantify our preferences and manage uncertainty. This book focuses on theories and tools to represent uncertainty using both data and expert judgment and the use of the resulting models to make informed choices when there is limited or no data. The presented concepts and methods are based on 300-years of research in probability theory and decision under uncertainty, and they are derived from few common-sense principles.

The book has three parts. Part 1 presents an overview of decision-making in engineering and business and explains the role of uncertainty in chapter 1. Chapter 2 reviews and compares alternative tools to represent uncertainty in a hierarchical fashion from the most general to specific. These theories are intervals, convex sets, objective and subjective probability, evidence theory and imprecise probability. It also describes how to make choices using these tools.

The second part focuses on objective probability (long term frequency). Chapter 3 presents the fundamental concepts of probability and methods to assess the performance and reliability of time invariant systems. Chapter 4 explains statistical methods for constructing probabilistic models from observed data. Chapter 5 focuses on dynamic systems.

Part 3 presents a structured approach to make decisions that are consistent with the decision maker's estimates of uncertainty and risk attitude. A main advantage of this approach is that it is derived from a few axioms (common sense principles). If one disputes this approach, then he has to explain which axiom is wrong.

In most practical decisions, uncertainty needs to be modeled using judgment. Chapter 6 defines the concept of subjective probability. It introduces a structured method to estimate a person's probability of an event, and update it when additional information (in the form of observations from an experiment or expert judgment) becomes available.

In addition, this chapter explains and demonstrates how to represent imprecision in these estimates.

Chapter 7 explains how to make risky decisions. It presents principles and tools for representing a decision-maker's attitude toward risk and for accounting for this attitude when making choices. Besides determining the best course of action among alternatives, the chapter presents a method to determine the value of additional information and shows how to investigate the effect of imprecision in probabilities on the optimum decision. Chapter 7 focuses on decisions in which a designer considers only one attribute. Chapter 8 extends the methods to multiple attribute decision making.

The presentation uses everyday language and real-life examples to elucidate concepts and methods for making engineering and business decisions in the presence of uncertainty.

The book audience includes:

1.   Advanced undergraduate and graduate students in mechanical, civil, industrial, aerospace and ocean engineering
2.   Students in business administration
3.   Managers and research and development engineers in the aerospace, automotive, civil, shipbuilding and power industries
4.   Researchers at universities and national labs

Readers will benefit from this book in the following ways:

1.   They will learn a structured approach for decision under uncertainty with limited information. In addition, they will learn what tools are available and understand their logical foundations.
2.   They will learn how to select the most suitable tool and apply it to a given engineering or business decision.
3.   They will understand how to improve the competitiveness of their organizations using a structured, risk-based approach for design decision making under uncertainty.
4.   They will also understand how to improve their personal and professional lives using the above approach.

Suggestions for instructors and students

Instructors can use this book in a senior or graduate level class on Probability and Statistics, Decision Analysis or Reliability Engineering. Chapters 1 and 3–5 are suitable for a class on Probability and Statistics. These chapters can also supplement a Reliability Engineering course. Instructors could cover chapters 1, 3, 4 and 6–8 in a Decision Analysis course. Chapter 2 on theories of uncertainty and the axiomatic definitions of subjective probability and utility in chapters 6 and 7 are suitable for a graduate class.

The book contains two types of examples. The first type consists of solved example problems that help the reader practice what he/she has already learned. We distinguish between the problem statement and the solution in these examples. The reader can try to solve these problems without looking at the solution. The second type presents a concept or method or explains their application in practice. There is no distinction between the problem statement and the solution for these examples.

About the Authors

Efstratios Nikolaidis is a Professor of Mechanical, Industrial and Manufacturing Engineering at the University of Toledo, Ohio. His research is on reliability analysis and optimization of aerospace, automotive and ocean structures, and on structural dynamics. He has published three books, three book chapters, and more than one hundred journal and conference papers, mostly on probabilistic methods, possibility, evidence theory and imprecise probability. Since the middle 1990's, he has focused on decision under uncertainty with limited information. Professor Nikolaidis and Dr. Pandey have developed an efficient experimental method for comparison of probabilistic and non probabilistic methods for decision under uncertainty with limited information.

Zissimos P. Mourelatos is a Professor of Mechanical Engineering at Oakland University in Rochester, Michigan. He conducts research in the areas of structural dynamics and reliability methods in engineering design. Before joining Oakland University, he spent 18 years at the General Motors Research and Development (GM R&D) Center. His research interests include design under uncertainty, structural reliability methods, reliability analysis with insufficient data, Reliability-Based Design Optimization (RBDO), vibrations and dynamics, and NVH (Noise, Vibration and Harshness). Professor Mourelatos has published five book chapters and over 120 journal and conference publications. He is the Editor-in-Chief of the *International Journal of Reliability and Safety*, an Associate Editor of the *ASME Journal of Mechanical Design and a SAE Fellow*.

Vijitashwa Pandey graduated with a PhD from the University of Illinois at Urbana-Champaign. He has also worked as a post-doctoral researcher with Dr. Mourelatos at Oakland University. His research revolves around decision based design, design optimization and uncertainty modeling. He is an active researcher in the field of mechanical engineering with peer reviewed publications in conferences and journals. He is a strong proponent of sustainability and interdisciplinary efforts in engineering design.

Chapter 1

Design Decision under Uncertainty

1.1 Decision under Uncertainty

A decision is an irrevocable allocation of resources to achieve a desirable payoff (Hazelrigg, 1996). It is irrevocable in the sense that reverting to the status quo will result in loss of time, money or something else tangible. It is an allocation of resources because a decision is a commitment to act and to invest resources such as money and manpower that could be invested elsewhere. A decision aims to achieve a desirable payoff for the decision-maker. The decision-maker must select that course of action among alternatives that will produce the best consequence. Below are some examples of decisions:

a)   A venture capitalist considers the proposal of an inventor to fund a project that will develop a new computer mouse. The inventor offers a fraction of the future revenue from the project to the venture capitalist in return for funding the project. The venture capitalist wants to decide whether to fund the project and what fraction of the revenue to demand.

b)   The mayor of a city threatened by a hurricane wants to decide quickly whether to order evacuation.

c)   The vice president of product development in an automotive company needs to decide whether to develop a hybrid engine for a new car model.

d)   An engineer considers two alternative designs of an automotive component. The engineer must decide whether to sign off one of these designs or develop a new design.

The above decisions are difficult and have significant consequences on the decision-maker and the society. A main challenge is that the decision-maker cannot readily select the best course action because he/she cannot predict the outcomes of uncertain events that affect the payoff of each alternative course of action. In the first example, the project revenue depends on the success of the project in developing a computer mouse, the performance of the product and on the demand for it. These outcomes are uncertain. To be successful, the decision-maker should be able to value each alternative course of action by considering its possible payoffs, the probabilities of the payoffs and the decision-maker's attitude toward risk.

To make good decisions, one must know all the important types of uncertainty and understand their characteristics in order to construct models of uncertainty that represent her beliefs. Decision-makers face different types of uncertainty, including uncertainty due to inherent randomness, lack of knowledge and human error.

Uncertainty is the state where a decision-maker cannot accurately predict the outcome of an event. For example, an investor who considers funding a risky project is uncertain about its success. Uncertainty is often categorized into aleatory (random) and epistemic. The first type is due to variability, which is an intrinsic property of natural phenomena or processes. We cannot reduce variability, unless we change the phenomenon or the process itself. For example, there is variability in the thickness of metal plates produced in an automotive supplier's factory. We cannot reduce variability by collecting data on plate thickness. We can only reduce variability by changing the manufacturing process, for example, by using better equipment or more skillful operators.

Epistemic uncertainty is due to lack of knowledge. As an example, the inventor in the first decision may not be able to predict the response of the computer mouse to input motions. This could be because the inventor uses an approximate model since the final mouse configuration is still undecided. Epistemic uncertainty is reducible; we can reduce it by collecting data or acquiring knowledge. The inventor can predict the behavior of the mouse better as the design matures. Another example could be that of an engineer facing uncertainty in predicting the stress in a structure due to deficiencies in the predictive models. Improving these models can reduce this uncertainty. Reducible uncertainty usually has a bias component. For example, crude finite element models of a structure tend to underestimate deflections and strains.

In order to manage uncertainty effectively, a decision-maker should know all possible outcomes of the uncertain events and assess the likelihood of these outcomes. For example, the venture capitalist in the first decision should assess how likely it is that the project will be successful in developing the computer mouse. Theories of uncertainty, including probability theory, evidence theory, and imprecise probability are available for quantifying uncertainty, but there is no consensus as to which theory is suitable in different situations. A main objective of this book is to help readers understand uncertainty and its causes, and the tools that are available for managing it.

1.1.1 *Good versus bad decisions*

When facing uncertainty, the decision-maker cannot determine the best course of action with certainty because the same course of action can have good or bad consequences depending on the outcome of an uncertain event. Consider that the city mayor in the above example orders evacuation after careful consideration of important facts such as weather forecasts, road conditions, and availability of transportation. This decision will result in considerable loss of revenue for the local businesses and the city, and can cause numerous traffic accidents. Yet, this does not mean that evacuating the city is a bad decision because evacuation will save many lives should the hurricane hit the city.

It is important to define the term "good decision." A good decision can have undesirable consequences and a bad decision can have desirable consequences. Suppose that after a decision-maker made a choice, he/she learned about the consequences. The decision-maker should not judge the decision only according to its consequences because these were unknowable at the time of the decision.

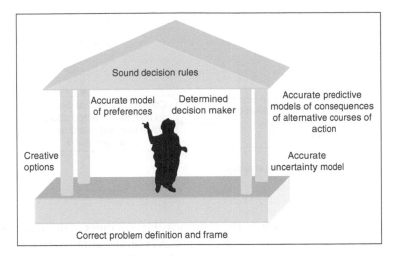

Figure 1.1 Ingredients of a good decision.

A good decision is one made after careful consideration of the problem, the decision-maker's objectives, preferences, the alternative courses of action and the available information about uncertainty. Looking back, the decision-maker should conclude that he/she would have made the same decision given what he/she knew at the time the decision was made.

Figure 1.1 shows the following important ingredients of a good decision (Howard),

a) A determined decision-maker willing to invest sufficient resources to define and frame the decision problem correctly, collect information and implement the decision
b) Accurate representation of the decision-maker's preferences
c) A comprehensive list of creative options (alternative courses of action)
d) A model that is representative of the decision-maker's beliefs about all important uncertainties
e) An accurate model for predicting the consequences of each course of action given the outcomes of the uncertain events
f) Sound decision rules based on right logic
g) Correct problem definition and frame

It is critical that a decision-maker define and solve the right problem and think carefully about what aspects of a decision to consider and what aspects to neglect. Many decision-makers rush through or skip this task and do not invest enough resources in it. Chapter 7 will present guidelines for defining and framing a decision problem. Items b) to d) stipulate that the decision-maker must collect and study evidence about the outcomes of the uncertain events and develop accurate predictive models. They also require that the decision-maker perform a careful introspection in order to understand

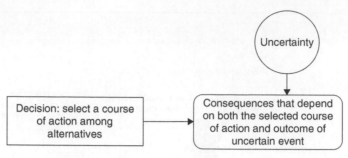

Figure 1.2 Decision Under Uncertainty.

his/her own preferences and beliefs about the possible outcomes of the uncertain events. Finally, rules about what option the decision-maker should choose given the decision-maker's preferences and beliefs are required. These rules must rely on common sense and they should not impose a decision on the decision-maker.

1.1.2 *Elements of a Decision*

Figure 1.2 depicts the elements of a typical decision under uncertainty and the way these elements influence each other. This representation of a decision is called "influence diagram." There are three elements in this diagram: 1) alternative courses of action (represented by rectangles), 2) uncertain events (represented by circles or ovals), and 3) consequences (represented by rounded rectangles). Arrows explain how one element influences others. For example, the two arrows originating from the rectangle and the circle and pointing toward the rounded box indicate that the consequences of a decision depend both on the selected course of action and on the outcome of the uncertain events. The decision-maker wants to select the course of action that will yield the most desirable consequence.

In another example, consider the vice president of product development in an automotive company who wants to select one of the following two programs: develop a hybrid-electric version of an existing car model (course of action A1), or develop a diesel version of same car model (course of action A2). The success of the car model depends primarily on its cost and the fuel prices when the new model will enter the market. The vice president is uncertain about both these factors. One reason is that he/she does not know to what extent the technologies for both projects will develop over the project duration. Another reason is that demand for both models depends on the gas price when the car will enter the market, and this price is uncertain.

Figure 1.3 shows an alternative way to represent the decision about the car program, called decision tree. The box on the left represents the decision to build a hybrid or a diesel version of the car. The four circles represent uncertainties (for example uncertainty in the price of gas). The branches originating from the circles represent the outcomes of the uncertain events (the price of gas and cost of the car when the car will enter the market). The consequences of each course of action are shown on the right of each branch.

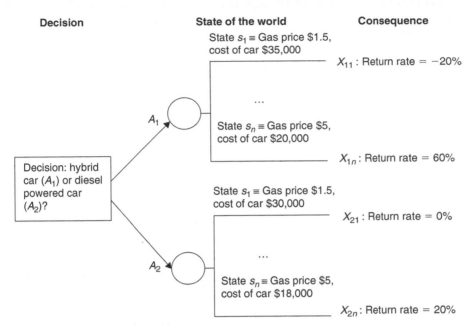

Figure 1.3 Example of a decision problem under uncertainty.

1.1.3 Limited Information

Uncertainty is the gap between the present state of knowledge and certainty. Information reduces this gap. Constructing models of uncertainty requires information. There is limited information about uncertainty in many important decisions because:

a) Most important decisions are made in the early stages of a project and the consequences depend on uncertain events that will be resolved only in the later stages or after the project completion. For example, the vice president in an automotive company should select the power train type and set performance targets at the early stages of the car program. At this stage, the vice president does not know if these targets are feasible and if they will generate high demand for the car, because she does not know how the design and the customer preferences will evolve.

b) Human actions affect the consequences of decisions and it is impractical to predict such actions. Investors, negotiators, and professors submitting technical proposals face such decisions frequently. For example, investors cannot predict how markets will react to a rise in the prime rate.

c) Many decisions involve development of new systems, novel materials, and geometries, about which there is scare or no data.

d) Constructing probabilistic models of uncertainty requires a very large amount of data. A complete probabilistic model of a set of random variables consists of the joint probability distribution of all the variables – not just their marginal probability distributions. For example, one needs to specify the joint probability distribution function of the Young's modulus and the yield stress of the steel plate of a car door because these properties are dependent. Estimation of the joint distribution requires a large amount of information, which is often unavailable. The common practice of assuming statistical independence for the variables is often erroneous and can result in serious errors such as severe overestimation of the reliability.

Unfortunately, often decision-makers invest little effort in quantifying uncertainty in most important decisions. They often make sweeping assumptions that are not supported by expert judgment or empirical evidence. For example, decision-makers often assume that random variables in a decision problem are normal and mutually independent. These assumptions simplify a problem but often result in poor decisions. For example, these assumptions can result in overestimation of the reliability of high-consequence systems (Ben Haim and Elishakoff, 1990).

One reason most people have difficulty modeling uncertainty when there is limited data is that they are not aware of or do not understand the notion of subjective probability. This is unfortunate because this is a very important tool for modeling uncertainty when making decisions and there is scarce empirical data. In addition, most people have difficulty eliciting subjective probabilities (O'Hagan et al. 2006).

There is an arsenal of theories and tools for modeling uncertainty, including subjective probability, imprecise probability, Dempster-Shafer evidence theory, and possibility theory. With the exception of probability, the literature on these tools focuses primarily on their mathematical development and on numerical tools for uncertainty propagation. Most publications do not explain the meaning of various measures of uncertainty and they do not present methods for estimating these measures in a way that is consistent with the definition of these measures (Cook, 2004). For example, few publications explain the meaning of the statement "the plausibility[1] of failure of a bridge is 0.5." Finally, very few applications of these theories in engineering design have been documented. The engineering community needs an integrated presentation of these tools that bridges the gap between theory and practice for methods for design under uncertainty.

1.2 The Role of Decision Analysis in Engineering Design

Engineering design is the process of applying the various techniques and scientific principles for defining a device or a system, which will satisfy a given need. A designer must define a system or process in sufficient detail to permit its realization (Norton, 1999). The design process consists of the following steps:

- Identification of the need
- Goal statement and specifications of product planning and performance attributes
- Ideation and invention of alternative concepts
- Analysis and evaluation of these concepts
- Selection of one or more promising concept

---

[1] Plausibility of a proposition is a measure of the degree to which the available evidence does not contradict the proposition in Depmster-Shafer evidence theory (see Chapter 2).

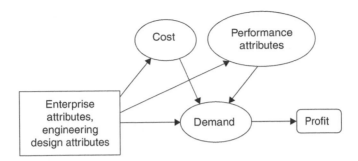

Figure 1.4 Decisions and uncertainties in product development and their impact on profit.

- Detailed design
- Prototyping, testing and sign off

Design is an iterative process. This process circulates throughout the above steps until it converges to a final design.

Engineering design involves a sequence of decisions from the product planning stage to the design sign off stage. These decisions involve significant uncertainty about the final product performance, cost, and demand. There is a growing recognition of the substantial role that decisions play in engineering design. Thurston (1991) proposed that the multiattribute decision analytic method should be used to assess design alternatives. Hazelrigg (1996) presented a simple framework in which he viewed product development as a decision with one objective: maximize the expected utility of the profit generated by the product. Product manufacturers have now understood that principles of decision making under uncertainty can help them increase profits by developing tools and processes for making good design decisions (Cafeo, 2005, Donndelinger, 2006). Many publications including the ASME *Journal of Mechanical Design, Research in Engineering Design, Mechanism and Machine Theory*, the *Journal of Engineering Design*, and a book on Decision Making in Engineering Design (Lewis et al. 2006) describe developments in this field.

Figure 1.4 presents an overview of the decisions involved in product development. First, a senior management group selects the product attributes. These attributes affect the product cost and demand, which in turn affect the manufacturer's profit. Product attributes belong to two broad categories: enterprise product planning attributes and performance attributes. Product planners select the first category of attributes, which include volume, price, warranty, and finance options. Engineering design attributes are quantifiable product properties that define the product configuration. For example, engineering attributes of a car include its wheelbase, and the main characteristics of the engine, transmission, and suspension. Engineering attributes influence the product performance attributes, such as comfort, handling, and acceleration of a car. Both enterprise and performance attributes influence demand that in turn influences profit. A main challenge is to set targets for the product attributes that are ambitious in order to stimulate demand, and yet affordable in order to keep cost low. In today's fiercely

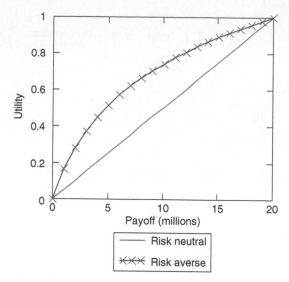

Figure 1.5 Utility functions of the payoff of risky a venture for two decision-makers with different attitude toward risk.

competitive environment, manufacturers must use the best available tools for making informed decisions about the product attributes.

A manufacturer can view product development as an optimization problem. A manufacturer seeks the optimum values of the enterprise attributes, X_{ent}, and engineering attributes, X_{eng}, to maximize profit,

Find X_{ent}, X_{eng}
to maximize Profit = Demand $(X_{perf}, \text{Price}) \cdot \text{Price} - \text{Cost}$
so that $g_{ent}(X_{ent}) \geq 0$
and $g_{eng}(X_{eng}) \geq 0$

(1.1)

Performance attributes X_{perf} are functions of the engineering attributes X_{eng} and $g_{ent}(X_{ent})$, $g_{eng}(X_{eng})$ are constraints on the enterprise and engineering attributes.

Product planning and development is a risky venture that requires a large investment and produces an uncertain payoff. Therefore, when a manufacturer makes decisions he/she should consider all possible scenarios and the profit corresponding to each scenario. In addition, the manufacturer should consider his/her attitude toward risk. In decision analysis, a utility function that measures the decision-maker's satisfaction with profit quantifies risk attitude. Figure 1.5 shows the shape of the utility function for a risk neutral, and a risk averse decision-maker. The utility function is a straight line for a risk neutral decision-maker. This indicates that the risky venture is equivalent to a safe investment with fixed payoff equal to the expected profit of the venture. However, most people are risk averse; their utility function is concave indicating that they prefer a sure payoff equal to the expected return of the risky venture to the opportunity of getting a higher payoff from this venture.

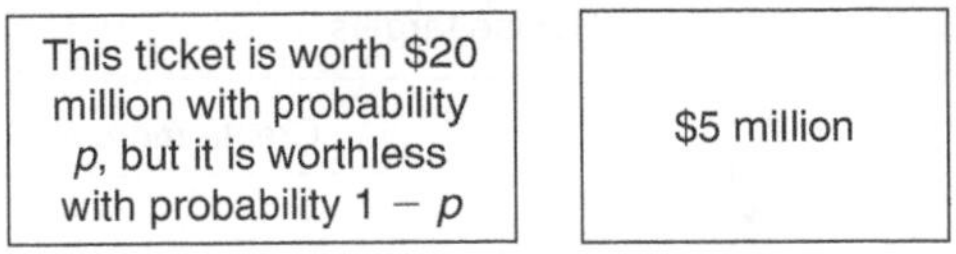

This ticket is worth $20 million with probability p, but it is worthless with probability $1 - p$	$5 million

Figure 1.6 If a decision-maker is indifferent between the lottery ticket on the left and the sure amount on the right, then the utility of $5 million is equal to p.

Often, we scale the utility function (without loss of generality when our satisfaction is monotonic in the payoff) so that the lowest possible payoff of a risky venture has utility zero and the highest one. Then, utility of a sure amount is equal to the probability p for which a decision-maker is indifferent between this amount and a gamble that pays the highest possible payoff with probability p and the lowest one with probability $1 - p$. For example, if the risk averse decision-maker is indifferent between a sure amount of $5 million and a gamble that pays off $20 million with probability $p = 0.5$ and zero otherwise, then the utility of $5 million is 0.5 (Figure 1.6).

Chapter 7 explains that expected utility is a suitable measure of the worth of the payoff of a risky venture to a designer. Therefore, ideally, the designer should seek the optimum values of the product attributes in order to maximize the expected utility,

$$\text{Find } X_{ent}, X_{eng} \qquad (1.2)$$

$$\text{to maximize } E[U(\text{Profit})]$$

where $U(.)$ is the utility and $E[.]$ the expected or mean value of a random variable. For most products, the relation between cost and product attributes is extremely complex. Therefore, designers do not have accurate models to calculate the terms in optimization problems (1.1) and (1.2). For example, it is impractical to develop an accurate model for predicting the effect of the sound pressure level at the driver's ear at 50 mph, the 0–60 mph acceleration time, or the maximum lateral acceleration on the cost of a car.

1.2.1 Sequential decisions in product development

The one-step decision based design formulation in Figure 1.4 is impractical for real-world systems. Design is an iterative process consisting of a sequence of tasks in which alternative design concepts are created, refined, evaluated, and screened out. Figure 1.7 illustrates this process on the design of a beach wheelchair. First, planners establish targets for the performance of the product. Table 1.1 shows the performance targets for the wheelchair. Then, designers create a set of alternative design concepts. In this step, creativity and ingenuity play an important role in coming up with promising design concepts. The performance attributes of the design alternatives are compared to the targets, and inferior design alternatives are screened out. Table 1.2 shows the results of the comparison of three design concepts for the beach wheelchair. The calculations shown in Table 1.2 show a simple 'weight and rate' approach in which each attribute has a relative weight which is multiplied by the score for that attribute. The final

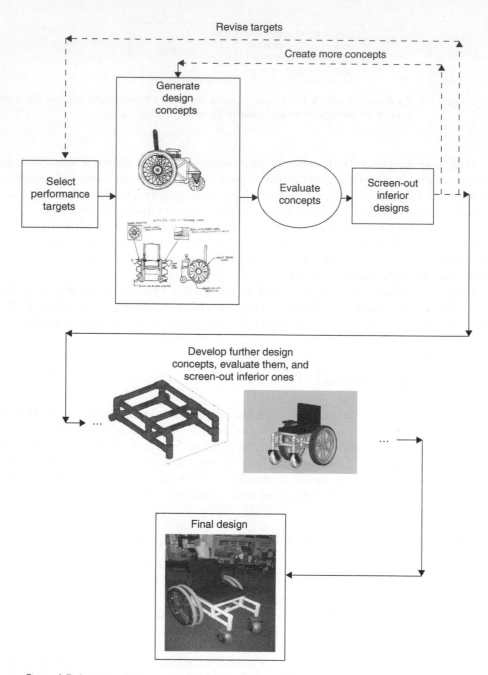

Figure 1.7 Iterative design consisting of a sequence of decisions.

Table 1.1 Performance targets for beach wheelchair.

Weight ≤ 30 pounds
Width less ≤ 26 inches.
Overall length ≤ 42 inches.
Displacement in sand ≤ 0.5 inch
Maximum safe weight of user = 300 lb
Height of center of gravity ≤ 30 in
Cost ≤ $1,350

Table 1.2 Evaluation of three design concepts for beach wheelchair.

Client Requirements	Importance	Concept 1	Concept 2	Concept 3
Mobility	7	10	8	8
Weight	9	1	3	8
Size	10	3	8	7
Physical Effort	7	10	4	4
Total Score	–	179	191	226

worth is just the sum of these values for each attribute[2]. A designer may eliminate design concept 1 based on the information in this table. If no satisfactory designs are found, designers return to the beginning of the process, create new concepts, and repeat the first step. Designers might also revise the performance targets if they realize that they are not achievable.

In the next step, information about the attributes of the system is used to define the attributes of the subsystems. For the wheelchair design example, the material properties and the topology of the frame are defined at this stage. Those designs concepts that passed the screening in the previous step are developed further and evaluated using refined analytical models and physical tests. The design process proceeds until one design is finally fully developed and approved.

In the initial design stages, there is significant uncertainty in the configuration and performance of the final design. Uncertainty is reduced as the design process evolves, because options are eliminated, the configuration of the final design is finalized, predictive models are refined and prototypes are built and tested.

Sometimes a set of alternative concepts are developed in parallel and the best design is selected downstream. This design approach is referred to as *set-based-design*. In this process, designers communicate in terms of sets. Imprecise information is used to identify and eliminate inferior alternatives (Reduc et al. 2006). Toyota has successfully employed this design approach and has demonstrated that a manufacturer can bring a

---

[2] A simplistic approach like this is usually acceptable for the initial stages of concept selection. Later in the book, we present rigorous approaches for evaluating and selecting among alternatives. More involved methods and the concomitant complexities are justified in the final stages of concept selection where only relatively good designs remain.

product faster to the market by postponing design decisions to the late stages of design (Sobek et al. 1999).

1.2.2 Challenges in design decision making under uncertainty and scope of this book

The following important challenges need to be addressed in order to develop an integrated decision based design methodology:

- Define the decision problem and make sure that you solve the right problem
- Frame the decision correctly
- Manage complexity
- Develop creative design concepts in the early design stages where there is significant uncertainty
- Identify the most promising concepts in the early design stages
- Model preferences of an organization
- Model uncertainty when there is limited information
- Evaluate and compare methods for modeling uncertainty and making decisions with limited information

Design is a risky venture consisting of a sequence of decisions with uncertain payoff. In each decision, the decision-maker considers alternative courses of action, selects one or more alternative design concepts for further development, revises his/her expectations about the product performance, collects more information about the performance of the alternative design concepts or cancels the venture. This book focuses on the following question: how should a designer, acting as a rational decision-maker, select the best course of action among a set of alternatives? For this purpose, the designer needs to assess the value of each alternative design concept and the value of additional information by considering both the uncertainty in the attributes of these design alternatives and his/her preferences. The book presents and demonstrates tools for making such decisions.

In principle, a decision-maker has only one objective: maximize profit. However, it is often impractical to predict profit as a function of the attributes of a design. Therefore, a decision-maker should determine the value of each design by considering different criteria about the performance of the design. Chapters 7 and 8 present a method for constructing a function that measures the performance of a design in terms of these criteria and returns a single value that depicts its merit.

The next chapter presents, compares some important theories for modeling uncertainty and explains what theories are most suitable for a given design problem. It discusses how a decision-maker should interpret different measures of uncertainty and presents operational definitions of these measures.

1.3 Conclusion

Design is a risky venture involving a sequence of decisions under uncertainty. Designers make important decisions, which greatly affect the success of a product, in the early stages of product development when uncertainty is large. A decision made under

Table 1.3 A perspective about how decision analysis tools can help designers make better decisions.

Can	Cannot
Provide the decision-maker with tools to elicit and quantify his/her beliefs about the likelihood of the outcomes of uncertain events	Determine the outcomes of the uncertain events
Help a decision-maker model his/her attitude toward risk	Tell the decision-maker whether he/she should or should not take risks
Assess the value of additional information	Tell the decision-maker what information to collect
Determine the best course of action given the likelihood of the outcomes of the uncertain events and the decision-maker's attitude toward risk	Determine that course of action that is guaranteed to maximize profit

uncertainty should not be judged based only on its payoff because the payoff depends on the outcomes of uncertain events. A good decision is one made after careful consideration of the problem, the decision-maker's objectives, preferences, the alternative courses of action and the available information about uncertainty. Looking back, the decision-maker should conclude that he/she would have made the same decision given what he/she knew at the time the decision was made.

Over the last 20 years, product manufacturers have understood that decision analysis tools can help them increase profits by making informed, rational choices under uncertainty. Decision analysis has a strong logical foundation and helps the decision-maker select the best course of action that is consistent with his/her beliefs about the likelihood of uncertain events that influence the payoff of a decision and his/her attitude toward risk. Table 1.3 explains how decision analysis tools can help designers and highlights some misconceptions about the capabilities of these tools. Thurston (2001) presented a comprehensive discussion of these limitations and capabilities.

Questions

1.1   Describe a decision that you made in the past. Describe the elements of this decision.

1.2   Describe the uncertainties involved in the decision in the above question. Explain how these uncertainties affected the outcome of this decision.

1.3   Which uncertainties in the above decision were random and which were epistemic?

1.4   Explain and comment on how you managed these uncertainties in the above questions.

1.5   Describe a bad decision with a good outcome and a good one with a bad outcome. Why do you think that the first decision was bad and the second good despite the outcome?

1.6 Describe three examples of random uncertainty and three examples of epistemic uncertainty.

1.7 Recruiters in Microsoft ask job candidates to make some challenging estimates during the interview. For example, recruiters ask candidates to estimate how many gas stations are in the US! Frequently, recruiters screen out those who fail to give a satisfactory estimate or justify it adequately (Thielen, 1999, pp. 25–33). Why do you think Microsoft recruiters seriously consider this type of questions in making hiring decisions?

1.8 Explain why most practical design decisions are broken down into a sequence of decisions.

1.9 Traditionally, designers take into account uncertainties using a worst-case scenario approach. Specifically, they represent the uncertainties with variables and use conservative *characteristic* or *design* values to account for the uncertainties. These values represent unfavorable scenarios that are unlikely to occur in real life.

For example in aircraft design, there are two main types of loads: maneuvering and gust loads. Characteristic maneuvering and gust loads are very high loads that occur rarely. Similarly, designers consider a characteristic value of the ultimate failure stress that is a small percentile of this stress. Then they make design decisions under the requirement that the airplane should be able to sustain these loads.

What are the strengths and weaknesses of this approach?

1.10 Design of a consumer product, such as a new car model, starts with setting performance targets for the most important attributes by examining competitors' cars and conducting customer surveys. For example, car manufacturers set targets for the sound pressure level at 70 mph, lateral acceleration on a test track, weight, and mpg. Then designers try to design the product to meet these targets.

What are the strengths and weaknesses of this approach?

1.11 Many people make everyday decisions and significant decisions, sometimes, considering only the most likely scenario. For example, a person takes an umbrella when leaving her house in the morning when the weather report predicts a higher than 50% chance of raining.

What are the strengths and weakness of this approach?

1.12 An author tries to sell you a book presenting a new decision making method. The author claims that if you follow this approach you will always enjoy the highest possible payoff. Comment on the author's claim.

Chapter 2

Overview of Theories of Uncertainty and Tools for Modeling Uncertainty

2.1 Introduction: Management of Uncertainty in Design

Design is a risky venture frequently involving a sequence of decisions under uncertainty (Chapter 1). The objective of a designer, who is also the decision-maker, is to select the product attributes in order to maximize the payoff of the venture. Designers need tools that will enable them to account for the uncertainty involved in design, and their own risk attitude.

The process of managing uncertainty consists of four steps: 1) *identify* the important uncertainties, 2) *model* uncertainties, 3) *infer* the payoffs of alternative courses of action, and 4) *choose* the best course of action. Figure 2.1 highlights these four steps in design decision-making under uncertainty. First, designers perform deterministic design in order to get a rough approximation of the final design. In this step, uncertainty is quantified using simple models such as design allowable values and safety factors. Designers identify important uncertainties by performing sensitivity analysis. For this purpose, they vary the uncertain variables in intervals, one at a time, and find the resulting intervals of the performance characteristics of the design. Tools, such as tornado diagrams, help designers identify the most important uncertainties (Howard, 1988). The second step is to model all the important sources of uncertainty. In the third step, a decision maker infers the uncertainty in the performance of each design alternative and the uncertainty in the payoff from building and selling each alternative design. The decision maker uses tools for propagating uncertainty through a system, such as probability calculus. Finally, the decision maker compares these alternatives and chooses the best one(s). This is an iterative procedure, as designers may want to create more alternatives, and obtain additional information from refined predictive models or tests.

Designers need tools for modeling and propagating uncertainty through a system, and for selecting the best design. This chapter presents an overview of the most important theories of uncertainty and the available tools for managing it. Each theory relies on a measure of the likelihood that an outcome or set of outcomes of an uncertain event will materialize. The presentation in this chapter focuses on the philosophy of each theory and the interpretation of the measure of likelihood. Tools for modeling uncertainty, making inferences and choices are presented. Readers will learn that each theory is suitable for a different decision problem, and the choice of the best tool depends of the type of uncertainties, and the amount and type of the available information.

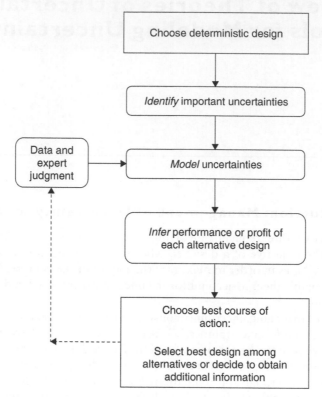

Figure 2.1 Four steps for managing uncertainty in design-decision making.

2.2 Theories of Uncertainty

Figure 2.2 summarizes all theories of uncertainty in this chapter and shows their relations. First, we explain two representations of uncertainty in which the designer specifies regions that contain all possible realizations of the uncertain variables: intervals and convex sets. These representations are suitable for problems with scarce information.

Then we present the objective and subjective interpretations of probability. Although most people are familiar with the former, this is only applicable to problems involving repeatable experiments. Subjective probability is more general than objective probability because it measures a decision maker's belief that a one-time event will occur.

Finally, we review theories of uncertainty that use bounds. These could be suitable for decisions in which a decision maker is ambiguous about the probability of an event. Imprecise probability is the most general theory of uncertainty. Probability, Dempster-Shafer Evidence Theory, p-boxes and Possibility are special cases of imprecise probability.

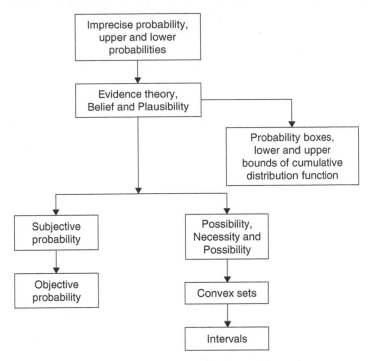

Figure 2.2 Families of the theories of uncertainty and the measures that they employ to quantify the likelihood of outcomes of uncertain events.

2.2.1   *Intervals*

The simplest model of uncertainty in a variable is an interval containing all of its possible realizations. For *n* variables, the extension of this model is a hypercube. Intervals are practical and popular because almost everyone can understand and use them to model uncertainty without making assumptions that are not supported by the available evidence.

Most people prefer estimating ranges of uncertain variables rather than probability distributions. There are well-established methods to propagate interval information through a system in order to estimate the range of its performance (Moore, 1979, Neumaier, 1990). Muhanna and Mullen (2001) and Mullen and Muhanna (1999a, b) developed interval-based finite element methods for structural analysis.

Interval calculations are easy and efficient when the relation between the input and output variables is monotonic. For example, the minimum and maximum stresses in a linear structure subjected to static loads correspond to the extreme values of the loads. Traditional structural design tools, such as design allowable values of strength, loads, and geometric imperfections are based of the concept of modeling uncertainty through intervals. These tools have been successfully used in design of aircraft, bridges, buildings, offshore platforms, and ships.

The premise of the interval approach for making choices is that each alternative is as good as its payoff under the worst-case scenario. This makes sense when there is limited information about the likelihood of the possible scenarios. At the same time, this conservative criterion may cause designers to reject alternatives that are very likely to produce a great payoff in almost all possible scenarios.

One criticism of an interval representation is that it does not provide information about the likelihood of a variable assuming a specific value. This leads to the following paradox. Consider two values that are close to one end of the interval, their location being such that one of the values lies inside the interval while the other one lies outside. The interval representation of uncertainty presumes that one value is possible, whereas the other is impossible to materialize regardless of how close the two values are, which is questionable.

Another disadvantage of the interval representation is that when a decision maker considers the most unfavorable combination of values of the uncertain variables in a decision it forces the decision maker to make decisions driven by very unlikely scenarios. For example, consider that the wind and earthquake loads on a building are modeled by intervals. It is too expensive to design a building that is expected to survive when the maximum values of both wind and earthquake loads occur simultaneously.

2.2.2 Convex Sets

Ben-Haim and Elishakoff (1990) introduced a representation of uncertainty based on convex sets. Convex sets in this context are regions in the space of random variables that contain all possible realizations of these variables. The philosophy of the convex set representation of uncertainty and decision-making is similar to that of intervals: "a course of action is as good as its payoff under the worst-case scenario, and this scenario occurs for some combination of the uncertain variables in a convex set." However, a convex set representation of uncertainty is less conservative than an interval representation because convex regions can be constructed to exclude highly unlikely realizations, which usually occur at a corner of a hypercube.

Ben-Haim (2001) developed information-gap models of uncertainty, which are convex sets whose size is controlled by an *uncertainty parameter*. This parameter measures the deviation of a state of the world from an expected state. One can control if highly unlikely events should be considered by adjusting this parameter. Ben-Haim's approach to decision-making is based on the following principle: "when there is a big information deficit, the decision maker should select the course of action that is more immune (in terms of having undesirable consequences) than all other alternatives to deviations of the state of the world from an expected state." The deviation is measured by the uncertainty parameter.

2.2.3 Objective Probability

Objective probability measures the likelihood of an outcome by its long-term relative frequency. Most people think of probability in this fashion. However, this view of probability is too narrow because it can describe uncertainty only in few practical problems.

Objective probability of the outcome is its long-term relative frequency (Table 2.1a). This definition assumes that a decision maker could perform a repeatable experiment[1],

---

[1] The term "repeatable" refers to the experiment, not its outcome.

Table 2.1a Overview of objective probability.

Measure of Likelihood	Interpretation	Underlying Assumptions	Strengths and Weaknesses
Probability	Long-term relative frequency in an infinitely repeatable experiment.	An infinitely repeatable experiment can be conducted.	**Strengths** • Most people easily understand the concept of relative frequency. • Standard procedures to measure and validate relative frequency are available. **Weakness** The theory has limited applicability.

while keeping all relevant conditions identical, in order to observe how frequently the outcome occurs. Flipping a coin and rolling a die are repeatable experiments. Objective probability is a property of the system observed, not of a particular decision maker. In principle, a decision maker can check an estimate of the objective probability by repeating the experiment many times until the relative frequency converges.

Most people think of probability as long-term relative frequency. Moreover, there is a well-developed framework for estimating probabilities from observations, as well as confidence intervals for these probabilities. In some engineering problems, a modest number of observations (e.g., few hundred) are sufficient in order to model uncertainty. The reason is that mathematicians have already derived from first principles the types of the probability distributions of some quantities, such as loads and strength. For example, extreme wave or wind loads often follow an Extreme Value distribution (see Ang and Tang, 1984, and section 3.2 of this book).

The theory has severe limitations though (French, 1986). There is no satisfactory definition of the term "infinitely repeatable experiment." Conditions that affect the outcome of an experiment change in each repetition, howsoever slightly. For example, each time an analyst performs an experiment in order to estimate some quantity, the problem changes, albeit slightly, and this affects the accuracy of the estimate. Even the flip of a fair coin is not an infinitely repeatable experiment because the initial position and velocity of the coin change in each flip. Strictly speaking, even the shape of the coin changes slightly during the experiments.

A second limitation of objective probability is that the concept of an infinitely repeatable experiment is meaningless for one-time events. Examples of such events are:

a) A given presidential election
b) The bankruptcy of a particular financial company within a given period
c) The S&P 500 index closing below 2,000 points on December 31, 2020

Events for which it seems that massive data exist from a repeated experiment are actually one-time events. Consider an analyst who calculated the maximum vibratory displacement on the door of a car by using a commercial code. The analyst wants to estimate the probability that the error in the displacement exceeds 15% of the estimated value. Although, usually there is data from previous applications of the same computer code, these are not results of a repeated experiment because each application of the code is unique. For example, the analyst, the structure, the applied loads, and the materials change in each application of the code.

It is clear that objective probability while immensely useful, suffers from the serious drawback that it cannot model uncertainty when experiments cannot be performed repeatedly. This drawback has led many researchers to consider subjective probability, which is presented later in this chapter.

2.2.3.1 Tools for managing uncertainty

Formulation of a Probabilistic Design Problem

In probabilistic design, a designer seeks the best configuration to maximize the average utility $U(\mathbf{X}, \mathbf{P})$ of a design. Utility depends on input variables that can be categorized into decision or design variables $\mathbf{X}$, and random variables $\mathbf{P}$. Designers can only change the design variables, which are considered deterministic here. In a design optimization problem, a designer seeks the optimum values of the design variables in order to maximize the expected utility,

Find $\mathbf{X}$ (2.1a)

To maximize $E[U(\mathbf{X}, \mathbf{P})]$

so that $\mathbf{x}_L \leq \mathbf{X} \leq \mathbf{x}_U$

In the above formulation, the design variables are constrained to vary between lower and upper bounds $\mathbf{x}_L$ and $\mathbf{x}_U$. A bold letter indicates a vector, an upper case letter indicates a variable, and a lower case letter indicates a realization of the variable. In subsequent formulations of the design problem, the side constraints on the design variables in the third line of Equation (2.1a) will be omitted.

Often, it is difficult to construct the utility function for a design. Therefore, alternative formulations of the above design problem are considered. A designer could minimize the average value of an attribute, $L(\mathbf{X}, \mathbf{P})$ such as the cost or weight, and impose constraints on other attributes. Failure of some systems is catastrophic, and the boundary between success and failure is crisp (for example, the failure of an aircraft wing due to aeroelastic divergence or the collapse of a bridge.) In this case, it is common practice to seek to minimize one attribute (for example the weight or cost) and satisfy minimum acceptable requirements for the remaining attributes. The system is idealized so that it fails under a finite number of failure modes, and the probability of system failure is calculated from the probabilities of the modes. The probability of system failure is required to be acceptably low. Let $I(\mathbf{X}, \mathbf{P})$ be the failure indicator function of the system, which is defined to be one if the system fails and zero if it survives. The design problem in Equation (2.1a) reduces to a system Reliability-Based Design Optimization (RBDO) problem that seeks the most efficient design whose system probability of failure (violation of the minimum requirements for the remaining attributes) does not exceed an allowable value. The formulation of the system RBDO problem is as follows,

Find $\mathbf{X}$ (2.1b)

To minimize $E[L(\mathbf{X}, \mathbf{P})]$

so that $p_{sys} = P[I(\mathbf{X}, \mathbf{P}) = 1] \leq p_{max}$

where p_{sys} is the system probability of failure. Often, the minimum expected value of the loss function $E[L(\mathbf{X}, \mathbf{P})]$ is approximated by the value of this loss function for the average values of the design variables and parameters $l[E(\mathbf{X}), E(\mathbf{P})]$.

In order to formulate one of the above problems, a designer completes the following tasks:

1. Model Uncertainty: Estimate the probability distribution of the random variables, $\mathbf{P}$.
2. Infer the Performance of Alternative Designs: Determine the probability distributions of the utility or loss functions by propagating the uncertainties through the system.
3. Choose the Best Design among Alternatives: Find the best design, that is, the design with the highest expected utility (formulation 2.1a) or lowest loss function (formulation 2.1b).

The following paragraphs highlight available tools for three steps of a design decision: modeling uncertainty, propagating uncertainty to infer the performance characteristics of alternative designs, and choosing the best design. Generally, modeling uncertainty is the most challenging task because, in many practical design decisions, there is insufficient information to model uncertainty. In addition, the computational cost for propagating uncertainty can be very high.

Modeling uncertainty
Decision makers have access to a rich collection of statistical tools for constructing models of the input variables to obtain meaningful estimates of the performance of a design. Many methods and software for estimating statistical summaries such as probabilities, mean values, standard deviations, and histograms from observed data (Fox, 2005) are available. Catalogs of standard probability distributions describing input variables encountered in practice, such as time to failure of a device, loads and strength of structures, waiting time in a queue, have also been published (Bury, 1999, Johnson et al. 1972, 1994). Both univariate and multivariate probability distributions are established from these summaries by fitting candidate distributions to the data and performing statistical tests to eliminate the ones that poorly fit the data.

Inferring the performance of a design by propagating uncertainty through it
In principle, a decision maker can derive the probability distribution of the performance variable of a system, such as its reliability, by using probability calculus (Chapter 3). However, this approach is too difficult and computationally expensive for problems involving more than about three random variables. Numerical methods,

such as Monte-Carlo simulation, second moment methods, dimension reduction, and stochastic finite elements are used in many applications. However, each of the above methods is suitable for a different set of problems. Moreover, the computational cost of propagating uncertainty, while considerably reduced, is still high for many practical problems. Development of computationally efficient tools to reduce this cost is an active research area.

Choosing among alternative courses of action

A decision maker can measure the worth of the payoff on a course of action by using expected utility. The utility can be a univariate or multivariate function of the performance attributes of a design. Utility theory has a well-established foundation (Clemen, 1997). However, often, designers think that it is impractical to work with utility functions. Instead, they compare alternative designs in terms of a single performance attribute such as cost or weight, while imposing minimum acceptable requirements on the remaining performance attributes, such as the reliability.

Chapters 3 and 4 (Probability Theory and Statistical Inference respectively) explain tools for constructing models of uncertainty and propagating uncertainty through a system. Chapter 7 (Decision Analysis) explains a method for making choices among alternatives with uncertain payoff.

2.2.4 *Subjective Probability*

The theory of subjective probability addresses the above limitations of objective probability. Unlike objective probability, subjective probability is a property of a person; not of the system of interest. Subjective probability of an outcome measures a person's belief that the outcome will occur. A fundamental assumption is that a person is always decisive when comparing alternative ventures whose payoff depends on the outcome of an uncertain event. Objective probability is a special case of subjective probability, and the latter converges to the objective probability of an event with the amount of data available.

Suppose that a facilitator wants to measure an expert's belief about the outcome. For this purpose, the facilitator observes the expert's willingness to bet on this outcome in a gamble. The stronger the expert believes that the outcome will occur, the more money he/she is inclined to bet on it. Consider the lottery ticket in Figure 2.3. This ticket is worth $1 if an outcome occurs and zero otherwise. The expert's probability of the outcome is the highest buying price for the ticket. For example, if the facilitator wants to elicit an expert's probability that a particular bank will be bankrupt within the next two years (outcome *A*), the facilitator asks the expert for his/her highest buying price of a ticket that will pay $1 if the bank goes bankrupt and zero otherwise. Jeffrey (2002) presents a thorough and readable review of subjective probability.

> This ticket is worth $1 if
> outcome *A* occurs;
> otherwise it is worthless.

Figure 2.3 Lottery ticket used to define subjective probability and imprecise probability.

Table 2.1b presents the subjective view of probability. Subjective probability theory presumes that a person is always decisive when comparing a sure amount with a lottery ticket; that is the person may prefer the lottery ticket to the sure amount or vice versa, or he/she be indifferent between the two. Indifference between the ticket and the sure amount means that the person would happily agree to exchange one for the other. An important implication of this assumption is that the highest buying price of the ticket is always equal to its lower selling price.

Figure 2.4 shows a procedure for eliciting an expert's subjective probability of some outcome. The expert is risk neutral. This means that the worth of a risky venture to the expert is equal to the mean value of the monetary return. First, a facilitator asks the expert to compare a trivial amount (e.g. a few cents) to the ticket. The expert should prefer the ticket to the sure amount (which means he/she would buy the ticket for this amount) because its price is trivial. The facilitator increases the amount incrementally and repeats the question. If the amount increases in small increments then it will reach a limit p for which the expert will become indifferent between the ticket and the sure amount. This means that the expert's probability of the event is equal to p.

In order to check this estimate of the expert's probability, the facilitator should repeat the comparison by reducing the amount from \$1 in small increments. If the expert is decisive when comparing options, the expert ought to be indifferent between the ticket and a sure amount that is also equal to p. That means that the maximum selling price and the minimum buying price of the ticket are both equal to p and this amount is equal to the expert's probability of outcome A.

Table 2.1b Overview of subjective probability.

Measure of Likelihood	Interpretation	Underlying Assumptions	Strengths and Weaknesses
Probability	A person's fair price for a lottery ticket that pays \$1 if the outcome occurs and zero otherwise. Fair price is the sure amount for which the person is indifferent between the ticket and the sure amount, i.e., he/she would happily agree to exchange the lottery ticket for the sure amount and vice-versa.	• The decision maker is decisive. If he/she is given the choice between the sure amount and the lottery ticket, the decision maker will prefer the ticket below a limit, and the sure amount above this limit. The decision maker is indifferent between the ticket and the sure amount exactly at the limit. • The definition presented here assumes that the decision maker is risk-neutral in the given range of outcomes for which the probability is assessed.	Strengths • Theory is applicable to all decision problems under uncertainty. • Subjective probability is a general case of objective probability. Weaknesses • It is challenging to understand the concept of subjective probability and to elicit this probability (O'Hagan and Oakley, 2004). • The definition of risk neutral decision-maker is cyclical; it uses the concept of subjective probability.

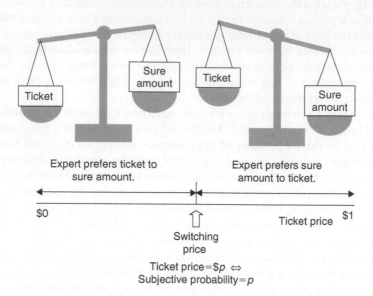

Figure 2.4 Comparing the lottery ticket in Figure 2.3 to sure amount in order to elicit an expert's subjective probability.

People frequently make decisions that are similar to the above example by buying or selling futures. There are a number of web-based event (or prediction) markets that provide a range of trading and gambling services, and pseudo markets (in which the participants trade virtual currency) such as Newsfutures.com and Ideosphere.com. Buying or selling prices in these markets reveal the aggregate belief of the players that some outcome of an important event will occur (for example a particular candidate will win a presidential election or a company's CEO will resign by a certain date.)

There is considerable empirical evidence that the probability estimates reflected by the prices of futures are accurate (Wolfers and Zitzewitz, 2004). Table 2.2 shows the average buying and selling prices of Standard and Poor's Future Price Securities on a gambling service. Prices in the second column are for tickets (futures) that pay $100 if the Standard and Poor 500 index finishes within certain ranges shown in the first column by the end of 2003. These prices reveal the players' aggregate probabilities for each range. For example, the market thinks that the index will close between 900 and 999 in the end of 2003 with probability 0.275. The table also compares these prices to actual settlement prices of options in the third column. In general, actual settlement prices in December 2003 agree reasonably well with future prices in July 2003. However, players overpaid and/or asked for too much for tickets whose payoffs depended on rare outcomes such as that S&P 500 index will close above 1200 or below 700. This means that players accurately estimated the probabilities of most outcomes in the table except for those corresponding to extreme deviations of the index that the players overestimated.

Subjective probability calculations must follow the same rules as objective probabilities. Probabilities that satisfy the above rules are internally consistent or *coherent*.

Table 2.2 Average prices of Standard and Poor futures vs. actual prices from the Chicago Mercantile exchange.

S&P 500 level on December 31, 2003	Average buying and selling prices	Actual settlement prices in December 2003
1200 and over	4	2.5
1100 to 1199	13.5	13.2
1000 to 1099	30.5	33.3
900 to 999	27.5	30.5
800 to 899	16.5	13
700 to 799	5.5	5
600 to 699	5.5	2
Under 600	6.5	1
S&P 500 on July 23, 2003:	985	

Table 2.3 Decision maker losses $0.80 for sure by betting $0.10 against both events "head up" and "tail up". The amounts shown are the profits and losses for all combinations bets and outcomes.

Bet/Outcome	Head up	Tail up
Against Head Up	($0.90)	$0.10
Against Tail Up	$0.10	($0.90)
Net gain (loss)	($0.80)	($0.80)

Coherence guarantees that a person uses all the available information to estimate these probabilities and avoids a gamble that leads to sure loss (Dutch book). For example, coherence requires that the probability of the union of disjoint outcomes must be equal to the sum of their probabilities. A person who violates this rule will accept gambles that lead to sure loss. Consider that an expert's probabilities for outcomes "head up" and "tail up" in a coin flip are both 0.1. These are incoherent probabilities because one of these outcomes will occur for sure and because the sum of the probabilities of these outcomes is not equal to one. This means that the expert will agree to bet $0.10 against both outcomes "head up" and "tail up" simultaneously (the third outcome, which does not exist). Betting $0.10 against the outcome "head up" is like selling insurance, that is, for an initial payment of $0.10, the expert assumes the responsibility to pay $1 if this outcome occurs. By placing these bets, the expert will lose a total $0.80 for sure as explained in Table 2.3. The expert can avoid sure loss by adjusting the probabilities of the two outcomes so that they add up to one.

In general, a facilitator cannot validate an expert's subjective probability. However, the facilitator could elicit the probabilities of different outcomes of an uncertain event and check if these probabilities cohere. If they do not, then the facilitator should explain this to the expert and ask her to revise these probabilities.

Objective probability is a special case of subjective probability. If a decision maker knows the long term frequency of an event, then his fair price of the ticket in Figure 2.3 should be equal to the relative frequency of this event. This means that the decision maker should agree to buy or sell the ticket for a price equal to the relative frequency of the event.

Suppose that first, an expert estimates his/her subjective probability about an outcome by using the comparison in Figure 2.4. The expert relies only on judgment because he/she has no data. If he/she obtains data from a repeatable experiment later, she/he can update her/his probability in view of the data using Bayes' rule (Chapters 3 and 6). Then the updated subjective probability should converge to the objective probability (long-term relative frequency) with increasing number of experiments.

2.2.4.1 Tools for managing uncertainty

The same tools for propagating uncertainty and making choices described in the section on objective probability are also applicable to subjective probability. However, different tools are suitable for modeling uncertainty. When little or no data is available, probabilities are elicited from experts. These subjective probabilities can be updated later if data from observations become available. Often elicitation involves a facilitator and an expert; the facilitator elicits the expert's probabilities or other statistical summaries, such as percentiles, by probing the expert's inclination to bet on outcomes of uncertain events. Chapter 6 explains procedures for elicitation and updating of subjective probabilities.

2.2.5 Imprecise Probability

Imprecise probability measures a person's belief that an outcome will occur by the person's inclination to bet on or against this outcome, like subjective probability. However, the theory allows for indecision when comparing risky ventures. This is the most general theory of uncertainty presented; all other theories in this chapter are special cases of imprecise probability.

In many decisions, experts cannot pinpoint the exact values of the probabilities of the relevant outcomes. One reason is that instead of a single probability value, a range of values could represent an expert's belief about the likelihood of an outcome. Most people's behavior violates the precepts of subjective probability in this case. Ellsberg (1961) demonstrated this by performing the following experiment. He asked subjects to pick up a ball from a bag that contained 90 red, blue, and yellow balls, without looking at the balls. He told the subjects that 30 balls were red and the rest blue and yellow, but did not specify the proportion of the latter balls. Then he asked each subject to select one of the following gambles:

Ticket A: Win \$1 if the ball chosen is red.

Ticket B: Win \$1 if the ball is blue.

Most subjects selected ticket A. Then, Ellsberg asked the subjects to select one of the following gambles:

Ticket C: Win \$1 if the ball is red or yellow.

Ticket D: Win \$1 if the ball is blue or yellow.

Most subjects preferred ticket *A* to *B* and ticket *D* to *C*. These two choices can be shown to be inconsistent as follows. If a subject prefers ticket *A* to *B*, then

$$P(red) > P(blue) \qquad (2.2)$$

where $P(red)$ and $P(blue)$ are the probabilities of outcomes "the ball chosen is red" and "the ball chosen is blue" respectively. Similarly, because the subject prefers ticket *D* to *C*,

$$P(blue \cup yellow) > P(red \cup yellow) \qquad (2.3)$$

where symbol "$\cup$" denotes the union. Now the second inequality implies that $P(blue) > P(red)$ because the probability of disjoint events equals the sum of the probabilities of these events. This latter inequality contradicts inequality (2.2). Therefore, the two choices of the subjects in Ellsberg's experiment are inconsistent.

In order to explain such inconsistencies, Ellsberg (1961), Knight (1921) and others argue that experts distinguish between risk (uncertainty in the outcomes of an experiment when the probabilities are known) and ambiguity (unknown probabilities), and are averse to ambiguity, just as they are averse to risk. Because of ambiguity aversion, people value a risky venture by using a more conservative estimate than its average payoff. Also, people increase the probability of an unfavorable outcome when they are ambiguous about its probability. In Ellsberg's experiment, subjects underestimate the probabilities of picking up a blue ball and picking up a red or yellow ball because they do not know the proportions of blue and yellow balls and they do not like betting on outcomes with ambiguous probabilities. Economists have demonstrated that ambiguity aversion explains the way markets behave better than subjective probability (Bossaerts, and Plott, 2004, Bossaerts, 2007).

Usually, when an expert is ambiguous about the probability of an outcome, the expert's minimum selling price of a lottery ticket, such as the one in Figure 2.3, exceeds its maximum buying price. The spread between the two prices increases with the expert's ambiguity about the probability of the outcome. For example, there is a spread between buying and selling prices of futures that reflects the lack of information about the likelihood of the relevant outcome.

Table 2.4 is an expanded version of Table 2.2, showing the maximum buying prices and minimum selling prices of Standard and Poor futures. A portion of the spread between buying and selling prices reflects the players' ambiguity about the probability distribution of the closing level of S&P 500 in the end of 2003. For example, according to the values in the second and third columns, the market believes that S&P 500 index could close in the end of 2003 between 900 and 999 with a probability as low as 0.25 and as high as 0.3. The spread is large for bets on rare outcomes such as the S&P 500 closing above 1200. This is reasonable because players are very uncertain about the probabilities of rare outcomes.

Consider again the facilitator who elicits an expert's probability of outcome *A*. The expert is highly uncertain about this outcome. First, the facilitator asks the expert to compare a small amount (e.g. a few cents) to the ticket in Figure 2.3. If the expert prefers the ticket, then the facilitator increases the amount incrementally and repeats

Table 2.4 Prices of Standard and Poor futures vs. actual prices from the Chicago Mercantile exchange.

S&P 500 level on December 31, 2003	Buying price	Selling price	Actual settlement prices in December 2003
1200 and over	2	6	2.5
1100 to 1199	11	16	13.2
1000 to 1099	28	33	33.3
900 to 999	25	30	30.5
800 to 899	14	19	13
700 to 799	3	8	5
600 to 699	4	7	2
Under 600	5	8	1
S&P 500 on July 23, 2003		985	

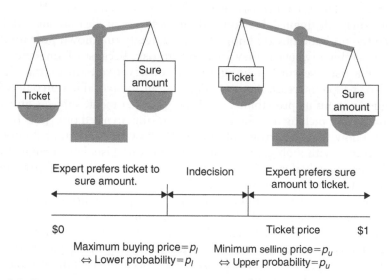

Figure 2.5 Elicitation of lower and upper probabilities.

the question. As Figure 2.5 shows, if the amount increases in small increments it will reach a limit $\$p_l$ above which the expert will be unable to decide which option is better. This is the expert's low probability of outcome A. The expert will pay up to $\$p_l$ for the ticket but he/she will not sell the ticket for a price higher than $\$p_l$, because the expert is ambiguous about the probability of outcome A and because he/she does not want to assume a high risk of paying \$1 if A occurs. Suppose that the facilitator keeps increasing the sure amount. Eventually, the sure amount will become high enough (equal to $\$p_u$ in Figure 2.5) so that the expert will decide that the sure amount is better than the ticket. This is the upper probability of A.

Imprecise probability refers to a family of theories that use both a lower and an upper estimate for the likelihood of an outcome. These are the maximum buying price and the minimum selling price of the ticket in Figure 2.3. Figure 2.2 shows this family of theories and the measures of likelihood that they employ. These include Dempster-Shaffer evidence theory, possibility theory, and subjective and objective probabilities. Imprecise probability differentiates between random uncertainty (uncertainty that is inherent in natural phenomena and processes) and epistemic uncertainty (uncertainty due to lack of knowledge). The gap between lower and upper probabilities reflects the latter uncertainty. Moreover, imprecise probability is more consistent than subjective probability with the way people reason and make decisions in the face of uncertainty. Aughenbaugh, and Paredis, (2006), and Pandey and Nikolaidis (2008) demonstrated that decision makers who use imprecise probability are protected from large losses because they are less inclined to gamble than those who use subjective probability when they are ambiguous about some probabilities. However, in many decisions, these theories cannot identify the best alternative course of action.

Table 2.5 summarizes the properties of imprecise probability. Imprecise probability follows different rules than subjective probability. Like subjective probabilities, imprecise probabilities must avoid sure loss and cohere. However, coherence does not guarantee avoidance of sure loss. The requirement for coherence implies that all available information is used to estimate the probabilities of outcomes. For example, coherence implies that the lower probability of the union of two disjoint outcomes must be greater than or equal to the sum of the lower probabilities of these outcomes.

In order to avoid sure loss a person should not pay too much for lottery tickets. For example, sure loss occurs when the sum of a person's lower probabilities of an outcome and its complement exceeds one. The reason is that the person will pay more than $1 for two lottery tickets that together are worth $1 for sure.

2.2.5.1 Tools for managing uncertainty

Modeling uncertainty
A facilitator can elicit an expert's lower and upper probabilities in a similar way as in subjective probability. The facilitator asks the expert for the maximum buying and minimum selling prices of a ticket such as the one in Figure 2.3 in order to estimate the lower and upper probabilities of outcome A. However, unlike in subjective probability, the facilitator does not require that the two prices be equal.

Inferring the performance of a design by propagating uncertainty through it
The principle of natural extension (Walley, 1991) can be used to estimate imprecise probabilities of outcomes. The lower and upper expected utilities of the uncertain payoff of a design decision can be calculated by solving a pair of optimization problems.

Making choices among alternative courses of action
In imprecise probability, alternative courses of action are compared in pairs. The minimum and maximum values of the difference between the upper and lower utilities are calculated for this purpose. This comparison may result in indecision as expected utilities form a range.

Table 2.5 Overview of imprecise probability.

Measure of likelihood	Interpretation	Underlying assumptions	Strengths and weaknesses
Lower and Upper Probabilities	Lower probability of an outcome is the maximum buying price of a lottery ticket that pays $1 if the outcome occurs and zero otherwise. Upper probability is the minimum selling price of the same ticket. Selling price is the minimum sure amount for which the decision maker agrees to assume the risk of paying $1 if the outcome occurs.	• The decision maker can be indecisive; he/she prefers the ticket to a certain amount below some limit p_l, and the sure amount above a higher limit p_u. The decision maker cannot decide when asked to choose between the ticket and a sure amount $p \in [p_l, p_u]$. • The definition presented here assumes that the decision maker is risk-neutral in the given range of outcomes for which the probability is assessed.	Strengths • Imprecise probability is a general case of subjective probability. • Allows for indecision • Describes better than subjective probability the way most people make choices in real life in the presence of uncertainty. • Reduces the frequency of gambling, thereby protecting a decision maker from large losses. Weaknesses • The theory is not mature and has not been demonstrated on the solution of practical engineering problems. • It is expensive to calculate imprecise probabilities. • The gap between lower and upper probability can be very large. • There is no information about how likely values between lower and upper probabilities are.

2.2.6   Dempster-Shafer Evidence Theory

Probabilistic models specify the probabilities of all individual outcomes (singletons) of an uncertain event. For example, the probability mass function of the number of heads, r, in n flips of a fair coin specifies the probabilities of all possible numbers of heads from 0 to n. Similarly, the PDF of the time to failure of a pump provides information about the probability that the pump will fail during any infinitesimal interval $[t, t + dt]$. Often, a decision-maker cannot estimate the probabilities of all singletons; he/she can only estimate the probabilities of events, such the probability that the pump will fail before 200 hours or between 150 and 250 hours.

Probability distributions are constructed from statistical summaries, such as, quantiles or credible intervals elicited from experts, by fitting a probability distribution to these estimates (O'Hagan et al. 2004). This task could be challenging when a decision maker has scarce data and does not understand well the underlying process.

In Dempster-Shafer Evidence Theory, the requirement of estimating the probabilities of all individual outcomes is relaxed. A decision maker can construct a model of uncertainty by using only estimated probabilities of events containing more than one outcome. Construction of a model is based on the body evidence, which consists of events, called *focal element* $A_i, i = 1, \ldots, n$, and a function $m(\cdot)$ called *basic probability assignment (BPA)*, or *basic belief assignment* or *mass function* that assigns a value from zero to one to each focal element. The values of the BPA function of all focal elements add up to one, but these values are not probabilities. Rather, the basic probability of focal element A_i is the portion of the evidence that supports event A_i and it cannot be assigned to any of its subsets. *Body of evidence* is the collection of all focal elements and the corresponding values of the BPA function.

The following are immediate consequences of the definition of a BPA function.

a) The value of the BPA function $m(A)$ of event A is equal to the part of the probability of A that a decision maker cannot distribute to any of its subsets. Consequently,

$$m(A) = P(A) - \sum_{j=1}^{n} P(A_j) \qquad (2.4)$$

where A_j are those subsets of A to which the decision maker can distribute a portion of $P(A)$.
b) In the special case where a decision maker knows the probability of A, but not the probabilities of its subsets, the BPA function of A is equal to its probability.

The following two examples illustrate the definition of the BPA.

Example 2.1:   This example is similar to the ones in Oberkampf and Helton, (2005) and Almond (1995). An exceptionally smart girl got ten Easter eggs as a present; four were black, five were white and the last was gray. The donor told the girl that black eggs contain a chocolate and white a cookie. She wants to estimate the probability that an egg selected at random contains a chocolate.

Solution:
Picking up an egg at random and checking what is inside is an experiment with two outcomes: find a chocolate and find a cookie. A cookie will be found inside a black egg and perhaps in a gray egg. Four-tenths of the evidence support outcome "chocolate" and five-tenths support "cookie." The girl cannot attribute the remaining one tenth of the evidence corresponding to the gray egg to any of these two outcomes. In order to express her ignorance, she assigns the remaining one tenth to the sample space "chocolate or cookie." Figure 2.6a shows the body of evidence in this example.

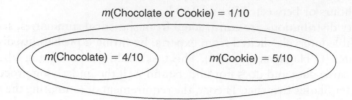

m(Chocolate or Cookie) = 1/10

m(Chocolate) = 4/10   m(Cookie) = 5/10

Figure 2.6a Evidence theory approach: Probabilities of outcomes in Easter egg example.

p(Chocolate) = 0.45   p(Cookie) = 0.55

Figure 2.6b Probabilistic approach: Probabilities of outcomes in same example.

A probabilistic approach to this problem assigns probabilities only to outcomes "chocolate" and "cookie." If the girl has no reason to believe that the gray egg is more likely to contain either a chocolate or cookie, then it is reasonable to split the remaining 1/10th of the evidence evenly and assign the two parts to outcomes "chocolate" and "cookie." Thus, the probabilistic approach assigns probabilities 0.45 and 0.55 to outcomes "chocolate" and "cookie."

Example 2.2:   Often, probabilities are estimated by asking experts to estimate statistical summaries such as quartiles. We want to estimate the cost of a project at its early stage. An expert estimates that the cost would not exceed $15, $18, and $21 million with probabilities 0.25, 0.5, and 0.75, respectively. He/she also told us that the cost is very unlikely to fall under $10 and to exceed $28 million. He/she explained that when he/she says that an outcome is very unlikely he/she means that its probability is 0.01. The expert also said that the cost would never exceed $35 million. Find the body of evidence that corresponds to this information.

Solution:
Based on the expert's judgment, each of the intervals [0, 10] and [28, 35] has probability 0.01. It is not known how these probabilities are distributed inside these intervals. Therefore, each of these intervals has basic probability 0.01.

From Equation (2.4), the basic probability of the interval [0, 15] is the difference of the probabilities of the intervals [0,15] and [0,10]. This difference is equal to 0.24. There is no evidence about how this probability is distributed inside this interval, therefore, the BPA of interval [0, 15] is also 0.24. We calculate the basic probabilities of the remaining intervals in Table 2.6 similarly. Figure 2.7 shows the body of evidence in this table.

In situations like the above two examples, if the decision-maker does not want to make additional assumptions or cannot afford to collect more data, then he/she has to specify the probability of an outcome up to an interval. The probability of another outcome *A* can be as low as the sum of the basic probabilities of the focal elements

Table 2.6 Body of evidence for cost of the project in example 2.2.

Interval (million)	Value of BPA function, m
[0,10]	0.01
[0,15]	0.24
[0,18]	0.25
[0,21]	0.25
[0,35]	0.24
[28,35]	0.01

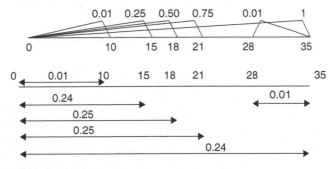

Figure 2.7 Judgments from Expert about the Project Cost.
Upper Panel: Each Triangle Shows a Range of the Cost (Base of Interval) and the Corresponding Probability of Enclosure (Number above the Peak).
Lower Panel: Focal Elements (Depicted by Arrows) and their values of the BPA Function.

that are contained in A, and as high as the sum of the basic probabilities of the focal elements that intersect A,

$$\underline{P(A)} = \sum_{A_i \subseteq A} m(A_i) \tag{2.5}$$

$$\overline{P(A)} = \sum_{A_i \cap A \neq \emptyset} m(A_i) \tag{2.6}$$

In the above equations, $\underline{P(A)}$ and $\overline{P(A)}$ are the lower and upper bounds of the probability of A. The above two bounds of the probability of A are called *Belief* and *Plausibility*. Belief measures the degree to which the evidence implies that A occurs, while plausibility measures the degree to which the evidence does not contradict that A occurs, respectively. The gap between Plausibility and Belief of an outcome measures the uncertainty in the probability of this outcome.

The following remarks help understand the interpretation of the BPA function.

a)   If a decision maker knows the probability of event A, and A does not contain any focal elements, then $m(A) = P(A)$ and $P(A) = \underline{P(A)} = \overline{P(A)}$.

b)   Suppose that focal element B contains focal element A and does not contain nor intersect other focal elements. Then, the value of the BPA function of B is equal to the Belief of B minus the Belief of A:

$$m(B) = \underline{P(B)} - \underline{P(A)} \tag{2.7}$$

This result can be generalized. The information that a body of evidence contains is equivalent to the information provided by the values of the Belief functions of the same elements. That is, if we know the values of the Belief function of all focal elements then we can calculate the values of their BPA function.

c)   Consider that the focal elements $A_i, i = 1, \ldots, n$ are singletons. Then the value of the BPA function of each element is equal to the probability of this element. A decision-maker is certain about the probabilities of the outcomes in this case.

d)   It is sufficient to specify one of the three measures (BPA, Belief and Plausibility) in order to compute the other two. For example, Equation (2.6) enables one to calculate the Plausibility of any event given the BPA function.

In Example 2.1, the Belief of outcome "Chocolate" is 0.40 and the Plausibility 0.50. In Example 2.2, the Belief that the cost will exceed \$25 million is equal to the basic probability of interval [28, 35], because this is the only focal element contained in the range [25, ∞]. The Plausibility of the same outcome is the sum of the basic probabilities of the ranges [0, 35] and [28, 35] because only these two ranges intersect range [25, ∞]. Therefore, the Belief and Plausibility that the cost can exceed \$25 million are 0.01 and 0.25. This result means that the probability of the cost exceeding \$25 million can be as low as 0.01 and as high as 0.25. The large gap between the two bounds shows that the decision maker is highly uncertain about the true probability.

The information consisting of focal elements and their basic probabilities is called a *random set*.

Example 2.3:   The probabilities of two outcomes A and B are 0.6 and 0.5 but the probability of their intersection is unknown. No information is available about other events in the sample space. Find the bounds of the probability of the intersection of A and B and the corresponding body of evidence.

Solution:
In this example, we calculate the Beliefs of all events and then derive the BPA functions using Equation (2.7).

The values of the belief functions of both A and B are equal to their known probabilities. The minimum value of the probability of the intersection is 0.1 (Figure 2.8). Therefore, the belief of the intersection is $P(A \cap B) = 0.1$. From Equation (2.7), $m(A \cap B) = P(A \cap B) = 0.1$. Then, the values of the BPA function and the focal elements are:

$$m(A) = \underline{P(A)} - \underline{P(A \cap B)} = 0.6 - 0.1 = 0.5$$
$$m(B) = \underline{P(B)} - \underline{P(A \cap B)} = 0.5 - 0.1 = 0.4$$

The BPA of the sample space S is zero.

An advantage of evidence theory is that it does not force a decision maker to estimate the probabilities of all the outcomes that affect the consequences of a decision when

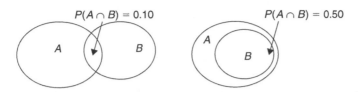

Figure 2.8 Lower and upper bounds of the probability of the intersection of A and B in Example 2.3.

there is not enough evidence to do so. On the other hand, the concept of the basic probability assignment is poorly defined, and often misunderstood. Unlike probability, there is no structured process to estimate the basic probability assignment of events that is consistent with the definition of this concept.

2.2.6.1 p-boxes

A *p*-box consists of a lower and upper bounds of the CDF of a random variable. Information for a basic probability assignment is equivalent to a *p*-box. That is, from the basic probability assignment for some variable, we can calculate its lower and upper CDFs and vice-versa.

2.2.6.2 Possibility

Possibility theory is a limiting case of Dempster-Shafer evidence theory, in which focal elements are nested (they can be ordered in a way that each element is a subset of the next one). Then the Plausibility of an outcome reduces to a measure called Possibility and the Belief to another measure called Necessity. Possibility is suitable for problems where uncertainty is very large. When the Possibility of an event is less than one then its Necessity is zero. When the Necessity of an event is greater than zero then its Possibility is one.

Interval and convex set representations of uncertainty are special cases of Possibility in which there is a single focal element (that is the interval or the convex set) with BPA value equal to one.

2.3 Conclusion

This chapter presented theories for managing uncertainty. Tools for modeling uncertainty, making inferences and choosing among alternative courses of action were explained. Imprecise probability is the most general theory, and all others are special cases of this theory.

Many people believe that it is difficult to estimate probabilities because they think of probability as a long-term relative frequency. Subjective probability is far more general and useful than its objective counterpart. According to this theory, the probability of an outcome measures a decision-maker's belief that this outcome will materialize. A facilitator can estimate this probability by probing the decision-maker's disposition toward betting on or against this outcome. This theory has a solid justification. Many well-tested tools are available for managing uncertainty in decision-making.

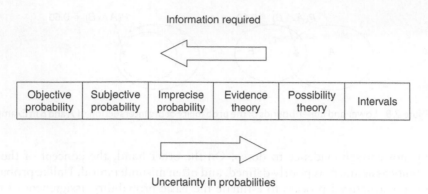

Figure 2.9 Alternative ways to represent uncertainty.

Often, designers are uncertain about probabilities. Most designers become indecisive (cannot choose among some of the alternative courses of action) when confronting uncertainty in probability. Probability does not account explicitly for this type of uncertainty and does not permit indecision. There is a family of theories that do account for the uncertainty in the probabilities of outcomes by using upper and lower bounds of these probabilities. Imprecise probability is the most general member of this family. However, there is no structured procedure to estimate the measures of uncertainty employed in Evidence Theory and Possibility Theory.

The choice of a theory and tools for managing uncertainty depends on the degree of accuracy to which we can afford to estimate probabilities. If probabilities can be estimated accurately from data or judgment, probability should be used. In the other extreme where we are unable to estimate probabilities, interval methods could be better. Figure 2.9 orders these theories in terms of the amount of required information and the degree of uncertainty in the probabilities.

Questions and Exercises

2.1   Show which statements below are true and false

   a) A decision maker tries to estimate the cost of a project that has not started yet. After doing some research, he concludes that the cost could be between $10 and $25 million. He cannot make any other judgments about the cost. An interval approach is suitable for modeling cost.

   b) A decision maker knows the functional relation between variables X and Y, $Y = g(X)$, but she is uncertain about variable X. She believes that the value of X is in the range $[x_l, x_u]$. Then variable Y should be in the range $[g(x_l), g(x_u)]$.

   c) Objective probability is a special case of subjective probability.

   d) A decision maker models uncertainty by using subjective probability. If the decision maker's subjective probabilities cohere, then the decision maker will avoid sure loss.

e) A decision maker estimates that the probabilities of an event and its complement are both 0.7. This decision maker is vulnerable to sure loss.

f) A risk neutral decision maker represents uncertainty by using imprecise probability. Consider two disjoint events A and B. The minimum buying price of a ticket that pays \$1 if and only if $A \cup B$ occurs is greater than or equal to the sum of the minimum buying prices of one ticket that pays \$1 if and only if A occurs and a similar ticket for event B.

g) If the Necessity of an event is strictly positive then its Possibility is one.

h) The Beliefs of two disjoint events and the Beliefs of their union satisfy the following equation,

$$\underline{P}(A \cup B) \geq \underline{P}(A) + \underline{P}(B)$$

i) The Plausibility of the events in the above statement satisfy the following equation,

$$\overline{P}(A \cup B) \geq \overline{P}(A) + \overline{P}(B)$$

j) If $A \subset B$ and A is the only focal element contained in B, then

$$\underline{P}(B) = m(A) + m(A^C \cap B)$$

k) Let $Y = f(X)$ be a non monotonic function of X. If the body of evidence of X consists of nested focal elements so does the body of evidence of Y.

2.2 Why do you think people are ambiguity averse? Is it justified?

2.3 What is the fair price of a ticket that is worth \$1 if you get heads up in one flip of a fair coin and zero otherwise to a risk neutral decision-maker?

2.4 A friend of your says that the fair price of the coin in the above question is \$0.2. Is this estimate rational? What will happen to your friend if he would agree to sell this ticket for \$0.2 and assume the risk of having to pay \$0.2 if the coin lands heads up in one flip.

2.5 You will flip a thumbtack once. Consider two lottery tickets which are worth \$1 if the thumbtack lands with its tip up or down, respectively. You do not know the long term frequency of these two outcomes. What is the fair price of this pair of tickets?

2.6 The worth of a ticket is determined as follows: A coin is selected blindly from a bag with bent and fair coins. If a fair coin is selected, then the bearer of the ticket gets a refund of the price that he paid. If a bent coin is selected, then it is flipped once. The ticket is worth \$1 if the outcome is heads up and \$0 otherwise.

A risk neutral player examines the bag and says that his fair price of the ticket is $0.4. What event's probability is equal to 0.4 on the basis of this assessment?

> • If a fair coin is selected, then the bearer of this ticket gets full refund of the price.
> • If a bent coin is selected, then this ticket is worth $1 for a heads up flip, and zero for a tails up flip.

2.7   Does the fair price of the ticket in the above problem depend of the proportion of bent and fair coins?

2.8   The compliance (displacement/force) of the steady state harmonic response of a single degree of freedom system without damping is,

$$x(r) = \frac{1}{|1 - r^2|}$$

where r is the ratio of the frequency of the excitation to the natural frequency of the system. The table below shows the BPA of the frequency ratio,

Frequency range, r	m
[1.6,1.8]	0.8
[1.5,2.0]	0.19
[1.2,2.5]	0.01

Find the BPA of the compliance.

2.9   If the system in the previous problem has damping the compliance is,

$$x(r) = \frac{1}{\sqrt{(1 - r^2)^2 + 4\zeta^2 r^2}}$$

where the damping ratio is $\zeta = 0.1$.

a. Is it true that the maximum and minimum values of the compliance occur when the frequency ratio r becomes equal to its minimum and maximum values?

b. Find the lower and upper bounds of the compliance if the frequency could be as low as 0.6 and as high as 1.5.

2.10   The table below shows the BPA of the frequency of the system in previous problem.

Frequency range, r	m
[0.9,1.2]	0.8
[0.75,1.35]	0.19
[0.6,1.5]	0.01

Find the BPA of the compliance.

2.11   You work in a civil engineering office. Your manager asks you to estimate the likelihoods that the cost of a custom house could exceed $800,000, $1,000,000 and $1,200,000. There is very limited data and experience about this project in your office. You enlist the help of an expert for this purpose. The expert believes that the project could cost from $600,000 to $1,000,000 or from $1,000,000 to $1,300,000, with probabilities 0.33 and 0.67, respectively. The expert cannot provide more refined estimates.

        Choose a suitable theory for representing uncertainty in the cost by using the expert's judgments and estimate the likelihoods of the cost exceeding the limits that the manager is interested in. Select the best way to present the results to your manager.

2.12   Suppose that you decide to use evidence theory in problem 2.11. What is the body of evidence? Find the Plausibility and Belief of the event that the cost will not exceed $900,000.

Chapter 3

Objective Probability

Overview of this Chapter

Design of a car, aircraft, or cell phone is a decision in which a designer selects the configuration that is likely to produce the most desirable outcome (e.g., highest profit). There are many sources of uncertainty in design, including variability in material properties and geometry, human errors and operating conditions.

Figure 3.1 explains the role of uncertainty in design and shows three important tasks for managing it: a) modeling uncertainty in the variables that drive the performance of a design or the profit from a risky venture, b) estimation of the resulting uncertainty in the performance, and c) selection of the best design. In this last step, the designer should account for his/her attitude toward risk.

This chapter presents probabilistic tools for the first two tasks for managing uncertainty; model sources of uncertainty and estimate the resulting uncertainty in the performance of a system (Figure 3.1). Probability is viewed as the long-term relative frequency of an event.

Objective probability deals with averages of mass phenomena such as arrival of customers in a bank, stamping of parts in a factory, telephone calls in a call center, and failure of a system. Certain averages converge to a constant value (long-term average) with the number of observations. For example, the average time to failure of a pump approaches a constant with the number of pumps tested.

Probability and statistics process knowledge about these averages. The term *statistics* refers to methods for estimating long-term averages from observations. For example, statistical methods estimate the average frequency of failure of a fuel pump from records. *Probabilistic methods* calculate long-term averages of some events from those of other events. For example, using probabilistic methods, we can find the probability that no more than three out of 100 pumps will fail over a certain period, given the probability of failure of a single pump.

This chapter explains the fundamental concepts of probability theory in three sections. First, the definition and properties of objective probability are presented. Probability distributions, which are tools for modeling uncertainty in experiments with numerical outcomes (such as the demand for a new product, the stress in a rod or the time to failure of a pump), are studied.

The second section reviews the properties of common probability distributions. Experimental and analytical studies have shown that some distributions describe well

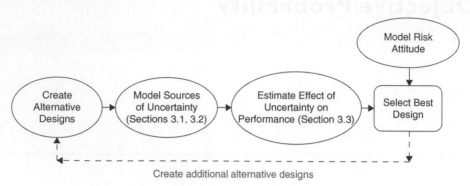

Figure 3.1 Tasks for managing uncertainty in a design decision. The sections of this chapter that explain each task are in parentheses.

some uncertain quantities. This section provides guidelines for selecting a distribution for a probabilistic analysis or a design problem.

The last section studies the fundamental problem of calculating the probability distribution of a function of random variables and presents tools for solving this problem. These tools enable a decision maker to quantify uncertainty in the payoff of a risky venture or the uncertainty in the performance of a design. This information is important for selecting the best course of action among alternatives.

This chapter also demonstrates through examples that probabilistic analysis can help people avoid bad decisions by enabling them to estimate the probabilities of the consequences of their actions. A bad decision is one that a person would avoid if an expert had shown him that it was likely to produce undesirable consequences.

3.1   Probability and Random Variables for Modeling Uncertainty

This section explains the definition and properties of objective probability and random variables. A detailed presentation of probabilities and random variables can be found in introductory books on probability (Daveport, 1970 and Papoulis 1965) or in advanced books (Loeve, 1963). The presentation here is focused towards engineering, business and statistics with particular applicability to decision making under uncertainty.

3.1.1   *Fundamentals of Objective Probability*

3.1.1.1   *Definition of probability*

An *experiment* is an activity with uncertain outcome. Flipping a coin or rolling a die are examples of experiments. Each time we flip a coin we do not know if we will get "heads" or "tails." *Event* is a collection of outcomes of the experiment. For example, consider that we roll a die. The collection of outcomes {1, 3, 5} is an event. The collection of all possible outcomes of an experiment is called a *sample space* or a *universal set*.

An experiment can have a finite or an infinite number of outcomes. In the latter case, the number of outcomes can be countable or uncountable. As we explained in Chapter 2, objective probability of an outcome is the long-term frequency of the outcome in an experiment that is repeated many times, while all relevant conditions remain the same. Formally, the probability of outcome A is,

$$P(A) = \lim_{n \to \infty} \frac{n_A}{n} \tag{3.1}$$

where n_A is the number of times outcome A occurred and n is the total number of replications of the experiment. Probability can be defined by using the counter or indicator function I_{A_i}, which is one if A occurs in the ith replication and zero otherwise, as follows,

$$P(A) = \lim_{n \to \infty} \frac{1}{n} \sum_{i=1}^{n} I_{A_i} \tag{3.2}$$

3.1.1.2 Axioms of probability

Probabilities of events obey three axioms that were introduced by Kolmogorov. Probability is a single number assigned to each member of a σ-algebra. This is the class of subsets of the sample space which is closed under union, intersection, and complementation. The collection of all possible events of an experiment is a σ-algebra. Probability satisfies the following three axioms.

1. Probability of an event is nonnegative,

$$P(A) \geq 0 \tag{3.3a}$$

2. The probability of the certain event is always equal to one,

$$P(S) = 1 \tag{3.3b}$$

3. The probability of the union of pairwise disjoint events is equal to the sum of the probabilities of these events,

$$P\left(\bigcup_{i=1}^{n} A_i\right) = \sum_{i=1}^{n} P(A_i) \tag{3.3c}$$

Although the above axioms look innocuous, people must be very careful when assigning probabilities to events, if they want to satisfy these axioms. If a decision maker has a very large number of observations from a repeated experiment on the relative frequency of an event, then the decision maker must assign a probability to the event that is equal to its relative frequency in order to satisfy the third axiom (see the law of large numbers in section 3.3.2.4).

There is controversy about the third axiom. When some people are ambiguous about the probability of an event they prefer assigning lower and upper bounds to its

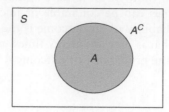

Figure 3.2 Probability of an event A and its complement A^C add up to 1.

probability instead of a single number. In Chapters 6 and 7, we will see that these bounds violate the third axiom.

The following properties of probability follow from Kolmogorov's axioms.

a. The probability of the complement of an event is equal to one minus the probability of the event,

$$P(A^C) = 1 - P(A) \tag{3.4}$$

where A^C is the complement of A.

Proof: The union of an event and its complement is equal to the entire sample space: $A \cup A^C = S$ (see Figure 3.2). Also, the probability of the union of an event and its complement is equal to the sum of their probabilities because the event and its complement are disjoint. Therefore, $P(A) + P(A^C) = P(S) = 1$. Equation (3.4) follows from this equality.

b. The probability of the empty event $\emptyset$ is always zero.

Proof: The empty event is the complement of the certain event. From equation (3.4) we conclude that the probability of the empty event is $P(\emptyset) = 1 - P(S)$, which is equal to zero.

c. The probability of any event A is no greater than 1.

Proof: The probabilities of an event and its complement add up to one, $P(A) + P(A^C) = 1$. Therefore, $P(A) = 1 - P(A^C)$. From this equation we conclude that $P(A) \leq 1$ since $P(A^C) \geq 0$.

d. The probability of the difference of two events is,

$$P(A - B) = P(A) - P(A \cap B) \tag{3.5}$$

where the difference of events A and B contains everything in A that is not in B.

Proof: The union of the difference of events A and B and the intersection of these two events is equal to A (see Figure 3.3) $(A - B) \cup (A \cap B) = A$. The difference and the intersection of A and B are disjoint. Therefore, $P(A - B) + P(A \cap B) = P(A)$ and $P(A - B) = P(A) - P(A \cap B)$.

In the special case where B is a subset of A, the probability of the difference of events A and B is equal to the difference of the probabilities of A and B.

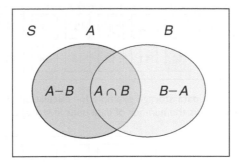

Figure 3.3 Probability of the difference of two events.

e. If event A contains B then the probability of A is greater or equal to the probability of B.

    Proof: The probability of A is greater than that of B because the difference of the probabilities of the two events is equal to $P(A - B)$ and because probability is always non negative.

f. The probability of the union of two events is equal to the sum of their probabilities minus the probability of their intersection,

    $$P(A \cup B) = P(A) + P(B) - P(A \cap B) \tag{3.6}$$

    Proof: First, we break down the union of A and B into disjoint events with known probabilities and then add these probabilities to find the probability of the union. Figure 3.3 shows that the union of A and B is equal to the union of the following disjoint events; the difference of A and B and event B, $A \cup B = (A - B) \cup B$. Therefore, $P(A \cup B) = P(A - B) + P(B)$. Equation (3.6) follows by substituting $P(A) - P(A \cap B)$ for $P(A - B)$ (see Equation 3.5).

An alternative justification of Equation (3.6) is that when we add the probabilities of A and B in order to find the probability of their union we add twice the probability of the intersection. Therefore, we subtract $P(A \cap B)$ from the sum of the probabilities of A and B in order to find the probability of their union.

We have already repeatedly used the approach of breaking down an event whose probability we want to calculate into the union of disjoint events, whose probabilities are easy to find. This *divide-and-conquer approach* is very useful in calculating probabilities of events from those of other events. The examples below demonstrate this approach.

Example 3.1: This example is similar to the one in Davenport, (1970). A data storage unit consists of 16 data blocks. Any number of blocks is equally likely to be in storage, including zero. Calculate the probabilities of the following events:

A: the unit is at least 1/8 full but no more than 5/8 full
B: the unit is at least 3/8 full but no more than 7/8 full
C: the union of events A and B above

Figure 3.4a Sample space in data storage example for calculation of the probability of event A: each square shows the number of data units in storage.

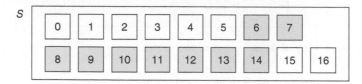

Figure 3.4b Sample space in data storage example for calculation of the probability of event B: each square shows the number of data units in storage.

Solution:
The sample space consists of 17 outcomes (0 to 16 data blocks in storage). These outcomes have probability 1/17. Figure 3.4a shows the sample space and event *A* (shaded squares). The probability of *A* is equal to the sum of the probabilities of these outcomes, which is 9/17.

Event *B* consists also consists of 9 outcomes (Figure 3.4b). Therefore, the probability of *B* is also 9/17.

The probability of the union of *A* and *B* is equal to the sum of the probabilities of these events minus the probability of the intersection. The intersection consists of outcomes: {6, 7, 8, 9, 10} and has probability 5/17. Therefore, the probability of the union of *A* and *B* is 13/17. This can be verified by observing that the union consists of outcomes {2, 3, ..., 14} each of which has probability 1/17.

Example 3.2:   Consider a drill in a factory. Blanks arrive at random intervals and if the drill is busy they are placed in a queue. Table 3.1 shows the results of 60 observations of the number of parts in the queue. Estimate the probabilities of a) more than 3 parts waiting, and b) no more than 8 parts waiting in the queue.

Solution:
We assume that the probability of each number of parts is equal to the relative frequency. The probability of more than three parts waiting in the queue is equal to the sum of the probabilities of 4, 5, ... parts in the queue. Therefore, this probability is $3/60 + 2/60 + 1/60 = 1/10$. The probability of no more than 8 parts waiting is the sum of the probabilities of zero to 8 parts waiting, which is equal to 1.

Note that the above numbers are estimates of the true probabilities (or long-term relative frequencies) of the above two events. These estimates will change if additional data become available. When we face an important decision, such as adding a second drill to the system in order to reduce the size of the queue, it is important to know the uncertainty in these estimates. In Chapter 4, we will learn how to estimate confidence

Table 3.1 Observations of the number of parts waiting in a drill queue.

Number of blanks	Number of observations	Relative frequency
0	3	3/60
1	15	15/60
2	21	21/60
3	15	15/60
4	3	3/60
5	2	2/60
6	1	1/60
7	0	0
8	0	0
. . .	0	0
Total	60	1

intervals, which are ranges containing the long term relative frequency with a high coverage probability (for example 0.95).

In many problems, we do not have observations such as the ones in example 3.2, but we have no reason to believe that the outcomes of the sample space have different likelihood. For example, in games of chance involving a roll of a die, the spin of a roulette disk or a wheel of fortune, all outcomes seem equally likely. In this case, we may estimate probabilities on the basis of the *principle of insufficient reason* (French, 1986, 217–220). This principle asserts that two events are equally likely if there is no reason, such as lack of symmetry, to assume that one is more likely than the others. In such problems, the probability of an event is equal to the number of outcomes that imply that the event occurs divided by the total number of outcomes in the sample space.

3.1.1.3 Conditional probability

In general, evidence that an event occurred changes the probability of another event. The reason is that occurrence of an event may change the sample space. For example, suppose that a box contains a fair coin and a biased coin. The coins look identical but the biased one has probability of landing heads up 0.9. We pick up one coin at random, flip it once, and get heads. This evidence suggests that it is more likely that we have selected the biased coin. Conditional probability of an event A is the probability of this event in view of the evidence that another event B occurred. This probability is given by the following equation,

$$P(A/B) = \frac{P(A \cap B)}{P(B)} \qquad (3.7)$$

where $P(A/B)$ is the conditional probability of A given the occurrence of B. The above equation is called Bayes' rule. Figure 3.5 shows that when event B occurs, then the sample space shrinks from S to B and this changes the probability of A. Equation (3.7) has the following relative frequency interpretation. We repeat an experiment in which B may or may not occur n times and we observe if A occurs in each trial. Then, the relative frequency of event A given the occurrence of B is equal to the number of

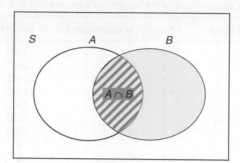

Figure 3.5 Explanation of the definition of the conditional probability of event A given the evidence that B occurred.

simultaneous occurrences of events A and B (represented by the cross hatched region in the Venn diagram in Figure 3.5) normalized by the number of occurrences of B (shaded region).

Equation (3.7) can be written as follows,

$$P(A/B) = \frac{P(B/A)P(A)}{P(B)} \tag{3.8}$$

In the above equation $P(B/A)$ is the probability of observing evidence B given that event A has occurred, while $P(A)$ is the probability of A before observing evidence B. $P(B/A)$ is called *likelihood of event A* and $P(A)$ is *prior probability* of A. In plain English, the above equation says:

> The probability of event A after observing evidence B is proportional to the likelihood that we will observe this evidence if event A occurs times the probability of A before observing the evidence.

The conditional probability of A given the evidence B is different than the conditional probability of B given A: $P(A/B) \neq P(B/A)$. For example, the probability of a patient having high fever given that he/she has strep throat is practically equal to one. However, the probability of a patient having strep throat given that he/she has high fever is considerably less than one.

Example 3.3:   A box contains a fair coin and a biased coin that look identical. The biased coin has a probability of 0.9 to land heads up in a flip. We pick up one coin and flip it once. If the coin lands heads up, find the probability it is the biased one.

Solution:
In this example, the evidence B consists of the observed head in a flip of the randomly selected coin. The likelihood of the evidence $P(B/A)$ is equal to the probability of getting a head given that the biased coin was selected. This is equal to 0.9. The prior probability of selecting the biased coin $P(A)$ is assumed 0.5 since there is no reason to

assume that selecting one coin is greater than the other. The probability of getting a head in one flip of a randomly selected coin is equal to the sum of the probabilities of selecting each coin and getting heads when flipping the selected coin,

$$P(B) = P(B/A)P(A) + P(B/A^C)P(A^C)$$

where $P(A^C)$ is the probability of selecting the fair coin. Putting all terms in Equation (3.8) we obtain,

$$P(A/B) = \frac{P(B/A)P(A)}{P(B/A)P(A) + P(B/A^C)P(A^C)} = \frac{0.9 \cdot 0.5}{0.9 \cdot 0.5 + 0.5 \cdot 0.5} = 0.643$$

As expected, the probability of having selected the biased coin increases in view of the observed evidence that the coin landed heads up. The probability of having selected the fair coin given the evidence is 0.357.

Two events are called independent if the probability of the intersection is equal to the product of the probabilities of these events $P(A \cap B) = P(A)P(B)$. The definition of independence and the equation for the conditional probability imply that, if two events are independent, knowing that one has occurred does not change the probability of the other.

Example 3.4:   An HIV test yields a false positive in 2/100 cases (false positive: the test indicates incorrectly that a non infected person is infected). The test has a negligible probability of yielding a false negative (i.e., indicating incorrectly that an infected person is not infected). In a low risk group, 1 per 100,000 members is HIV infected. Consider that a low risk member tests positive. What is the probability that the result of the test is correct and the member is actually infected?

Solution:
The prior probability of a person in a low risk group being HIV infected is 10^{-5}. Using Bayes' rule we obtain the following equation for the probability of infection given the positive test result,

$$P(HIV/Pos) = \frac{P(Pos/HIV)P(HIV)}{P(Pos)} \tag{3.9}$$

where $P(HIV)$ is the prior probability of a person in a low risk group being infected, $P(HIV/Pos)$ is the probability of the same event given that the person tested positive, $P(Pos/HIV)$ is the probability that an infected person tests positive, and $P(Pos)$ is the probability that a person selected randomly from a low risk group (which is either infected or healthy) tests positive. The latter probability is,

$$P(Pos) = P(Pos/HIV)P(HIV) + P(Pos/HIV^C)P(HIV^C)$$

From the problem statement,

$$P(HIV) = 10^{-5}, \quad P(Pos/HIV) = 1, \quad P(Pos/HIV^C) = 0.02$$

Plugging the above probabilities in Equation (3.10) we find that the probability that a person selected randomly from the low risk group will test positive is 0.02. This is identical to the probability of a false positive. This result is reasonable because all but one person in a group of 100,000 low risk persons is HIV infected on average. By plugging this result in Equation (3.9) we find that the probability of the person who tested positive is actually infected is only 5/10,000.

This result seems counterintuitive; almost everyone would be very scared if he/she tested HIV positive. Here is an intuitive explanation for the result. In a group of 100,000 low risk people, only one is infected on average. If we test 100,000 people in a low risk group, 2,000 will test positive. Out of the two thousand only one is actually infected and the rest 1,999 are not. Therefore, a person in this group has chance of 1 per 2,000 to be infected. We can assess the severity of this risk by comparing it with the probability of dying in a car accident is one year, which is about one per 7,000.

Two general conclusions can be drawn from this example: a) The test has very little value to a member of a low risk group. b) Knowing how to calculate probabilities can help a person make better decisions by interpreting evidence from the test correctly. For example, suppose that one who is in a low risk group tests HIV positive. If he/she knows the odds of being actually infected, he/she will avoid making drastic and very costly changes in his/her lifestyle.

3.1.1.4 Combined experiments

Often, a decision maker deals with multiple experiments or a repeated experiment. When the experiments are independent, one can calculate probabilities of events efficiently by combining these experiments into one. Two experiments are independent if knowing the outcome of one does not change the probabilities of the outcomes of the other. For example, when we flip a fair coin twice we perform two experiments, each with outcomes Heads (H) and Tails (T). We can combine the two experiments into one experiment with outcomes all possible combinations of the outcomes of the two flips: $\{HH, HT, TH, TT\}$. The two coin flips are independent experiments.

The sample space of a combined experiment is the Cartesian product of the sample spaces of the individual experiments,

$$S = S_1 \times S_2 \cdots S_n$$

where S is the sample space of the combined experiment and $S_i, i = 1, \ldots, n$ the sample space of the ith sub-experiment. For example, if that we flip a coin and roll a die the sample space of the combined experiment consists of all pairs of the outcomes of a coin flip $S_1 = \{H, T\}$ and a roll of a die $S_2 = \{1, 2, \ldots, 6\}$ (Table 3.2).

If a combined experiment consists of independent sub-experiments then the probabilities of the outcomes of the combined experiment are equal to the products of the probabilities of the corresponding outcomes of the sub-experiments. For example, if the coin and the die in Table 3.2 are unbiased, then the probabilities of the outcomes in Table 3.2 are all equal to $\frac{1}{2} \times \frac{1}{6}$. Therefore, information about the probabilities of the outcomes of the sub-experiments is sufficient in order to determine the probabilities of the outcomes of the combined experiment. However, if the sub-experiments

Table 3.2 Outcomes of the combined experiment of flipping a coin and rolling a die. The first column lists the outcomes of the coin flip and the first row the outcomes of the roll of a die.

	1	2	3	4	5	6
H	HI	H2	H3	H4	H5	H6
T	TI	T2	T3	T4	T5	T6

are dependent, then we need to estimate the probabilities of all of the outcomes of the combined experiment.

Bernoulli process

An experiment with two possible outcomes is called a *Bernoulli trial*. A series of such experiments is called a *Bernoulli process*.

Example 3.5: This example involves the calculation of the probabilities of the outcomes of a Bernoulli trial: We flip a fair coin 3 times. Find the probabilities of each of the events of getting a total of 0 to 3 heads.

Solution:
The process of flipping a coin three times is a combination of three experiments, each of the three coin flips. We assume that the three experiments are independent. This is reasonable because the probability of getting heads is known and it is fixed from one trial to the next – it does not depend on what happened in the previous trials.

The space of all possible outcomes of the combined experiment consists of all possible triplets of outcomes of the three experiments in which we flip the coin. There are 2^3 such triplets, which are the rows of Table 3.3.

According to the principle of insufficient reason, the outcome in each row of the table has probability 1/8. We get no heads only in the first outcome. Therefore, the probability of getting no heads is 1/8. Outcomes "one head in three flips" and "two heads in three flips" have probability 3/8. Finally, the probability of getting three heads is 1/8.

Example 3.6: We flip biased coin n times. The probability of getting heads is $P(H) = p$. Find the probability of k heads.

Solution:
Consider an outcome in which we get k heads and $(n - k)$ tails in a particular sequence. Since the n trials are independent, the probability of this outcome is the product of the probabilities of the outcomes of the trials which is equal to $p^k(1 - p)^{n-k}$. The event "k heads in n flips" comprises $N_n(k)$ outcomes, where $N_n(k)$ is given by the following equation (Papoulis, 1965, p. 58),

$$N_n(k) = \binom{n}{k} = \frac{n!}{k!(n-k)!} \tag{3.10}$$

Table 3.3 Sample space of the Bernoulli trial consisting of the three coin flips.

Outcome #	First flip	Second flip	Third flip	Total number of heads in three flips
1	T	T	T	0
2	T	T	H	1
3	T	H	T	1
4	T	H	H	2
5	H	T	T	1
6	H	T	H	2
7	H	H	T	2
8	H	H	H	3

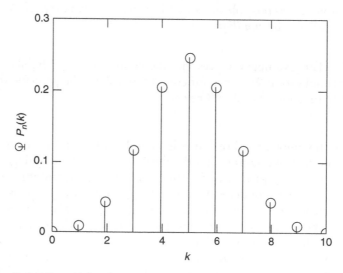

Figure 3.6a Probability of k heads in 10 flips of a fair coin.

The appendix shows the proof of this equation. Therefore, the probability of getting k heads in n flips of the coin is,

$$P_n(k) = \binom{n}{k} p^k (1-p)^{n-k} \tag{3.11}$$

Figures 3.6a to 3.6c show the probabilities of the number of heads in 10 flips of three coins with probabilities of heads equal to 0.5, 0.3 and 0.7 obtained from Equation (3.11).

Example 3.7: A redundant system consists of 100 components that fail independently. The system fails if two or more components fail. The probability of failure of a component is 10^{-4}. Find the reliability of the system, that is, the probability that the system will not fail.

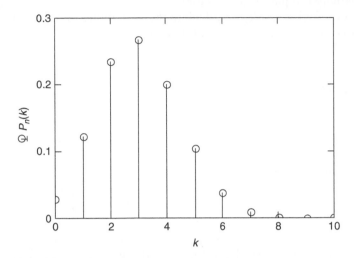

Figure 3.6b Probability of k heads in 10 flips of a coin with probability of heads 0.3.

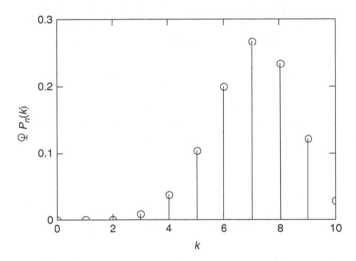

Figure 3.6c Probability of k heads in 10 flips of a coin with probability of heads 0.7.

Solution:
Operating a system can be thought as a Bernoulli process with 100 trials in each of which a component can survive or fail. The probability of k components failing is obtained from Equation (3.11) where $n = 100$ and $p = 10^{-4}$. The probability of survival is equal to the sum of the probabilities that none of the components or one component will fail,

$$R = \sum_{k=0}^{1} \binom{n}{k} p^k (1-p)^{n-k}$$

This probability is equal to 0.9999508.

In order to verify this result, we calculate the probability of failure and check if the probabilities of failure and survival add up to 1.

The probability of failure is:

$$P(F) = \sum_{k=2}^{100} \binom{n}{k} p^k (1-p)^{n-k}$$

which is equal to $4.918 \cdot 10^{-5}$ and the two probabilities add up to one.

Sequential experiments

In some applications, we perform a sequence of experiments in which the sample space is determined from the outcome of the previous experiment. This sequence of experiments is a combined experiment with sample space the union of the sample spaces of the two experiments. For example, suppose that we pick up at random a die from a box that contains two dice with six and four sides, respectively. The sample space of the first experiment in which we select a die is {"six-sided die", "four-sided die"}. If the six-sided die is selected, then the sample space of the second experiment will be $S_1 = \{1, 2, 3, 4, 5, 6\}$, while if the four-sided die is selected the sample space will be $S_2 = \{1, 2, 3, 4\}$. The sample space of the combined experiment is the union of $S_1 \cup S_2$.

Consider the general case where the first experiment has n outcomes, and for each outcome a different experiment is conducted. The probability of an outcome A of the combined experiment is calculated from the probabilities of the outcomes of the first experiment E_i, $i = 1, 2 \ldots n$ and the conditional probabilities of the outcomes second experiment, given the outcome of the first as follows,

$$P(A) = \sum_{i=1}^{n} P(A/E_i)P(E_i) \tag{3.12}$$

Equation (3.12) is called the *total probability theorem*.

Example 3.8: If the probabilities of picking up a four-sided and a six-sided dice are 0.5 and the dice are fair, find the probabilities of getting each possible number of dots in a roll of the selected die.

Solution:
The probability of one dot is calculated from the total probability theorem, $P(1 \text{ dot}) = P(1 \text{ dot/six-sided die})P(\text{six-sided die}) + P(1 \text{ dot/four-sided die})P(\text{four-sided die})$ which is $\frac{1}{6} \times \frac{1}{2} + \frac{1}{4} \times \frac{1}{2} = \frac{5}{24}$. Similarly, each of the probabilities of getting two to four dots are equal to $\frac{5}{24}$. Finally, the probabilities of five and six dots are both $\frac{1}{12}$. For example, $P(5 \text{ dots}) = P(5 \text{ dots/six-sided die})P(\text{six-sided die})$ which is $\frac{1}{6} \times \frac{1}{2} = \frac{1}{12}$.

Example 3.9: Most structural systems such as buildings, bridges, cars and offshore platforms are redundant, which means that they can survive failure of one or more

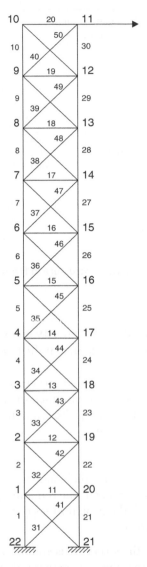

Figure 3.7 Redundant structure that fails under different scenarios triggered by failure of one member.

structural members. If a component fails, the load is redistributed to the surviving members. This increases the probabilities of failure of these members.

For example, the tower in Figure 3.7 can fail under 50 failure scenarios each triggered by failure of one of its 50 members (Figure 3.8). Application of the horizontal load on the top is an experiment whose sample space consists of 51 outcomes: {all members survive, member 1 fails, ..., member 50 fails}. If one of the 50 members fails, a second experiment is performed with 50 outcomes: {survival of the partially damaged

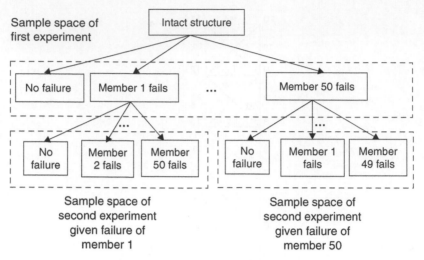

Figure 3.8 Failure scenarios of redundant structure in Figure 3.7.

structure, failures of each one of the remaining 49 members}. The probabilities of these outcomes change, depending on the outcome of the first experiment, i.e. depending on which member failed first.

Members 1, 2 and 50 of the above 50-bar structure are the most vulnerable. When the intact structure is loaded, their probabilities of failure are 0.03, 0.01 and 0.02 respectively. The probabilities of failure of the remaining members are negligible. Find the probability of failure of the system.

Solution:
If one of members 1, 2 or 50 fails, the partially damaged structure may be able to sustain the applied load without failure or one of the remaining members may fail, which will lead to collapse. Table 3.3 shows the probabilities of failure of the remaining members of the partially damaged structure.

The probability of failure of the structure is equal the sum of the conditional failure probabilities of each surviving members given that one member has already failed times the probability of failure of that member.

$$
\begin{aligned}
P(F) = &\ [P(\text{Member 2 fails/Member 1 fails}) + P(\text{Member 50 fails/}\\
&\ \text{Member 1 fails})]P(\text{Member 1 fails})\\
&\ + [P(\text{Member 1 fails/Member 2 fails}) + P(\text{Member 50 fails/}\\
&\ \text{Member 2 fails})]P(\text{Member 2 fails})\\
&\ + [P(\text{Member 1 fails/Member 50 fails}) + P(\text{Member 2 fails/}\\
&\ \text{Member 50 fails})]P(\text{Member 50 fails})
\end{aligned}
$$

Using the data in Table 3.4 we find that the probability of system failure is 0.01.

Table 3.4 Sequence of two experiments for 50-bar structure; in the first experiment the intact structure is loaded, in the second the partially load structure is loaded. The sample space of the second experiment depends of the outcome of the first.

First Experiment: Outcomes and Probabilities

All members survive	Member 1 fails	Member 2 fails	Member 50 fails
0.94	0.03	0.01	0.02

Second Experiment: Outcomes and Probabilities

No failure	Member 2 fails	Member 50 fails	No failure	Member 1 fails	Member 50 fails	No failure	Member 1 fails	Member 2 fails
0.87	0.05	0.08	0.75	0.05	0.2	0.8	0.1	0.1

We calculate the probability of survival in order to check the above probability,

$$P(S) = P(\text{All members survive}) + P(\text{No other failure/Member 1 fails})P(\text{Member 1 fails})$$
$$+ P(\text{No other failure/Member 2 fails})P(\text{Member 2 fails})$$
$$+ P(\text{No other failure/Member 50 fails})P(\text{Member 50 fails})$$
$$= 0.94 + 0.87 \cdot 0.03 + 0.75 \cdot 0.01 + 0.8 \cdot 0.02 = 0.99$$

3.1.2 Random variables

The outcomes of many physical phenomena are associated with numbers. Examples of such outcomes are the number of customers waiting in a queue in a bank, the demand for and the profit from a new product, the maximum stress at failure in a beam and the number of cycles to failure of a crankshaft. Even if the outcome is not numerical, it can be associated with a number. For example, a failure indicator function (1 failure, 0 survival) can describe the state of a system.

A random variable is a function defined over the sample space of an experiment, which assigns to every outcome a real value. In order to distinguish between random variables and their values, we use capital letters for variables and lower case letters for their values. For example, variable K denotes the number of heads in two flips of a coin and k the number of heads in a particular pair of flips. There are discrete and continuous random variables. The former assume a list of possible values while the latter assume a continuum of values.

3.1.2.1 Discrete random variables

The probability mass function (PMF) of a discrete random variable X, $f_X(x)$, assigns a probability to each value of the variable

$$f_X(x) = P(X = x) \tag{3.13}$$

The probability mass function can also be written as follows,
$f_X(x) = \sum_{i=1}^{n} \delta(x - x_i)P(X = x)$ where $\delta(x)$ is the unit impulse function and x_i are all possible values of variable X.

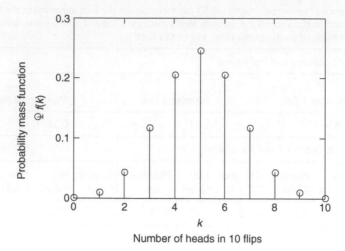

Figure 3.9 PMF of the number of heads in 10 flips of a fair coin.

The PMF of the variable "number of heads in 10 flips of a fair coin" is, $f_K(k) = \binom{n}{k} p^k (1-p)^{n-k}$, where $p = 0.5$ (Equation 3.11). Figure 3.9 shows this PMF.

The PMF is nonnegative. The sum of the values of the probability mass function for all possible values of the variable is always one.

Cumulative Distribution Function (CDF) of a variable is the function that assigns to each value of the variable the probability that the variable is less than or equal to the value.

$$F_X(x) = P(X \leq x) \tag{3.14}$$

The CDF of the number of heads in 10 flips of a fair coin is, $F_K(k) = \sum_{i=0}^{Floor(k)} \binom{n}{i} \times p^i (1-p)^{n-i}$ where $p = 0.5$. Note that k is a real number in this expression. Floor function of a real number is the largest integer that is less than or equal to the real number. Figure 3.10 shows the CDF of the same variable in Figure 3.9.

The CDF has the following properties:

1. It is non negative. The reason is that the CDF is a probability.
2. It is non decreasing, $F_X(x_1) \leq F_X(x_2)$ for $x_1 < x_2$. The reason is that if $x_2 > x_1$ then the probability of $X \leq x_2$ is greater than or equal to $X \leq x_1$. For example, the probability of getting no more than 6 heads is always greater than or equal to the probability of getting no more than 5.
3. The CDF assumes a value of zero at $-\infty$ and one at $+\infty$, because a number cannot be less than or equal to minus infinity and it is always less than or equal to infinity.

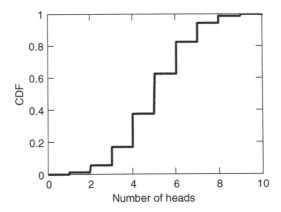

Figure 3.10 CDF of the number of heads in 10 flips.

4. The CDF is continuous on the right: $\lim_{\varepsilon \to 0} F_X(x+\varepsilon) = F_X(x)$ where ε is a small positive number. This is obvious from Equation (3.14), and the fact that $\lim_{\varepsilon \to 0} P(X \le x+\varepsilon) = P(X \le x)$.

The CDF of a variable for $X = x$ can be obtained by adding the values of the PMF for all values of x that are less than or equal to X

$$F_X(x) = \sum_{x_i \le x} f_X(x_i) \tag{3.15}$$

If we know the CDF of a variable, then we can calculate the probability of the variable assuming any given value or being in any interval or collection of intervals. For example, $P(x_1 < X \le x_2) = F_X(x_2) - F_X(x_1)$.

One can obtain the CDF of a variable from its PMF and vice versa. For example, the CDF of the number of heads in 10 flips of the fair coin $F_K(k)$, for $k = 3$ is, $F_K(3) = f_K(0) + f_K(1) + f_K(2) + f_K(3)$ where $f_K(k)$ is the PMF of the number of heads. The PMF $f_X(x)$ can be calculated by subtracting from the CDF at $X = x$, the CDF at a value that is smaller than x by an infinitesimal amount,

$$f_X(x) = F_X(x) - \lim_{\varepsilon \to 0} F_X(x - \varepsilon)$$

For example, the PMF of the number of 3 heads in Figure 3.9 is,

$$f_K(3) = F_K(3) - F_K(2.999...) = 0.171 - 0.054 = 0.117$$

Shape measures: mean value and variance

The PMF or the CDF define completely the probabilistic structure of a random variable. The PMF provides sufficient information to calculate the probability of any event defined in terms of the random variable. We will define some shape measures that give a decision maker an idea about the properties of the PMF. Estimation of these measures requires less data than the PMF. There are four important shape measures: the mean value, variance, skewness and kurtosis. The mean value and the variance describe the location of the center of the PMF and its spread. The skewness indicates the degree of the skew, and the kurtosis the sharpness of the peak of the PMF. The latter two measures will be defined only for continuous random variables because they are more useful for these variables.

The mean (expected) value of a discrete variable $E(X)$ can be thought as the average of a large sample of values of this variable. The mean value is the center of gravity of the probability mass function. The following equation defines the mean value,

$$E(X) = \sum_{i=1}^{n} x_i f_X(x_i) = \sum_{i=1}^{n} x_i P(X = x_i) \tag{3.16}$$

The variance $Var(X)$ is a measure of the dispersion of the samples of a random variable. It is the moment of inertia of the PMF,

$$Var(X) = \sum_{i=1}^{n} [x_i - E(X)]^2 f_X(x_i) \tag{3.17}$$

The variance is also denoted by σ_X^2. The variance can be also calculated from the following equation, which is equivalent to Equation (3.17),

$$Var(X) = \sum_{i=1}^{n} x_i^2 f_X(x_i) - E(X)^2 \tag{3.18}$$

The standard deviation σ_X is the square root of the variance. The standard deviation normalized by the mean value is called coefficient of variation.

Example 3.10:   Find the mean value, variance and standard deviation of the number of heads in 10 flips of a fair coin in Figure 3.9.

Solution:
The mean value is: $E(K) = \sum_{k=0}^{10} k \cdot f_K(k) = \sum_{k=0}^{10} k \cdot \binom{n}{k} p^k (1-p)^{n-k} = 5$, where $p = 0.5$.
The variance is calculated from Equation (3.17):

$$Var(K) = \sum_{k=0}^{10} [k - E(K)]^2 \cdot f_K(k) = \sum_{k=0}^{10} [k - E(K)]^2 \cdot \binom{n}{k} p^k (1-p)^{n-k} = 2.5$$

Equation (3.18) yields the same result.
The standard deviation is 1.581.

3.1.2.2 Continuous random variables

A random variable, such as a person's height or the temperature, may assume values in a continuous range. It does not make sense to assign probabilities to values of the random variable, because the probability of a particular value is generally zero. For example, the probability of a person's height being equal *exactly* to 1.8 meters (not 1.79999999 or 1.80000001) is zero because there are infinite possible values of the height in an interval. In this case, we assign probabilities to intervals. The likelihood of a specific value (for example the height of a person is 1.8 meters) is quantified by a probability density (e.g. probability per meter) instead of a probability.

The CDF of a continuous variable is defined the same way as for a discrete one; it is the probability of a variable being less or equal to a real value. Formally, a random variable is continuous if its CDF is differentiable everywhere except for a countable number of values. The probability mass function of a continuous variable is not defined. Instead, we define the probability density function (PDF) of a variable, $f_X(x)$, which is the derivative of the CDF,

$$f_X(x) = \frac{dF_X(x)}{dx} \tag{3.19}$$

The PDF is defined at all values where the derivative of the CDF exists.
Properties of the PDF

1. The PDF is non-negative because it is the derivative of a non-decreasing function.
2. The integral of the PDF from minus infinity to plus infinity is always one,

$$\int_{-\infty}^{+\infty} f_X(x)\, dx = 1 \tag{3.20}$$

The reason is that the above integral is equal to the difference of the values of the CDF of X at plus infinity and minus infinity, which is one. The above integral is equal to the area under the PDF.
3. The integral of the PDF from a to b is equal to the probability of the variable being in the interval $(a, b]$.

$$\int_{a}^{b} f_X(x)\, dx = F_X(b) - F_X(a) = P\{X \in (a,b]\} \tag{3.21}$$

Note that if the CDF of a random variable is differentiable everywhere, then $P\{X \in (a,b]\} = P\{X \in [a,b]\}$
From the above equation, by setting $a = -\infty$ and $b = x$, we obtain the CDF of a random variable in terms of the PDF of this variable:

$$F_X(x) = \int_{-\infty}^{x} f_X(x')\, dx' \tag{3.22}$$

4. The probability of a continuous variable X assuming a value x is zero, if $F_X(x)$ is differentiable at $X = x$. For an infinitesimal interval, $(x, x + \delta x]$,

$f_X(x)\delta x \approx P\{X \in (x, x + \delta x]\}$. As the length of the interval, δx, tends to zero, so does $\lim_{\delta x \to 0} P\{X \in (x, x + \delta x]\} = P(X = x)$.

Example 3.11:   We put in operation a water pump and we observe the time that it fails. This is an experiment, whose outcome is the time to failure. We know that the pump will fail between 0 and 2000 hrs of operation, and there is no reason to assume that one value is more likely than another in this interval. Find a) the PDF and the CDF of the time to failure, b) the probability that the pump will fail before 1,000 hrs of operation, and c) the probability of failure between 500 and 700 hours.

Solution:

a)   The PDF of the pump is constant between 0 and 2000 hrs and zero elsewhere:

$$f_T(t) = \begin{cases} c & \text{for } 0 \le t \le 2000 \\ 0 & \text{for } t > 2000 \end{cases}$$

The area under the PDF must be one. Therefore,

$$\int_0^{2000} f_T(t)\, dt = 1 \Rightarrow c = \frac{1}{2000}$$

and the PDF of the time to failure is:

$$f_T(t) = \begin{cases} \dfrac{1}{2000} & \text{for } 0 \le t \le 2000 \\ 0 & \text{for } t > 2000 \end{cases}$$

The CDF of the time to failure is:

$$F_T(t) = \int_0^t f_T(t')\, dt' = \begin{cases} \dfrac{t}{2000} & \text{for } 0 \le t \le 2000 \\ 1 & \text{for } t > 2000 \end{cases}$$

Figures 3.11a and 3.11b show the PDF and CDF of the time to failure.

b)   The probability that the pump will fail before 1,000 hours is equal to the CDF of the time to failure at $t = 1000$, $P(T < 1000) = 0.5$.

c)   The probability of failure between 500 and 700 hours is:

$$P(500 < T \le 700) = F_T(700) - F_T(500) = \frac{700}{2000} - \frac{500}{2000} = 0.1$$

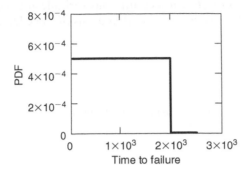

Figure 3.11a PDF of the time to failure.

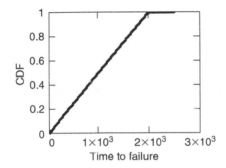

Figure 3.11b CDF of the time to failure.

Moments of a continuous random variable

The mean value of a discrete variable is the sum of each value that the variable can assume scaled by the corresponding probability (Equation 3.16). We can extend this definition to continuous variables by dividing the x axis into differential elements and multiplying the value of the variable for each differential element by the probability of this vale,

$$E(X) \approx \sum_{i=-\infty}^{\infty} x_i P(x_i < X \le x_i + \Delta x) \tag{3.23}$$

In the limit, as the integral length tends to zero the equation for the mean value becomes,

$$E(X) = \int_S x \, dP \tag{3.24}$$

where $dP = \lim_{\Delta x \to 0} P(x < X \le x + \Delta x)$ and S is the sample space. This equation is the counterpart of Equation (3.16) for a continuous variable and it is called the *Lebesgue integral*. Equation (3.24) can be expressed in terms of the CDF as follows,

$$E(X) = \int_{-\infty}^{\infty} x \, dF_X(x) \tag{3.25}$$

The mean value of a continuous variable is therefore,

$$E(X) = \int_{-\infty}^{+\infty} x f_X(x) \, dx \tag{3.26}$$

The ith moment of a continuous variable is the mean value of the ith power of the variable,

$$m_i' = E(X^i) = \int_{-\infty}^{\infty} x^i f_X(x) \, dx \tag{3.27}$$

The ith central moment is the mean value of the ith power of the difference of the variable and its mean value,

$$m_i = E\{[X - E(X)]^i\} = \int_{-\infty}^{\infty} [x - E(X)]^i f_X(x) \, dx \tag{3.28}$$

The variance is,

$$Var(X) = \sigma_X^2 = \int_{-\infty}^{+\infty} [x - E(X)]^2 f_X(x) \, dx \tag{3.29}$$

The following is an alternative way to calculate the variance,

$$Var(X) = \sigma_X^2 = \int_{-\infty}^{+\infty} x^2 f_X(x) \, dx - [E(X)]^2 \tag{3.30}$$

Standard deviation σ_X is the square root of the variance. Equations (3.29) and (3.30) are the counterparts of (3.17) and (3.18). It is observed from Equations (3.26) and (3.29) that the mean value and variance are equal to the first and second central moments, respectively.

Skewness is a measure of the skew of the PDF. It is equal to the third central moment normalized by the third power of the standard deviation,

$$skew(X) = \frac{m_3}{\sigma_X^3} \tag{3.31}$$

Positive values of the skewness indicate that the right tail is longer than the left. Negative values indicate that the left tail is longer than the right tail. A symmetric PDF, such as the one for a normal distribution, has zero skewness.

Kurtosis is a measure of the sharpness of the peak of the PDF and it is equal to the fourth moment normalized by the square of the variance,

$$kurt(X) = \frac{m_4}{\sigma_X^4} \tag{3.32}$$

The normal PDF has kurtosis equal to three. A value of the kurtosis greater than three indicates that the peak of the PDF is sharper than the normal. This PDF has heavier tail than the normal, that is, the variable assumes extreme values that are several standard deviations away from the mean with higher probability than the a normal PDF with same mean and standard deviation.

Example 3.12:   The PDF of the time between arrivals (inter-arrival time) of customers in a bank during lunch time is: $f_T(t) = \frac{1}{\lambda}e^{-\frac{t}{\lambda}}$ for $t \geq 0$, where T is the inter-arrival time. Parameter λ is equal to 5 minutes.

a)   Plot this PDF
     Also, find the following quantities,
b)   The mean value and standard deviation of the inter-arrival time
c)   The CDF of the inter arrival time
d)   The skewness and kurtosis
e)   The probability that the inter arrival is less or equal to 20 minutes.
f)   The probability that the inter arrival time is between 5 and 10 minutes.

Solution:

a)   Figure 3.12 shows the PDF of the inter-arrival time.
b)   The mean value of the inter-arrival time is:

$$E(T) = \int_0^{+\infty} t f_T(t)\, dt = \frac{1}{\lambda} \int_0^{+\infty} t e^{-\frac{t}{\lambda}}\, dt = \lambda = 5 \text{ seconds} \tag{3.33}$$

The standard deviation is the square root of the variance. The variance is:

$$Var(T) = \int_0^{+\infty} t^2 f_T(t)\, dt - [E(T)]^2 = \frac{1}{\lambda} \int_0^{+\infty} t^2 e^{-\frac{t}{\lambda}}\, dt - \lambda^2 = \lambda^2 \tag{3.34}$$

Therefore, the standard deviation is: $\sigma_T = \lambda = 5$ seconds.

c)   The CDF is obtained from Equation (3.22),

$$F_T(t) = \int_{-\infty}^{t} f_T(t')\, dt' = \int_0^{t} \frac{1}{\lambda} e^{-\frac{t'}{\lambda}}\, dt' = 1 - e^{-\frac{t}{\lambda}} \tag{3.35}$$

Figure 3.13 shows this CDF.

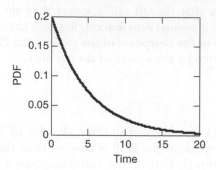

Figure 3.12 PDF of inter arrival time.

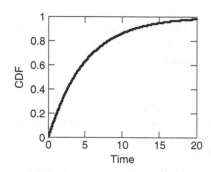

Figure 3.13 CDF of inter arrival time.

d) The central third moment of T is,

$$m_3 = \int_0^\infty (t - \lambda)^3 f_T(t) \, dt = \frac{1}{\lambda} \int_0^\infty (t - \lambda)^3 e^{-\frac{t}{\lambda}} \, dt = -e^{-\frac{t}{\lambda}} (2\lambda^3 + 3\lambda^2 t + t^3) \Big|_0^\infty = 2\lambda^3$$

Therefore, from Equation (3.31) the skewness is equal to 2.
Skewness is positive because the PDF has a longer right tail than the left.
The fourth moment is,

$$m_4 = \int_0^\infty (t - \lambda)^4 f_T(t) \, dt = \frac{1}{\lambda} \int_0^\infty (t - \lambda)^4 e^{-\frac{t}{\lambda}} \, dt$$

$$= -e^{-\frac{t}{\lambda}} (9\lambda^4 + 8\lambda^3 t + 6\lambda^2 t^2 + t^4) \Big|_0^\infty = 9\lambda^4$$

From Equation (3.32) the Kurtosis is equal to 9. This indicates that the peak of the distribution is sharp and the interarrival time can assume extreme values with high probability.

e)   The probability of the inter arrival time being less or equal to 20 minutes is equal to the value of the CDF at 20 minutes: $F_T(20) = 0.982$. Thus after the arrival of a customer, the next customer arrives in less than 20 minutes after the previous customer 98.2% of the time. Consequently, the next customer will arrive after 20 minutes 1.8% of the time. For comparison, if the interarrival time were normal with same mean value and standard deviation as the exponential, a customer would arrive 20 minutes after the previous only 0.14% of the time.

f) $P(5 \leq t \leq 10) = F_T(10) - F_T(5) = 0.233$.

3.1.2.3   Conditional Probability Distribution and Density Functions

In section 3.1.1.3 the conditional probability of an event was defined to be the probability of the event in view of evidence that another event has occurred. According to this definition and Equation (3.7), the conditional CDF of a variable X is the conditional probability of the variable being no greater than some value x, given the occurrence of another event E,

$$F_X(x/E) = P(X \leq x/E) = \frac{P[(X \leq x) \cap E]}{P(E)} \tag{3.36}$$

where $P[(X \leq x) \cap E]$ is the probability of the intersection of events $X \leq x$ and E. A conditional CDF has the properties of an ordinary CDF listed in section 3.1.2.1. For example, $F_X(x/E)$ is a non decreasing function of x, that starts from zero at $x = -\infty$ and increases monotonically to one at $x = +\infty$. In addition, the probability of $X \in (x_1, x_2]$ given E is equal to $F_X(x_2/E) - F_X(x_1/E)$.

If the conditional CDF is differentiable, then the conditional PDF of X given E is equal to the derivative of the conditional CDF:

$$f_X(x/E) = \frac{dF_X(x/E)}{dx} \tag{3.37}$$

A conditional PDF has the properties of ordinary PDFs.

Example 3.13:   This is a continuation of the Example 3.12. The PDF of the time between arrivals (inter-arrival time) of customers in a bank during lunch time is: $f_T(t) = \frac{1}{\lambda}e^{-\frac{t}{\lambda}}$ for $t \geq 0$, where T is the inter arrival time. Parameter λ is equal to 5 minutes. Find the conditional CDF and PDF of the inter arrival time of a customer given that the customer arrives 5 minutes after the previous customer.

Solution:

From Equation (3.36), for time t greater or equal to 5 minutes, the conditional CDF of the inter-arrival time is the probability of the customer arriving after 5 minutes but before or at time t, normalized by the probability of the customer arriving after 5 minutes. For t less than 5 minutes, the conditional probability is zero. Therefore,

$$F_T(t/T > 5) = \begin{cases} 0 & \text{if } t < 5 \\ \dfrac{P(T \leq t \cap T > 5)}{P(T > 5)} & \text{if } t \geq 5 \end{cases}$$

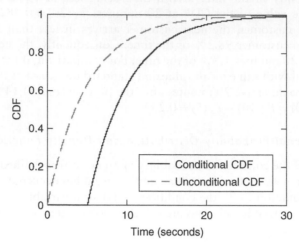

Figure 3.14 Conditional CDF of the inter arrival time of a customer given that the customer arrives in the bank at least 5 minutes after the previous customer.

The probability of the customer arriving between 5 and T minutes is: $P(T \leq t \cap T > 5) = F_T(t) - F_T(5)$, while the probability of the customer arriving after 5 minutes is $P(T > 5) = 1 - F_T(5)$. Therefore,

$$
F_T(t/T > 5) = \begin{cases} 0 & \text{if } t < 5 \\[2mm] \dfrac{F_T(t) - F_T(5)}{1 - F_T(5)} & \text{if } t \geq 5 \end{cases} \tag{3.38}
$$

Figure 3.14 shows this conditional CDF and compares it to the unconditional CDF of the inter-arrival time.

The conditional PDF of the inter-arrival time is the derivative of the conditional CDF,

$$
f_T(t/T > 5) = \begin{cases} 0 & \text{if } t < 5 \\[2mm] \dfrac{f_T(t)}{1 - F_T(5)} & \text{if } t \geq 5 \end{cases} \tag{3.39}
$$

According to this equation, the conditional PDF is obtained by truncating the unconditional PDF of the inter-arrival time in the range $t \geq 5$ and normalizing it by the probability of the customer arriving after 5 minutes. The latter probability is equal to 0.368. Figure 3.15 shows both conditional and unconditional PDFs.

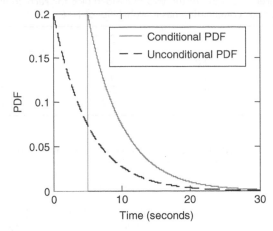

Figure 3.15 Conditional PDF of a customer inter-arrival time given that the customer arrives at least 5 minutes after the previous customer.

3.1.3 *Multiple random variables*

When a decision problem involves multiple random variables, the individual (marginal) probability distributions of these variables do not completely define the probabilistic structure of these variables. The reason is that marginal probability distributions do not account for possible dependencies among these variables. This subsection explains the notion of joint probability distributions. For simplicity, we consider two random variables only; however all concepts can be extrapolated readily to higher dimensions.

3.1.3.1 *Discrete random variables*

The joint probability mass function (PMF) of two discrete random variables X and Y shows the probability of the two variables being equal to each pair of possible values of these variables x_k and y_l,

$$f_{X,Y}(x_k, y_l) = P(X = x_k, Y = y_l) = p_{kl} \tag{3.40}$$

A plot of the joint PMF consists of spikes with height p_{kl} at all combinations of values of the two random variables, $X = x_k$ and $Y = y_l$.

Example 3.14: We roll two fair, four-sided dice. a) Find the joint PMF of the number of dots on each die. b) Find the joint PMF of the sum, Z and the difference W of the number of dots of the first and second die.

Solution:

a) Let random variables X and Y represent the number dots of the first and second dice respectively. There are 16 possible pairs of values of X and Y:$\{x_1 = 1, y_1 = 1\}$, $\{x_1 = 1, y_2 = 2\}, \ldots, \{x_4 = 4, y_4 = 4\}$, and each pair has probability 1/16.

Table 3.5 Sum and difference of dots of the two four-sided dice for each combination of numbers of dots X and Y. Each combination has a probability of $1/16$.

Number of dots $X \backslash Y$	1	2	3	4
1	2,0	3,–1	4,–2	5,–3
2	3,1	4,0	5,–1	6,–2
3	4,2	5,1	6,0	7,–1
4	5,3	6,2	7,1	8,0

Table 3.6 Joint PMF of the sum, Z, and the difference, W, of the number of dots.

$Z \backslash W$	−3	−2	−1	0	1	2	3	$f_Z(z)$
2	0	0	0	1/16	0	0	0	1/16
3	0	0	1/16	0	1/16	0	0	2/16
4	0	1/16	0	1/16	0	1/16	0	3/16
5	1/16	0	1/16	0	1/16	0	1/16	4/16
6	0	1/16	0	1/16	0	1/16	0	3/16
7	0	0	1/16	0	1/16	0	0	2/16
8	0	0	0	1/16	0	0	0	1/16
$f_W(w)$	1/16	2/16	3/16	4/16	3/16	2/16	1/16	1

b) The sum of the number of dots ranges from 2 to 8 and the difference ranges from −3 to 3. Table 3.5 shows the sum and the difference of the number of dots for each combination of the number of dots of each die. Each cell of this table has probability 1/16. From the information in Table 3.5, we can find the probabilities of all combinations of values of the sum and the difference, which are shown in Table 3.6. The last row and column in this table show the marginal PMFs of the sum and difference of the number of dots.

The joint PMF has the following properties:

a) It is non-negative.
b) The probabilities of all possible values of the random variables add up to one,

$$\sum_{k,l} f_{X,Y}(x_k, y_l) = 1 \qquad (3.41)$$

The marginal PMF of one random variable, for example X, $f_X(x_k)$ is obtained by adding up the values of the joint PMF for all possible values of the other variable while the first variable X is kept equal to x_k,

$$f_X(x_k) = \sum_l f_{XY}(x_k, y_l) \qquad (3.42)$$

Similarly: $f_Y(y_l) = \sum_k f_{XY}(x_k, y_l)$

In the above example with the two four-sided dice, Equation (3.42) was used to calculate the marginal PMFs of the sum and difference of the number of dots in Table 3.6. For example, the marginal PMF of the sum of the dots Z is,

$$f_Z(z) = \sum_{w=-3}^{3} f_{ZW}(z, w)$$

For $Z = 4$, this equation yields,

$$f_Z(4) = \sum_{w=-3}^{3} f_{ZW}(z, w_i) = 0 + 1/16 + 0 + 1/16 + 0 + 1/16 + 0 = 3/16$$

Two discrete variables are independent if knowing the value of one does not change the PMF of the other. The joint PMF on two independent variables is equal to the product of the marginal PMFs: $f_{X,Y}(x, y) = f_X(x)f_Y(y)$.

Example 3.15: In Example 3.14 determine if the sum and the difference of the number of dots are independent.

Solution:
The joint PMF of X and Y is equal $1/16$ and it is equal to the product of the two marginal PMFs. Therefore, the two variables are independent. This is reasonable because the outcome of the roll of the first die does not say anything about what will happen in the second roll.

On the other hand, the sum and difference of the number of dots are dependent because knowing the sum of the dots in a roll of the two dice affects the probability of the difference. For example if we learn that the sum is two, we will conclude that each die landed with the face with one dot up. In Table 3.6, the joint PMF is different than the product of the marginal PMFs because the two variables are dependent.

The joint cumulative distribution (CDF) of two discrete random variables X and Y shows the probability that these variables are less than or equal to x_k and y_l,

$$F_{X,Y}(x_k, y_l) = P(X \leq x_k, Y \leq y_l) \tag{3.43}$$

The joint CDF of two independent random variables is equal to the product of the marginal CDFs of these variables.

Example 3.16: In the two four-sided dice example (Example 3.14) find the joint CDFs of the following pairs of random variables: a) of the number of dots X and Y, and b) the joint CDF of the sum Z and the difference W of the number of dots.

Solution:
According to the definition in Equation (3.43), the joint CDF for $X = x$ and $Y = y$ is the sum of the probabilities of all pairs on outcomes for which the number of dots in

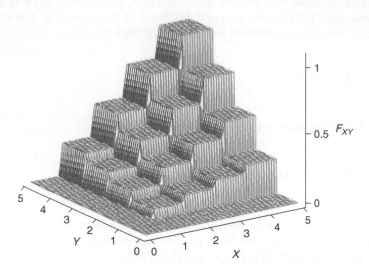

Figure 3.16   Joint CDF of the number of dots of the two four-sided dice, X and Y.

the first die is less than or equal to x and the number of dots in the second is less or equal to y,

$$F_{X,Y}(x,y) = \sum_{l=1}^{4} \sum_{k=1}^{4} f_{XY}(x_k, x_l) \cdot I(x_k \le x) \cdot I(y_l \le y)$$

where $I(\cdot)$ is an indicator function that is equal to 1 if the event in the parenthesis occurs and zero otherwise. Figure 3.16 shows the CDF obtained from the above equation. The joint CDF consists of steps at those locations where the PMF is nonzero.

The above joint CDF is equal to the product of the marginal CDFs of the two random variables because the outcome of a roll of one die does not affect the probability of the outcomes of a roll of the other.

In order to find the joint CDF of the sum and the difference of the number of dots for $Z = z$ and $W = w$ we add up the probabilities of those pairs of numbers of dots for which their sum is less than or equal to z and the difference less than or equal to w,

$$F_{Z,W}(z,w) = \sum_{l=1}^{4} \sum_{k=1}^{4} f_{X,Y}(x_k, y_l) \cdot I(x_k + y_k \le z) \cdot I(x_k - y_l \le w)$$

For example, the joint CDF for $Z = 4$ and $W = -1$ is,

$$F_{Z,W}(4, -1) = f_{X,Y}(1,2) + f_{X,Y}(1,3) = 2/16$$

Similarly, the joint CDF for $Z = 8$ and $W = 2$ is the sum of the values of $f_{X,Y}(x,y)$ for all combinations of x and y but $(4,1)$,

$$F_{Z,W}(8,2) = 1 - f_{X,Y}(4,1) = 15/16$$

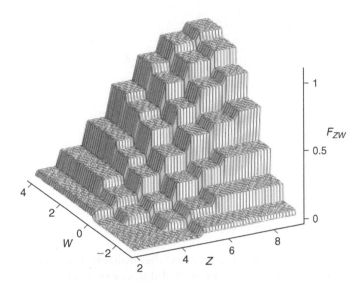

Figure 3.17 CDF of the sum Z and the difference W of the number of dots.

Figure 3.17 shows the joint CDF calculated from the above equation.

The joint CDF ranges from zero to one. In addition it has the following properties:

1. It is a monotonically increasing function of x and y[1]:

$$F_{X,Y}(x_i, y_j) \geq F_{X,Y}(x_k, y_l) \quad \text{for } x_i \geq x_k \text{ and } y_j \geq y_l \tag{3.44}$$

2. It is zero if one variable is equal to minus infinity,

$$F_{X,Y}(-\infty, y) = F_{X,Y}(x, -\infty) = 0 \tag{3.45}$$

3. It is equal to one if both variables are equal to infinity,

$$F_{X,Y}(\infty, \infty) = 1 \tag{3.46}$$

4. The marginal CDF of one variable is obtained from the joint CDF by setting the other variable equal to infinity,

$$F_X(x) = F_{X,Y}(x, \infty) \tag{3.47}$$

If we know the joint CDF then we can compute probability of any event associated with these two variables. For example,

$$P(x_1 < X \leq x_2 \cap y_1 < Y \leq y_2) = F_{X,Y}(x_2, y_2) - F_{X,Y}(x_1, y_2)$$
$$- F_{X,Y}(x_2, y_1) + F_{X,Y}(x_1, y_1) \tag{3.48}$$

---

[1] A stricter condition is defined for continuous variables called the *n-increasing* property. It requires that the mixed derivative of the CDF function should be non-negative everywhere. Equation (3.48) is implied by the *n*-increasing property for the two-variable case.

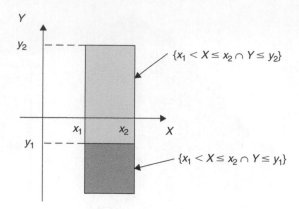

Figure 3.18 Explanation of Equation (3.49).

Proof: In the following, we use the fact that the probability of an event and a subset thereof is equal to the difference of the probabilities of these events.

Event $\{x_1 < X \leq x_2 \cap y_1 < Y \leq y_2\}$ is equal to the difference of events $\{x_1 < X \leq x_2 \cap Y \leq y_2\}$ and $\{x_1 < X \leq x_2 \cap Y \leq y_1\}$ (see Figure 3.18). Since the second event is a subset of the first, the probability of the difference is equal to the difference of the two probabilities,

$$P(x_1 < X \leq x_2 \cap y_1 < Y \leq y_2) = P(x_1 < X \leq x_2 \cap Y \leq y_2)$$
$$- P(x_1 < X \leq x_2 \cap Y \leq y_1) \qquad (3.49)$$

Now

$$P(x_1 < X \leq x_2 \cap y_1 < Y \leq y_2) = P(x_1 < X \leq x_2 \cap Y \leq y_2) - P(x_1 < X \leq x_2 \cap Y \leq y_1)$$
$$= P(X \leq x_2 \cap Y \leq y_2) - P(X \leq x_1 \cap Y \leq y_2)$$
$$- P(X \leq x_2 \cap Y \leq y_1) + P(X \leq x_1 \cap Y \leq y_1)$$

Remembering the definition of the CDF we conclude that the last equation is equivalent to (3.48).

Although, the marginal PMF and CDF of two variables can be obtained from their joint PMF and CDF, respectively, the reverse is not true. The joint PMF of two variables contains more information than their marginal PMFs.

Covariance and correlation coefficient

The covariance and the correlation coefficient are measures of the dependence of two variables. Positive (negative) covariance indicates that the variables move in the same direction (opposite directions).

The covariance of variables X and Y is:

$$\text{cov}(X, Y) = \sum_x \sum_y P(X = x_i, Y = y_j)\{x_i - E(X)\}\{y_j - E(Y)\} \qquad (3.50)$$

where the two sums are for all possible combinations of values of the two variables. The correlation coefficient is the covariance normalized by the product of the standard deviations of the variables,

$$\rho_{XY} = \frac{\text{cov}(X, Y)}{\sigma_X \sigma_Y} \tag{3.51}$$

The correlation coefficient varies between -1 and 1. If the covariance (and hence the correlation coefficient) of two random variables is zero, then the variables are uncorrelated.

If two random variables are independent, then their covariance is zero. However, zero covariance does not always imply independence.

Example 3.17:   Table 3.7 shows the joint PMF of two variables X and Y. Find a) the joint CDF of X and Y, and b) the covariance and correlation coefficient of these variables.

X\Y	0	1
0	0	1/4
1	1/4	1/2

Solution:
We use the following notation for the values of variables X and Y, $x_1 = 0, x_2 = 1$ and $y_1 = 0, y_2 = 1$.

a)   The joint CDF is calculated by the following equation,

$$F_{X,Y}(x,y) = \sum_{l=1}^{2} \sum_{k=1}^{2} f_{XY}(x_k, y_l) \cdot I(x_k \le x) \cdot I(y_l \le y)$$

where $I(\cdot)$ is the indicator function of the event in the parenthesis. Figure 3.19 shows this CDF.

b)   First, we find the marginal PMFs of X and Y. Then we find the mean values and standard deviations of these variables, and the covariance and correlation coefficient.

From Equation (3.42) we find that these marginal PMFs are $f_X(0) = 1/4$ and $f_X(1) = 3/4$ and $f_Y(0) = 1/4$ and $f_Y(1) = 3/4$. The mean values of X and Y are both 0.75 from Equation (3.16). Then, using Equation (3.17) we find that each variable has variance, $\sigma_X^2 = \sigma_Y^2 = (0 - 0.75)^2 \cdot 0.25 + (1 - 0.75)^2 \cdot 0.75 = 0.187$. Therefore, the standard deviations of these variables are $\sigma_X = \sigma_Y = 0.433$.

The covariance is calculated from Equation (3.50),

$$\text{cov}(X, Y) = (0 - 0.75)^2 \cdot 0 + 2(1 - 0.75)(0 - 0.75) \cdot 0.25 + (1 - 0.75)^2 \cdot 0.5$$

$$= -0.063$$

The correlation coefficient is -0.333 from Equation (3.51).

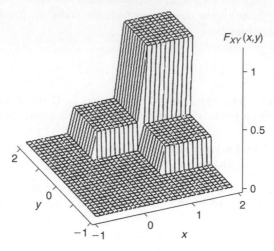

Figure 3.19 Joint CDF of variables X and Y.

3.1.3.2 *Continuous random variables*

The definition of the joint CDF of two continuous random variables is identical with that for discrete variables (see Equation 43). Properties 1–4 and Equations (3.44 to 3.48) for discrete variable CDFs apply to continuous variables too. However, the CDF is continuous as opposed to the CDF of discrete variables, which consists of steps (see Figures 3.16 and 3.17).

The joint PDF of two random variables is the second derivative of the CDF with respect to these variables:

$$f_{X,Y}(x,y) = \frac{\partial^2 F_{X,Y}(x,y)}{\partial x \partial y} \tag{3.52}$$

This equation can be interpreted as follows: the probability of a pair of random variables being in the rectangle with sides $[x, x+dx]$ and $[y, y+dy]$ is equal to the joint PDF of these variables times the area of the rectangle.

The joint PDF has the following properties:

a. It is non-negative, $f_{X,Y}(x,y) \geq 0$
b. The volume under the joint PDF is one, $\int_{-\infty}^{\infty} \int_{-\infty}^{\infty} f_{X,Y}(x,y)\, dx\, dy = 1$
c. The marginal PDF of variable X is obtained from the joint PDF of X and Y by integrating out Y,

$$f_X(x) = \int\limits_{-\infty}^{\infty} f_{X,Y}(x,y)\, dy \tag{3.53}$$

The last equation is the counterpart of Equation (3.42) for discrete variables.

The probability of variables X and Y being in a range D is the volume under the joint PDF with support D,

$$P\{(X, Y) \in D\} = \iint_D f_{X,Y}(x, y) \, dx \, dy \tag{3.54}$$

Equation (3.53) is important because it enables a decision maker to compute the probability of an event, such as failure of a structure.

From Equation (3.54) it follows that the joint CDF and PDF are related by the equation,

$$F_{X,Y}(x, y) = \int_{-\infty}^{y} \int_{-\infty}^{x} f_{X,Y}(x, y) \, dx \, dy \tag{3.55}$$

Two continuous random variables are independent if their joint PDF is equal to the product of the marginal PDFs of these variables,

$$f_{X,Y}(x, y) = f_X(x) f_Y(y) \tag{3.56}$$

Example 3.18: A rod with strength (stress at failure) Y is subjected to stress X. The strength is a random variable with Weibull PDF,

$$f_Y(y) = \frac{\beta}{\theta - Y_{min}} \left(\frac{y - Y_{min}}{\theta - Y_{min}} \right)^{\beta-1} \exp\left[-\left(\frac{y - Y_{min}}{\theta - Y_{min}} \right)^{\beta} \right] \quad \text{for } y \geq Y_{min}$$

where Y_{min} is the minimum value of the strength, and β and θ are the shape and scale parameters. The values of these parameters are, $Y_{min} = 90{,}000$ psi, $\beta = 2$ and $\theta = 130{,}000$ psi.

The stress follows a Rayleigh PDF,

$$f_X(x) = \frac{x - X_{min}}{\sigma_R^2} \exp\left\{ -\frac{1}{2} \left(\frac{x - X_{min}}{\sigma_R} \right)^2 \right\}$$

where X_{min} is the minimum value of the stress. The values of these parameters are $X_{min} = 70{,}868$ psi, and $\sigma_R = 15{,}260$ psi. See section 3.2.1.2 for the properties of the Weibull and Rayleigh distributions. The stress and strength are independent. Find the probability of failure, that is, the probability of the strength being less than the stress.

Solution:
Figure 3.20 shows the PDFs of the strength and stress. The two PDFs overlap in the region from 90,000 to about 130,000 psi. This means that the stress can exceed the stress.

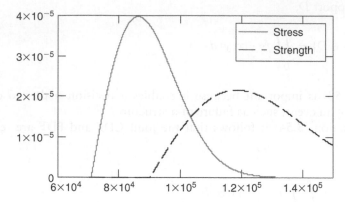

Figure 3.20 PDFs of strength and stress.

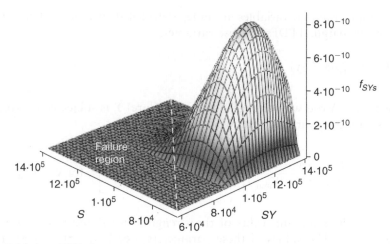

Figure 3.21 Joint PDF of stress and strength.

The joint PDF of the stress X and strength Y is equal to their product because X and Y are independent, $f_{X,Y}(x, y) = f_X(x)f_Y(y)$. Figure 3.21 shows this joint PDF while Figure 3.22 shows the same PDF but truncated in the failure region.

According to Equation (3.54), the probability of failure is the volume under this joint PDF supported by the region in which the stress exceeds the strength,

$$PF = \iint\limits_{X>Y} f_{X,Y}(x, y)\, dx\, dy = \int\limits_{Y_{min}}^{\infty} f_Y(y) \left[\int\limits_{y}^{\infty} f_X(x)\, dx \right] dy$$

The failure probability was found to be 0.032 by numerical integration. The upper limit of the inner integral for the stress was set equal to 140,000 psi in order to reduce the computational cost. This did not affect the accuracy of integration because the stress has a very low probability to exceed this limit (Figure 3.20).

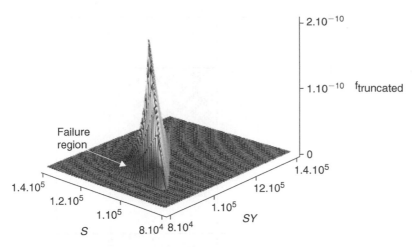

Figure 3.22 Joint PDF of stress and strength truncated in the failure region where the strength falls short of the stress.

Covariance of two continuous random variables

The counterpart of Equation (3.50) for the covariance of two random variables is,

$$Cov_{XY} = \iint\limits_{x,y} \{x - E(X)\}\{y - E(Y)\} f_{X,Y}(x,y)\, dx\, dy \tag{3.57}$$

The covariance can be calculated by the following equation, which is equivalent to (3.57),

$$Cov_{XY} = \iint\limits_{x,y} x \cdot y \cdot f_{X,Y}(x,y)\, dx\, dy - E(X)E(Y) \tag{3.58}$$

Conditional CDF and PDF

Equation (3.36) in Section 3.1.2.3 defined the conditional CDF of a random variable, Y, given that event E has occurred,

$$F_Y(y/E) = \frac{P(Y \leq y \cap E)}{P(E)}.$$

Here, we study the special case where conditioning event E refers to a random variable X, for example $E = [x_1 < X \leq x_2]$. Then,

$$F_Y(y/E) = \frac{P(Y \leq y \cap x_1 < X \leq x_2)}{P(x_1 < X \leq x_2)} = \frac{F_{X,Y}(x_2, y) - F_{X,Y}(x_1, y)}{F_X(x_2) - F_X(x_1)} \tag{3.59}$$

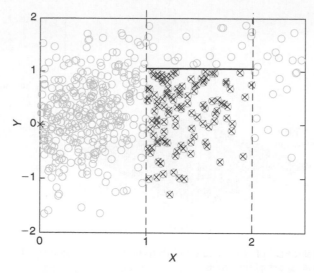

Figure 3.23 Relative frequency interpretation of the conditional probability of Y being less than or equal to 1 given that X is in the range $[1, 2]$. This probability is approximately equal to the fraction of observations in the strip enclosed by the dashed lines that are under the solid horizontal line.

Equation (3.59) has the following relative frequency interpretation. We collect a large number of pairs of observed values of the two variables, X and Y (gray circles in Figure 3.23). Conditional CDF $F_Y(1/1 < X \leq 2)$ is approximately equal to the fraction of the number of observations in the strip for which $1 < X \leq 2$ that are also under the solid horizontal line (black crosses).

The conditional PDF of Y given E is obtained by differentiating Equation (3.59) with respect to y,

$$f_Y(y/E) = \frac{\frac{\partial}{\partial y}\left(\int\limits_{-\infty}^{y}\int\limits_{x_1}^{x_2} f_{X,Y}(x,y')\, dx\, dy'\right)}{F_X(x_2) - F_X(x_1)} = \frac{\int\limits_{x_1}^{x_2} f_{X,Y}(x,y)\, dx}{F_X(x_2) - F_X(x_1)} \tag{3.60}$$

If two variables are independent, then the conditional of Y given event E is equal to the marginal CDF of Y.

If the conditioning event is $E: X = x$ then the conditional probability of Y can be found from Equation (3.60) by setting $x_2 = x_1 + \delta$, where δ is infinitesimal,

$$f_{Y/X}(y/X = x) = \lim_{\delta \to 0} \frac{\int\limits_{x_1}^{x_1+\delta} f_{X,Y}(x,y)\, dx}{\int\limits_{x_1}^{x_1+\delta} f_X(x)\, dx} = \lim_{\delta \to 0} \frac{f_{X,Y}(x,y) \cdot \delta}{f_X(x) \cdot \delta} = \frac{f_{X,Y}(x,y)}{f_X(x)} \tag{3.61}$$

In the above equation the PDF of x in the denominator is obtained from the joint PDF of X and Y by integrating out variable y, $f_X(x) = \int_{-\infty}^{\infty} f_{X,Y}(x,y)\, dy$.

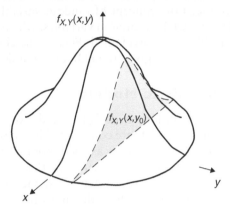

Figure 3.24 Joint PDF of random variables X and Y, $f_{X,Y}(x,y)$.

From the above equation, we observe that the marginal PDF of Y is obtained by multiplying the conditional probability of Y given $X = x$ by the marginal PDF of X and by integrating out x.

$$f_Y(y) = \int_X f_{Y/X}(y/X = x)f_X(x)\, dx. \qquad (3.62)$$

The above equation is the counterpart of Equation (3.12) for the total probability theorem for events.

Similarly, the conditional PDF of X given $Y = y$ is,

$$f_{X/Y}(x/Y = y) = \frac{f_{X,Y}(x,y)}{f_Y(y)} = \frac{f_{X,Y}(x,y)}{\int_{-\infty}^{\infty} f_{X,Y}(x,y)\, dx} \qquad (3.63)$$

Figure 3.24 illustrates the calculation of $f_{X/Y}(x/y_0)$ in Equation (3.63). The conditional PDF of X given $Y = y_0$ is equal to the joint PDF of X and Y at y_0 normalized by the shaded area of slice $f_{X,Y(x,y_0)}$.

$$f_{X/Y}(x/y_0) = \frac{f_{X,Y}(x,y_0)}{\int_{-\infty}^{+\infty} f_{X,Y}(x,y_0)\, dx}$$

The conditional mean value of Y given $X = x$ is $E(Y/X = x) = \int_Y y \cdot f_Y(y/X = x)\, dy$. Then the unconditional mean value of Y is,

$$E(Y) = \int_X E(Y/X = x)f_X(x)\, dx \qquad (3.64)$$

The first part of this chapter explained the definition and properties of objective probability. It also introduced the concept of random variables. The definitions and the properties of the PDF and CDF of random variables, which define the probabilistic structure of random variables, were explained. Finally, the chapter explained how to calculate the probabilities of events given the PDF of the underlying variables.

3.2 Common Probabilistic Models

In order to make informed decisions a designer needs to construct models of the random variables. Probabilistic models are constructed based on data and expert judgment. Some probability distributions, such as the binomial, lognormal and Gumbel distributions accurately describe quantities such as the number of voltage spikes within a time period, the strength of a system, and the applied loads on an off-shore platform. Other distributions, such as the uniform, exponential and beta, are useful when there is little data or the understanding of the underlying process is poor. The reason is that these distributions are flexible and they provide conservative estimates of probability of an extreme event that can cause failure. This part presents some common probability distributions and illustrates their use for modeling uncertainty.

A comprehensive description of families of probability distributions of a single variable that are useful in engineering analysis and design is found in Bury (1999). Todinov (2005) presents probability distributions that are useful in system reliability analysis, such as distributions of strength of materials (stress to failure). Probabilistic models of quantities that are useful in structural reliability, such as extreme loads, and strength of materials can be found in Ang and Tang (1975, chapters 1, 2 and 4), Madsen et al. (1986), Ochi (1990), Nikolaidis and Kaplan (1992), Melchers (1999) and Haldar and Mahadevan (2000). Fox (2005) presented guidelines for selecting the most suitable distribution depending on the amount of available information and the degree of understanding of a phenomenon.

3.2.1 Distributions of a single random variable

3.2.1.1 Discrete variables

Bernoulli distribution
This is the distribution of a binary variable that assumes values of 0 and 1. The binary variable characterizes the state of a system with two states: failure (the variable is equal to 0) and success or survival (the variable is equal to 1). The mean value of this variable is p and the variance $p(1-p)$, where p is the probability of success.

Geometric distribution
This distribution characterizes the number of successful Bernoulli trials (Section 3.1.1.4), x, before failure in $x+1$th trial. If the probability of success is p then the PMF of the number of successful trials before failure is,

$$f_X(x) = p^x(1-p) \tag{3.65}$$

This PMF has mean value $\frac{1-p}{p}$ and variance $\frac{1-p}{p^2}$

Binomial distribution
This distribution characterizes the number of successful Bernoulli trials x out of a total of n trials. For example, when a quality control manager inspects n nominally identical parts, the manager performs a Bernoulli trial with outcomes "defective" and "good part" in each inspection. Assume that these trials are independent. Then both numbers of defective and good parts follow the Binomial distribution.

The PMF of a variable following a Binomial distribution (Binomial variable) is:

$$f_X(x) = \binom{n}{x} p^x (1-p)^{n-x} = \frac{n!}{(n-x)!x!} p^x (1-p)^{n-x} \tag{3.66}$$

where p is the probability of a successful trial. This variable has mean value np and variance $np(1-p)$. The corresponding CDF is:

$$F_X(x) = \sum_{i=0}^{x} \frac{n!}{(n-i)!i!} p^i (1-p)^{n-i} \tag{3.67}$$

Example 3.19: On average, there is one defective part out of 1,000 nominally identical parts produced in a factory. Find the probability that there is no more than one defective part in a batch of $n=100$ parts.

Solution:
The probability of no more than one defective part is equal to the CDF of the number of defective parts evaluated at $x=1$. This probability is found to be 0.9954 by plugging $p=10^{-3}$, $n=100$ and $x=1$ in Equation (3.67).

Poisson distribution
This distribution characterizes the number of random events occurring in a fixed interval of time. If the time between successive events is exponential, then the number of occurrences in time interval t, X, is Poisson,

$$f_X(x) = \frac{(vt)^x e^{-vt}}{x!} \tag{3.68}$$

where v is the occurrence rate of events (events per unit time). Both the mean value and variance of X are equal to vt.

Example 3.20: An accident occurs at a busy intersection once a week, on average. The time between successive accidents follows an exponential distribution. A) Determine the PMF of the number of accidents in 28 days. B) Suppose that 11 accidents happened in 28 days. Should one be surprised by this evidence?

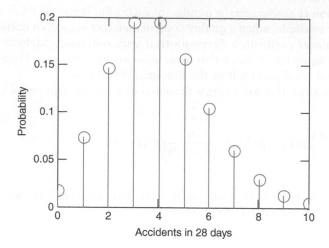

Figure 3.25 Probabilities of 0 to 10 accidents at an intersection in four weeks.

Solution:
The occurrence rate of accidents is $v = \frac{1}{7}$ accidents per day. Let N be the number of accidents in $t = 28$ days. Then, the probability of $N = n$ is obtained from Equation (3.68) by plugging N for X and n for x,

$$f_N(n) = \frac{(vt)^n e^{-vt}}{n!}$$

Figure 3.25 shows the probability of the number of accidents. It is observed that the most likely numbers of accidents are 3 and 4.

The probability of no more than 10 accidents is the sum of the probabilities from zero to 10 accidents. This probability is 0.997. The probability of having more than 10 accidents is only 0.003. Therefore, it is surprising that 11 accidents occurred in four weeks[2]. One may question the assumptions that one accident occurs per week on average, or that the time between accidents follows the exponential distribution based on the result.

Example 3.21: A friend of yours decided not to buy flood insurance for his house because, according to government data, a catastrophic flood in his neighborhood occurs once every 500 years. Your friend plans to stay in his house for 30 years. A major flood would cause at least $200,000 damage to the house, which would devastate your friend. Flood insurance costs $200 per year and it is not likely to change much in the future. Did you friend make an informed decision?

---

[2] It is generally not correct to call a probability small or large just by looking at its value. Judgment is required as well as some knowledge of the outcomes. In the given example, one can question the assumptions as they appear likely to be incorrect since from our experience, traffic statistics data is reflected in reality.

Solution:
To find out if your friend made an informed decision we could calculate the probability of a major flood in the 30 years that he plans to stay in his house. Assume that the time between floods is an exponential random variable. The mean value of this variable is 500 years. We will calculate the probability of one or more floods in 30 years. The probability of n floods in 30 years is calculated from Equation (3.68) by substituting 1/500 for v and 30 for t. The probability of one flood is 0.057. The probability of one or more floods is $1 - f_N(0)$ which is 0.058.

In order to avert the risk of a financial disaster, your friend would have to pay $30 \cdot \$200 = \$6,000$. In conclusion, your friend assumed the risk of a financial disaster which may happen with 6% probability (expected loss of $\$200,000 \cdot 0.058 = \$11,600$) in order to save $6,000. He/she should reconsider this decision in view of the above calculations.

3.2.1.2 Continuous variables

This subsection presents nine families of probability distributions. You should represent quantities for which you have limited or no data with the first two families. In addition, these two families could represent an unknown quantity for which you cannot collect data from a repeatable experiment. Examples are the distance between two planets and the value of a market index in the future.

The remaining families of distributions are suitable for specific types of variables such as measurement errors, proportions and number of times an extreme event occurs in a long period. You can observe sample values of these variables from a repeated experiment, but it is impractical or impossible to determine their true value.

Uniform distribution
This distribution should be used as a rough model in the absence of data. The uniform distribution characterizes a variable X when a decision maker knows the range of this variable $[x_l, x_u]$, and he/she has no reason to believe that one value is more likely than any other in this range. For example, the uniform distribution could model the error of a predictive model. The uniform PDF is constant in the range of the variable,

$$
f_X(x) = \begin{cases} \dfrac{1}{x_u - x_l} & \text{if } x_l \leq x \leq x_u \\[2mm] 0 & \text{otherwise} \end{cases} \tag{3.69}
$$

The CDF of this variable is,

$$
F_X(x) = \begin{cases} \dfrac{x - x_l}{x_u - x_l} & \text{if } x_l \leq x \leq x_u \\[2mm] 0 & \text{if } x < x_l \\[2mm] 1 & \text{if } x > x_u \end{cases} \tag{3.70}
$$

The mean value and variance of X are: $E(X) = \frac{x_l + x_u}{2}, \sigma_X^2 = \frac{(x_u - x_l)^2}{12}$.

Triangular distribution

This distribution is used when there limited or no data, like the uniform distribution. A decision maker should consider this distribution when she can estimate the bounds and the most likely value of an uncertain quantity. The PDF of a triangular distribution resembles a triangle with the peak at the most likely value and the ends at the lower and upper limits. Convolution of two independent uniform distributions is a triangular distribution.

Poisson process and the exponential distribution

Rare events, such as a severe earthquake, a voltage spike or a shock occur randomly in space or in time, at a constant rate. Occurrences of such events satisfy the following conditions,

- The probability of the event occurring in a short interval Δt is proportional to the length of the interval, i.e. the probability is approximately equal to $v \cdot \Delta t$ where v is the rate of occurrence of the event (occurrences per unit time).
- The probability of more than one event occurring in the short interval above is zero.
- The number of events occurring in one interval is independent of what happened in any other non overlapping interval. For example, the probability that a customer could arrive at a bank in a short time interval is independent of the previous customers' arrivals.

The occurrence of events that satisfy the above conditions is called Poisson process.

The interarrival time between events in a Poisson process follows the exponential probability distribution. The PDF of this distribution is,

$$f_X(x) = \begin{cases} \dfrac{1}{\lambda} e^{-\frac{x-x_l}{\lambda}} & \text{if } x \geq x_l \\ 0 & \text{if } x < x_l \end{cases} \tag{3.71}$$

The CDF of this variable is,

$$F_X(x) = \begin{cases} 1 - e^{-\frac{x-x_l}{\lambda}} & \text{if } x \geq x_l \\ 0 & \text{if } x < x_l \end{cases} \tag{3.72}$$

The exponential PDF has two parameters; the mean interarrival time $\lambda + x_l$ and its minimum value x_l. The standard deviation of this PDF is equal to λ. The inverse of parameter λ is equal to the interarrival rate (occurrences of the underlying event per unit time) v. Figure 3.26 shows the PDF of an exponentially distributed random variable with minimum value zero and standard deviation 5.

The exponential distribution is suitable for variables about which the decision maker only knows their mean values and lower bounds. Often, this distribution represents the interarrival time of customers in a bank or at a post office, and the time to complete a task. The exponential distribution has a heavy tail, which means that the underlying variable can exceed a limit with a considerable probability even if this limit is three

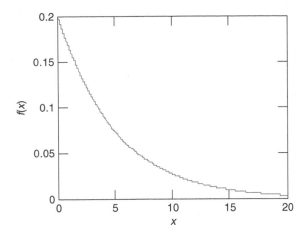

Figure 3.26 PDF of an exponentially distributed random variable with mean value and standard deviation 5 and zero minimum value.

or more standard deviations away from the mean value. For example, the probability of an exponential variable exceeding its mean value by three standard deviations is 0.049, while the same probability for a normally distributed random variable is only 0.00135.

Example 3.22:   A major flood occurs in Cedar Rapids, Iowa every 500 years, on average. Two major floods occurred in 1993 and 2008. People were surprised about this occurrence of two major floods within such a short period. Is the surprise justifiable?

Solution:
To find out if people should be surprised, we will calculate the probability that two major floods can occur in less than 15 years. People's surprise could be justifiable if this event has very low probability (for example 1%).

The time between occurrences of very rare events is usually exponential. We assume that the time between major floods is exponential with a mean value of 500 years. The probability of two major floods in 15 years is equal to the probability of the time between floods being less or equal to 15. This probability is equal to the CDF of the interarrival time in Equation (3.72) for $\lambda = 500$ and $x_l = 0$, which is 0.03. People's reaction could be reasonable because this is a low probability. However, the author thinks that people would not be that surprised if they knew that two major floods can happen in 15 years with 3% probability. This is because people react differently based on how uncertainty is presented to them.

Normal distribution
The normal (or Gaussian) distribution is one of the most commonly used distributions. Many statistical and probabilistic tools revolve around this distribution. The normal distribution is suitable for characterizing a quantity that is the result of many random causes. Examples are noise signals, the lifetime of a system of standby components

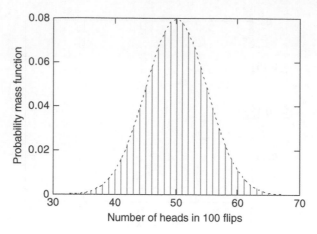

Figure 3.27 Approximation of a binomial PDF by a normal with same mean and standard deviation.

and perfect switching devices, measurement errors and scores of IQ tests. The normal distribution approximates other distributions, such as the binomial, Poisson and Student's distributions. Figure 3.27 shows that the PMF of the number of heads in 100 flips of a fair coin represented by binomial distribution, is very close to a normal PDF with same mean value (50) and standard deviation (5).

The family of normal distributions is closed with respect to linear transformations i.e. the sum of normal random variables is normal, as is a normal random variable scaled by a constant.

The normal probability distribution is defined in terms of the mean value μ and standard deviation σ,

$$f_X(x) = \frac{1}{\sqrt{2\pi}\sigma} e^{-\frac{(x-\mu)^2}{2\sigma^2}} \tag{3.73}$$

This distribution is symmetric and has a bell shape (Figure 3.27). The mass is clustered around the mean value and the likelihood of deviations from the mean diminishes rapidly with the distance from the mean. For example, the probabilities that a normal random variable deviates no more than one, two and three standard deviations from its mean value are 68.3%, 95.4% and 99.7%, respectively. A *standard* normal random variable has zero mean and unit standard deviation. Since the set of normal distributions is closed under linear transformations, a normal random variable can be transformed into a standard normal by subtracting from the normal variable its mean value and normalizing the difference by its standard deviation. $Z = \frac{X-\mu}{\sigma}$. The PDF of a standard normal random variable, Z, is,

$$\phi_Z(z) = \frac{1}{\sqrt{2\pi}} e^{-\frac{z^2}{2}} \tag{3.74}$$

There is no closed form expression for the normal CDF. Fortunately, calculations involving random variables can be performed very efficiently and conveniently because

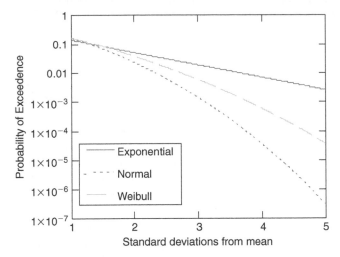

Figure 3.28 Probability of three types of random variables exceeding thresholds located between one and five standard deviations from the mean value.

percentiles of the standard normal CDF $\Phi(z)$ have been tabulated and approximated by closed form functions.

Example 3.23: The stress in a rod S is normal with mean value $\mu_S = 90{,}000$ psi and standard deviation $\sigma_S = 5{,}000$ psi. The strength is $S_u = 110{,}000$ psi, deterministic. Find the probability of failure, i.e. the probability of the stress exceeding the strength.

Solution:
The probability of failure is $P(F) = P(S > S_u)$. We write this expression in an equivalent form by subtracting the mean value of the stress from both sides and normalizing by the standard deviation of the stress $P(F) = P\left(\frac{S - \mu_S}{\sigma_S} > \frac{S_u - \mu_S}{\sigma_S}\right)$. The left hand side of the inequality in the parenthesis is a standard normal variable, and the right hand side is equal to 4. Therefore, the probability of failure is equal to the probability of a standard normal variable exceeding 4, which is equal to one minus the value of the CDF of a standard normal variable at 4. This probability is found to be $3.2 \cdot 10^{-5}$ by using an approximation of the standard normal CDF or tables of a standard normal distribution.

The normal distribution has a light tail, i.e. the probability that a normal random variable exceeds a threshold located a certain number of standard deviations from its mean value is small compared to the corresponding probability for other distributions. Figure 3.28 compares the probability that an exponential, a normal and a Weibull variable exceed a threshold that is between one to five standard deviations larger than the mean value. The exceedance probability for a normal random variable diminishes faster than for the exponential and Weibull.

Approximations of the loads and/or strength by a normal distribution may lead to underestimation of the probability of failure of a system. Therefore, use of the normal distribution may lead to designs that look safe on the paper but are likely to fail in practice. Therefore, decision makers should use the normal approximation only if they have strong evidence that supports this approximation (Fox, 2005).

The central limit theorem

The central limit theory gives prominence to the normal distribution because it suggests that many random variables in practical applications are normal. The central limit theorem states that the distribution of the sum of many random variables, none of which dominates, tends to a normal distribution, even if the random variables are not normal. For example, if the error in some measurement is the result of many contributing factors, then the distribution of the error approximates the normal distribution. The average of many independent realizations of a random variable (e.g., ten uniform random variables) is approximately normal (Papoulis, 1965, p. 266).

Beta distribution

The beta distribution characterizes quantities that are bounded from both sides and whose most likely values are known, such as an unknown probability or the cost of a project. Because this distribution has exceptional flexibility (see Figure 3.29) it is useful as a rough model in the absence of data. There are two versions of the beta distributions with two and four parameters respectively.

The two-parameter beta PDF is,

$$f_X(x, \beta_1, \beta_2) = \frac{\Gamma(\beta_1 + \beta_2)}{\Gamma(\beta_1)\Gamma(\beta_2)} x^{\beta_1 - 1} \cdot (1 - x)^{\beta_2 - 1} \quad \text{for } 0 < x < 1 \tag{3.75}$$

where $\Gamma(\cdot)$ is the Gamma function defined as $\Gamma(\beta) = \int_0^\infty x^{\beta - 1} e^{-x} dx$. Parameters β_1, β_2 are the shape parameters of the beta PDF. For integer values of the argument, the Gamma function is equal to the factorial function $\Gamma(\beta) = (\beta - 1)!$ The PDF of the Beta distribution in equation (3.75) is zero for x outside the interval [0, 1]. This PDF is useful for characterizing variables restricted in the range from 0 to 1 whose most likely value is known. For example, a decision maker can use the beta PDF to characterize the proportion of defective parts in a lot.

The CDF of the beta distribution is,

$$F_X(x) = \frac{B_x(\beta_1, \beta_2)}{B(\beta_1, \beta_2)} \tag{3.76}$$

where $B(\beta_1, \beta_2)$ is the beta function, $B(\beta_1, \beta_2) = \frac{\Gamma(\beta_1)\Gamma(\beta_2)}{\Gamma(\beta_1 + \beta_2)}$ and $B_x(\beta_1, \beta_2) = \int_0^x x'^{\beta_1 - 1}(1 - x')^{\beta_2 - 1} dx'$ is the incomplete beta function.

The mean value of the two-parameter beta PDF is,

$$E(X) = \frac{\beta_1}{\beta_1 + \beta_2} \tag{3.77}$$

and its variance is,

$$\sigma_X^2 = \frac{\beta_1 \beta_2}{(\beta_1 + \beta_2)^2 (\beta_1 + \beta_2 + 1)} \tag{3.78}$$

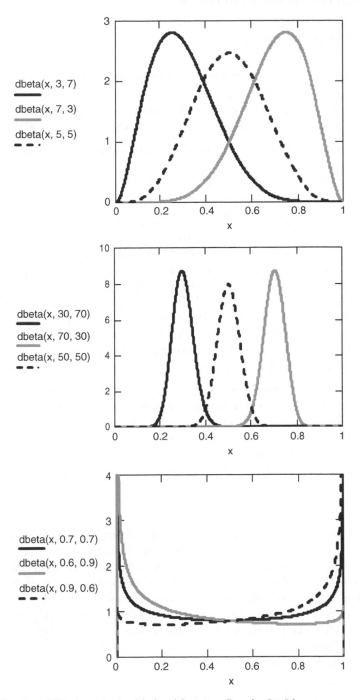

Figure 3.29 Beta PDF plotted using Mathcad function dbeta(x, β_1, β_2).

The coefficient of variation of this PDF is,

$$COV_X = \sqrt{\frac{\beta_2}{\beta_1(\beta_1 + \beta_2 + 1)}} \tag{3.79}$$

The most likely value or mode of the Beta distribution is given by $x_m = \frac{\beta_1 - 1}{\beta_1 + \beta_2 - 2}$. This equation is valid for $\beta_1 \geq 1$ and $\beta_1 + \beta_2 > 2$.

Figure 3.29 shows the effect of the shape parameters on the beta PDF.

The following effects of the shape parameters on the shape of the PDF are observed:

- The mean value and median of X are controlled by the ratio of the shape parameters.
- Increasing the shape parameters decreases the dispersion of the PDF.
- The mean and the median approach each other as the shape parameters increase.

The uniform PDF is a special case of the beta distribution for $\beta_1 = \beta_2 = 1$.

Winkler (1967) proposed to estimate the shape parameters of the beta distribution using the *Equivalent Prior Sample* method. The size of the equivalent prior sample reflects one's confidence in her estimate of the mean value of the proportion. The values of the shape parameters are derived from the equivalent prior sample size $\hat{n}$ and the mean value of the proportion p as follows, $\beta_1 = p\hat{n}$ and $\beta_2 = (1 - p)\hat{n}$.

For example, an expert estimates that the probability of an event is 0.3 and the equivalent prior sample size is 100. This means that the expert is as confident in her subjective estimate of the probability as if she had conducted 100 experiments and observed 30 occurrences of the event. The shape parameters are $\beta_1 = 30$ and $\beta_2 = 70$ in this case.

Example 3.24: Nominally identical parts are delivered to a manufacturer from two factories, A and B. We want to estimate the proportion of parts from these two factories. An expert says that 20 percent of parts are from factory A. When asked to quantify his/her confidence in this estimate the expert says that it is very unlikely that the percent of parts from factory A is less than 15 or more than 25. We want to estimate the PDF of the percentage of parts from factory A based on the evidence from the expert.

Solution:

When we do not know a quantity such as a probability, we can treat it as a random variable, p. The expert gave us an idea about the bounds and the mean value of the proportion of parts from factory A. We assume that the probability of a part being from factory A follows the beta distribution. The shape parameters of the distribution of the number of parts are estimated as follows. The mean value of this variable is 0.2 based on the expert's evidence. The statement that the percentage of parts from factory A is very unlikely to be less than 15% or more than 25% is interpreted as follows: p is in the range from 0.15 and 0.25 with probability 0.95.

The expert's judgment is expressed by the following equations,

$$E(p) = \frac{\beta_1}{\beta_1 + \beta_2} = 0.2$$

$$P(0.15 < p \leq 0.25) = F_P(0.25) - F_P(0.15) = 0.95$$

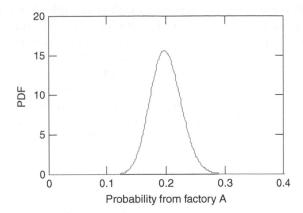

Figure 3.30 PDF of probability that a part selected at random is from factory A reflecting the expert's judgment.

where $F_P(p)$ is the CDF of the probability of a part being from factory A, which is obtained from Equation (3.76). Solving the two equations for shape parameters β_1 and β_2 we find that $\beta_1 = 48.64$ and $\beta_2 = 194.56$. The PDF of the probability p is shown in Figure 3.30.

The coefficient of variation of p is equal to 0.4 from Equation (3.79). This indicates that the expert is somewhat uncertain about the true value probability p.

The four-parameter beta distribution is useful for quantities that assume values in a known range. Examples are the cost and duration of a project, and the strength and modulus of elasticity of some material. The four-parameter PDF is,

$$f_X(x, x_l, x_u, \beta_1, \beta_2) = \frac{\Gamma(\beta_1 + \beta_2)}{\Gamma(\beta_1)\Gamma(\beta_2)} \left(\frac{x - x_l}{x_u - x_l} \right)^{\beta_1 - 1} \cdot \left(1 - \frac{x - x_l}{x_u - x_l} \right)^{\beta_2 - 1} \frac{1}{x_u - x_l}$$

$$\text{for } x_l < x < x_u \qquad (3.80)$$

The PDF is zero outside the interval $[x_l, x_u]$. If variable x is relocated by $\mu = x_l$ and scaled by $\theta = x_u - x_l$, then the reduced variable $x' = \frac{x - \mu}{\theta}$ follows the two-parameter beta distribution.

The mean value of the four-parameter beta PDF is,

$$E(X) = \frac{\beta_1}{\beta_1 + \beta_2}(x_u - x_l) + x_l \qquad (3.81)$$

and its variance is,

$$\sigma_X^2 = \frac{\beta_1 \beta_2 (x_u - x_l)^2}{(\beta_1 + \beta_2)^2 (\beta_1 + \beta_2 + 1)} \qquad (3.82)$$

The most likely value or mode of the distribution is $x_m = x_l + (x_u - x_l)\frac{\beta_1 - 1}{\beta_1 + \beta_2 - 2}$. This equation is valid for $\beta_1 \geq 1$ and $\beta_1 + \beta_2 > 2$.

Lognormal distribution

The lognormal distribution characterizes a random variable that is the product of a large number of random variables. This distribution is a good candidate for modeling positive random variables whose histograms have positive skewness. The lognormal distribution fits well fatigue strength data, including the endurance strength of steel. This distribution is a popular model of material strength.

Various probabilistic models of fatigue give raise to the lognormal distribution. An important model is the theory of proportionate effect (Parzen, 1967). Consider some material that is under cyclic loading. The increase of the length of a crack in each cycle is approximately proportional to the length of the same crack that that has been already created in the beginning of the cycle,

$$A_k - A_{k-1} = \varepsilon_k A_{k-1} \tag{3.83}$$

where A_k is the length of the crack in the end of the kth cycle and ε_k is a random variable. It is assumed that random variables $\varepsilon_i, i = 1, \ldots, k$ are independent. Then, the crack length is the product of k independent random variables,

$$A_k = \prod_{i=1}^{k} (1 + \varepsilon_i) A_1 \tag{3.84}$$

Therefore, for a large number of load cycles, the cumulative damage is lognormal. The lognormal distribution is also used to model the time to completion of a task.

The lognormal PDF is related to the normal as follows. If Y is a normal random variable, then variable $X = e^Y$ is lognormal. Conversely, if a variable X is lognormal, then its natural logarithm $\ln(X)$ is normal.

The PDF of this distribution is,

$$f_X(x) = \frac{1}{x\sigma\sqrt{2\pi}} e^{-\frac{1}{2}\left[\frac{\ln(x)-\mu}{\sigma}\right]^2} \quad \text{for } x > 0 \tag{3.85}$$

where μ and σ are the mean value a standard deviation of the natural logarithm of X. The PDF is zero for x less or equal to zero. The mean value and standard deviation of the lognormal distribution are,

$$E(X) = e^{\mu + \frac{\sigma^2}{2}} \tag{3.86}$$

$$\sigma_X^2 = e^{(2\mu + \sigma^2)}(e^{\sigma^2} - 1) \tag{3.87}$$

Extreme probability distributions

Extreme distributions deal with the maximum and the minimum values of observations of a random quantity (for example the 50-year maximum wind speed, the maximum ground acceleration during an earthquake and the maximum 500-year volume flood). Information about these extreme values is important for design of a system that fails the first time the applied load exceeds its capacity. Gumbel (1958) studied and systemized extreme distributions and advocated their use in practical problems.

Consider a sample of n values $(x_1, x_2, \ldots, x_n)$ of a random quantity. These values can be thought of as realizations of n random variables $X_i, i = 1, 2, \ldots, n$, which follow the same parent PDF $f_X(x)$. For example, X_i, $i = 1, 2, \ldots, n$, can be the annual maximum wave heights at a location in the Atlantic Ocean. The objective is to determine the probability distribution of the maximum and minimum of the above n values. Random variables X_i, $i = 1, 2, \ldots, n$, are independent if each variable represents the maximum value of a process over a long period, such as the annual maximum wave height, the ground acceleration or wind speed.

If sample $(x_1, x_2, \ldots, x_n)$ is rearranged in ascending order so that $y_1 \leq y_2 \leq \cdots \leq y_n$, then set $(y_1, y_2, \ldots, y_n)$ is called the *ordered sample* of size n and y_j is called the jth *order statistic*. Random variables Y_1, and Y_n have their own PDFs, which depend only on the shape of the tail of the parent PDF. For different parent distributions, which share some common characteristics, extreme random variables Y_1 and Y_n converge with n to the same distribution called *extreme distribution*.

There are three types of asymptotic extreme distributions (Gumbel, 1958, Ang and Tang, 1984, and Ochi, 1990). These types are explained below.

Extreme I (Gumbel distribution)

Extreme type I (Gumbel) is the extreme distribution whose original PDF is unbounded toward infinity and decreases exponentially as x tends to infinity. The extreme values of variables whose original distribution is exponential, normal distribution, log-normal, chi-square, and Gamma follow the extreme type I distribution with CDF,

$$F_{Y_n}(y) = \exp\left[-e^{-\alpha_n(y - w_n)}\right] \tag{3.88}$$

where w_n is a location parameter, called the characteristic maximum value of the original variable X. Scale parameter α_n is an inverse measure of the dispersion of the maximum value. These parameters can be estimated using information about the probability distribution of the original random variables X, measured values of X or measured extreme values of X.

The PDF of the Gumbel distribution is,

$$f_{Y_n}(y) = \alpha_n e^{-\alpha_n(y - w_n)} \exp\left[-e^{-\alpha_n(y - w_n)}\right] \tag{3.89}$$

The characteristic maximum value w_n has probability to be exceeded by the original variable X equal to $1/n$, that is, w_n is the $1 - \frac{1}{n}$ quantile of X,

$$P(X > w_n) = \frac{1}{n} \Leftrightarrow F_X(w_n) = 1 - \frac{1}{n} \tag{3.90}$$

The characteristic maximum value is also the mode of the extreme type I PDF. The probability that the maximum value Y_n will exceed the characteristic maximum value is approximately $1 - F_{Y_n}(w_n) = 1 - e^{-1} = 0.632$ for large values of n.

The mean value of the extreme type I distribution is

$$E(Y_n) = w_n + \frac{\gamma}{a_n} \tag{3.91}$$

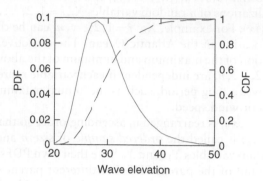

Figure 3.31 PDF of 100-year elevation.

where $\gamma = 0.577216$ is the Euler number. The variance is,

$$\sigma_{Y_n}^2 = \frac{\pi^2}{6a_n^2} \tag{3.92}$$

The extreme type I distribution has skewness equal to 1.1414. This means that the right tail of the distribution is longer than the left (Figure 3.31).

Example 3.25: The 100-year maximum wave elevation in the North Sea has mean value 30 m and coefficient of variation 0.16 (Guedes, Soares and Moan, 1982). This maximum value follows the extreme I type distribution. Find the parameters of this distribution, sketch the PDF and CDF and calculate the probability that the wave elevation will exceed 40 m.

Solution:
Let Y_{100} be the 100-year wave elevation. The variance of this variable is 23.04 m^2. The scale parameter α_n is found to be 0.267 from Equation (3.92). The characteristic value is obtained from Equation (3.91), $w_n = 27.84$ m. Figure 3.31 shows the PDF and CDF of the 100-year wave elevation. The mode of the PDF in this figure is 27.84 m.

The probability of the 100-year wave height exceeding 40 m is equal to one minus the value of the CDF for this wave height. The latter CDF is found from Equation (3.88). This probability is 0.038.

Example 3.26: Find the scale and location parameters of the largest value of n independent, identically distributed variables that follow the exponential distribution with mean value λ,

$$f_X(x) = \begin{cases} \dfrac{1}{\lambda}e^{-\frac{x}{\lambda}} & \text{if } x \geq 0 \\ 0 & \text{if } x < 0 \end{cases}$$

Assume n is large.

Solution:
The characteristic value w_n can be found from the CDF of X by solving Equation (3.90) for w_n,

$$F_X(w_n) = 1 - \frac{1}{n} \Leftrightarrow$$

$$1 - e^{-\frac{w_n}{\lambda}} = 1 - \frac{1}{n} \Leftrightarrow$$

$$w_n = \lambda \ln(n)$$

From Equation (3.89) if $y = w_n$ then $f_{Y_n}(w_n) \cdot e = \alpha_n$. But $f_{Y_n}(w_n) \cdot e = n f_X(w_n)$ for large values of n (see proof below). Therefore, $\alpha_n = n \frac{1}{\lambda} e^{-\frac{w_n}{\lambda}} = \frac{n}{\lambda} e^{-\ln(n)} = \frac{1}{\lambda}$.

Proof that $f_{Y_n}(w_n) \cdot e = n f_X(w_n)$:
Since $X_i, i = 1 \ldots n$ are independent, $F_{Y_n}(x) = [F_X(x)]^n$. Therefore, the PDF of the maximum value is related to the PDF of the original variable as follows,
$f_{Y_n}(x) = n[F_X(x)]^{n-1} f_X(x)$. Substituting w_n for x we get,

$$f_{Y_n}(w_n) = \frac{n}{F_X(w_n)} F_X(w_n)^n f_X(w_n) = \frac{n}{F_X(w_n)} F_{Y_n}(w_n) f_X(w_n)$$

But $F_X(w_n) = 1 - \frac{1}{n}$ and $F_{Y_n}(w_n) = e^{-1}$.

Therefore, for large values on n, $f_{Y_n}(w_n) = n e^{-1} f_X(w_n)$ or $f_{Y_n}(w_n) \cdot e = n f_X(w_n)$. Q.E.D.

The Gumbel distribution has the following reproductive property. If k independent, identically distributed variables follow the Gumbel distribution, then the maximum of these variables also follows the Gumbel distribution with the same scale parameter but with location parameter given by,

$$w_{kn} = w_n + \ln(k)/\alpha_n \tag{3.93}$$

Equation (3.93) is called stability postulate.

Example 3.27:   Find the PDF of the maximum 200-year and 500-year wave heights in example 3.25.

Solution:
Based on the previous paragraph, the location parameters of the 200-year and 500-year wave heights are 30.43 m and 33.86 m from Equation (3.93). The scale parameters are equal to the scale parameter of the 100-year wave height, which is 0.267. Figure 3.32 shows the PDFs of these heights.

The PDF of the smallest value Y_1 is,

$$F_{Y_1}(y) = \exp\left[-e^{-\alpha_n(y - w_1)}\right] \tag{3.94}$$

where w_1 is the characteristic smallest value of original variable X. The probability of X not exceeding this characteristic value is $1/n$.

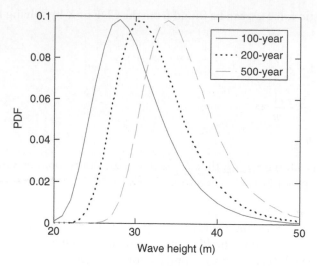

Figure 3.32 PDFs of 100-year, 200-year and 500-year wave heights.

Extreme II distribution

Type II distribution (Frechet) is associated with the original distribution that has unlimited value but it only has a finite number of moments. Cauchy distribution and Pareto distributions are examples of original distributions whose extreme value follows the type II extreme distribution (Bury, 1999, p. 294). For the largest value, the CDF of extreme type distribution is,

$$F_{Y_n(y)} = e^{-\left(\frac{u_n}{y}\right)^{\beta}} \tag{3.95}$$

$$f_{Y_n}(y) = \frac{\beta}{u_n}\left(\frac{u_n}{y}\right)^{\beta+1} e^{-\left(\frac{u_n}{y}\right)^{\beta}} \tag{3.96}$$

where u_n is a location parameter, also called characteristic largest value, and β is a shape parameter.

Extreme III distribution

The extreme type III distribution characterizes the extreme values of variables with known upper or lower bounds. These distributions have three parameters; the shape parameter β, the upper or lower limits $Y_{\max}$ or $Y_{\min}$, respectively, and the characteristic minimum or maximum values w_1 and w_n. The shape parameter controls the skewness, and the characteristic value the dispersion of the distribution. For values of the shape parameter less than four the PDF is skewed to the left, and for values greater than four is skewed to the right. The PDF is symmetric for $\gamma = 4$.

For the maximum case, the cumulative probability distribution function, CDF is,

$$F_{Y_n}(y) = \exp\left[-\left(\frac{Y_{\max} - y}{Y_{\max} - w_n}\right)^{\beta}\right] \quad y \le Y_{\max} \tag{3.97}$$

The corresponding PDF is,

$$f_{Y_n}(y) = \frac{\beta}{Y_{\max} - w_n}\left(\frac{Y_{\max} - y}{Y_{\max} - w_n}\right)^{\beta - 1}\exp\left[-\left(\frac{Y_{\max} - y}{Y_{\max} - w_n}\right)^{\beta}\right] \quad y \le Y_{\max} \tag{3.98}$$

The characteristic value is the $1 - \frac{1}{n}$ quantile of the original distribution. The mean value, $E(Y_n)$, variance, $\sigma_{Y_n}^2$, and skewness, γ are related to the parameters w_n, and $Y_{\max}$ as follows,

$$E(Y_n) = Y_{\max} - (Y_{\max} - w_n)\Gamma\left(1 + \frac{1}{\beta}\right) \tag{3.99}$$

$$\sigma_{Y_n}^2 = (Y_{\max} - w_n)^2\left\{\Gamma\left(1 + \frac{2}{\beta}\right) - \left[\Gamma\left(1 + \frac{1}{\beta}\right)\right]^2\right\} \tag{3.100}$$

$$\gamma = \frac{\Gamma\left(1 + \frac{3}{\beta}\right) + 2\left[\Gamma\left(1 + \frac{1}{\beta}\right)\right]^3 - 3\Gamma\left(1 + \frac{2}{\beta}\right)\Gamma\left(1 + \frac{1}{\beta}\right)}{\left\{\Gamma\left(1 + \frac{2}{\beta}\right) - \left[\Gamma\left(1 + \frac{1}{\beta}\right)\right]^2\right\}^{3/2}} \tag{3.101}$$

The CDF of the minimum is called Weibull distribution. This distribution is important in modeling the strength (stress at failure) of materials, the loading capacity of series systems, and their time to failure. For example, the Weibull distribution models the strength of a chain consisting of many links, which fail independently. The weakest link dictates the strength of the chain. Then the distribution of the strength of the chain approximates the Weibull distribution as the number of links increases to infinity. Finally, the Weibull distribution is used to model the time to complete a task.

The CDF of the Weibull distribution is,

$$F_{Y_1}(y) = 1 - \exp\left[-\left(\frac{y - Y_{\min}}{w_1 - Y_{\min}}\right)^{\beta}\right] \quad \text{for } y \ge Y_{\min} \tag{3.102}$$

where $Y_{\min}$ is the lower bound value of the distribution, w_1 is the characteristic smallest value of X, and β is the shape parameter. The characteristic smallest value is the $1/n$ quantile of the distribution.

Its corresponding PDF is,

$$f_{Y_1}(y) = \frac{\beta}{w_1 - Y_{\min}}\left(\frac{y - Y_{\min}}{w_1 - Y_{\min}}\right)^{\beta - 1}\exp\left[-\left(\frac{y - Y_{\min}}{w_1 - Y_{\min}}\right)^{\beta}\right] \quad \text{for } y \ge Y_{\min} \tag{3.103}$$

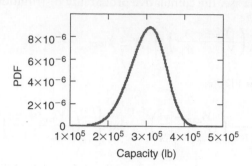

Figure 3.33 Weibull PDF of the capacity of a structural member.

In order to estimate the parameters of the distribution, first, an individual could estimate statistical summaries from observations and/or from expert judgment, and then estimate the distribution parameters based on these summaries.

The mean value, $E(Y_1)$, variance, $\sigma_{Y_1}^2$, and skewness, γ, of the Weibull distribution are given below in terms of the lower bound, characteristic value and shape parameter,

$$E(Y_1) = Y_{\min} + (w_1 - Y_{\min})\Gamma\left(1 + \frac{1}{\beta}\right) \tag{3.104}$$

$$\sigma_{Y_1}^2 = (w_1 - Y_{\min})^2 \left\{\Gamma\left(1 + \frac{2}{\beta}\right) - \left[\Gamma\left(1 + \frac{1}{\beta}\right)\right]^2\right\} \tag{3.105}$$

$$\gamma = \frac{\Gamma\left(1 + \frac{3}{\beta}\right) + 2\left[\Gamma\left(1 + \frac{1}{\beta}\right)\right]^3 - 3\Gamma\left(1 + \frac{2}{\beta}\right)\Gamma\left(1 + \frac{1}{\beta}\right)}{\left\{\Gamma\left(1 + \frac{2}{\beta}\right) - \left[\Gamma\left(1 + \frac{1}{\beta}\right)\right]^2\right\}^{3/2}} \tag{3.106}$$

The parameters of the distribution are found by solving equations (3.104) to (3.106).

Example 3.28: The minimum capacity of the structural member of an offshore platform follows the Weibull distribution with mean value 300,000 lb, and coefficient of variation 0.15. The lower bound is 100,000 lb. Calculate the characteristic value, shape parameter and skewness of the distribution, and estimate the probability that the capacity is no greater than 200,000 lb.

Solution:
The variance of the distribution is $2.025 \cdot 10^9$ lb. The characteristic value and the shape parameter are found to be $w_1 = 317{,}600$ lb and $\beta = 5.09$ by solving Equations (3.104) and (3.105). The skewness is found to be $\gamma = -0.27$ from Equation (3.106). Figure 3.33 shows the PDF of the minimum capacity. It is observed that the PDF is skewed to the left.

Special cases

The Weibull distribution in equation (3.102) reduces to the exponential distribution in Equation (3.72) when shape parameter $\beta = 1$. The Weibull distribution becomes the Rayleigh distribution for $\beta = 2$, $Y_{min} = 0$ and $w_1 = \sqrt{2}\sigma_R$. This distribution characterizes the square root of the sum of the squares of two normal random variables with zero mean values and common variance. For example, if the x and y coordinates of a point are normal with zero mean and equal variance, then the distance of this point from the origin is Rayleigh.

According to the *reproductive* property of the extreme type III distribution, the maximum of a sample drawn from this distribution also follows the same distribution. For example, of k independent identically distributed variables following the distribution in equation (3.97), its maximum also follows the extreme type III distribution with same shape parameter β. The numerator of the exponent becomes $(Y_{max} - y)k^{1/\beta}$. This means that if we determine the distribution of the extreme values of a sample of size n, then we can easily determine the distribution of samples of size equal to a multiple of n.

3.2.2 *Joint normal distribution*

A complete probabilistic model of multiple random variables is their joint probability distribution. There are a few common joint probability distributions. Random variables are assumed independent in many applications, but this assumption may result in significant errors in the estimation of the probability of an event. It could also lead to error in the probability distribution and the mean value of a variable that quantifies the performance of a system, such as the mean value of the utility of the profit of a risky venture.

Copulas are practical tools to characterize the dependence of random variables. A copula is a parametric relation between the joint CDF of two variables and their marginal CDFs. These models are very flexible and their parameters are estimated on the basis of expert judgment and observations. Copulas for more than two random variables are also available. We will explain these tools in Chapter 4.

The joint probability distribution of a set of normal random variables has been studied extensively and is well documented. The normal PDF of two random variables X and Y is,

$$f_{X,Y}(x,y) = \frac{1}{2\pi\sigma_X\sigma_Y\sqrt{1-\rho^2}} \exp\left[-\frac{1}{2(1-\rho^2)}\left\{\left(\frac{x-\mu_X}{\sigma_X}\right)^2 + \left(\frac{y-\mu_Y}{\sigma_Y}\right)^2\right.\right.$$

$$\left.\left. -2\rho\left(\frac{x-\mu_X}{\sigma_X}\right)\left(\frac{y-\mu_Y}{\sigma_Y}\right)\right\}\right] \qquad (3.107)$$

where μ_X, and μ_Y are the mean values of X and Y, σ_X, and σ_Y are their standard deviations and ρ their correlation coefficient.

The conditional PDF of Y given that $X = x$ is normal with mean value $\mu_Y + \rho \left(\frac{\sigma_Y}{\sigma_X} \right)(x - \mu_X)$ and standard deviation $\sigma_Y \sqrt{1 - \rho^2}$,

$$f_{X/Y}(x/y) = \frac{f_{X,Y}(x,y)}{f_Y(y)} = \frac{1}{\sqrt{2\pi}\sigma_Y\sqrt{1-\rho^2}} \exp\left[-\frac{1}{2} \left\{ \frac{y - \mu_Y - \rho\left(\frac{\sigma_Y}{\sigma_X}\right)(x - \mu_X)}{\sigma_Y\sqrt{1-\rho^2}} \right\}^2 \right]$$

(3.108)

The marginal PDFs of each of the two variables are obtained by integrating out the other variable. These marginal PDFs are also normal.

When the correlation coefficient is zero, the joint normal PDF reduces to the product of the marginal PDFs,

$$f_{X,Y}(x,y) = f_X(x)f_Y(y)$$

(3.109)

Therefore, two uncorrelated normal random variables are independent. This conclusion is not true for random variables that follow other distributions that are not normal.

Example 3.29: This example, demonstrates the importance of the dependence of random variables on the calculation of the failure probability of a system.

Loads S_1 and S_2 are applied on a cantilever beam at two points with distance 1 m and 2 m from the clamped end. The loads follow a joint normal distribution with mean value 1 kN and standard deviation 0.2. The beam can sustain a bending moment equal to 4 kN m. Find the failure probability when the correlation coefficient of the loads is 0, 0.3 and 0.6.

Solution:
Figures 3.34 and 3.35 show the joint PDF of these loads for correlation coefficients of 0.6 and zero, respectively. The failure region is near the upper right corner of these figures because failure occurs when both loads become large simultaneously. The probability of failure is the volume under the joint PDF of the loads over the failure region. Therefore, the probability of failure increases with the correlation coefficient.

The failure region defined by the inequality $S_2 \geq \frac{4-S_1}{2}$. Therefore, the probability of failure is,

$$PF = \int_{-\infty}^{+\infty} \int_{\frac{4-s_1}{2}}^{+\infty} f_{S_1,S_2}(s_1,s_2) \, ds_2 \, ds_1$$

The probability of failure can be estimated accurately by narrowing the limits of the first integration from $[-\infty, \infty]$ to $[-0.2, 2.2]$, which corresponds to the mean value minus $\pm$ six standard deviations, because the probability of a normal variable being outside this range is only $2 \cdot 10^{-9}$. Similarly, we changed the upper limit in the second

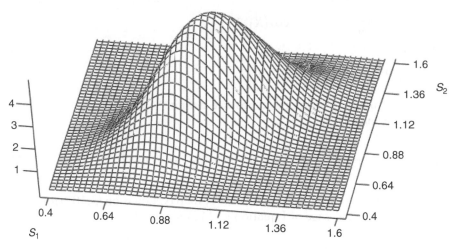

Figure 3.34 Joint PDF of loads on the example beam for a correlation coefficient of 0.6.

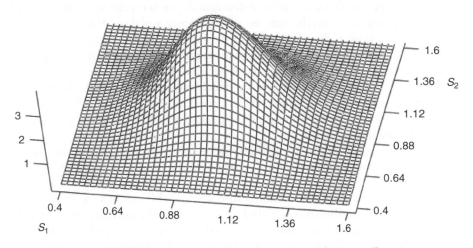

Figure 3.35 Joint PDF of loads on example beam for zero correlation coefficient.

integration to 2.2. Using numerical integration, we found that this probability is 0.013, 0.022 and 0.033 for correlation coefficients of 0, 0.3 and 0.6. This shows that the results of the calculations in probabilistic analysis are affected by the dependence of the random variables.

In section 3.3 we will see that when random variables are normal, we can calculate probabilities of failure without performing numerical integration.

Consider random vector $\mathbf{X}$ consisting of n jointly normal random variables, $\mathbf{X} = \{X_1, X_2, \ldots, X_n\}^T$. The joint PDF of these variables is defined by the vector of the mean values,

$$\boldsymbol{\mu}_{\mathbf{X}} = \{\mu_{X_1}, \ldots, \mu_{X_n}\}^T \tag{3.110}$$

and the covariance matrix

$$\mathbf{\Lambda_X} = \begin{bmatrix} \sigma^2_{X_1} & \cdots & \mathrm{cov}(X_1, X_n) \\ \vdots & \vdots & \vdots \\ \mathrm{cov}(X_n, X_1) & \cdots & \sigma^2_{X_n} \end{bmatrix} \tag{3.111}$$

The covariance matrix is symmetric, positive definite[3].
   The joint normal PDF is,

$$f_{\mathbf{X}}(\mathbf{x}) = \frac{1}{(2\pi)^{n/2}|\mathbf{\Lambda_X}|^{1/2}} \exp\left[-\frac{1}{2}(\mathbf{x} - \boldsymbol{\mu_X})^T \mathbf{\Lambda_X^{-1}}(\mathbf{x} - \boldsymbol{\mu_X})\right] \tag{3.112}$$

where $|\mathbf{\Lambda_X}|$ is the determinant of the covariance matrix. Equation (3.107) for a bivariate normal PDF is a special case of Equation (3.112) for $n = 2$.

Summary of section 3.2

The choice of a probabilistic model in a decision depends on the understanding of the underlying phenomenon and the amount of experience and data. If the decision maker knows that a particular distribution describes a physical quantity well then he/she should consider using this distribution. For example, the Gumbel distribution is a good candidate for modeling the 100-year extreme wave elevation or flood level, and the lognormal distribution can be suitable for modeling fatigue strength. If observations are available, then the decision-maker should check the fit of the candidate distribution to these observations. Tests for checking the goodness of fit of a candidate distribution will be explained in Chapter 4.

   In many decisions, limited data is available. As a result, the decision-maker should consider the following guidelines in order to choose the proper probability distribution,

a)   The distribution should not underestimate probabilities of rare events that can cause failure. Distributions with light tails, such as the normal, are inappropriate for modeling uncertainty with limited data because they tend to underestimate the probabilities of ranges of values away from the mean.

b)   The distribution should be flexible so that the decision maker will be able to adjust its shape if more data become available.

c)   The parameters of the distribution should be estimated easily based on experience or judgment. For example, bounds and most likely values are easier to estimate than moments.

   Uniform, exponential and beta distributions are good candidates in cases where limited data is present because they satisfy the above criteria.
   Dependence of the random variables can significantly influence the results of probabilistic analysis. A complete probabilistic model requires the joint PDF of the random variables. A decision maker faces two challenges when building such a model,

---

[3] A matrix $\mathbf{M}$ is positive definite if the product $\mathbf{x}^T\mathbf{M}\mathbf{x}$ is positive for all vectors $\mathbf{x}$.

a)   There are few families of common multivariate distributions and there is limited amount of expertise with these distributions.

b)   A very large amount of data is required to construct a joint probability distribution.

In Chapters 4 and 5, we will study how to construct models of uncertainty using observations and expert judgment.

3.3   Probability Calculations

As we explained in Chapter 1, design is a risky venture with uncertain payoff. The designer synthesizes, evaluates and compares alternative designs in order to select the design with the highest payoff (Figure 3.1). For this purpose, the designer needs to estimate the performance attributes of a system given a probability model of the uncertain input variables that influence the system's performance. For example, a car manufacturer needs to estimate the probability distribution of the sound pressure at the driver's ear given the probability distribution of the road excitation, the plate gages, and the material properties.

The above task is an input-output problem: given function $Y = g(X)$ and a probabilistic model of input variables X, determine the probabilistic structure of the output Y. This section presents tools for time invariant problems where variables X do not vary with time. Chapter 5 deals with time-varying problems.

3.3.1   *Probability distributions of a function of one random variable*

3.3.1.1   *Probability distribution*

The objective is to find the CDF $F_Y(y)$ of dependent variable $Y = g(X)$ given the CDF of independent variable X, $F_X(x)$. The CDF of dependent variable Y is the probability that this variable is less than or equal to a value y, $F_Y(y) = P(Y \leq y)$. From Figure 3.36, the probability that $Y \leq y$ is equal to the probability that X assumes a value within range I_y. This range is called the inverse image of $[-\infty, y]$ under function $g(x)$. Therefore,

$$F_Y(y) = F_X(x_u) - F_X(x_l) \tag{3.113}$$

If function $g(x)$ is monotonically increasing then the inverse image of $[-\infty, y]$ reduces to the interval $[-\infty, x_u]$ and Equation (3.113) becomes,

$$F_Y(y) = F_X(x_u) \tag{3.114}$$

If the inverse image of $[y, \infty]$ under function $g(x)$ consists of intervals, $[x_{l_i}, x_{u_i})$ $i = 1, \ldots, m$ then Equation (3.113) is generalized to,

$$F_Y(y) = \sum_{i=1}^{m} \left[F_X(x_{u_i}) - F_X(x_{l_i}) \right] \tag{3.115}$$

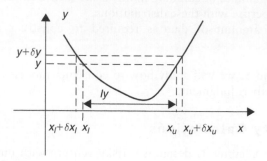

Figure 3.36 Finding the CDF and PDF of dependent variable *Y* from the CDF and PDF of independent variable *X*.

Example 3.30:   Find the CDF of $Y = \sin(X)$ where X is a uniformly distributed random variable in the range $[0, 2\pi]$.

Solution:
The CDF of X is,

$$F_X(x) = \begin{cases} 0 & x < 0 \\ \dfrac{x}{2\pi} & 0 \le x < 2\pi \\ 1 & x \ge 2\pi \end{cases}$$

If dependent variable Y is positive, then $\sin(X) \le y$ if X is in range $[0, \arcsin(y)]$ or $[\pi - \arcsin(y), 2\pi]$, where $\arcsin(y)$ is the principal value of the inverse sine function (Figure 3.37). For negative values of Y, $\sin(X) \le y$ if X is in range $[\pi - \arcsin(y), 2\pi + \arcsin(y)]$.

The CDF of sine is zero for values of y less than -1. The CDF is 1 for values if y greater than or equal to 1. Between -1 and 1 the CDF of $\sin(X)$ is,

$$F_Y(y) = \begin{cases} F_X[2\pi + \arcsin(y)] - F_X[\pi - \arcsin(y)] & -1 \le y < 0 \\ F_X[\arcsin(y)] + 1 - F_X[\pi - \arcsin(y)] & 0 \le y \le 1 \end{cases}$$

Figure 3.38 shows the CDF of $\sin(X)$ in the range from -1 to 1. It is observed that the CDF increases very steeply for $Y = -1$ and 1. The reason is that function $Y = \sin(X)$ is flat when Y assumes these two values (Figure 3.37). This means that, if we sample values of X from a uniform distribution in $[0, 2\pi]$, many values of $Y = \sin(X)$ will be clustered near -1 and 1. Therefore, the probability of $Y \le y$ will increase very steeply near -1 and 1.

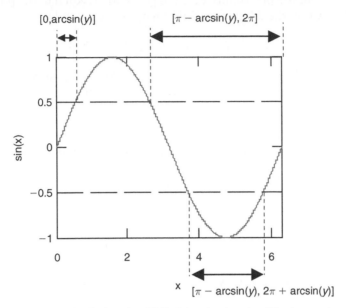

Figure 3.37 Explanation for finding the CDF of the sine function. The inverse image of sin(X) ≤ y depends on the sign of y.

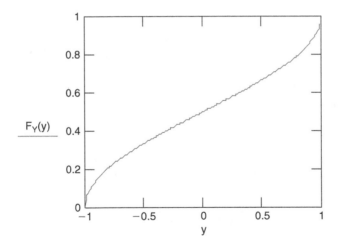

Figure 3.38 CDF of Y = sin(X) if X is uniform in the range [0, 2π].

3.3.1.2 *Probability density function*

We want to find the PDF $f_Y(y)$ of random variable $Y = g(X)$ given the PDF of X, $f_X(x)$, where function $g(X)$ is differentiable. First, we find the probability that $Y \in [y, y + \delta y]$ where δy is a small interval, and then find the limit, $\dfrac{P(y \leq Y \leq y + \delta y)}{\delta y}$ as δy tends to zero. In Figure 3.36, the probability that $Y \in [y, y + \delta y]$ is equal to the probability that X is in any interval in the domain of X that maps to $[y, y + \delta y]$,

$$P(y \leq Y \leq y + \delta y) = P(x_l + \delta x_l \leq X \leq x_l) + P(x_u \leq X \leq x_u + \delta x_u)$$

Note that δx_l is negative because function $f_Y(y)$ is decreasing at $X = x_l$. If the above two intervals are infinitesimal, the above equation becomes,

$$f_Y(y)\delta y = f_X(x_l)(-\delta x_l) + f_X(x_u)(\delta x_u)$$

The PDF of Y is found by dividing the left hand side by δy and finding the limit of the right hand side as δy tends to zero,

$$f_Y(y) = \lim_{\delta y \to 0} \frac{f_X(x_l)}{-\dfrac{\delta y}{\delta x_l}} + \frac{f_X(x_u)}{\dfrac{\delta y}{\delta x_u}} = \frac{f_X(x)}{\left|\dfrac{dy}{dx}\right|}\Bigg|_{x=x_l} + \frac{f_X(x)}{\left|\dfrac{dy}{dx}\right|}\Bigg|_{x=x_u}$$

This result is generalized as follows,

$$f_Y(y) = \sum \frac{f_X(x)}{\left|\dfrac{dy}{dx}\right|}\Bigg|_{x_i = g^{-1}(y)} \tag{3.116}$$

where $g^{-1}(y)$ is the inverse function of $y = g(x)$.

If the derivative of $g(x)$ is zero for some value x_0, then the PDF of Y is equal to infinity for $y_0 = g^{-1}(x_0)$. This means that if function $g(x)$ has zero slope at $X = x_0$ then many sample values of y are clustered at $Y = y_0$.

Example 3.31:   Find the PDF of $Y = g(X) = \sin(X)$ given that X is uniform from $[0, 2\pi]$.

Solution:
The PDF of $\sin(X)$ is obtained from Equation (3.116),

$$f_Y(y) = \frac{1}{2\pi}\left[\frac{1}{|\cos(x_1)|} + \frac{1}{|\cos(x_2)|}\right]$$

From Figure 3.37,

$$x_1 = \arcsin(y), \quad \text{and} \quad x_2 = \pi - \arcsin(y) \quad \text{if } y \geq 0$$
$$x_1 = \pi - \arcsin(y), \quad \text{and} \quad x_2 = 2\pi + \arcsin(y) \quad \text{if } y < 0$$

Figure 3.39 shows the PDF of $\sin(x)$. It is observed that the $\sin(x)$ is much more likely to be almost equal to -1 or 1 than equal to other values (for example zero). This observation could surprise someone because angle x is equally likely to assume any value between 0 and 2π. The justification is that function $y = \sin(x)$ is flat for

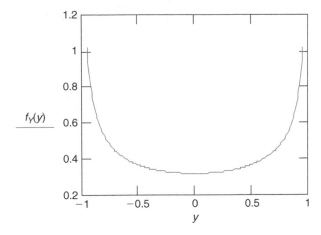

$f_Y(y)$

Figure 3.39 PDF of $y = \sin(x)$.

angles $\frac{\pi}{2}$ and $\frac{3\pi}{2}$ (Figure 3.37), for which the sine is equal to 1 and -1, respectively. If we sample from a uniform distribution between 0 and 2π we will find many values of the angle x with sine almost 1 or -1.

Example 3.32:   The steering column of a car is modeled as a single degree of freedom system subjected to a harmonic force. The steering column can vibrate excessively because, occasionally, the frequency of excitation resonates with the natural frequency of the column. Attaching a Dynamic Vibration Absorber (DVA) to the system is a practical solution to this problem. If the DVA is perfectly tuned to the excitation frequency, then the original system does not vibrate. However, the system is very sensitive to variations of the natural frequency of the DVA if the mass of the DVA is small (e.g. one percent of the mass of the original system). Figure 3.40 shows a two degree of freedom model comprising the original systems and the DVA.

An automotive manufacturer asked a supplier to design a DVA for the steering column. The manufacturer wants to minimize the mass of the DVA and keep the displacement of the system (normalized by the quasi-static displacement) under 5. The normalized natural frequency (natural frequency divided by the excitation frequency) of the DVA attached to the column is normal with standard deviation 0.025. The damping ratio of the original system is 1%. The variability in the frequency of excitation and the natural frequency of the original system are negligible compared to the variability in the normalized frequency of the DVA. The supplier designed two alternative DVA's with mass 1% and 2% of the system. Compare these designs in terms of the manufacturer's specifications.

Solution:
We will compare the alternative DVA designs in terms of the probability that the displacement of the original system will exceed the allowable limit of 5. The steady

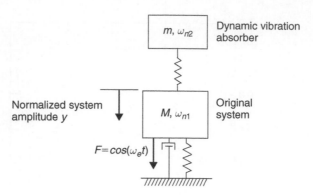

Figure 3.40 Original system with a dynamic vibration absorber.

state amplitude y (normalized by the quasi-static response of mass M) of the original system to a harmonic excitation is calculated using the following equation:

$$y(\beta_1, \beta_2, r, \zeta) = \frac{1 - (1/\beta_2)^2}{\{[1 - r(1/\beta_1)^2 - (1/\beta_1)^2 - (1/\beta_2)^2 + (1/\beta_1\beta_2)^2]^2}{\qquad\qquad + 4\zeta^2[1/\beta_1 - 1/(\beta_1\beta_2^2)]^2\}^{1/2}} \qquad (3.117a)$$

A negative value of the displacement shows that the system vibrates out of phase relative to the excitation. Only the absolute value of the displacement is important.

In the above equation β_1 is the ratio of the original system natural frequency to the excitation frequency, β_2 is the ratio of the absorber natural frequency to the excitation frequency, r the mass ratio of the absorber to the original system, and ζ the damping ratio of the original system.

According to the data in the problem definition $\beta_1 = 1$ and $\zeta = 0.01$. The normalized frequency of the DVA β_2 is normal with a mean value of one and a standard deviation of 0.025. The mass ratios of the two alternative DVA designs are 0.01 and 0.02, respectively.

Figure 3.41 shows the variation of the displacement with the natural frequency of the DVA for mass ratios equal to 0.01 and 0.02. The same figure shows the PDF of the frequency of the DVA. The displacement of the design with mass ratio 0.02 is less sensitive to variations in the frequency than that of the design with mass ratio 0.01. However, we should keep the mass low in addition to the displacement. We will calculate the PDFs of the displacements of both DVA designs and the probabilities of the displacement exceeding 5 in order to help the manufacturer select a design.

The PDF of the displacement $f_Y(y)$ is obtained from the following equation:

$$f_Y(y) = \frac{f_{B2}(\beta_2)}{\left|\dfrac{dy}{d\beta_2}\right|} \qquad (3.117b)$$

where $f_{B2}(\beta_2)$ is the PDF of the frequency of the DVA, and $\frac{dy}{d\beta_2}$ is the derivative of the displacement with respect to the frequency of the DVA. The right hand side of

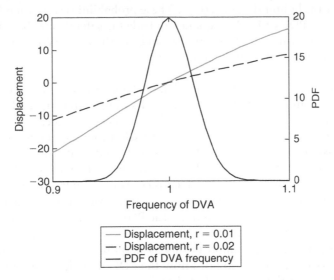

Figure 3.41 Variation of the displacement of the original system with the frequency of the DVA for mass ratios 0.01 and 0.2. The PDF of the frequency of the DVA is also shown.

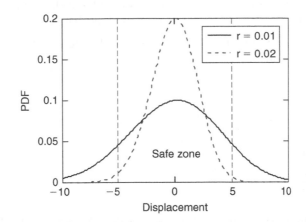

Figure 3.42 PDFs of the displacement of the original system for the two alternative designs with mass ratios 0.01 and 0.02.

Equation (3.117b) is calculated for the value of the frequency of the DVA for which $Y = y$. This value was found by solving numerically Equation (3.117a) for β_2.

Figure 3.42 shows the PDFs of the displacements of the steering column for the two alternative DVA designs with mass ratios 0.01 and 0.02. A safe zone containing the values of β_2 for which the original system has acceptable amplitude is shown. The absolute value of the displacement of the design with the lighter DVA is likely to exceed the threshold of 5. The probability that the absolute value of the displacement of the heavier DVA design can exceed the threshold of 5 is 0.013. The corresponding probability for the lighter design is 0.21. These probabilities were found by numerical integration of the PDF of DVA frequency outside the safety zone. Based on the calculated probabilities of unacceptable performance of the two DVA designs, the heavier design should be better than the lighter design.

Examples 3.31 and 3.32 demonstrate that probabilistic analysis is a valuable tool for making decisions under uncertainty. The results of probabilistic analysis provide the designer with the probabilities of all possible outcomes of a decision. This is valuable information because it enables the designer to select the best course of action among alternatives based on his preferences. Interval methods do not provide such information because they only tell the designer what the possible outcomes of a decision are; they do not show the likelihood of these outcomes. For example, application of interval arithmetic to example 3.31 only tells us that $\sin(x)$ varies in the range $[-1, 1]$.

However, probabilistic methods require the PDF of the random variables that affect the performance of a system. It is dangerous for a designer to assume a PDF if he/she has no evidence that supports this assumption. Consider again the problem in Example 3.31 but now we do not know the PDF of angle x. Assume that x is uniform from 0 to 2π on the grounds that there is no reason that one angle is more likely than another. This seemingly innocuous assumption leads to the following paradox. If angle x is uniform then its sine is more likely to be -1 or 1 than 0. But if we cannot decide if one value of angle x in the range from 0 to 2π is more likely than another value, then the same should be true for the values of $\sin(x)$. Therefore, $\sin(x)$ should also be uniform from -1 to 1. This is at odds with probability calculus.

Example 3.33:   Find the PDF of $Y = g(X) = aX + b$ in terms of the PDF of X, where a and b are known constants.

Solution:
Equation (3.116) yields,

$$f_Y(y) = \frac{f_X(x)}{\left|\dfrac{dy}{dx}\right|}\Bigg|_{x=\frac{y-b}{a}} = \frac{f_X\left[\dfrac{y-b}{a}\right]}{|a|} \tag{3.118}$$

Figure 3.43 shows the PDFs of random variables X and $Y = 2 \cdot (X - 1) = 2X - 2$, where X is normal with mean value 1 and standard deviation 0.5. The PDF of Y is obtained by setting $a = 2$ and $b = -2$. The PDF of Y is more dispersed than that of X and this PDF is centered on zero.

3.3.1.3   *Mean value and standard deviation of a function of one variable*

The mean value of dependent variable $Y = g(X)$ is,

$$E(Y) = \int\limits_{-\infty}^{\infty} y f_Y(y) \, dy \tag{3.119}$$

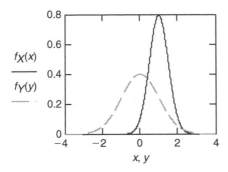

Figure 3.43 PDF of normal random variable X, which has mean value 1 and standard deviation 0.5 and PDF of dependent variable $Y = 2X - 2$.

We can calculate the mean value of Y without calculating the PDF of this dependent variable. From Equation (3.116),

$$f_Y(y)\, dy = \sum_i f_X(x_i)|dx_i| \quad \text{where } x_i = g^{-1}(y)$$

In the above equation intervals dx_i do not overlap. Therefore, Equation (3.119) becomes,

$$E(Y) = \int_{-\infty}^{\infty} g(x) f_X(x)\, dx \tag{3.120}$$

The mean value is a linear operator; if $Y = aX + b$ then $E(Y) = aE(X) + b$. This result can be seen by substituting $ax + b$ for $g(x)$ in Equation (3.120).

The variance of $g(x)$ is the mean value of the square of $g(X) - E[g(X)]$. Therefore, the variance of $g(X)$ is calculated from either one of the following equations,

$$\sigma_{g(X)}^2 = \int_{-\infty}^{\infty} \{g(x) - E[g(x)]\}^2 f_X(x)\, dx \tag{3.121}$$

$$\sigma_{g(X)}^2 = \int_{-\infty}^{\infty} [g(x)]^2 f_X(x)\, dx - \{E[g(x)]\}^2 \tag{3.122}$$

From Equations (3.121) and (3.122), we observe that when we add a constant to a random variable, we shift its mean value by the same constant, without changing the variance. Moreover, scaling the variable by a constant scales both the mean value and standard deviation by the same constant. Consequently, transformation $Z = \frac{X - E(X)}{\sigma_X}$ yields variable Z that has zero mean value and unit standard deviation.

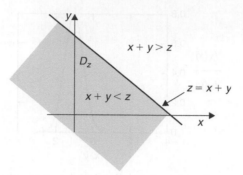

Figure 3.44 Region of values of two random variables in which their sum is less than or equal to z.

3.3.2 *Distribution of functions of multiple random variables*

3.3.2.1 *One function of two variables*

We will consider two random variables. The results are generalized easily for more variables. The objective is to find the CDF of a known function of two random variables $Z = g(X, Y)$ given the joint PDF of X and Y $f_{X,Y}(x, y)$.

The CDF of dependent variable Z is,

$$F_Z(z) = P[Z \leq g(x, y)] = P[(X, Y) \in D_z]$$

where D_z is the region in the space of variables X and Y for which every pair of values x and y map into a value $g(x, y)$ that is less than or equal to z.

Then,

$$F_Z(z) = \iint\limits_{D_z} f_{X,Y}(x, y)\, dx\, dy \tag{3.123}$$

The PDF of Z is the derivative of the CDF in Equation (3.123).

Example 3.34: Find the CDF and PDF of the sum of two variables, $Z = X + Y$.

Solution:
Figure 3.44 shows the region that contains the values (x, y) whose sum is less than or equal to z.

Then,

$$F_Z(z) = \int\limits_{-\infty}^{+\infty} \int\limits_{-\infty}^{z-x} f_{X,Y}(x, y)\, dy\, dx \tag{3.124}$$

According to Equation (3.124), the probability of $Z \leq z$ is the sum of the conditional probabilities of $Z \leq z$ given that $X = x$, times the probability of $X = x$ for all possible values of X. If we change the order of integration in the above equation, then the CDF of Z is also,

$$F_Z(z) = \int_{-\infty}^{+\infty} \int_{-\infty}^{z-y} f_{X,Y}(x,y) \, dx \, dy \tag{3.125}$$

The PDF of Z is the derivative of the CDF of Z with respect to z,

$$f_Z(z) = \frac{dF_Z(z)}{dz} = \int_{-\infty}^{\infty} f_{X,Y}(z-y,y) \, dy = \int_{-\infty}^{\infty} f_{X,Y}(x, z-x) \, dx \tag{3.126}$$

If variables X and Y are independent then the PDF of their sum becomes,

$$f_Z(z) = \int_{-\infty}^{\infty} f_X(x)f_Y(z-x) \, dx = \int_{-\infty}^{\infty} f_X(z-y)f_Y(y) \, dy \tag{3.127}$$

According to the above equation, the PDF of the sum is equal to the convolution integral of the PDFs of X and Y.

Example 3.35: Find the probability of failure of a system with load capacity S_u subjected to a load S. The PDFs of the capacity and the load are $f_{S_u}(s_u)$ and $f_S(s)$, respectively. Assume that the load and capacity are independent.

Solution:
Failure occurs when the load is greater than or equal to the capacity. Therefore, this probability is equal to the probability that the difference of the capacity minus the load is negative. This probability is equal to the CDF of this difference evaluated at zero.

Let $Z = S_u - S$. The CDF of Z is $F_Z(z) = \int_{-\infty}^{\infty} f_S(s)P(Z \le z/S = s) \, ds$, where $P(Z \le z/S = s)$ is the conditional probability that the difference of the strength and the load is less than or equal to z, given that the stress is s. The latter conditional probability is equal to $F_{S_u}(z+s)$.

Therefore, $F_Z(z) = \int_{-\infty}^{\infty} f_S(s)F_{S_u}(z+s) \, ds$. The probability of failure is the CDF of the difference of the strength minus the load evaluated at $z = 0$,

$$P(F) = F_Z(0) = \int_{-\infty}^{\infty} f_S(s)F_{S_u}(s) \, ds \tag{3.128}$$

Equation (3.128) is interpreted as follows: the probability of failure is the integral of the conditional probability of failure given that the stress is equal to s weighted by the PDF of the stress, for all possible values of the stress.

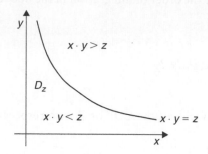

Figure 3.45 Region of values of two random variables for which the product of these variables is less than or equal to z.

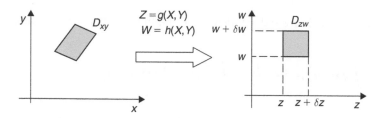

Figure 3.46 Mapping of region D_{xy} to region D_{zw} under transformation (3.129).

Example 3.36: Find the CDF of the product of two variables, $Z = X \cdot Y$ where X and Y are positive. X and Y are independent.

Figure 3.45 shows the region in the space of variables x and y, where $x \cdot y \leq z$. The probability that Z is less than or equal to z is,

$$P(Z \leq z) = \int_0^\infty \int_0^{\frac{z}{x}} f_Y(y) f_X(x) \, dy \, dx = \int_0^\infty f_X(x) \left[\int_0^{\frac{z}{x}} f_Y(y) \, dy \right] dx = \int_0^\infty f_X(x) F_Y\left(\frac{z}{x}\right) dx$$

3.3.2.2 Two functions of two random variables

The objective of this subsection is to find the joint PDF of two dependent variables,

$$Z = g(X, Y) \quad \text{and} \quad W = h(X, Y) \tag{3.129}$$

given the joint PDF of independent variables X and Y. First, we will solve the problem for one-to-one transformations and then consider general transformations.

The probability that dependent variables z and w are in region D_{zw} is equal to the probability that x and y are in region D_{xy} (Figure 3.46),

$$P[(x, y) \in D_{xy}] = P[(z, w) \in D_{zw}] = f_{Z,W}(z, w) \, dz \, dw \tag{3.130}$$

Assume that the areas of both regions are infinitesimal.

The probability of variables X and Y being in region D_{xy} is the joint PDF of these variables times the area of D_{xy},

$$P[(x,y) \in D_{xy}] = f_{X,Y}(x,y)|D_{xy}| \tag{3.131}$$

where $|D_{xy}|$ denotes the area of region D_{xy}. Consequently the joint PDF of dependent variables Z and W is equal to the joint PDF of X and Y scaled by the ratio of the areas of D_{xy} and D_{zw},

$$f_{Z,W}(z,w) = f_{X,Y}(x,y)\frac{|D_{xy}|}{|D_{zw}|} \tag{3.132}$$

or

$$f_{Z,W}(z,w) = \frac{f_{X,Y}(x,y)}{\dfrac{|D_{zw}|}{|D_{xy}|}} \tag{3.133}$$

The ratio of the areas in the denominator of Equation (3.133) is the Jacobian of transformation $|J(x,y)|$,

$$|J(x,y)| = \begin{vmatrix} \dfrac{\partial g(x,y)}{\partial x} & \dfrac{\partial g(x,y)}{\partial y} \\ \dfrac{\partial h(x,y)}{\partial x} & \dfrac{\partial h(x,y)}{\partial y} \end{vmatrix} \tag{3.134}$$

Therefore, the joint PDF of dependent variables Z and W is obtained from the following equation,

$$f_{Z,W}(z,w) = \frac{f_{X,Y}(x,y)}{|J(x,y)|}\bigg|_{\substack{x=g^{-1}(z,w) \\ y=h^{-1}(z,w)}} \tag{3.135}$$

If multiple pairs of values of variables X and Y map to the same pair of values of Z and W, we consider the contributions of all these pairs in the right hand side of Equation (3.135),

$$f_{Z,W}(z,w) = \sum_{i=1}^{m} \frac{f_{XY}(x_i,y_i)}{|J(x_i,y_i)|}\bigg|_{\substack{x_i=g^{-1}(z_i,w_i) \\ y_i=h^{-1}(z_i,w_i)}} \tag{3.136}$$

The joint PDF of n dependent variables that are functions of n independent variables can be obtained by plugging the joint PDF and the Jacobian of the transformation from the independent to dependent variables in Equation (3.136).

3.3.2.3 *The method of auxiliary variables*

We can use the method described in section 3.3.2.2, for finding the PDF of a function of many independent variables, by using auxiliary variables. Suppose that we want to find the PDF of function $Z = g(X, Y)$. Here we present an alternative method to that in subsection 3.3.2.1.

Consider auxiliary variable $W = Y$. Note that the transformation from variables (X, Y) to (Z, W) is one-to-one. Then from Equation (3.136) the joint PDF of Z and W is,

$$f_{Z,W}(z, w) = \left. \frac{f_{X,Y}(x, y)}{|J(x, y)|} \right|_{x = g^{-1}(z, y)} \tag{3.137}$$

We find the PDF of Z by integrating out variable w,

$$f_Z(z) = \int_{-\infty}^{\infty} f_{Z,W}(z, w)\, dw = \int_{-\infty}^{\infty} \left. \frac{f_{X,Y}(x, y)}{|J(x, y)|} \right|_{x = g^{-1}(z, y)} dy \tag{3.138}$$

Example 3.37: Find the PDF of the sum of two variables, $Z = X + Y$.

Solution:
Introduce auxiliary variable $W = Y$. Then, the inverse transformation from (z, w) to (x, y) is,

$$x = g^{-1}(z, w) = z - y$$
$$y = h^{-1}(z, w) = y$$

The Jacobian of the transformation is,

$$|J(x, y)| = \begin{vmatrix} \dfrac{\partial z}{\partial x} & \dfrac{\partial z}{\partial y} \\[2mm] \dfrac{\partial w}{\partial x} & \dfrac{\partial z}{\partial y} \end{vmatrix} = \begin{vmatrix} 1 & 1 \\ 0 & 1 \end{vmatrix} = 1$$

The joint PDF of variables Z and Y is (Equation 3.137),

$$f_{Z,W}(z, y) = f_{X,Y}(z - y, y)$$

The PDF of the sum is obtained by integrating out y,

$$f_Z(z) = \int_{-\infty}^{\infty} f_{X,Y}(z - y, y)\, dy$$

This equation is identical to Equation (3.126).

Discussion

In many practical design problems, we need to find the PDF of a function of many random variables. For example, reliability assessment of structural systems requires calculation of the PDF of the difference of the load capacity and the applied load. These two quantities are functions of dimensions, material properties and load variables (wind speed, wave elevation, and current speed). In this case, the auxiliary variable method yields an equation similar to (3.138) that requires the calculation of an $n-1$ fold integral, where n is the number of random variables. The integration must be performed numerically because there is usually no closed form expression for this integral. This calculation is impractical for more than about four variables.

The auxiliary variable method cannot be used directly to calculate the reliability of such a system. Numerical methods such as Monte-Carlo simulation, first and second order methods and dimension reduction are popular for these problems.

3.3.2.4   Mean value and standard deviation of a function of many variables

Here we extend the results in section 3.3.1.3 to the mean value of a function of many variables. The mean value of $Y = g(X_1, \ldots, X_n)$ is,

$$E(Y) = \int\limits_{-\infty}^{\infty} \cdots \int\limits_{-\infty}^{\infty} g(x_1, \ldots, x_n) f_{X_1,\ldots,X_n}(x_1, \ldots, x_n)\, dx_1 \ldots dx_n \tag{3.139}$$

The above equation, which is the counterpart of equation (3.120) for one variable, implies that we can calculate the mean value of a function of random variables without calculating the joint PDF of these variables.

From Equation (3.139) we can show that the mean value is a linear operator; the mean value of a linear function of random variables,

$$Y = \sum_{i=1}^{n} a_i X_i \tag{3.140}$$

is,

$$E(Y) = \sum_{i=1}^{n} a_i E(X_i) \tag{3.141}$$

The variance of Y is,

$$\sigma_Y^2 = \sum_{i=1}^{n} a_i^2 \sigma_{X_i}^2 + 2 \sum_{i=1}^{n} \sum_{j>i}^{n} a_i a_j \rho_{ij} \sigma_{X_i} \sigma_{X_j} \tag{3.142}$$

For uncorrelated independent random variables the above equation becomes,

$$\sigma_Y^2 = \sum_{i=1}^{n} a_i^2 \sigma_{X_i}^2 \tag{3.143}$$

Thus, the variance of the sum of uncorrelated random variables $Y = \sum_{i=1}^{n} X_i$ is equal to the sum of their variances.

$$\sigma_Y^2 = \sum_{i=1}^{n} \sigma_{X_i}^2 \tag{3.144}$$

The standard deviation of the sum is equal to the square root of the sum of the squares of the standard deviations,

$$\sigma_Y = \left(\sum_{i=1}^{n} \sigma_{X_i}^2 \right)^{1/2} \tag{3.145}$$

A consequence of the above equation is that the standard deviation of the average of n independent, identically distributed random variables $Y = \frac{1}{n} \sum_{i=1}^{n} X_i$ tends to zero with n converging to infinity,

$$\lim_{n \to \infty} \sigma_Y = 0 \tag{3.146}$$

The sum $\hat{x} = \frac{1}{n} \sum_{i=1}^{n} x_i$ is called *sample mean of X* because it is the average of n realizations (or sample values of X).

Figure 3.47 shows the PDFs of the average of n independent, normal random variables with zero mean and standard deviation 1, for $n = 1, 10, 40, 160$. The standard deviation of the average tends to zero while the mean value remains zero with n increasing. The PDF tends to a spike at zero as n tends to infinity.

Equation (3.146) is very important because it lays the foundation for estimation of mean values of variables from observations. This estimation is based on the Law of Large numbers. There are many versions of the this law including Bernoulli's, Borel's, Tchebycheff's, Markov's and Kinchin's (Papoulis, 1965, pp. 263–266). Here we explain Tchebycheff's version of the law of large numbers.

If variables $X_1, \dots, X_n$ are independent, identically distributed with finite variance, then the average of these variables,

$$\overline{X} = \frac{1}{n} \sum_{i=1}^{n} X_i \tag{3.147}$$

converges to the mean value of these variables in the mean square sense. This means that the mean square difference $E\{[\overline{X} - E(X)]^2\}$ tends to zero with n converging to infinity,

$$\lim_{n \to \infty} E\left\{ \left[\frac{1}{n} \sum_{i=1}^{n} X_i - E(X) \right]^2 \right\} = 0 \tag{3.148}$$

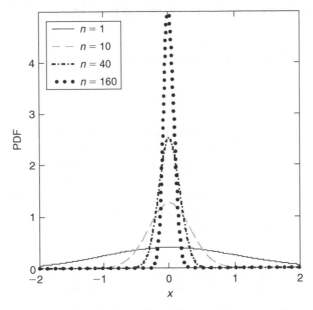

Figure 3.17 Demonstration that the dispersion of the sample mean tends to zero with the sample size converging to infinity, while the central value does not change.

According to the above equation, the average of n independent observations of a random variable converges to the mean value of this variable with the number of observations converging to infinity.

According to the law of large numbers, the relative frequency of an event tends to the probability of this event with the number of observations converging to infinity. This is true because the relative frequency of an event is the average of a counter or indicator function that is one if the event occurs and zero otherwise, and the mean value of the counter function is the probability of the event.

3.3.2.5   *Calculations involving normal random variables*

Calculations involving variables that are linear combinations of normal random variables are very easy because of the following properties of normal variables,

a)   Linear combinations of normal variables are normal.
b)   The normal PDF is defined in terms of its mean value and standard deviation.
c)   The mean value and standard deviation of a linear combination of variables can be readily calculated from Equations (3.141) and (3.142).

Example 3.38:   The stress in a rod, S, is normal with mean value $\mu_S = 90,000\,\text{psi}$ and standard deviation $\sigma_S = 5,000\,\text{psi}$. The strength is also normal with mean value

$S_u = 110,000$ psi, and standard deviation $\sigma_{S_u} = 10,000$ psi. Find the probability of failure. Assume that the stress and strength are independent.

Solution:
The rod fails if the difference of strength and stress becomes negative,

$$g(S, S_u) = S_u - S < 0 \tag{3.149}$$

Function $g(S, S_u)$ is called performance function. The difference is also normal with mean value equal to the difference of the mean values of the strength and stress. The standard deviation is equal to the square root of the sum of the variances of these variables (Equations 3.141 and 3.142). Therefore, the difference of the strength and stress is normal with mean value $E[g(S, S_u)] = 20,000$ psi and standard deviation $\sigma_{g(S,S_u)} = 11,180$ psi. The failure criterion can be expressed in terms of a standard normal random variable as follows,

$$P[g(S, S_u) < 0] = P\left\{ \frac{g(S, S_u) - E[g(S, S_u)]}{\sigma_{g(S,S_u)}} < \frac{-E[g(S, S_u)]}{\sigma_{g(S,S_u)}} \right\}$$

Therefore,

$$P(F) = P(Z < -\beta)$$

where

$$\beta = \frac{E[g(S, S_u)]}{\sigma_{g(S,S_u)}}$$

Constant β is called safety index or reliability index and it is the margin between the strength and the applied load normalized by the standard deviation of this difference. In this example, the safety index is 1.789, which means that the strength is 1.789 standard deviations of the difference greater than the stress. Therefore, the probability of failure is equal to the probability that a standard normal standard random variable is greater than the safety index or equivalently, that a standard normal random variable is less than $-\beta$ (Figure 3.48). This probability is found to be equal to 0.0368 from tables of the CDF of a standard normal distribution or from a computer program such as Mathcad.

If the random variables were not normal, we would need to numerically integrate the joint PDF of the random variables over the failure region in order to calculate the probability of failure. This is an expensive computation. In this example, we used a closed form solution because the variables are normal.

Example 3.39: This example is same as the previous example but the stress and strength are jointly normal with correlation coefficient $\rho = 0.5$.

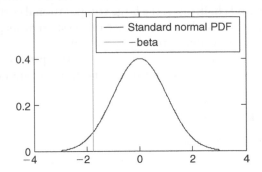

Figure 3.48 The probability of failure is the probability of a standard normal random variable being less than the negative safety index β.

Solution:
The margin between the mean values of the strength and stress does not change but the standard deviation decreases to 8,660 psi. The safety index becomes $\beta = 2.30$ and the probability of failure is 0.011.

These two examples demonstrate that,

- It is much easier to perform probabilistic analysis when the performance function is linear and the random variables are jointly normal than if one of these conditions is violated.
- Dependence of the random variables affects the calculations significantly.
- Assuming independence of the random variables can lead to significant under or overestimation of the probability of failure.

3.4   Concluding Remarks

Two major tasks of a designer are to model the uncertainty in the input variables that drive the performance of a system or the payoff of a risky venture, and to estimate the resulting uncertainty in the system's performance or the payoff of the venture. This chapter presented probabilistic tools to accomplish both the tasks. These tools are based on the notion that probability is the long-term relative frequency of an event.

Objective probability has a solid theoretical justification. Cumulative probability distributions and probability density functions characterize uncertainty. In this chapter, we reviewed common families of probability distributions and guidelines for modeling important uncertain quantities, such as, the time to failure, the lifetime load, and the capacity of a system. We also learned that, when we have limited data and experience about an uncertain quantity, we should select a PDF that has a flexible shape, does not underestimate the probability of extreme events, and its parameters can be estimated easily. The uniform, exponential and beta distributions are good candidates in this case. Finally, we studied computational tools for propagating uncertainty through a system and for quantifying the uncertainty in its performance.

In practical design decisions, it is often difficult to construct a complete probabilistic model of the random variables. The reason is that such a model consists of the joint probability distribution of these variables, and many observations are required to build such a model. The second challenge is the large computational cost required to estimate the probability distribution of the performance of a system or its probability of failure from the probability distribution of the input variables.

Chapters 6 and 7 will present the concept of subjective probability that circumvents the requirement for a very large number of observations for modeling uncertainty.

Questions and Exercises

Fundamentals of Objective Probability

3.1   Answer the following questions below. Mark T or F for True-False questions.

a) The probability of the union of two events E and F, which may intersect (see figure below), is $P(E \cup F) = P(E) + P(F) - P(E \cap F)$.   (T-F)
Hint: Sketch in a Venn diagram the union and intersection of E and F, $P(E \cup F)$ and $P(E \cap F)$.

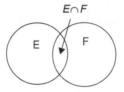

Problem 3.1b

b) In question a, $P(E) = 0.2$, $P(F) = 0.1$ and $P(E \cap F) = 0.05$. Sketch on a Venn diagram event $E \cap F^C$ where F^C denotes the complement of F. How much is $P(E \cap F^C)$?

c) In question b, $P(E \cap F) + P(E \cap F^C) = P(E)$ (T-F)

d) If two events are disjoint (see figure below), then they are independent. (T-F)

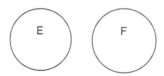

Problem 3.1d

e) Events A and B are independent. Then their intersection is always not empty (T-F)

f) The conditional probability of getting 1 in a roll of a fair die given that the outcome of the roll is less than or equal to 3 is 1/3.   (T-F)

g) If two events are independent, then the probability of the union is equal to the product of the probabilities of these events, $P(E \cup F) = P(E) \cdot P(F)$. (T-F)

h) If two events are independent, then the probability of the intersection is equal to the product of the probabilities of these events, $P(E \cap F) = P(E) \cdot P(F)$. (T-F)

3.2   Find the probabilities of the following events in terms of $P(A), P(B)$ and $P(A \cap B)$,

a) $(A \cup B) \cap (A \cup B^C)$
b) $(A \cup B) \cap (A^C \cup B) \cap (A \cup B^C)$
c) $A^C \cup B^C$

3.3   Show that if two events are disjoint, then the probability of each event is no greater than the probability of the complement of the other,

If $A \cap B = \emptyset$ then $P(A) \leq P(B^C)$

3.4   What is the probability of getting *exactly* two heads in five independent flips of a fair coin?

3.5   A call center of an internet provider company processes both technical help calls and sales calls. Equal numbers of technical and sales calls arrive every day on average. Fifty percent of technical calls are completed in less than two minutes, while the remaining fifty percent take more than two minutes. Ninety five percent of sales calls are completed in less than two minutes, while the remaining five percent call are completed in more than two minutes. An operator tells you that he just processed a phone call in more than two minutes without telling you if it was a sales or a technical call. Given that the phone call took more than two minutes what is the probability of being a technical help call?

3.6   A family has two children. You learn that one child is a girl. Find the probability that the other child is also a girl conditioned upon this information. Assume that the unconditional probability of a girl is 0.5.

3.7   If two fair dice are rolled once, what is the probability that the total number of dots shown is,

a) Equal to 5
b) Divisible by 3

3.8   A state conservation department has announced the opening of a manmade lake. The statistics that they publish are as follows:

a) 30% of the fish are of legal size
b) 10% of the trout are of legal size
c) 40% of the fish are either trout or of legal size

Find the probability that a fish is a trout and of legal size.

3.9   A slot machine has four separate wheels that rotate independently. On each wheel are four pictures of a lemon and one picture of a cherry. Each time the slot machine level is pulled, one picture on each wheel will appear in the window. In order to win a cash prize, three or more pictures of a cherry must appear in the window. Find the probability of winning a prize when the lever is pulled once.

Random Variables
3.10   Explain which of the following functions are probability distributions,

$$a)\ F_X(x) = \begin{cases} 0 & x < 0 \\ x^2 & 0 \leq x < 1 \\ 1 & x > 1 \end{cases}$$

b) $F_X(x) = \begin{cases} 0 & x \le 4 \\ 1 & x > 4 \end{cases}$

c) $F_X(x) = \begin{cases} 0 & x < 2 \\ x - 1 & 2 \le x < 3 \\ 1 & x \ge 3 \end{cases}$

d) $F_X(x) = \begin{cases} 0 & x < 0 \\ 1 - e^{-x} & 0 \le x \end{cases}$

3.11 The distance from Ann Arbor to Detroit is 40 miles. You will drive from Ann Arbor to Detroit at a constant speed of 60 mph, and your departure time is uniform from 12 noon to 1 PM. Calculate the CDF of the distance of your car from Ann Arbor at 1 PM.

3.12 The probability density function of variable X is uniform from 0 to 10. Find the probability that the solutions of equation in variable z (which is not considered random) $z^2 + xz + x = 0$ are real.

3.13 Let X be a random variable with CDF, $F_X(x)$, and PDF, $f_X(x)$. Find the conditional CDF and PDF of X conditioned upon the event A where,

a) $A = \{X > a\}$
b) $A = \{a_2 \ge X > a_1\}$

3.14 The median of random variable X is the real number x_m such that the probability of X being less than or equal to the median is 0.5. Find the median of the following random variables with given probability density functions,

a) $f_X(x) = \frac{1}{b} e^{-x/b}$ for $x \in [0, \infty)$ (Hint: First find the value of b so that you get a valid PDF.)

b) $f_X(x) = xe^{-\frac{x^2}{2}}$ for $x \in [0, \infty)$

3.15 A random variable X has the probability function of the form,

$$f_X(x) = \begin{cases} 2x & x \in [0, x_1] \\ 0 & \text{elsewhere} \end{cases}$$

Find the value of x_1 and the probability that $X > 1$.

Multiple random variables

3.16 Determine if the statements below are true or false, where $F_{XY}(x, y)$ and $f_{XY}(x, y)$ are the joint CDF and PDF of two random variables X and Y.

a) $F_{XY}(-\infty, y) = 0$
b) $F_{XY}(\infty, y) = 1$
c) $f_{XY}(x, y)$ can be greater than 1.

d) The marginal PDF of variable X is equal to the ratio of the joint PDF of X and Y and the marginal PDF of Y,

$$f_X(x) = \frac{f_{X,Y}(x,y)}{f_Y(y)}$$

e) $f_X(x) = f_{XY}(\infty, y)$

f) $f_X(x) = \int\limits_{-\infty}^{+\infty} f_{XY}(x,y)\, dy$

g) $F_X(x/Y \in A) = \int\limits_{-\infty}^{x} f_X(x'/Y \in A)\, dx'$, where A is a range of values of Y.

h) $F_X(x) = F_{XY}(x, \infty)$

i) The conditional PDF of X conditioned upon $Y = y$ is $f_{X/Y}(x) = \dfrac{f_{X,Y}(x,y)}{f_Y(y)}$.

j) The conditional CDF of X conditioned upon $Y = y$ is $F_{X/Y}(x) = \dfrac{F_{X,Y}(x,y)}{F_Y(y)}$.

3.17   Determine the constant b so that each of the following functions are valid joint PDFs,

a) $f_{XY}(x,y) = \begin{cases} 3xy & \text{for } 1 \geq x \geq 0,\ b \geq y \geq 0 \\ 0 & \text{elsewhere} \end{cases}$

b) $f_{XY}(x,y) = \begin{cases} bx(1-y) & \text{for } 0.5 \geq x \geq 0,\ 1 \geq y \geq 0 \\ 0 & \text{elsewhere} \end{cases}$

c) $f_{XY}(x,y) = \begin{cases} b(x^2 + 4y^2) & \text{for } 1 \geq x \geq -1,\ 2 \geq y \geq 0 \\ 0 & \text{elsewhere} \end{cases}$

3.18   Which of the following are bivariate CDFs?

a) $F_{XY}(x,y) = \begin{cases} 1 - e^{-(x+y)} & \text{for } x \geq 0, y \geq 0 \\ 0 & \text{elsewhere} \end{cases}$

b) $F_{XY}(x,y) = \begin{cases} 0 & \text{for } x < y \\ 1 & \text{for } x \geq y \end{cases}$

3.19   The joint PDF of random variables X and Y is,

$$f_{XY}(x,y) = \begin{cases} e^{-(x+y)} & \text{for } x \geq 0,\ y \geq 0 \\ 0 & \text{elsewhere} \end{cases}$$

Calculate the marginal CDFs of X and Y and the probability that X is less than Y.

3.20   If $f_{XY}(x,y) = g(x)h(y)$, find the marginal PDFs of the two variables.

3.21   Let X and Y be two random variables. Show that,

$$F_X(-\infty/Y \in A) = 0$$

where A is some set of values and the probability of the conditioning event is nonzero.

3.22   Consider the random variables X and Y have the following CDF,

$$
F_{XY}(x,y) = \begin{cases}
1 & \text{for } x \geq 1,\ y \geq 1 \\
x & \text{for } 1 > x \geq 0,\ y \geq 1 \\
y & \text{for } 1 > y \geq 0,\ x \geq 1 \\
xy & \text{for } 1 > x \geq 0,\ 1 > y \geq 0 \\
0 & \text{otherwise}
\end{cases}
$$

a) Find the marginal CDFs of these variables.
b) Determine if the two variables are independent.

Common probabilistic models

3.23   A random variable has a uniform PDF (Equation 3.69). Calculate the probability that

$$
X \leq \frac{2x_l + x_u}{3}
$$

3.24   The time to failure of a particular class of computer hard disks follows the exponential probability distribution (Equation 3.72). Test results show that the probability of a hard disk failing less than 2,000 hours of operation is e^{-1}.

a) Calculate the mean time to failure and write expressions for the CDF and PDF of the time to failure. Assume that $x_l = 0$.
b) Calculate the probability that a hard disk could fail before 1,000 hrs of operation.

3.25   One earthquake of magnitude greater than 6.5 in the Richter scale occurs in 10 years, in a particular city, on average. Assume that earthquakes of such magnitude occur independently.

a) What family of probability distributions represents the number of earthquakes of such magnitude in a given period?
b) Find the parameters of the probability distribution in question a) from the information provided.
c) Find the probability that more than one earthquakes could occur in 5 years.

3.26   An automotive manufacturer estimated the following parameters of the time to failure of a timing belt,
   Mean value = 80,000 miles
   Minimum value = 30,000 miles
   Standard deviation = 20,000 miles
   The manufacturer's engineers believe that the Weibull distribution represents the time to failure.

a) Estimate the parameters of the Weibull distribution.

b) The manufacturer recommends replacing timing belts every 60,000 miles. Calculate the probability of failure before this limit.

3.27   Same problem as 3.26 but represent the time to failure by the Gumbel distribution.

3.28   You want to construct probabilistic models of the quantities presented below. Specify one or more families of probability distributions that you would consider for each quantity.

Quantity	Probability distribution(s)
Data from repeated measurements of the diameter of the same tube taken at the same location	
Time to complete a phone call	
Time between two earthquakes of magnitude greater than 8 in Richter scale in Chile	
Number of earthquakes of magnitude greater than 8 in Richter scale in Chile over the next 100 years	
Number of heads in 10 flips of a fair coin	
Time to failure of a piece of equipment	
Proportion of defective parts in a shipment	
50-year wind speed at a location in the North Sea	
Stress at failure of a series system (that is, a system that fails if any of its components fail)	
Length of the longest crack in an aircraft wing.	

3.29   Specify one or more families of probability distributions that you would consider for each quantity presented in the table below.

Quantity	Probability distribution(s)
Error in the stress in a particular structure calculated by finite element analysis by a particular analyst.	
Height of Sears Tower	
Population of Toledo, Ohio	
Mathematical constant e	

3.30   What are the fundamental differences between the quantities in problems 3.28 and 3.29 and the uncertainty associated with these quantities?

Probability calculations

3.31   The CDF of a continuous random variable X, $F_X(x)$, is a random variable itself because it is a function of a x. Prove that the probability distribution of $F_X(x)$ is uniform from 0 to 1.

3.32   Let X and Y be independent random variables with probability density functions $f_X(x)$ and $f_Y(y)$, respectively. Show that the probability density function of their product is,

$$f_Z(z) = \int\limits_{-\infty}^{+\infty} \frac{1}{|x|} f_X(x) f_Y\left(\frac{z}{x}\right) dx$$

3.33   Let $F_X(x)$ be the CDF of a random variable X. Assume a new random variable $Y = g(X)$. Find the CDF of Y if $g(X)$ has the following forms,

$$g(x) = \begin{cases} x - c & x \geq c \\ x & c > x > -c \\ x + c & x \leq -c \end{cases}$$

$$g(x) = \begin{cases} b & x \geq b \\ x & b > x > -b \\ -b & x \leq -b \end{cases}$$

$$g(x) = \begin{cases} b & x \geq 0 \\ -b & x < 0 \end{cases}$$

3.34   Let X be a random variable with PDF $f_X(x)$. Assume a new random variable $Y = g(X)$. Find the PDF of Y if $g(X)$ has the following forms,

$$g(X) = a \tan(X) \quad X \text{ is uniform in the range } \left[-\frac{\pi}{2}, \frac{\pi}{2}\right]$$

$$g(X) = \begin{cases} e^{-aX} & \text{for } X \geq 0 \\ 0 & \text{otherwise} \end{cases}$$

3.35   Show that if X and Y are independent identically distributed Gaussian variables then the square root of the sum of the squares of these variables has Raleigh PDF.

3.36   Let X and Y be independent identically distributed Gaussian variables, with mean values 1 and 3 and standard deviations 0.5 and 1.5. Find the probability distribution of $Z = Y - X$ and the probability that this variable could be negative.

3.37   In problem 3.36, find the probability that $Z < 0$ when X and Y have correlation coefficient equal to 0.5.

3.38   A table's top surface has sides equal to a and b, both normally distributed with mean 1 m and standard deviation 0.2 m. Find the probability that the surface is a square. Also find the probability that the area of the surface is less than 1.1 m^2.

3.39   Find the distribution of the radius of a circle with area equal to the area of the table's top surface in the previous problem.

3.40   An aircraft's take-off run is triangularly distributed with lower limit of 6000 ft, upper limit of 11000 ft and most likely value of 7500 ft. Its stopping distance from touch down to complete halt is also triangularly distributed with values of 5000 ft, 9000 ft and 6000 ft respectively. Find the probability that the aircraft can land on the same runway as it can take-off. Hint: Find the probability that the take-off distance is longer than the landing distance.

Appendix

Calculation of the number possible ways of getting an outcome k times in a Bernoulli process with n trials.

We flip a coin n times. We need to find the number of ways we can get heads k times. There are 2^n possible outcomes of the Bernoulli process,

TTT...T
HTT...T
...
HHH...H

We want to count the outcomes in which H appears k times. We use the following notation for the outcomes of a Bernoulli process: We represent each trial by a number according to the order it was performed and frame the number if the outcome of this trial was heads. For illustration, we show the outcomes for the special case where H appears twice,

$\boxed{1}\boxed{2}\,3\,...n$

$\boxed{1}\,2\,\boxed{3}\,...n$

...

$1\,2\,3\,\boxed{n-1}\,\boxed{n}$

The first row represents the outcome in which we get heads in the first two trials, the second the outcome in which we get heads in the first and third trials and the last the outcome in which we get heads in the last two. We observe from this representation that the number of ways in which we can get two heads in n trials is equal to the number of distinct pairs drawn from a pool of n objects. In general, the number of outcomes in which we get k heads in n trials is equal to the number of distinct k-tuples drawn from a pool of n objects, where two k-tuples are distinct if they differ at least by one object. This number is equal to: $\frac{n(n-1)\cdots(n-k+1)}{2\cdot3\cdots k} = \frac{n!}{k!(n-k)!}$ and it is denoted by $\binom{n}{k}$.

Chapter 4

Statistical Inference – Constructing Probabilistic Models from Observations

4.1 Introduction

Chapter 3 explained how to calculate the probability of an event from the probabilities of other events. Two examples where probability calculations are required are,

- Find the probability that at least 2 out of 100 pumps from a factory could fail during the first year of operation.
- Find the likelihood that an earthquake with magnitude 8 or higher on the Richter scale could occur in Southern California in the next 30-years.

The above calculations will require the probability of failure of one pump and the probability distribution of the strongest earthquake in one year. A decision maker can estimate these quantities from observations or expert judgments. This chapter will explain how to construct probabilistic models and draw conclusions from observed data by using statistics. Chapter 6 will explain how to construct a probabilistic model by using expert judgments and by combining these judgments with observations.

The science of *Statistics* deals with methods of recording, organizing, analyzing and reporting quantitative information. The branch of *descriptive* statistics focuses on collecting, summarizing and presenting data. The branch of *inferential* statistics deals with methods to analyze the data, construct probabilistic models and draw conclusions. One uses these conclusions to make forecasts and improve a system. The three examples below demonstrate how statistics can help a decision maker.

- An engineer constructs and tests 30 prototypes of a pump and observes one failure. The engineer wants to estimate the probability of failure of a pump and quantify the accuracy of this estimate. In addition, she wants to test the hypothesis that the probability of failure exceeds 0.1.
- The productivity of an electronic assembly and test department in a factory is low. In this department, parts produced by another department enter a preparation area where they are machined, deburred and cleaned. Then they are assembled and sealed at a second station. Finally, the parts are inspected at a third station. These parts are shipped or repaired at a rework station. The department manager records and estimates the waiting times at the four stations in order to identify what slows down the flow of parts. This statistic is very high at the rework station compared to those of the other stations. This suggests that the manager can improve

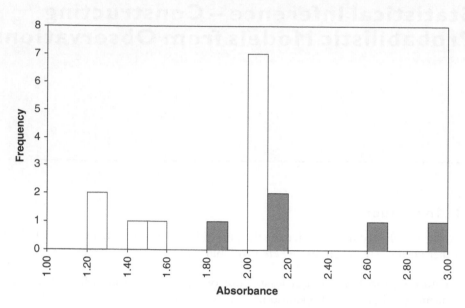

Figure 4.1 Cleaned Plates (White Bars) vs. Control Plates (Grey Bars).

productivity significantly by adding operators to the rework area. Before implementing this change, the manager simulates the operations in the factory using a program for manufacturing system simulation, such as Arena (Kelton et al. 2010, Chapter 4), and compares the statistics of the performance attributes of the original and a modified system. The results confirm that the latter system is much more efficient than the original.

• In the final example, a biologist wants to decide if a new cleaning procedure could clean surfaces infected by bacteria. For this purpose, he grows bacteria on 16 glass plates, and tries to clean 11 of them by using this procedure. Then he compares the amount of bacteria left on these 11 plates to the amount of bacteria on five control plates. The amount of bacteria left on a plate is quantified by a quantity called absorbance. The higher the absorbance, the more bacteria are on the plate.

Figure 4.1 compares the conditions of the plates that the biologist tried to clean to those of the control plates. The processed plates seem to have fewer bacteria than the control plates. However, it is unsafe to conclude that the cleaning procedure is effective from the results in Figure 4.1 because the results could be due to luck. The biologist could collect more data but this would be very expensive. Statistics provide the biologist with formal procedures to test the hypotheses that the cleaning procedure is effective by using the limited data in Figure 4.1.

It is important to define the terms *population* and *sample*. A population consists of all members in a group about which we want to draw a conclusion. Sample is a part of the population. In the first example, the population comprises all pumps that the engineer's factory will make. The set of 30 pumps is a sample drawn from

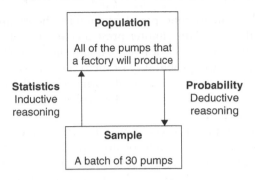

Figure 4.2 The relation between statistics and probability.

this population. The engineer wants to determine the properties of the population of pumps from those of the sample of 30 pumps.

It is impractical to calculate the properties of a population. For example, the engineer cannot test all the pumps that the factory will produce in order to estimate their reliability. The engineer should infer the probability of failure of the population of the pumps from the observed number of failures in the sample of the 30 pumps. The process of inferring the properties of a population from the properties of the sample is called statistical inference. This is an inductive procedure in that it makes inferences about the population from a sample.

Statistical methods are valuable tools for reducing cost and improving quality in design and manufacturing. In the above example, an engineer has a test budget and wants to determine the properties of the population. This is challenging because tests are expensive. In addition, even a modest improvement in the accuracy of the estimated failure probability often requires too many additional observations. For example, in order to decrease the standard error in the estimated failure probability of the pumps to one half, the engineer needs to increase the number of observations by four.

Statistical methods enable the engineer to use resources effectively in order to obtain the maximum amount of information about a population. For example, these methods enable her to extract the maximum amount of information from a sample and to design the tests to obtain the required information at a low cost.

After a decision maker infers the population properties, she can predict the properties of another sample by using probability. For example, the engineer calculates the probability that more than 10% of the pumps could fail during the warranty period using the methods in chapter 3.

Figure 4.2 depicts the relation between statistics and probability. When we use statistics, we proceed from the specific to the general, that is, we infer the properties of the population from those of a sample. This is inductive reasoning. When we use probability, we proceed from the general to the specific, which is deductive reasoning.

Statistical summaries describe the properties of a population. These are tools that describe the key features of the population (for example the central tendency, and the scatter). For numerical data, these summaries include the mean, variance, and probability distribution.

4.1.1   *Objective, scope and summary of this chapter*

Chapter 3 explained the main concepts of probability theory and presented methods for calculating probabilities. This chapter presents concepts of statistical inference of probabilistic models from data. Section 4.2 explains how to estimate the mean value of a random variable and the probability of an event and assess the accuracy of this estimate. Section 4.3 explains the concept of testing a hypothesis about the properties of a population from data. Sections 4.4 and 4.5 explain how to estimate the probability distribution of random variables.

The presentation in Chapter 4 is confined to inferences made by using data. In many decisions, a decision maker has limited or no data. Then the decision maker should rely on judgments from experts, besides observed data, in order to infer the properties of a population. Chapter 6 will explain the concept of subjective probability. It will present and demonstrate methods for estimating a probabilistic model by using expert judgments and by combining such judgments with data.

4.2   Estimating Mean Values of Random Variables and Probabilities of Events

4.2.1   *Sample mean*

We want to estimate the mean value of a random variable from a sample of observations. First, we will present equations for estimating the statistics of the population from which the sample was drawn when the observations are statistically independent. Then we will develop equations when the sample values are correlated in section 4.2.3.

Observations $X_1, \ldots, X_n$ are considered to be random variables in the sense that each time we draw a different sample from the population, we get different values. Let μ and σ be the unknown mean value and standard deviation of the population. An estimator of the mean value is the mean or average of the sample,

$$\hat{\mu}_n = \frac{1}{n}\sum_{i=1}^{n} X_i \tag{4.1}$$

Note that the sample mean is a random variable, because it assumes a different value each time we draw a new sample. This estimator is useful because it has three desirable properties,

a)   It is unbiased: On average, it is equal to the mean value of the population.
b)   It is consistent: It gets better as the sample size increases, in the sense that its variance decreases with n.
c)   It has minimum variance among all linear estimators: A linear estimator has the following form:

$$\hat{\mu}_n = \sum_{i=1}^{n} a_i X_i \tag{4.2}$$

where a_i are arbitrary constants.

We prove these properties below.

a)   Estimator $\hat{\mu}_n$ in Equation (4.1) is unbiased. The mean value of $\hat{\mu}_n$ is equal to the sum of the mean values of random variables X_i because the mean of the sum is equal to the sum of the mean values. Equation (4.1) follows from the fact that $E(X_i) = \mu$. Q.E.D.

b)   Estimator $\hat{\mu}_n$ is consistent. The variance of the sample mean $\hat{\mu}_n$ is,

$$Var(\hat{\mu}_n) = \frac{1}{n^2} Var\left(\sum_{i=1}^{n} X_i\right)$$

The variance of the sum of independent random variables is equal to the sum of their variances. Therefore,

$$Var(\hat{\mu}_n) = \frac{1}{n^2} \sum_{i=1}^{n} Var(X_i) = \frac{\sigma^2}{n} \tag{4.3}$$

where $\sigma^2 = Var(X_i)$. Equation (4.3), shows that the variance of the estimator of the mean decreases with the sample size n. Therefore, this estimator is consistent. Q.E.D.

   The variance of the sample mean converges to zero with the sample size according to equation (4.3). Hence, $E\{(\hat{\mu}_n - \mu)^2\}$ converges to zero with the sample size. We say that the sample mean converges to the population mean, in the mean square sense.

   The linear estimator (Equation 4.2) is unbiased. This means that, if we use a sufficiently large sample in this equation, then we will accurately estimate the mean value of random variable X from this estimator. This is true for any nonzero combination of values of the weighting constants $a_i, i = 1, \ldots, n$ that add up to one. However, we will need a very large sample in order to estimate the mean value accurately, for some combinations of values of these coefficients. For a given sample size, the weighting constants should be all equal to $1/n$, in order to obtain the most accurate estimate of the mean value from Equation (4.2). Most accurate estimate is the one with minimum variance.

c)   Finally, we prove that the sample mean has minimum variance. For this purpose we will prove that, of all linear estimators that have the form of Equation (4.2), the one that has minimum variance satisfies the condition, $a_i = 1/n$.

Proof: The mean value of estimator (4.2) is equal to the sum of coefficients a_i. Therefore, from the requirement that the estimator (4.2) must be unbiased, it follows that these coefficients must add up to 1.

   The variance of the linear estimator (4.2) is,

$$Var(\hat{\mu}_n) = Var\left(\sum_{i=1}^{n} a_i X_i\right) = \sum_{i=1}^{n} a_i^2 Var(X_i) = \sigma^2 \sum_{i=1}^{n} a_i^2 \tag{4.4}$$

The values of a_i that minimize (4.4) are the solution to the following constrained optimization problem,

Find $a_1, \ldots, a_n$

To minimize $\sum_{i=1}^{n} a_i^2$ (4.5)

So that $\sum_{i=1}^{n} a_i = 1$.

The Lagrangian of this problem is,

$$L(a_1, \ldots, a_n) = \sum_{i=1}^{n} a_i^2 + \lambda \cdot \left(\sum_{i=1}^{n} a_i - 1 \right)$$

The derivative of the Lagrangian with respect to constants a_i must be zero,

$$\frac{\partial L}{\partial a_i} = 2a_i + \lambda = 0 \quad \text{for } i = 1, \ldots, n$$

$$\text{Therefore, } a_i = -\frac{\lambda}{2} \quad \text{for } i = 1, \ldots, n$$

Since coefficients a_i must add up to one, it follows that $a_i = \frac{1}{n}$ for $i = 1, \ldots, n$, and the minimum variance estimator is the sample mean in Equation (4.1).
Q.E.D.

Equation (4.3) is impractical for assessing the accuracy of the estimator of the mean value of X in Equation (4.1) because this equation involves the variance of the population, σ^2, that the user does not know. The user can replace σ^2 by its estimator $\hat{\sigma}_n^2$ in Equation (4.9)

$$\hat{V}ar(\hat{\mu}_n) = \frac{\hat{\sigma}_n^2}{n} = \frac{\sum_{i=1}^{n} (X_i - \hat{\mu}_n)^2}{n(n-1)}$$ (4.6)

An equivalent expression for the sample variance is obtained by expanding the square in the numerator of the right hand side of Equation (4.6),

$$\hat{V}ar(\hat{\mu}_n) = \frac{1}{n(n-1)} \left[\sum_{i=1}^{n} X_i^2 + n\hat{\mu}_n^2 - 2 \cdot \hat{\mu}_n \sum_{i=1}^{n} X_i \right]$$ (4.7)

Therefore,

$$\hat{V}ar(\hat{\mu}_n) = \frac{1}{n(n-1)} \left[\sum_{i=1}^{n} X_i^2 - n \cdot \hat{\mu}_n^2 \right]$$ (4.8)

Equation (4.8) is equivalent to (4.6) but it allows for a more efficient calculation of the sample variance than (4.6). Note that the variance of the sample mean measures its sample-to-sample variability.

Table 4.1 Interarrival times of 10 customers at a bank during lunch time.

Customer	Interarrival Time (min)
1	33.35
2	8.22
3	2.69
4	5.25
5	0.97
6	8.74
7	1.71
8	5.95
9	11.96
10	9.58

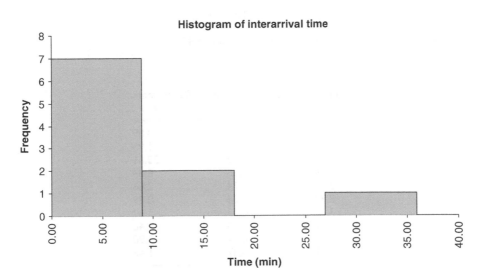

Figure 4.3 Histogram of interarrival times in example 4.1.

Example 4.1: A bank manager wants to estimate the mean value of the interarrival time of customers during lunch time. The manager recorded the interarrival times of 10 customers, which are shown in Table 4.1. Calculate the sample mean value of the interarrival time. Calculate the variance of the sample mean from both Equations (4.6) and (4.8) and compare the results.

Solution:
Figure 4.3 displays the data in Table 4.1 classified into four bins. Each bin is the base of one bar with height equal to the number of values that the bin contains. This type of plot is called histogram. The width of each bin is 9 min. According to this figure, 7 values are in the range from 0 to 9 minutes, 2 values in the range from 9 to 18 minutes and one value between 27 and 36 minutes. There is large dispersion in the data.

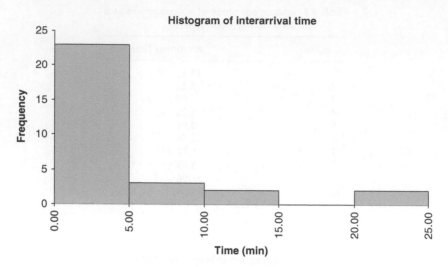

Figure 4.4 Histogram of interarrival times in example 4.2.

Table 4.2 Interarrival times of 30 customers at the bank during lunch time.

0.06	0.66	4.89
10.64	0.23	1.95
23.60	3.09	23.66
3.16	3.86	6.44
2.54	0.74	2.66
8.97	1.24	0.89
3.98	0.02	3.62
14.32	2.46	1.48
1.22	6.62	3.90
3.27	0.87	1.48

The sample mean of the interarrival time is 8.84 min from Equation (4.1). Both equations (4.6) and (4.8) yield 8.7 min² for the variance of the sample mean. Therefore, the standard deviation of the sample mean is 2.95 min.

Example 4.2: The manager recorded 30 values of the interarrival time shown in Table 4.2. Calculate the sample mean and the variance of the sample mean.

Solution:
The histogram in Figure 4.4 displays the data. It is observed that 23 of the 30 values in Table 4.2 are between 0 and 5 min. Three values are between 5 and 10 minutes. The sample mean of the interarrival time is 4.75 min. The variance of the sample mean is 1.23 min² and the standard deviation of the sample mean is 1.11 min.

4.2.2 Sample variance

The sample variance of X is an estimator of the variance of this variable,

$$\hat{\sigma}_n^2 = \frac{1}{n-1}\sum_{i=1}^{n}(X_i - \hat{\mu}_n)^2 \tag{4.9}$$

This estimator is unbiased as we prove below. We expand the square in the sum in the above equation,

$$E(\hat{\sigma}_n^2) = \frac{1}{n-1}\sum_{i=1}^{n}\{E(X_i^2) + \hat{\mu}_n^2 - 2E(X_i\hat{\mu}_n)\} \tag{4.10}$$

We will express each of the three terms on the right hand side in terms of the mean value and variance of the population. The mean square of variable X is equal to the square of its mean value plus its variance (see exercise 4.6 in the end of this section). Therefore, the first term of the sum is,

$$E(X_i^2) = \mu^2 + \sigma^2 \tag{4.11}$$

The mean square of the sample mean is equal to the sum of the square of the mean value of this estimator plus its variance,

$$E(\hat{\mu}_n^2) = \mu^2 + \frac{\sigma^2}{n} \tag{4.12}$$

The third term is,

$$E\{X_i\hat{\mu}_n\} = \frac{1}{n}E\left(\sum_{\substack{j=1,\ldots,n \\ i \neq j}}^{n} X_j X_i + X_i^2\right) = \frac{1}{n}\{(n-1)\mu^2 + \mu^2 + \sigma^2\} = \mu^2 + \frac{\sigma^2}{n} \tag{4.13}$$

Plugging Equations (4.11) to (4.13) into (4.10) yields,

$$E\{\hat{\sigma}_n^2\} = \frac{1}{n-1}\sum_{i=1}^{n}\left[\mu^2 + \sigma^2 + \mu^2 + \frac{\sigma^2}{n} - 2\mu^2 - 2\frac{\sigma^2}{n}\right] = \sigma^2$$

Q.E.D.

An equivalent estimator of the variance to that in Equation (4.9) is,

$$\hat{\sigma}_n^2 = \frac{1}{n-1}\left[\sum_{i=1}^{n}X_i^2 - n\cdot\hat{\mu}_n^2\right] \tag{4.14}$$

It is more efficient to compute the above estimator than the one in Equation (4.9).

Do not confuse the variance of the sample mean (Equation 4.6) with the sample variance (Equation 4.14). The variance of the sample mean measures the variation of the mean from one sample to another, while the sample variance measures the variation of the observations in a sample. Note that the variance of the sample mean (Equation 4.8) is equal to the sample variance (Equation 4.14) divided by the sample size,

$$\hat{V}ar(\hat{\mu}_n) = \frac{\hat{\sigma}_n^2}{n} \qquad (4.15)$$

The square root of the variance of the sample mean is called standard error of the mean.

Example 4.3: Find the variance of the sample for the two examples above by using both Equations (4.9) and (4.14).

Solution:
The sample variance is $87.03\,min^2$ and $36.91\,min^2$ for examples 4.1 and 4.2, respectively. The corresponding values of the standard deviation are $9.33\,min$ and $6.08\,min$.

Example 4.4: We can quantify the accuracy of the sample mean of the interarrival time by the coefficient of variation (the standard deviation of the interarrival time normalized by the mean value). The coefficient of variation of the sample mean from the 30 observations in Table 4.2 is $\frac{1.11}{4.75} = 0.23$. Therefore, the standard deviation of the sample mean is 23% of its value. The manager wants to reduce the coefficient of variation by one half. How many observations of the interarrival time does the manager need?

Solution:
The coefficient of variation (COV) of the sample mean is,

$$COV(\hat{\mu}_n) = \frac{Var(\hat{\mu}_n)^{1/2}}{\hat{\mu}_n} = \frac{\hat{\sigma}_n}{\hat{\mu}_n \cdot n^{1/2}} \qquad (4.16)$$

According to the above equation the coefficient of variation of the sample mean is inversely proportional to the square root of the sample size. Therefore, the manager needs four times as many observations (i.e., 120) in order to reduce the coefficient of variation to one half, i.e. 0.115. The number of observations as a function of the desired value of the coefficient of variation is,

$$n = \frac{\left(\frac{\hat{\sigma}_n}{\hat{\mu}_n}\right)^2}{COV(\hat{\mu}_n)^2} \qquad (4.17)$$

Figure 4.5 shows the coefficient of variation of the sample mean vs. the number of observations. The results in this figure were calculated from Equation (4.16) by using the sample mean and standard deviation from 30 observations. Increasing the sample size drastically improves the accuracy of the estimated mean value, for up to 30 observations. For larger sample sizes the rate of improvement decreases. The point of diminishing returns is reached after about 100 observations.

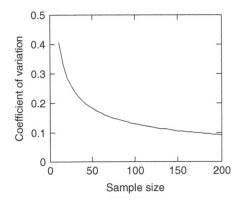

Figure 4.5 Coefficient of variation of the estimated mean interarrival time vs. the sample size.

4.2.3 Covariance and Correlation

Observed data are often dependent. For example, the waiting times of customers in a bank or callers in a call center have positive dependence. If a customer waits too long in a queue, then the next customer is likely to do the same. Simulation results are almost always dependent. Equations (4.8) and (4.15) for the variance of the sample mean are not applicable to dependent observations. First, we present equations for estimating the covariance and correlation coefficient from data. Then we derive an equation for the variance of the sample mean for correlated data.

Consider a sample of ordered observations $X_1, \ldots, X_n$ drawn from the same probability distribution. For example, X_i is the waiting time of the ith customer in a bank. The covariance of two observations X_i and X_{i+j} depends only on the separation of the values, j. An estimator of the covariance of these variables is,

$$\hat{C}_j = \frac{1}{n-j} \sum_{i=1}^{n-j} (X_i - \hat{\mu}_n)(X_{i+j} - \hat{\mu}_n) \tag{4.18}$$

Suppose that we observe the wait times of n customers every day for m days. Consider that the statistical properties of the waiting time do not change from one day to another. Then, we have m independent sequences of observations, $X_{i,k}$, where $i = 1, \ldots, n$ and $k = 1, \ldots, m$. We use the following estimator of the covariance,

$$\hat{C}_j = \frac{1}{(n-j)m} \sum_{k=1}^{m} \sum_{i=1}^{n-j} (X_{i,k} - \hat{\mu}_{mn})(X_{i+j,k} - \hat{\mu}_{mn}) \tag{4.19}$$

where $\hat{\mu}_{mn}$ is the sample mean of the waiting times from the data for all m days. The estimator in Equation (4.19) is more accurate than the one in Equation (4.18) because the m sequences are independent and because positive dependence reduces accuracy.

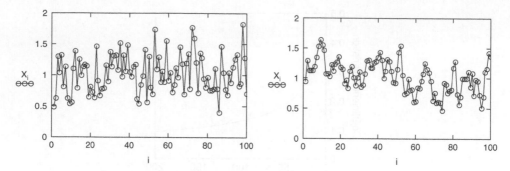

Figure 4.6   Simulated sample values from an ARMA model for correlation coefficient 0.1 (left panel) and 0.75 (right panel).

The correlation coefficient of the waiting time is the covariance normalized by the variance,

$$\hat{\rho}_j = \frac{\hat{C}_j}{\hat{\sigma}_n^2} \tag{4.20}$$

Example 4.5:   One way to simulate a sequence of dependent values drawn from the same probability distribution is by using an autoregressive moving average (ARMA) model,

$$X_i = \mu + \phi(X_{i-1} - \mu) + \varepsilon_i \tag{4.21}$$

where X_{i-1} and X_i are the $(i-1)$th and ith variables of the sequence, and ε_i are independent, identically distributed normal random variables. The mean value of ε_i is zero. Parameter μ is equal to the mean value of X_i. Usually, the variance of X_i, σ^2, is prescribed. Then, the variance of ε_i, σ_ε^2 is determined as follows,

$$\sigma_\varepsilon^2 = \sigma^2(1 - \phi^2) \tag{4.22}$$

Constant ϕ is equal to the correlation coefficient of X_{i-1} and X_i, ρ. The correlation coefficient of two values X_i and X_{i+j}, is equal to ρ^j.

Assume that $\mu = 1$ and $\sigma = 0.3$. Simulate four samples with $N = 100$, 1,000, 10,000, and 100,000 values from Equation (4.21) for four cases where the correlation coefficient of two subsequent values is $\rho = 0.1$, 0.3, 0.5 and 0.75. Estimate the correlation coefficient from these sample values.

Solution:
The left and right panels in Figure 4.6 show two sequences consisting of 100 sample values for $\rho = 0.1$ and $\rho = 0.75$. The time history is noisy for $\rho = 0.1$ because adjacent values change independently. Adjacent values tend to move together when the correlation coefficient is 0.75.

Table 4.3 Estimates of the correlation coefficient of two subsequent values X_i and X_{i+j}.

N	$\rho=0.1$	$\rho=0.3$	$\rho=0.5$	$\rho=0.75$
100	−0.04	0.267	0.389	0.717
1,000	0.067	0.314	0.478	0.758
10,000	0.108	0.299	0.499	0.746
100,000	0.1	0.299	0.497	0.753

Table 4.3 presents the estimated values of the correlation coefficients of two subsequent values X_i and X_{i+1} from Equation (4.20). We need at least 10,000 sample values in order to estimate accurately the correlation coefficient.

Ignoring correlation in a sample could lead to serious errors in the estimated properties of a population. Although we can still use Equation (4.1) for the sample mean, we need a new equation for its variance. We develop such an equation below.

Lemma 1: Consider n independent identically distributed variables $X_1, X_2, \ldots, X_n$. The variance of the variables is σ^2. The correlation coefficient of X_i and X_{i+j} is ρ_j. Then the variance of the sum is,

$$Var(X_1 + \cdots + X_n) = n\sigma^2 \left(1 + 2 \sum_{j=1}^{n-1} \frac{n-j}{n} \rho_j \right) \tag{4.23}$$

Proof: The variance of the sum is,

$$Var(X_1 + \cdots + X_n) = \sum_{i=1}^{n} Var(X_i) + 2\sum_{i=1}^{n-1} Cov(X_i X_{i+1}) + 2\sum_{i=1}^{n-2} Cov(X_i X_{i+2})$$
$$+ \cdots + 2 Cov(X_1 X_n)$$

where $Cov(X_i X_{i+j})$ is the covariance of two sample values in the sequence separated by $j-1$ terms.

The equation for the variance becomes,

$$Var(X_1 + \cdots + X_n) = n\sigma^2 \left\{1 + 2\frac{n-1}{n}\rho_1 + 2\frac{n-2}{n}\rho_2 + \cdots + \frac{2}{n}\rho_{n-1}\right\}$$

where

$$\sigma^2 = Var(X_i)$$
$$\rho_1 \sigma^2 = Cov(X_i X_{i+1})$$
$$\rho_j \sigma^2 = Cov(X_i X_j)$$

The above equation is equivalent to Equation (4.23). Q.E.D.

The variance of the sample mean $\hat{\mu}_n = \frac{1}{n}(X_1 + \cdots + X_n)$ is

$$Var(\hat{\mu}_n) = \frac{1}{n^2} Var(X_1 + \cdots + X_n) \tag{4.24}$$

We obtain the variance of the sample mean by plugging Equation (4.23) to (4.24),

$$Var(\hat{\mu}_n) = \frac{\sigma^2}{n}\left\{1 + 2\sum_{j=1}^{n-1}\left(1 - \frac{j}{n}\right)\rho_j\right\} \tag{4.25}$$

Usually we do not know the population variance σ^2. Therefore, we substitute the sample variance for the population variance in Equation (4.25),

$$Var(\hat{\mu}_n) = \frac{\hat{\sigma}_n^2}{n}\left\{1 + 2\sum_{j=1}^{n-1}\left(1 - \frac{j}{n}\right)\rho_j\right\} \tag{4.26}$$

Equation (4.26) reduces to Equation (4.15) for independent samples ($\rho = 0$). When sample values have positive correlation, Equation (4.15), which is for independent samples, underestimates the variance. The accuracy of the estimator of the mean deteriorates with the correlation coefficient. Therefore, it is more expensive to estimate the mean value of a population from a sample containing positively correlated observations than from a sample containing independent observations. Conversely negative correlation in a sample reduces the cost of estimating the populations mean.

Example 4.6:   This is a continuation of example 4.5. We estimate the standard deviation of the sample mean by taking the square root of the right hand side of Equation (4.26) when we have 100, 1,000, 10,000 and 100,000 observations. We consider four cases where the correlation coefficient is 0.1, 0.3, 0.5 and 0.75 and one case where the observations are independent.

The accuracy of the sample mean deteriorates with the correlation coefficient of the sample. Figure 4.7 shows the standard deviation of the sample mean vs. the sample size for the above last three values of the correlation coefficient. The gray line at the bottom shows the standard deviation for independent observations. The standard deviation for a correlation coefficient of 0.1 is practically identical to that for uncorrelated observations. The standard deviation increases with the correlation coefficient and it becomes very high when the correlation coefficient is 0.75. For example, the standard deviation of the sample mean when using 100,000 observations is 0.0065 for $\rho = 0.75$ and 0.001 for $\rho = 0.1$. For 100 observations, the standard deviation is 0.151 for $\rho = 0.75$ and only 0.03 for $\rho = 0$.

4.2.4   Confidence Interval for Mean Value

A decision maker can assess the accuracy of an estimate of the population mean by constructing the $100(1 - \alpha)$ percent confidence interval. This interval has the following interpretation in objective (frequency) probability theory. If the decision maker

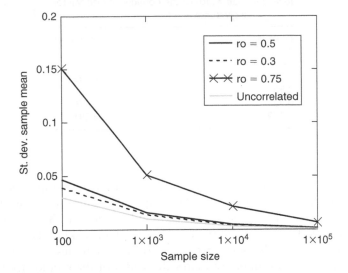

Figure 4.7 Effect of correlation on the accuracy of the sample mean.

calculates the sample mean and constructs the 95% confidence intervals from different samples, then these confidence intervals will contain the true mean value 95% of the time. The accuracy of the estimator increases with the width of the confidence interval decreasing.

According to the central limit theorem, for large samples, the sample mean follows the normal distribution (see Section 3.2.1.2). Then the standardized sample mean $\frac{\hat{\mu}_n - \mu}{\hat{\sigma}_n/\sqrt{n}}$ follows the standard normal distribution. Note that the denominator is equal to the standard deviation of the sample mean. Then, for a small probability level α,

$$-z_{1-\alpha/2} \leq \frac{\hat{\mu}_n - \mu}{\frac{\hat{\sigma}_n}{\sqrt{n}}} \leq z_{1-\alpha/2} \tag{4.27}$$

where $z_{1-\alpha/2}$ is the $1 - \alpha/2$ quantile of a standard normal variable. Therefore, the unknown mean value satisfies the following inequality with probability $1 - \alpha$,

$$\hat{\mu}_n - z_{1-\alpha/2} \cdot \frac{\hat{\sigma}_n}{\sqrt{n}} \leq \mu \leq \hat{\mu}_n + z_{1-\alpha/2} \cdot \frac{\hat{\sigma}_n}{\sqrt{n}} \tag{4.28}$$

The interval defined in Equation (4.28) is called $1 - \alpha$ percent confidence interval for the mean, and probability $1 - \alpha$ is called coverage. The quantity that is subtracted and added to the sample mean is called half-width of the confidence interval. Table 4.4 shows the half width of the confidence interval, in standard deviations of the sample mean, for different coverage probabilities. For example, for 0.95 coverage, the ends of the confidence interval should be 1.96 standard deviations of the sample mean away from it.

Table 4.4 Half width of the confidence interval for
a given coverage.

Coverage $1 - \alpha$	Half width $z_{1-\alpha/2}$
0.9	1.645
0.95	1.96
0.99	2.576
0.999	3.291

Equation (4.28) relies on two approximations: First, the sample mean is normal. Second, the standard deviation of the underlying random variable is equal to the sample standard deviation $\hat{\sigma}_n$. The confidence interval in Equation (4.28) is accurate for large samples. If the underlying probability distribution is highly skewed or has a heavy tail, then Equation (4.28) will be inaccurate even for large samples (Law 2007, pp. 232–237). Often, this equation underestimates the width of the confidence interval for a given coverage.

We develop an alternative confidence interval for small samples. If the sampling distribution is normal, then standardized variable $\frac{\hat{\mu}_n - \mu}{\hat{\sigma}_n/\sqrt{n}}$ has t-distribution (or Student distribution) with $n - 1$ degrees of freedom (Bury, 1999, pp. 124–126). Then, the equation for the $1 - \alpha$ percent confidence interval for the mean becomes,

$$\hat{\mu}_n - t_{n-1,1-\alpha/2} \cdot \frac{\hat{\sigma}_n}{\sqrt{n}} \leq \mu \leq \hat{\mu}_n + t_{n-1,1-\alpha/2} \cdot \frac{\hat{\sigma}_n}{\sqrt{n}} \qquad (4.29)$$

In the above equation, $t_{n-1,1-\alpha/2}$ is the $1 - \alpha/2$ quantile of a variable that follows the t-distribution with $n - 1$ degrees of freedom. Figure 4.8 compares the PDFs of a standard normal variable and three variables that follow the t-distribution with 5, 10 and 30 degrees of freedom. We observe the following,

a)   The t-distribution has heavier tails than the normal. This means that a variable that follows the t-distribution is more likely than a normal variable to assume a value that is more than a few standard deviations away from the mean. Therefore, for the same coverage, the confidence interval obtained from the t-distribution (Equation 4.29) is wider than that for the normal.

b)   The tails of the t-distribution become lighter with the number of degrees of freedom increasing.

c)   For 30 degrees of freedom, the t- and normal distributions almost coincide.

The values of the probability density function (PDF) and cumulative distribution function (CDF) of the t-distribution can be calculated by using built-in functions in software such as Mathcad, Excel or StatTools 5.5.1 for Excel.

Many decision makers use the bounds in Equation (4.29) even for non normal variables. Equation (4.29) is better than (4.28), because it yields intervals with coverage

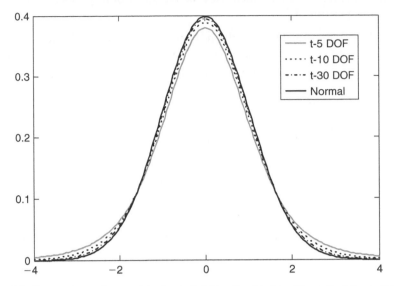

Figure 4.8 Comparison of normal and *t*-probability distributions. The tail of the *t*-distribution is heavier than that of the normal distribution but the difference is small for more than 30 degrees of freedom.

that is closer to the theoretical one. For large samples, both equations yield almost identical results.

Example 4.7: Calculate the 95% confidence intervals for the population mean by using the samples in Examples 4.1 and 4.2.

Solution:
In example 4.1, there are $n = 10$ observations. The sample mean is 8.84 min and the standard deviation of the sample mean, $\frac{\hat{\sigma}_n}{\sqrt{n}}$, is 2.95 min. The 0.975-quantile of a standard normal variable is 1.96 min. Therefore, from Equation (4.28),

$$\hat{\mu}_n - z_{1-\alpha/2} \cdot \frac{\hat{\sigma}_n}{\sqrt{n}} = 8.84 - 1.96 \cdot 2.95 = 3.06$$

$$\hat{\mu}_n + z_{1-\alpha/2} \cdot \frac{\hat{\sigma}_n}{\sqrt{n}} = 8.84 + 1.96 \cdot 2.95 = 14.63$$

The 0.975-quantile for the *t*-distribution for 9 degrees of freedom is 2.262. Equation (4.29) yields,

$$\hat{\mu}_n - t_{n-1,1-\alpha/2} \cdot \frac{\hat{\sigma}_n}{\sqrt{n}} = 8.84 - 2.262 \cdot 2.95 = 2.17$$

$$\hat{\mu}_n + t_{n-1,1-\alpha/2} \cdot \frac{\hat{\sigma}_n}{\sqrt{n}} = 8.84 + 2.262 \cdot 2.95 = 15.51$$

Note that Equation (4.29), which uses the t-distribution, yields a wider confidence interval than Equation (4.28).

In example 4.7, $n = 30$. The sample mean was 4.75 min and the standard deviation of the sample mean, $\frac{\hat{\sigma}_n}{\sqrt{n}}$, is 1.11 min. The 0.975-quantile of a standard normal variable is still 1.96 min but the same quantile of a t-distribution with 29 degrees of freedom is 2.045.

Therefore, from Equation (4.28),

$$\hat{\mu}_n - z_{1-\alpha/2} \cdot \frac{\hat{\sigma}_n}{\sqrt{n}} = 4.75 - 1.96 \cdot 1.11 = 2.57$$

$$\hat{\mu}_n + z_{1-\alpha/2} \cdot \frac{\hat{\sigma}_n}{\sqrt{n}} = 4.75 + 1.96 \cdot 1.11 = 6.93$$

The 0.975-quantile for the t-distribution for 29 degrees of freedom is 2.045. Equation (4.29) yields,

$$\hat{\mu}_n - t_{n-1,1-\alpha/2} \cdot \frac{\hat{\sigma}_n}{\sqrt{n}} = 4.75 - 2.045 \cdot 1.11 = 2.48$$

$$\hat{\mu}_n + t_{n-1,1-\alpha/2} \cdot \frac{\hat{\sigma}_n}{\sqrt{n}} = 4.75 + 2.045 \cdot 1.11 = 7.02$$

Note that the confidence bounds are considerably narrower for 30 observations than for 10 observations. The two methods that use the 0.975 quantile of the normal and t-distributions yield almost the same results.

4.2.4.1 Accuracy of confidence interval for mean value

We derived Equation (4.29) by assuming that the sampling distribution is normal. In order to check the accuracy of the confidence intervals from this equation we performed the following numerical experiment. We drew samples from normal, exponential and lognormal probability distributions, calculated 90%, 95% and 99% confidence intervals and checked if these intervals contained the true mean value of each probability distribution. Both the normal and lognormal distributions had mean value 0.5 and standard deviation 0.3. The exponential distribution had mean value 0.5. Samples with 5, 10 and 30 values were drawn from each distribution. We repeated this procedure 10,000 times for each distribution, coverage and sample size. Then we calculated the relative frequency that each confidence interval contained the mean value for all combinations of probability distributions, sample sizes, and confidence probabilities. This process yielded 27 observed coverage probabilities corresponding to all combinations of 3 probability distributions × 3 coverage probabilities × 3 sample sizes.

Figure 4.9 compares the anticipated coverage from Equation (4.29) to the observed coverage calculated from the experiment with 10,000 simulations for the normal and lognormal probability distributions. Table 4.5 presents the observed probabilities that the confidence intervals contained the mean value of the sampling distribution. Equation (4.29) underestimated the coverage for all probability distributions and sample

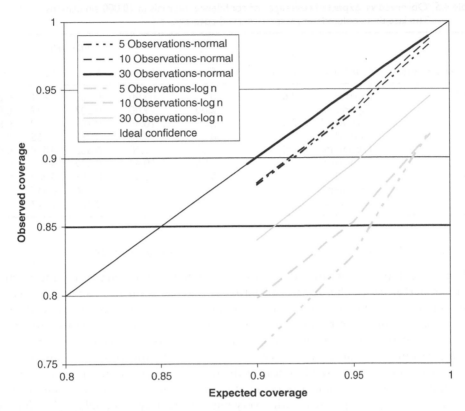

Figure 4.9 Observed vs. expected coverages for samples drawn from normal and lognormal probability distributions. Equation (4.29) underestimates coverage for the lognormal distribution

sizes. For example, the 95% confidence intervals for the mean value of the lognormal distribution from 10 observations contained the mean value only 85.3% of the time.

The following trends are observed from Table 4.5:

- Equation (4.29) considerably underestimated coverage. This is because this equation assumes that the sampling distribution is normal, which has light tails.
- The error in coverage is large for right skewed probability distributions with heavy tails.
- The error in coverage decreases with the number of observations, as expected.

If we had used Equation (4.28) instead of (4.29), then we would have underestimated coverage even more. The reason is that Equation (4.28) defines the confidence interval by using the standard normal distribution instead of the t-distribution, and because the normal distribution has lighter tails than the t-distribution. Therefore, it is better to use the t-distribution instead of the normal. These results are consistent with those that Law (2007, pp. 235–236) observed in a similar study.

Table 4.5 Observed vs. expected coverages of confidence intervals in 10,000 simulations. The last three columns show the observed coverage.

Expected coverage	Distribution	Coefficient of variation	Skewness	Kurtosis	Sample size 5	10	30
90%	Normal	0.6	0	3.00	0.88	0.881	0.895
	Exponential	1.0	2.00	9.00	0.807	0.837	0.875
	Lognormal	0.6	2.06	10.44	0.76	0.798	0.84
95%	Normal	0.6	0	3.00	0.932	0.935	0.945
	Exponential	1.0	2.00	9.00	0.862	0.888	0.922
	Lognormal	0.6	2.06	10.44	0.829	0.853	0.895
99%	Normal	0.6	0	3.00	0.983	0.987	0.989
	Exponential	1.0	2.00	9.00	0.936	0.947	0968
	Lognormal	0.6	2.06	10.44	0.919	0.917	0.945

4.2.4.2 Confidence interval for difference of two mean values

When we compare alternative designs we examine the difference of the mean values of their performance characteristics. Suppose that we want to compare the average number of parts that two alternative configurations (configurations A and B) of an electronic assembly line produce per hour. For this purpose, we need to estimate the difference of the mean values of the numbers of parts produced per hour by each configuration. Configuration A is better than B if the difference of the mean values is positive. Since there is randomness in the numbers of parts produced, we can only ascertain that configuration A is better if the difference of the mean values of parts produced, $\mu_A - \mu_B$, is positive with a very high probability (e.g., 0.95). Therefore, we must estimate confidence intervals for the difference of $\mu_A - \mu_B$. We can decide that configuration A is better than B only if the lower bound of this interval is greater than zero. We cannot decide which configuration is better if the confidence interval contains zero.

Assume that the samples from two populations are independent. First, consider two samples with equal numbers of observations, n. Then the confidence interval of the difference is,

$$\hat{\mu}_A - \hat{\mu}_B - t_{n-1,1-\alpha/2} \cdot \sqrt{\frac{\hat{\sigma}_A^2 + \hat{\sigma}_B^2}{n}} \leq \mu_A - \mu_B \leq \hat{\mu}_A - \hat{\mu}_B + t_{n-1,1-\alpha/2} \cdot \sqrt{\frac{\hat{\sigma}_A^2 + \hat{\sigma}_B^2}{n}}$$

(4.30)

where $\mu_A - \mu_B$ is the difference of the mean values, $\hat{\mu}_A - \hat{\mu}_B$ is the difference of the sample means, and $\hat{\sigma}_A^2$ and $\hat{\sigma}_B^2$ are the sample variances of the two variables.

The confidence interval of the difference of the mean values for unequal sample sizes is (Snedecor and Cohran, 1989),

$$\hat{\mu}_A - \hat{\mu}_B - t_{v,1-\alpha/2} \cdot \sqrt{\frac{\hat{\sigma}_A^2}{n_A} + \frac{\hat{\sigma}_B^2}{n_B}} \leq \mu_A - \mu_B \leq \hat{\mu}_A - \hat{\mu}_B + t_{v,1-\alpha/2} \cdot \sqrt{\frac{\hat{\sigma}_A^2}{n_A} + \frac{\hat{\sigma}_B^2}{n_B}}$$

(4.31)

Table 4.6 Absorbance of plates.

Control	Sonicated
2.118	1.207
3.000	1.500
2.140	1.544
2.623	1.274
1.896	2.087
	2.090
	2.087
	2.084
	2.084
	2.078
	2.087

The number of degrees of freedom in Equation (4.31) is,

$$\nu = \frac{\hat{\sigma}_{AB}^4}{\left(\dfrac{\hat{\sigma}_A^2}{n_A}\right)^2 /(n_A - 1) + \left(\dfrac{\hat{\sigma}_B^2}{n_B}\right)^2 /(n_B - 1)} \tag{4.32}$$

where $\hat{\sigma}_{AB} = \sqrt{\dfrac{\hat{\sigma}_A^2}{n_A} + \dfrac{\hat{\sigma}_B^2}{n_B}}$ is the standard deviation of the difference.

In the special case where the variances of the two populations are known to be equal, the number of degrees of freedom is,

$$\nu = n_A + n_B - 2 \tag{4.33}$$

Example 4.8:   Table 4.6 shows the data for the absorbance values of the control and cleaned plates. Calculate the 90% confidence interval of the difference of the mean values of the absorbance of the two groups of plates.

Solution:
The sample means of the control and cleaned plates are $\hat{\mu}_A = 2.355$ and $\hat{\mu}_B = 1.830$. The sample variances are, $\hat{\sigma}_A^2 = 0.2$ and $\hat{\sigma}_B^2 = 0.134$. There is no reason to assume that the sample variances are equal based on these numbers. The standard deviation of the difference is,

$$\hat{\sigma}_{AB} = \sqrt{\frac{\hat{\sigma}_A^2}{n_A} + \frac{\hat{\sigma}_B^2}{n_B}} = 0.229$$

From Equation (4.32), the number of degrees of freedom is 7. Therefore, the critical value of the Student t distribution of $\alpha = 0.1$ is $t_{7,0.95} = 1.895$. The 90% confidence interval of the difference of the mean values is [0.093, 0.96]. The lower bound of the

Table 4.7 Thirty measurements of the one-year
impact force (MN).

2.406	9.743	2.034	3.187
2.118	4.671	4.401	5.943
2.363	5.130	2.759	2.089
1.833	4.806	2.157	3.042
1.240	4.944	2.069	5.475
3.110	4.346	2.309	4.884
2.850	1.744	4.088	
4.085	3.153	2.668	

confidence interval is greater than zero. Therefore, the biologist can ascertain that the control plates have more bacteria than the cleaned ones, on average. This is evidence that the cleaning process is effective.

4.2.5 Confidence Interval for Variance

The $1 - \alpha\%$ confidence interval of the variance a normal random variable is,

$$\frac{(n-1)\hat{\sigma}_n^2}{\chi^2_{n-1,1-\alpha/2}} \leq \sigma^2 \leq \frac{(n-1)\hat{\sigma}_n^2}{\chi^2_{n-1,\alpha/2}} \tag{4.34}$$

where $\chi^2_{n-1,1-\alpha/2}$ and $\chi^2_{n-1,\alpha/2}$ are the $1 - \alpha/2$- and $\alpha/2$-quantiles of the Chi-squared probability distribution with $n - 1$ degrees of freedom. Equation (4.34) is also used for non-normal variables. The $1 - \alpha\%$ confidence interval for the standard deviation is obtained by calculating the square roots of the terms in this equation.

Example 4.9: Table 4.7 shows 30 observations of the one-year ice impact force on a platform in North Atlantic. Estimate the variance of the force and the 95%-confidence interval of the variance.

Solution:
The sample variance of the data in Table 4.7 is $\hat{\sigma}_n^2 = 3.028$. The $1 - \alpha/2$ and $1 - \alpha/2$ quartiles of the Chi-squared probability distribution with $n - 1 = 29$ degrees of freedom are: $\chi^2_{29,0.025} = 16.047$, and $\chi^2_{29,0.975} = 45.722$. From Equation (4.34) the lower and upper bounds of the 95% confidence interval of the variance are 1.921 and 5.472, respectively. The corresponding bounds for the standard deviation are 1.386 and 2.339. The confidence interval for the standard deviation is wide, which means that there is large uncertainty in the true value of the standard deviation. We need a larger sample in order to estimate the standard deviation with higher confidence.

4.2.6 Probability of an Event

The probability of an event A is the mean value of an indicator function, I_A, which is one if event A occurs and zero otherwise. Suppose that we repeat an experiment in which event A could or could not occur n times. Then the probability of A is the mean value of indicator function A (Section 3.1.1.1),

$$P(A) = \lim_{n \to \infty} \frac{1}{n} \sum_{i=1}^{n} I_{A_i} \tag{4.35}$$

The sample mean of the indicator function for n observations $I_{A_1}, \ldots, I_{A_n}$ is an unbiased estimator of the probability of A,

$$\hat{P}(A) = \frac{1}{n} \sum_{i=1}^{n} I_{A_i} \tag{4.36}$$

Theorem: An unbiased estimator of the variance of the probability of A, $\hat{P}(A)$ is,

$$\hat{Var}_{\hat{P}(A)} = \frac{\hat{P}(A)\{1 - \hat{P}(A)\}}{n} \tag{4.37}$$

Proof: The number of times k that an event A occurs in n independent repetitions of an experiment follows the binomial probability distribution (Section 3.2.1.1). This distribution has variance $P(A) \cdot \{1 - P(A)\} \cdot n$. Therefore, the variance of $P(A)$ is,

$$Var_{\hat{P}(A)} = \frac{P(A)\{1 - P(A)\}}{n} \tag{4.38}$$

We obtain the estimator of the variance of $\hat{P}(A)$ in Equation (4.38) by substituting $\hat{P}(A)$ for the probability of A in Equation (4.38). Q.E.D.

The standard deviation of $\hat{P}(A)$ is,

$$\hat{\sigma}_{\hat{P}(A)} = \sqrt{\frac{\hat{P}(A)\{1 - \hat{P}(A)\}}{n}} \tag{4.39}$$

The $(1 - \alpha) \cdot 100$ percent confidence interval of the probability can be estimated from Equations (4.28) or (4.29). The former equation yields,

$$\hat{P}(A) - z_{1-\alpha/2} \cdot \hat{\sigma}_{\hat{P}(A)} \leq P(A) \leq \hat{P}(A) + z_{1-\alpha/2} \cdot \hat{\sigma}_{\hat{P}(A)} \tag{4.40}$$

This is called the Wald interval. Equation (4.29) yields the following interval,

$$\hat{P}(A) - t_{n-1,1-\alpha/2} \cdot \hat{\sigma}_{\hat{P}(A)} \leq P(A) \leq \hat{P}(A) + t_{n-1,1-\alpha/2} \cdot \hat{\sigma}_{\hat{P}(A)} \tag{4.41}$$

From Equations (4.39–4.41), we derive the following rule of thumb:
The required number of observations in order to reduce the width of a confidence interval increases quadratically with the rate of decrease of the interval's width. If we want to reduce the width of a confidence interval to one half of the original width, then we need four times as many observations as those for the original interval.

The Wald confidence interval is accurate only when the number occurrences of the underlying event and its complement are not too small. Brown (2001) summarized rules of thumb from textbooks as to when the Wald interval may be used. One rule is that the predicted coverage of the confidence intervals is accurate only when an event and its compliment occur at least 5 (or 10) times in a sample, e.g., $n\hat{P}(A) \geq 5$ (or 10) and $n\{1 - \hat{P}(A)\} \geq 5$ (or 10).

In some applications, it is important to calculate the confidence interval for the difference between the probabilities of two events. For example, we need to know the confidence interval of the difference between the failure probabilities of two alternative designs in order to decide which one is safer. If the probability that the difference $P(F_1) - P(F_2)$ is positive is very high (e.g., 0.99), then system 2 is likely to be safer than 1 and vice versa. Assume that the probabilities of two events A and B are estimated from two independent samples of the same size. Then, the $(1 - \alpha) \cdot 100$ percent confidence interval for the difference of the probabilities of these events is,

$$\hat{P}(A) - \hat{P}(B) - z_{1-\alpha/2} \cdot \hat{\sigma}_{\hat{P}(A) - \hat{P}(B)} \leq P(A) - P(B) \leq \hat{P}(A) - \hat{P}(B) + z_{1-\alpha/2} \cdot \hat{\sigma}_{\hat{P}(A) - \hat{P}(B)}$$

$$\text{where } \hat{\sigma}_{\hat{P}(A) - \hat{P}(B)} = \sqrt{\hat{\sigma}^2_{\hat{P}(A)} + \hat{\sigma}^2_{\hat{P}(B)}} \tag{4.42}$$

Example 4.10:   We tested 500 nominally identical systems and found 25 defectives. The estimated probability of failure is 0.05 based on these results. Calculate the standard deviation and the 95% confidence interval of the probability of failure from Equation (4.39).

Solution:
From Equation (4.39), the standard deviation of the probability of failure is $\hat{\sigma}_{\hat{P}(A)} = 9.74 \times 10^{-3}$. This value is approximately equal to 20% of the estimated probability of failure. Equation (4.40) yields a 95% confidence interval of [0.031, 0.069].

Example 4.11:   In Example 4.10, how many tests do we need if the half width of the 95% should be equal to 10% of the estimated probability of failure?

Solution:
First we will estimate the number of tests by the rule of thumb above. Then we calculate the same number from Equations (4.39) and (4.40).

Approximate calculation:
The half width of the confidence interval in Example 4.9 is 0.019. Since the estimated probability of failure is 0.05, we should reduce the half width to 0.005, which is a reduction to one fourth of the original interval, approximately. A 50% reduction of the width of the confidence interval requires to increase the number of observations by four. Therefore, we need 16 times the original number of observations, i.e. $n = 8,000$.

Exact calculation:
In order to find the number of tests, we need the probability of failure. Since we do not know the true probability of failure we will use its estimated value from

Example 4.10, which is 0.05. The half width of the 95% confidence interval is $z_{1-\alpha/2} \cdot \hat{\sigma}_{\hat{P}(A)}$ and it should be equal to the specified width, w,

$$z_{1-\alpha/2} \cdot \hat{\sigma}_{\hat{P}(A)} = w \Leftrightarrow$$

$$\hat{\sigma}_{\hat{P}(A)} = \frac{w}{z_{1-\alpha/2}} \Leftrightarrow$$

$$\sqrt{\frac{\hat{P}(A)\{1 - \hat{P}(A)\}}{n}} = \frac{w}{z_{1-\alpha/2}}$$

Therefore, the required number of tests is,

$$n = \frac{z_{1-\alpha/2}^2}{w^2} \hat{P}(A)\{1 - \hat{P}(A)\}$$

For $w = 0.005$ and $\hat{P}(A) = 0.05$, the above equation yields, $n = 7{,}299$. This value is close to that given by the rule of thumb.

4.2.6.1 Accuracy of confidence intervals for probability

In order to assess the accuracy of the Wald interval in Equation (4.40) we need to calculate the true coverage of this interval. For a given true probability $P(A)$ and sample size n, we use the procedure below,

1. Calculate the probability that event A could occur k times in n experiments and determine the confidence interval of the apparent probability $\hat{P}(A) = \frac{k}{n}$, for each value of $k \in [0, n]$. This step yields $k + 1$ confidence intervals and their probabilities.
2. Identify those confidence intervals in step 1 that contain $P(A)$.
3. The coverage is the sum of the probabilities of the confidence intervals that contain $P(A)$.

Suppose that the true probability of an event is $P(A)$. The number of times K that event A will occur in n experiments follows the binomial distribution,

$$f_K(k) = \binom{n}{k} P(A)^k \{1 - P(A)\}^{n-k} \tag{4.43}$$

where $k \in [0, n]$.

From this equation, the probability mass function of the apparent probability of A is,

$$f_{\hat{P}(A)}\{\hat{P}(A)\} = \frac{n!}{\{n - \hat{P}(A)n\}!\{\hat{P}(A)n\}!} P(A)^{n\hat{P}(A)} \{1 - P(A)\}^{n\{1-\hat{P}(A)\}}$$

$$\text{for } \hat{P}(A) = 0, \ 1/n, \ldots, 1 \tag{4.44}$$

In the above equation, $\hat{P}(A) = \dfrac{\text{number of observed occurrences of } A}{\text{number of trials}} = \dfrac{k}{n}$.

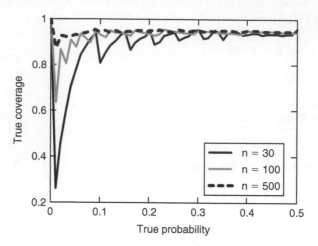

Figure 4.10 True coverage of the wald 95% confidence interval.

Equation (4.44) shows the probability that if we perform k experiments, then we will estimate that the probability of A is $\hat{P}(A)$.

Coverage associated with the Wald confidence interval is the probability that this interval could contain the true probability $P(A)$. The bounds of the confidence interval with coverage α are,

$$[\hat{P}(A) - z_{1-\alpha/2} \cdot \hat{\sigma}_{\hat{P}(A)}, \hat{P}(A) + z_{1-\alpha/2} \cdot \hat{\sigma}_{\hat{P}(A)}] \tag{4.45}$$

This confidence interval may or may not contain the true probability, depending on the value of $\hat{P}(A)$. The coverage is the sum of the probabilities of obtaining an apparent probability such that the confidence interval in Equation (4.45) contains probability $P(A)$,

$$p(\text{coverage}) = \sum_{\hat{P}(A)=0,1/n,\dots,1} f_{\hat{P}(A)}\{\hat{P}(A)\} \cdot I\{P(A) \in [p_{\min}, p_{\max}]\} \tag{4.46}$$

where $I\{P(A) \in [p_{\min}, p_{\max}]\}$ is a counter function, which is one if the confidence interval contains $P(A)$ and zero otherwise. Note that the coverage from Equation (4.46) is a function of the true probability and the number of experiments. Also the coverage is a discontinuous function of the true probability p because the binomial distribution (4.43) is discrete.

Figures 4.10 and 4.11 show the true coverage for the 95% and 90% Wald confidence intervals. We observe the following:

- The true coverage of the Wald confidence intervals differs significantly from the predicted coverage, especially for small probabilities. For $n = 30$ and $p = 0.02$ the 95% confidence interval contains the true probability only 45% of the time! For

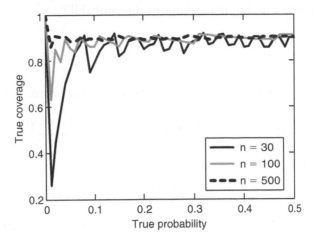

Figure 4.11 True coverage of the wald 90% confidence interval.

$n = 100$ and $p = 0.01$ the 95% confidence interval contains the true probability 63% of the time. Even for $n = 1,000$ and $p = 0.005$ the 95% confidence interval contains the true probability 87% of the time.

- The accuracy of the confidence intervals is poor for small probabilities and small samples.
- The true coverage of the Wald confidence interval displays an erratic behavior. For $n = 30$, the true coverage of the 95% interval is 0.937 for probability 0.09 but only 0.81 for probability 0.1.

4.2.6.2 *Alternative and more accurate confidence intervals*

Brown et al. (2001) and Henderson and Meyer (2001) investigated the accuracy of the Wald confidence interval and demonstrated its shortcomings. Brown et al. (2001) recommended using the Wilson, the Agresti-Coull and Jeffreys (Bayesian) confidence intervals, which we present below.

The Wilson "Score" confidence interval
The *Wilson "Score" confidence interval* is derived as follows. According to the central limit theorem, random variable $\frac{\hat{P}(A) - P(A)}{\sqrt{P(A)\{1 - P(A)\}/n}}$ converges to a standard normal variable with the sample size n. Therefore, if we calculate apparent probability $\hat{P}(A)$ from n experiments, then this probability will satisfy the inequality below with probability $1 - a$,

$$-z_{1-\alpha/2} \leq \frac{\hat{P}(A) - P(A)}{\sqrt{P(A)\{1 - P(A)\}/n}} \leq z_{1-\alpha/2} \tag{4.47}$$

For a given apparent probability $\hat{P}(A) = k/n$, we solve the following equations for $P(A)$ to find the lower and upper bounds of this probability,

$$\frac{\hat{P}(A) - P(A)}{\sqrt{P(A)\{1 - P(A)\}/n}} = \pm z_{1-\alpha/2} \tag{4.48}$$

The two solutions are the lower and upper bounds of the Wilson confidence interval,

$$CI_{\text{Wilson}} = \tilde{P}(A) \pm \frac{z_{1-\alpha/2} \cdot \sqrt{n}}{n + z_{1-\alpha/2}^2}\left[\hat{P}(A) \cdot \{1 - \hat{P}(A)\} + \frac{z_{1-\alpha/2}}{4n}\right]^{1/2} \tag{4.49}$$

where $\tilde{P}(A) = \dfrac{\tilde{k}}{\tilde{n}}$, $\tilde{k} = k + \dfrac{z_{1-\alpha/2}^2}{2}$ and $\tilde{n} = n + z_{1-\alpha/2}^2$ $\tag{4.50}$

The center of the Wilson interval (which is the first term in Equation (4.49)) is different than the relative frequency of event A.

Wilson interpreted the confidence interval as follows (Wilson, 1927): "If the true value of the probability $P(A)$ lies outside this interval, then the chance of having such bad luck as to have made an observation as bad as $\hat{P}(A) = k/n$ is less than or equal to α."

Example 4.12: We tested 500 nominally identical systems and found 25 defectives. The estimated probability of failure is 0.05 based on these results. Find the 95% confidence interval of the probability of failure from Equations (4.49) and (4.50).

Solution:
From Equation (4.50) $\tilde{n} = 504$, $\tilde{k} = 26.92$ and $\tilde{P}(A) = 0.053$. The Wilson confidence bounds from Equation (4.49) are 0.034 and 0.073. In Example 4.10 we found that the corresponding Wald bounds are 0.031 and 0.069. Wilson probability bounds are higher than the corresponding Wald bounds.

Example 4.13: Same as Example 4.12 but the numbers of tests and defectives are 8,000 and 400, respectively.

Solution:
From Equation (4.50) $\tilde{n} = 8004$, $\tilde{k} = 402$ and $\tilde{P}(A) = 0.05$. The confidence bounds from Equation (4.49) are 0.045 and 0.055. The half width of the confidence interval is 0.005, which is 10% of the estimated probability.

Example 4.14: Calculate the true coverage of the 95% Wilson confidence interval for $n = 30$, 100 and 500 experiments. Plot this probability as a function of the true probability. Consider true probabilities in the range from 0 to 0.2. Use the procedure in subsection "Accuracy of confidence intervals for probability".

Solution:
Figure 4.12 shows the true coverage of the Wilson interval. The true coverage is very close to 0.95 even for small probabilities. However, the confidence intervals are

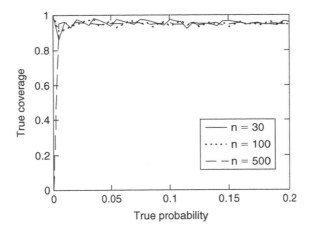

Figure 4.12 True coverage of Wilson score 95% confidence interval.

inaccurate for true probability equal to zero, where the coverage is 1 for $n = 100$ and 0 for $n = 500$.

The Agresti-Coull interval

The *Agresti-Coull interval* is obtained by replacing the apparent probability $\hat{P}(A)$ in Equations (4.39) and (4.40) with the center of the Wilson interval $\tilde{P}(A) = \frac{\tilde{k}}{\tilde{n}}$, and the number of experiments n with $\tilde{n} = n + z_{1-\alpha/2}^2$. The bounds of this interval are never shorter than those of the Wilson interval.

$$CI_{\text{Agresti}} = \tilde{P}(A) \pm z_{1-\alpha/2}\left[\tilde{P}(A) \cdot \{1 - \tilde{P}(A)\}/\tilde{n}\right]^{1/2} \tag{4.51}$$

For the 95% confidence interval $z_{1-\alpha/2}$ is 1.96 which is very close to 2. In this case, $\tilde{k} = k + 2$ and $\tilde{n} = n + 4$. Thus, the Agresti-Coull 95% confidence interval is derived from the Wald interval by increasing the number of experiments by four and number of occurrences of event A by 2.

Example 4.15: Find the Agresti-Coull 95% confidence interval for $n = 100$, and $k = 5$.

Solution:
From the observations, $\tilde{n} = 104$, $\tilde{k} = 6.92$ and $\tilde{P}(A) = 0.067$. The confidence interval from Equation (4.51) is [0.019, 0.115]. The bounds of this interval are significantly higher that the corresponding Wald bounds, [0.0073 , 0.093].

Figure 4.13 shows the true coverage of the Agresti-Coull 5% confidence interval. This interval is more accurate than the Wald interval especially for few observations. An additional advantage is that, unlike the Wilson interval, its coverage is not zero when the true failure probability is zero. The Agresti-Coull confidence interval could

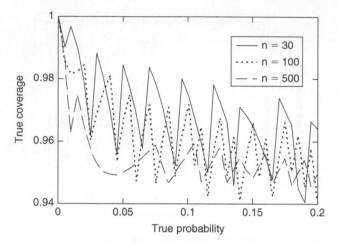

Figure 4.13 True coverage of Agresti-Coull 95% confidence interval.

be biased toward high probabilities. Suppose that the true probability of an event is 0.1 and this event occurs once in 10 experiments. Then the center of the Agresti-Coull 95% confidence interval will be 0.21, which is about two times the true probability. However, use of this interval protects a decision maker from underestimating the probability of a catastrophic event when observations from only a few experiments are available.

Bayesian credible interval
Section 6.4 will explain that the Bayesian approach combines prior information about the probability of an event with observations. Prior information is based usually on expert judgment. The Bayesian approach considers the probability of an event as a random variable with its own probability distribution. A prior probability distribution is estimated based on expert judgment. When data from observations become available, this probability distribution is updated by using Bayes' rule. A decision maker can assess the accuracy of the estimated probability by calculating the $1 - \alpha\%$ credible interval, which is the counterpart of the confidence interval in objective probability. This interval contains the probability of failure with probability $1 - \alpha$, so as it is equally likely that the true probability could be lower than the low bound and higher than the upper bound.

The meaning of a *credible interval* is different than that of a *confidence* interval.

- In objective probability theory, probability is the long-term relative frequency of an event. It does not make sense to talk about the relative frequency of one time events, such as "a given interval contains the value of a distribution parameter." Confidence interval has the following frequency interpretation. If a decision maker estimates the 95% confidence interval many times, each time using a different sample, then the confidence interval will contain the true probability 95% of the time.

- In subjective probability theory, the interpretation of a credible interval is as follows: the 95% credible interval contains the true probability of an event with probability 0.95. This subjective probability reflects a decision maker's belief that this credible interval contains the true probability. This means that the decision maker is indifferent between a lottery ticket that pays some prize (e.g. $100) with long term relative frequency 0.95 and a second ticket that pays the same prize if the credible interval contains the true probability (Chapter 6, section 6.2).

Chapter 6 explains the concepts of subjective probability and credible intervals. In this chapter, section 6.4 explains and demonstrates how to estimate credible intervals of an event.

4.2.7 *How to get the maximum return from your budget for data collection*

Statistical methods enable a decision maker to use resources effectively in order to obtain the maximum amount of information for a population from data. Here we present two applications where a decision maker extracts the maximum amount of information from tests by using statistical methods.

4.2.7.1 *Estimating the probability of failure of a series system efficiently*

Consider a series system consisting of n independently operating components. Suppose that we can perform a total of k tests on these components, to validate the system reliability. An approach that minimizes the system probability of failure by optimizing the number of tests, k_i, performed on each component will be presented.

The probability of failure of the series system is approximately equal to,

$$P(F_S) = P(F_1) + P(F_2) + \cdots + P(F_n) \tag{4.52}$$

where $P(F_i)$ is the probability of failure of the ith component and $P(F_S)$ the probability of failure of the system. Equation (4.52) is valid for small probabilities of failure (e.g., less than 0.01).

If the probability of failure of the ith component is estimated by testing a sample of k_i components, then the variance of the component probability of failure is,

$$\sigma^2_{P(F_i)} = \frac{P(F_i) \cdot \{1 - P(F_i)\}}{k_i} \tag{4.53}$$

Then the variance of the estimated system probability is,

$$\sigma^2_{P(F_S)} = \sigma^2_{P(F_1)} + \sigma^2_{P(F_2)} + \cdots + \sigma^2_{P(F_n)} \tag{4.54}$$

The objective is to reduce the variance of the system probability of failure. We develop a test plan for components for the following two cases: (1) the total number of tests is a constant (k); (2) the total testing cost is a constant (c). Two formulations are studied for the above cases. Both formulations minimize the variance of the system probability of failure by finding number of tests for each component.

Optimization Formulation I:

Find $k_1, k_2, \ldots, k_n$

To minimize $\sigma^2_{P(F_S)} = \sum_{i=1}^{n} \sigma^2_{P(F_i)}$ (4.55)

So that $\sum_{i=1}^{n} k_i = k$

The Lagrangian of the above problem is,

$$L = \sigma^2_{P(F_S)} + \lambda \cdot \left(k - \sum_{i=1}^{n} k_i \right)$$ (4.56)

The minimum of the Lagrangian, which satisfies the equality constraints, minimizes the variance [Equation (4.55)]. At the minimum, the gradient of the Lagrangian is zero,

$$\vec{\nabla} L = 0$$ (4.57)

$$\frac{\partial \sigma^2_{P(F_S)}}{\partial k_i} - \lambda = 0 \Rightarrow \frac{\partial \sigma^2_{P(F_i)}}{\partial k_i} = \lambda$$

$$\Rightarrow \frac{\partial \sigma^2_{P(F_1)}}{\partial k_1} = \frac{\partial \sigma^2_{P(F_2)}}{\partial k_2} = \cdots = \frac{\partial \sigma^2_{P(F_n)}}{\partial k_n} = \text{constant}$$ (4.58)

From Equation (4.53),

$$\frac{P(F_1) \cdot \{1 - P(F_1)\}}{k_1^2} = \frac{P(F_2) \cdot \{1 - P(F_2)\}}{k_2^2} = \cdots = \frac{P(F_n) \cdot \{1 - P(F_n)\}}{k_n^2}$$ (4.59)

We know that the total number of tests is equal to the sum of the tests on the individual components,

$$k_1 + k_2 + \cdots + k_n = k$$ (4.60)

Equations (4.59) and (4.60) are the optimality conditions, and can be solved to obtain the values of $k_1, k_2, \ldots, k_n$ in terms of preliminary estimates of the probabilities of failure of the components. We solve an example to present the optimality condition graphically.

Example 4.16:   Consider a two-component system. We have preliminary estimates of failures probabilities $P(F_1)$ and $P(F_2)$. Figure 4.14 shows the optimality condition,

$$\frac{\partial \sigma^2_{P(F_1)}}{\partial k_1} = \frac{\partial \sigma^2_{P(F_2)}}{\partial k_2} = \text{constant for minimizing the variance } \sigma^2_{P(F_S)}$$

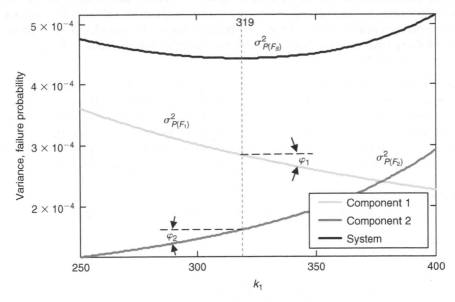

Figure 4.14 Optimality condition for a two-component series system: $\varphi_1 = \varphi_2$.
We can afford to test $k = 500$ components. We should test $k_1 = 319$ examples of the
first component and $k_2 = k - k_1 = 181$ of the second, to minimize the variance of the
system failure probability. Note that $\frac{\partial \sigma^2_{P(F_2)}}{\partial k_2} = -\frac{\partial \sigma^2_{P(F_2)}}{\partial k_1}$.

If the component probabilities of failure are very small then equation (4.59)
reduces to,

$$\sqrt{\frac{P(F_i)}{P(F_j)}} = \frac{k_i}{k_j} \quad \text{when } i = 1 \text{ and } j = 2, \ldots, n \tag{4.61}$$

Optimization Formulation 2:
This formulation will consider the testing costs explicitly. The cost of testing each
component is c_i and the total budget for testing all the components is c. Following is
the formulation of the optimization problem,

Find $k_1, k_2, \ldots, k_n$

To minimize $\sigma^2_{P(F_S)} = \displaystyle\sum_{i=1}^{n} \sigma^2_{P(F_i)}$ $\qquad\qquad$ (4.62)

So that $\displaystyle\sum_{i=1}^{n} c_i k_i = c$

The optimality condition can be derived using the same procedure as in formulation 1,

$$\frac{P(F_1) \cdot \{1 - P(F_1)\}}{k_1^2} = \frac{P(F_2) \cdot \{1 - P(F_2)\}}{k_2^2} = \cdots = \frac{P(F_n) \cdot \{1 - P(F_n)\}}{k_n^2} \tag{4.63}$$

$$c_1 \cdot k_1 + c_2 \cdot k_2 + \cdots + c_n \cdot k_n = c \tag{4.64}$$

In both formulations, we will obtain n equations with n unknowns. We determine how many tests on each component we should conduct by solving these equations. A numerical example with a three- component system is considered. The proposed methodology finds an optimum test plan to minimize the variance in the estimate of the system reliability.

Example 4.17: Consider a product line that produces systems made of three major components. We want to estimate the probability of failure of the system by testing samples of the three components. The budget allows testing a total of 100 components. The estimated probabilities of failure of the three components are 0.08, 0.05 and 0.02. Find the sizes of the samples of the three components to be tested using the variance reduction approach for determining an optimal testing plan for systems.

Solution:
The following equations will be solved to find the optimum sample sizes,

$$\frac{P(F_1) \cdot \{1 - P(F_1)\}}{k_1^2} = \frac{P(F_2) \cdot \{1 - P(F_2)\}}{k_2^2} \tag{4.65}$$

$$\frac{P(F_1) \cdot \{1 - P(F_1)\}}{k_1^2} = \frac{P(F_3) \cdot \{1 - P(F_3)\}}{k_3^2} \tag{4.66}$$

$$k_1 + k_2 + k_3 = 100 \tag{4.67}$$

Solving the above set of equations yields the optimal values for k_1, k_2 and k_3: $k_1 = 43$, $k_2 = 35$ and $k_3 = 22$. The variance of the system reliability is $3.96 \cdot 10^{-3}$. If we used the traditional approach of testing equal numbers of individual components ($k_1 = k_2 = k_3 = 33$) then the variance of system reliability would be $4.26 \cdot 10^{-3}$.

For the same example we consider different numerical values, $c = 10$ and probabilities of failure 0.13, 0.01 and 0.01 for the 3 components, respectively. The sample sizes for each component to be tested are 6, 2 and 2 respectively. The variance of the system using the optimal test plan is $2.87 \cdot 10^{-2}$. If three examples of each component were tested the variance would be $4.4 \cdot 10^{-2}$.

It is observed that following the optimum test plan we can estimate the system failure probability with higher confidence than testing equal numbers of components. The test plan indicates that we should test a larger number of components whose failure probabilities are higher.

4.2.7.2　Combining estimates of a mean value from different sets of observations

Often, we are given the mean value of a quantity but not the observations from which it was estimated. Consider a scenario, where a decision maker has multiple estimates of the mean value of a variable from different sets of data but he/she does not have the data. The decision maker only has estimates of the variance of the sample mean for each set. He/she wants to combine the estimates of the mean value in order obtain a more accurate estimate.

We present a method for combining these estimates so as to minimize the variance of the estimate of the sample mean when the observations are statistically independent. This approach can be applied for estimation of a probability, since probability is the mean value of a counter function. For example, an engineer calculated two estimates of the probability of failure of a system from two Monte-Carlo simulations by using importance sampling (Melchers, 1999, pp. 73–81). The engineer did not save the data from the two simulations. The engineer wants a more accurate estimate of the probability of failure. The approach below enables him/her to do so by combining the two estimates.

Let $\hat{\mu}_i, i = 1, \ldots, m$ be the sample means from Equation (4.1), and $\hat{V}ar(\hat{\mu}_i), i = 1, \ldots, m$ their variances from Equation (4.8). The decision maker combines these estimates by using the following equation,

$$\hat{\mu} = \sum_{i=1}^{m} a_i \hat{\mu}_i \tag{4.68}$$

where weighting factors $a_i, i = 1, \ldots, m$ are to be determined.

We require that estimator $\hat{\mu}$ be unbiased and have minimum variance among all estimators that satisfy Equation (4.68). Therefore, $a_i, i = 1, \ldots, m$ are solutions to the following constrained optimization problem,

Find $a_i, i = 1, \ldots, m$

To minimize $Var(\hat{\mu})$ $\tag{4.69}$

So that $E(\hat{\mu}) = \mu$

Factors $a_i, i = 1, \ldots, m$ must add up to one so that the estimator of the mean is unbiased. The Lagrangian of the above optimization problem is,

$$L(a_1, \ldots, a_m) = \sum_{i=1}^{m} a_i^2 Var(\hat{\mu}_i) + \lambda \left(1 - \sum_{i=1}^{m} a_i \right) \tag{4.70}$$

The derivative of the Lagrangian with respect to the factors must be zero,

$$\frac{\partial L}{\partial a_i} = 2a_i Var(\hat{\mu}_i) - \lambda = 0 \Leftrightarrow$$

$$a_i = \frac{\lambda}{2 \cdot Var(\hat{\mu}_i)} \tag{4.71}$$

Since factors $a_i, i = 1, \ldots, m$ add up to one, the Lagrange multiplier is,

$$\lambda = \frac{2}{\sum\limits_{i=1}^{m} \dfrac{1}{Var(\hat{\mu}_i)}} \tag{4.72}$$

Therefore, the optimum values of the factors are,

$$a_i = \left(Var(\hat{\mu}_i) \cdot \sum_{i=1}^{m} \frac{1}{Var(\hat{\mu}_i)} \right)^{-1} \tag{4.73}$$

The variance of the combined sample mean in Equation (4.67) is calculated from the following equation,

$$Var(\hat{\mu}) = \sum_{i=1}^{m} a_i^2 \, Var(\hat{\mu}_i) \tag{4.74}$$

where factors a_i are obtained from Equation (4.73).

Example 4.18: The means of two samples with sizes $n_1 = 20$ and $n_2 = 50$ are $\hat{\mu}_1 = 9.796$, $\hat{\mu}_2 = 10.827$, and the corresponding variances are, $Var(\hat{\mu}_1) = 0.217$, $Var(\hat{\mu}_2) = 0.074$. These are two samples drawn from the same population. Find a more accurate estimate of the mean by combining the two estimates.

Solution:
We will combine the two estimates by using Equation (4.68). From Equation (4.73) the weighting factors are $a_1 = 0.255$ and $a_2 = 0.745$. These factors add up to one. The weighting factor for the second sample is greater than that of the first because the second sample is bigger.

The combined estimate of the mean value is obtained from Equation (4.68), $\hat{\mu} = 10.564$. From Equation (4.74), the variance of this estimate is $Var(\hat{\mu}) = 0.055$. A crude estimate of the mean value is the average of the mean values of $\hat{\mu}_1$ and $\hat{\mu}_2$, which is equal to 10.311. The variance of this estimate is 0.073. The method in this section is more accurate than calculating the average of the two estimates of the mean value.

Conclusion

Statistical methods are powerful tools for presenting data and making inferences from them. A decision maker can use these tools to make forecasts and improve a system.

Statistical summaries, such as the mean value and variance of a variable and the probability of an event describe the properties of a population. It is impractical to determine the exact values of these quantities. Therefore, we estimate these summaries by the statistics of a sample of observations. This section presented the sample mean and variance, which approximate the corresponding values of a population. It also presented an estimator of the probability of an event. These estimators have three desirable properties; they are unbiased (they are on target, on average), consistent (converge to the corresponding statistical summary of the population with the sample size) and efficient (converge faster than all other competing estimators).

We assess the accuracy of an estimator by a confidence interval, which contains the underlying population statistic with some high probability (coverage), such as 95%. Confidence intervals for the mean value and the probability of an event were presented. Although these intervals are popular, they are often inaccurate, especially for highly skewed distributions. Confidence intervals for the mean value overestimate coverage because they rely on the assumption that the distribution of the variable is normal, which has light tails. This section presented three alternative confidence intervals for probabilities that are simple and more accurate than the commonly used Wald interval.

One must consider the correlation in a sample. Positive correlation can significantly reduce the accuracy of the estimator of the probability of an event. A much larger sample is needed to estimate this probability than when the sample is uncorrelated. On the other hand, negative correlation increases accuracy for a given sample size.

It is expensive to collect data. This section demonstrated with examples how to use available resources efficiently in order to estimate the probability of an event or the mean value of a variable accurately.

4.3 Statistical Hypothesis Testing

Science progresses in two ways: a) Scientists establish a set of axioms (propositions that are obviously true) and derive models that represent physical phenomena or human behavior on the basis of these axioms. An example of an axiom is the proposition that if a decision maker prefers chocolate milk to 2% milk and 2% milk to skim milk, then she prefers chocolate to skim milk. b) Scientists make a hypothesis in order to explain a physical phenomenon, collect observations and test this hypothesis. A hypothesis that gains credibility after passing many tests becomes a theory and is used to analyze systems, and make predictions and decisions.

This section presents a structured, consistent procedure to test a claim about a probabilistic model by using data drawn from the model. This procedure is called *statistical hypothesis testing*. The hypothesis to be tested is called *null hypothesis*. If this hypothesis is false, then the *alternative hypothesis* is true. The null hypothesis, denoted by H_0, usually represents the status quo. It is a specific statement about a probabilistic model. The alternative hypothesis is denoted by H_1.

Example 4.19: In section 4.1, the biologist wants to decide if a new cleaning procedure can clean surfaces infected by bacteria. The biologist compares the amounts of bacteria on 11 plates that have been subjected to the procedure to those on 5 control plates. The null hypothesis states that the mean values of the amounts of bacteria on the two plates are equal. The alternative hypothesis is that these amounts are different.

Example 4.20: An engineer wants to determine the probability distribution of the time to failure of a population of water pumps. The engineer makes the hypothesis that this time is exponentially distributed, with mean value 3,000 hours. This is the null hypothesis. He/she tests 30 pumps and obtains 30 values of the time to failure. The engineer wants to test the null hypothesis in view of the evidence from the tests.

The key idea of a statistical hypothesis test is that, if its results are surprising given that the null hypothesis is true, then the hypothesis should be rejected. Thus, a hypothesis test can only falsify a hypothesis – it cannot prove it.

Example 4.21: Suppose that the biologist in example 4.19 assumes that the mean absorbance values of the cleaned and control plates are identical. When the biologist examines Figure 4.1 he/she will be surprised because the absorbance values of the

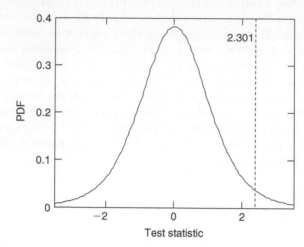

Figure 4.15   Probability density function of the test statistic in the biologist example conditioned on the hypothesis that the mean absorbance values of control and cleaned plates are equal.

control plates seem to be larger than those of the cleaned plates, on average. In view of these results, he/she will suspect that the mean absorbance values are different.

Example 4.22:   Suppose that all pumps in example 4.20 fail in less than 3,000 hours of operation. The engineer will be surprised and conclude that the mean value of the time to failure ought to be less than 3,000 hours.

We need a *decision rule* in order to decide if a hypothesis is false. A decision rule tells you whether to reject a hypothesis on the basis of a *test statistic*. This is a value determined from a sample drawn from the population. The probability distribution of the test statistic, conditioned on the null hypothesis H_0 being true is known. The test results suggest that the hypothesis is false if the test statistic assumes an unlikely value.

Example 4.23:   The biologist in examples 4.19 and 4.21 should use the following test statistic,

$$T = \frac{\hat{\mu}_A - \hat{\mu}_B}{\sqrt{\dfrac{\hat{\sigma}_A^2}{n_A} + \dfrac{\hat{\sigma}_B^2}{n_B}}}$$

(4.75)

This is a random variable that follows the Student distribution with $\nu - 1$ degrees of freedom. Equation (4.32) specifies the number of degrees of freedom (Snedecor and Cochran, 1989). In this example, the value of the test statistic is equal to 2.301. Figure 4.15 shows the PDF of the test statistic, given that the mean absorbance values of the two groups of plates are equal. The test statistic (marked by the dotted line) lies on the right tail of this distribution, away from the mean. The probability of drawing a value that is located so far from the mean on either side is only 0.061, which is low.

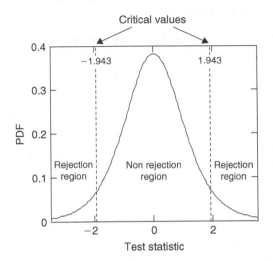

Figure 4.16 Rejection and non rejection regions for the hypothesis test about the difference of the mean absorbance values. The probability of the rejection region is $\alpha = 0.1$. Decision rule: Reject the hypothesis if the test statistic falls in the rejection region.

This surprising result suggests that the null hypothesis is false, that is, the mean values of the control and cleaned plates are different.

In order to develop a decision rule, we specify rejection regions in the range of the values of the test statistic. These regions are the extreme portions of one or both tails, while the non rejection region in the main body of the distribution. The value(s) that separate the rejection and non rejection regions are called *critical values*. We reject the null hypothesis if the test statistic falls within the rejection region. The decision rule is expressed in two equivalent forms:

1) For one rejection region, reject the hypothesis if and only the observed test statistic exceeds the critical value. For two rejection regions, reject the hypothesis if and only if the observed test statistic is smaller than the lower critical value or larger than the upper value.
2) Reject the null hypothesis if and only if the probability of obtaining a more extreme value of the test statistic (called *p-value*) is less than a level (called *level of significance*). The *p-value* is also called *observed level of significance*.

Example 4.24:   Figure 4.16 shows the rejection region for the test of the hypothesis that the mean values of the absorbance of the control and cleaned plates are equal in example 4.23. This region has a significance probability of $\alpha = 0.1$. The critical values are -1.943 and 1.943. The biologist should reject the null hypothesis because the observed value of the test statistic (which is 2.301) lies in the rejection region.

The p-value is equal to $P(T < -t_{observed} \cup T > t_{observed})$, where T is the test statistic and $t_{observed}$ is the observed value of this variable in this particular test. In this example,

p-value $= 0.061$. The null hypothesis should be rejected because this value is less than the level of significance: $p - \text{value} < \alpha$.

There are two types of errors in hypothesis testing: false rejection of a true null hypothesis (Type I error) and failure to reject a false one (Type II error). We commit type I error if we are unlucky enough to observe a value of the test statistic that lies in the rejection region, despite the fact that the null hypothesis is true. The probability that this can happen is equal to the significance level α (Figure 4.16). To lower this probability we reduce the significance level. However, this increases the probability of type II error because it makes it less likely to reject the null hypothesis. The choice of the probability of type I error depends on the consequences of committing it. Typical values of this probability are 0.01, 0.05 and 0.1.

Models are approximations of reality. Therefore, a test will practically reject all models when it uses a very large number of observations.

The probability of type II error is denoted by β. It is difficult to determine this probability because, usually, we do not know the probability distribution of the test statistic if the null hypothesis is false. When the number of observations is small, it is easy for a hypothesis to pass a test, even if it is false. One way to reduce β, without affecting the probability of type I error, is to increase the sample size.

In example 4.24, there is a probability of 0.1 to conclude that the mean values of the absorbance levels of the control and the cleaned plates are different, while these values are actually equal.

Hypothesis tests can be classified into tests of hypotheses involving one or two populations. Example 4.19 involves two populations (control and cleaned plates), while example 4.20 involves one (pumps). In addition, hypothesis tests can be classified into one-tailed and two-tailed tests. The alternative hypothesis in one-tailed tests can be expressed in terms of an inequality, e.g., $\mu_1 > \mu_2$. In two-tailed tests, the alternative hypothesis can be expressed in terms of the inequality, $\mu_1 \neq \mu_2$. Example 4.22 is a two-tailed test because the alternative hypothesis is that the mean values of the control and cleaned plates are different.

A hypothesis test involves the following steps

1.   State the null and alternative hypotheses.
2.   Choose the acceptable probability of type I error α (level of significance) and the sample size(s).
3.   Select the test statistic and determine its probability distribution.
4.   Determine the critical values that separate the rejection and non-rejection regions.
5.   Collect data.
6.   Calculate the observed value of the test statistic (or p-value).
7.   Reject the null hypothesis if the observed value of the test statistic lies in the rejection region (or the p-value is smaller than α). Do not reject the hypothesis otherwise.

Example 4.25:   Test the hypothesis that the mean value of the probability distribution from which the data in Table 4.8 was drawn is equal to 0.9.

Solution:
We complete steps 1–7 above to solve this problem.

Table 4.8 Data for example 4.25.

0.90993	0.364621
0.616695	0.829623
1.073277	0.878786
1.382942	1.040456
1.359505	0.890352
1.51994	0.901903
0.344924	0.888928
0.929746	1.402792
1.328507	0.974415
0.67399	0.944153
0.792939	0.846038
0.49287	1.591664
0.445927	1.259702
0.706711	1.712696
0.767948	0.803528

Step 1. The null hypothesis is that the mean value is equal to 0.9. The alternative hypothesis is that the mean value is different than 0.9. Therefore, this is a two-tailed test.

Step 2. We select an acceptable probability of type I error (significance probability) $\alpha = 0.05$. This means that we accept the possibility of concluding erroneously that the mean value is different than 0.9 if its probability is only 0.05. Also a sample size of 30 is acceptable.

Step 3. The test statistic in this test is the difference between the sample mean and the assumed population mean, normalized by the standard deviation of the sample mean,

$$T = \frac{\hat{\mu}_n - \mu}{\hat{\sigma}_n / \sqrt{n}} \tag{4.76}$$

This random variable follows the t-distribution with $n - 1 = 29$ degrees of freedom.

Step 4. The significance probability is equal to 0.05. The critical values of the test are the 2.5 and 97.5 percentiles of the t-distribution, which are equal to -2.045 and 2.045, respectively (Figure 4.17).

Step 5. The data collected is in Table 4.8.

Step 6. The sample mean is, $\hat{\mu}_n = 0.9559$ and the sample standard deviation is $\hat{\sigma}_n = 0.3562$. From Equation (4.76), the test statistic is equal to 0.859. The observed significance value (p-value) is equal to 0.398.

Step 7. We do not reject the hypothesis because the test statistic is in the non-rejection region.

Example 4.26: A goodness-of-fit test: Statistical tests that help a decision maker decide if a sample is statistically consistent with a theoretical distribution are called goodness-of-fit tests. The Chi-square, Kolmogorov-Smirnov, and Anderson-Darling are few of these tests (Law, 2007, pp. 340–353). Test the hypothesis that the data in

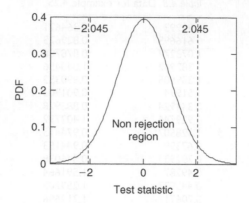

Figure 4.17 Critical values of the test, separating the non-rejection and rejection regions.

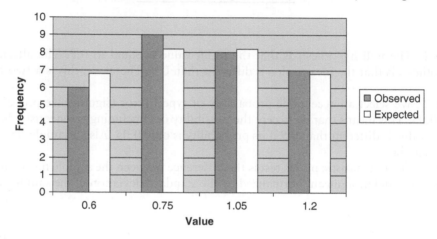

Figure 4.18 Observed vs. expected data used to test for normality of data.

Table 4.8 were drawn from a normal distribution with mean value equal to 0.9 and standard deviation 0.4.

Solution:
Figure 4.18 shows the histogram of the data with the expected number of values in four bins. The figure shows that the data agree with the histogram corresponding to the assumed distribution. Table 4.9 shows the observed and expected frequencies in the four bins corresponding to the histogram in Figure 4.18.

In this problem, we will use the Chi-square test. In this test, we divide the data into bins where the observed frequency of occurrence in each bin is f_i^o and the expected frequency (based on the assumed distribution) is f_i^e. Then, variable,

$$\chi^2 = \sum_{i=1}^{n} \frac{(f_i^o - f_i^e)^2}{f_i^e}$$

(4.77)

Table 4.9 Observed and expected frequencies in
Figure 4.18.

Bin	Observed freq., f_i^o	Expected freq., f_i^e
$[-1.0, 0.6]$	6	6.8
$[0.6, 0.9]$	9	8.2
$[0.9, 1.2]$	8	8.2
$[1.2, 3.0]$	7	6.8

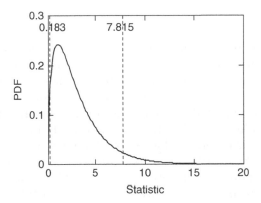

Figure 4.19 PDF of test statistic, critical value (7.815) and observed value of test statistic (0.183).

follows the Chi-square distribution. Usually, the parameters of the distribution are estimated from the data. The number of degrees of freedom of this distribution is in the range $[n-1-m, \; n-1]$, where n is the number of bins and m the number of parameters of the distribution. We recommend using the upper bound of this interval in order to reduce the risk of incorrectly rejecting the hypothesis that the data was drawn from the hypothesized distribution.

In the following we demonstrate the steps of the goodness-of-fit test in this example:

Step 1: The null hypothesis is that the observed data were drawn from a normal probability distribution with mean value equal to 0.9 and standard deviation 0.4.

Step 2. We select a significance probability of $\alpha = 0.05$. The sample size is 30.

Step 3: The test statistic is the value of variable χ^2, and it is calculated from Equation (4.77).

Step 4. The significance probability is equal to 0.05. Figure 4.19 shows the PDF of the Chi-square distribution with 3 degrees of freedom. The critical value of the test is 7.815. The rejection region is to the right of this value, and the non rejection region to it left.

Step 5. Table 4.9 shows the data.

Step 6. The value of the test statistic is equal to 0.183 (Equation 4.77), and the *p*-value is 0.98.

Step 7: Since the observed value of the test statistic is less than the critical value, there is no reason to suspect that the data is not consistent with the assumed probability distribution.

Table 4.10 Observed grade distribution of students according to their working habits.

Grade	Diligent	Lazy	Total
A	10	5	15
B	30	14	44
C	5	12	17
D	5	7	12
Total	50	38	88

Example 4.27:   Test for independence: The Chi-Square test can be used to determine if two random variables are independent. We divide the observed values of these variables into bins and compare the observed number of values with the expected number of values in each bin if the variables were independent. The sum of the square differences between observed and expected frequencies normalized by the expected frequencies in each bin follows the Chi-square distribution with $(n_1 - 1) \cdot (n_2 - 1)$ degrees of freedom, where n_1 and n_2 are the number of bins for each random variable.

Table 4.10 shows the distribution of the grades of 88 students in a statistics class. Students are divided into two categories (diligent and lazy) based on their homework grades and class attendance. Test the hypothesis that student grades are independent of work habits.

Solution:
The null hypothesis is that student grades are independent of work habits. In order to test this hypothesis, we need to calculate the expected number of students for every possible combination of grade and work habits if grades were independent of working habits. For this purpose, we estimate the unconditional probability distribution of student grades and the probability that a student could be diligent and lazy, based on the data in Table 4.10. The probabilities distribution of the grades is,

$$P(A) = 0.170, \quad P(B) = 0.5, \quad P(C) = 0.193, \quad P(D) = 0.136$$

The probabilities of the students being diligent and lazy are,

$$P(diligent) = 0.568, \quad P(lazy) = 0.432$$

Assume that student grades are independent of work habits. Then the joint probability of each combination is the product of the probabilities of each category. Table 4.11 presents the expected grade distribution for each student type.

The test statistic is equal to the sum of the square differences of the observed and expected numbers of students normalized by the latter in the eight cells in Tables 4.10 and 4.11. The value of this test statistic is equal to 9.236. The test statistic follows the Chi-square distribution with $(n_1 - 1) \cdot (n_2 - 1)$ degrees of freedom where n_1 and n_2 are the number of bins for variables "Grade" and "Work habit". This number is equal

Table 4.11 Expected grade distribution of students under the assumption that grades were independent of work habits.

Grade	Diligent	Lazy	Total
A	8.52	6.48	15
B	25	19	44
C	9.66	7.34	17
D	6.82	5.18	12
Total	50	38	88

to 3. The p-value corresponding to the test statistic is equal to $1 - F_{\chi^2}(9.236) = 0.026$. This very low value suggests that the null hypothesis is wrong.

Conclusion

Statistical hypothesis testing is a systematic procedure to test a hypothesis about the statistics of a population on the basis of observed data. This procedure can only falsify the hypothesis; it cannot prove nor estimate the degree of confidence in this hypothesis.

It is easy for a statistical model to pass a test, when few observations are available. For example, many types of probability distributions pass the Chi-square test when we have few observations (e.g., less than 30). In the other extreme, when there is a very large number of observations it is unlikely that any probabilistic model will pass a statistical hypothesis test because all models involve approximations of reality.

The limitations of hypothesis testing stem from the definition of probability as a long term frequency. This definition is too restrictive in the opinion of the authors. According to this definition, it is meaningless to think of the probability that a hypothesis (such as the probability of a disaster in a space shuttle flight is less than 0.01) is true. It is inconceivable to repeat an experiment in order to estimate the relative frequency that the hypothesis is true because the hypothesis is either true or false. Subjective probability (Chapter 6) does not suffer from the above limitations. A decision maker can estimate his/her probability that the hypothesis is true. This probability represents his/her belief that the hypothesis is true. When the decision maker obtains observations from space shuttle flights, he/she can update this probability.

4.4 Selecting Input Probability Distributions

Probabilistic analysis of a system requires the joint probability distribution of the input variables. Selection of this probability distribution significantly affects the conclusions of a probabilistic analysis. For example, the estimated probability of failure of an offshore platform in the North Sea during a 20-year period depends on the joint probability distribution of the significant wave height and the wind speed. This section presents an approach to estimate the probability distribution of one random variable. This approach is sufficient for modeling input variables that are statistically independent. Section 4.5 explains how to model dependent variables.

In order to construct a probabilistic model, decision makers use both observed data and expert judgment. This chapter focuses on estimation of probabilistic models by using data. Chapter 6 explains how to construct a model by combining data and expert judgment.

We can specify an empirical or a theoretical probability distribution. An empirical distribution is a histogram of the data, such as the ones shown in Figures 4.3 and 4.4. Construction of an empirical model does not require strong assumptions. On the other hand, an empirical distribution is discontinuous (see Figures 4.3 and 4.4). We cannot calculate the probabilities of extreme values that lie outside the range of the observed data. For example, we cannot calculate the probability that the interarrival time in Figure 4.4 could exceed 30 min from the histogram.

A decision maker selects a probability distribution for a random variable in three steps.

1. The decision maker chooses families of probability distributions that could represent the random variable. These distributions are consistent with the process associated with the random variable and amount of data.
2. He/she estimates the parameters of each family.
3. Finally, the decision maker assesses the conformity of the selected probability distributions to the available data. A goodness-of-fit test rejects probability distributions that do not fit the data in this step.

The above steps could be arduous. Fortunately, computer programs, such as @Risk, Minitab, and ExpertFit perform these tasks efficiently. However, a decision maker should study the properties of the uncertain quantity that he/she wants to model, the available information and the outputs of these programs carefully, in order to select a model that represents the data well.

The decision maker can choose one distribution or a set of admissible distributions from the results of the third step. When little data is available, many probability distributions seem to fit the data. In this case, the decision maker could select and use one or more distributions according to the guidelines in the next subsection, analyze the system for each distribution and draw conclusions based on the results. The following subsections explain each step.

4.4.1   Step 1: Select families of probability distributions

First, decision makers select some families of probability distributions that could represent the quantity of interest, considering the process associated with the random variable that they want to model, the amount of available data, and the consequences of the decisions that they plan to make. When assessing the probability of failure of a high consequence system, such as an offshore platform or a nuclear power plant, decision makers prefer to err on the conservative side. Often, they use distributions with heavy tails, such as beta and Weibull in order to avoid underestimating the likelihood of extreme events that could cause failure. Guidelines for selecting a probability distribution for a particular quantity are available in several books (e.g., Ang and Tang, 1975, chapter 5, Bury, 1999, Fox, 2005, Jordaan, 2005, Appendix, 1, Kelton et al. 2010, Appendix C, Law, 2007, section 6.2) and in section 3.2 of this book. Table 4.12 summarizes some of these guidelines.

Table 4.12 Selection of a probabilistic model given the amount of data and amount of experience.

Distribution	Suggested uses
Uniform	Small amounts of data or no data, good quantification of bounds
Triangular	Small amounts of data or no data, good quantification of bounds and most likely value
beta	Small amounts of data or no data, good quantification of bounds and most likely value. Suitable for the distribution of a proportion
Second-order probabilistic model	Small amounts of data or no data, poor quantification of bounds, large uncertainty in the most likely value
Weibull	Small amounts of data. Suitable for the time to failure, the stress to failure of a series system, and the maximum load over a long period
Lognormal	Small amounts of data, good quantification of lower bound
Exponential	Interarrival time between events that occur independently, at a constant rate. Suitable for the time between failures of a piece of equipment, and the time between severe earthquakes
Normal	Large amounts of data, large amount of experience. Suitable for errors of various types, and quantities that are the sum of large numbers of other large quantities
Extreme I, (Gumbel)	Maximum load over a long period, small amounts of data, large amounts of experience
Pareto	Maximum load over a long period, large amounts of experience

There are two sources of uncertainty in a probabilistic model; model and parameter uncertainty. The first type is the uncertainty about the family of distributions to which the model belongs. The second is uncertainty in the parameters of the distribution. It is important to quantify these sources of uncertainty and account for them when making design decisions. A decision maker can quantify uncertainty in a parameter by using a second order probabilistic model. That is a probability distribution whose parameters are random variables with their own probability distributions. The decision maker should perform sensitivity analysis in order to assess the effect of the model uncertainty on the conclusions of probabilistic analysis. This approach analyzes the same system by representing the random variables with different families of probability distributions that are consistent with the data. If the choice of the family of probability distributions significantly affects the probability of failure of a system, then the decision maker should collect additional information to reduce this uncertainty or use the most conservative estimate of the probability of failure.

When we have little data (for example 30 observed values) and we do not understand well the physics of the process from which the data is derived we should be careful not to assume more than what we actually know and not to underestimate probabilities of extreme catastrophic events. Probability distributions with light (or short) tails, such as the normal, are unsuitable for this case because they tend to underestimate probabilities of events corresponding to the tails of the distributions. The uniform, Weibull, beta distribution and a second-order probabilistic model are better choices.

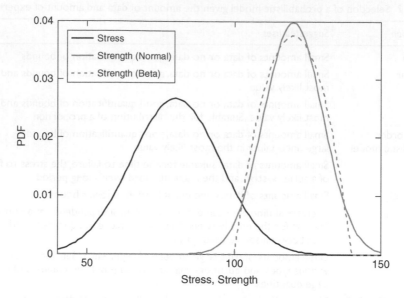

Figure 4.20 Probabilistic model of stress and two alternative probabilistic models of strength.

Example 4.28: Effect of the choice of the probabilistic model of the strength on the estimated probability of failure.

To illustrate the differences which can be obtained when care is not taken to correctly specify the statistical distribution of an input variable, consider a simple two-variable example that compares the tensile stress which a component sees during operation to its load carrying capability or ultimate tensile strength. This example has been adapted from (Fox, 2005). The interference region between these two variables is related to the failure probability. The distribution on the left (Figure 4.20) represents the tensile stress distribution, and the one on the right is the distribution of ultimate tensile strength. Thus, a failure will occur if the stress exceeds the tensile strength. In this example, suppose that the tensile stress distribution is truly normal with a mean $\mu = 80$ ksi and a standard deviation $\sigma = 15$ ksi and the tensile strength distribution is truly beta with a minimum and maximum values of $s_{min} = 100$ ksi and $s_{max} = 140$ ksi, respectively, with parameters $\beta_1 = 2$ and $\beta_2 = 2$. The resulting true probability is,

$$P(F) = \int_{100}^{140} \left\{ 1 - \Phi \left(\frac{s - \mu}{\sigma} \right) \right\} \frac{B_x(\beta_1, \beta_2)}{B(\beta_1, \beta_2)} ds$$

where $B_x(\beta_1, \beta_2)$ is the incomplete beta function and $x = \frac{s - s_{min}}{s_{max} - s_{min}}$. The above probability was found to be equal to 0.06 by using numerical integration. This is the true failure probability if the stress and strength distributions given above are correct.

But suppose that the tensile strength distribution was incorrectly modeled as being normally distributed with a mean of 120 ksi and a standard deviation 10 ksi. The

Table 4.13 30 measured one-year values of the impact force (MN).

2.406	9.743	2.034	3.187
2.118	4.671	4.401	5.943
2.363	5.130	2.759	2.089
1.833	4.806	2.157	3.042
1.240	4.944	2.069	5.475
3.110	4.346	2.309	4.884
2.850	1.744	4.088	
4.085	3.153	2.668	

resulting failure probability can be calculated in this case to be 0.00234. The first correct result yields a failure probability of 1 in 17 components – the later of 1 in 425 components. This is roughly a difference of nearly 26 times. This difference may be surprising when both models appear to model roughly the same variation pattern in tensile strength. This underscores the importance of the distributional form. This simple example demonstrates that small differences in input distributions can lead to large differences in the resulting answer – in this case, the failure probability.

It is dangerous to rely on an estimate of the probabilities of a critical event obtained from a small sample. The decision maker should select a set of probability distributions that could be representative of the data and calculate the probability of failure for each distribution. He/she could follow the guidelines in Table 4.12 in this task. In addition, the decision maker should quantify the uncertainty in the estimated probability of failure arising from statistical error in the estimates of the parameters.

Finally, the decision maker should use expert judgment, besides the data. The decision maker could consider the parameters of the distribution random and estimate their probability distributions on the basis of judgment. He/she will update these probability distributions in view of the observations by using Bayesian updating. Chapter 6 presents this approach.

4.4.2 *Step 2: Estimate the distribution parameters*

After selecting a set of probability distributions for a random variable, we estimate their parameters. This section presents three methods for doing so; the methods of moments, maximum likelihood and least squares.

4.4.2.1 *Method of moments*

The method of moments estimates some of the distribution parameters from estimates of the moments of the distribution. A decision maker determines the remaining parameters, such as the bounds of the variable, based on experience and judgment. The moments are estimated from observed values of the random variables by using the estimators in section 4.2.

Example 4.29: Table 4.13 presents 30 observations of one-year impact force on an offshore platform in the North Sea. It is known that this force is lognormal. The data

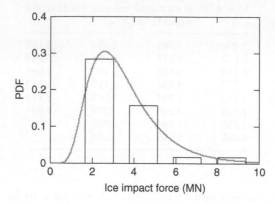

Figure 4.21   Histogram of impact forces and fitted PDF. The parameters of the PDF are $\mu = 1.15$ and $\sigma = 0.467$.

in this table are identical to that in Table 4.7. Figure 4.21 shows a histogram of these forces.

a. Estimate the parameters of the lognormal distribution.
b. Calculate the 95% confidence intervals of the mean value and standard deviation of the impact force.
c. Find the probability that the impact force could exceed 8 MN when the mean value of the force is equal to the sample mean. Also, find the same probability when the mean value is equal to the 95%-confidence bounds.

Solution:
a. The sample mean value and standard deviation of the impact force are 3.522 and 1.74 MN. We can find the parameters of the lognormal distribution by solving Equations (3.86) and (3.87) in Chapter 3 for μ and σ,

$$\mu = \ln\left(\frac{E^2(X)}{\sqrt{\sigma_X^2 + E^2(X)}}\right)$$

$$\sigma = \sqrt{\ln\left(1 + \frac{\sigma_X^2}{E^2(X)}\right)}$$

In the above equations, we substitute the sample mean and standard deviation for $E(X)$ and σ_X, respectively. Then the parameters of the lognormal distribution are, $\mu = 1.15$ and $\sigma = 0.467$. Figure 4.21 shows the fitted PDF of the impact force for the estimated values of the distribution parameters together with the histogram of the measured values. The two models in this figure agree well.

b.   The lower and upper 95%-bounds of the mean value of the force are calculated from Equation (4.29), where $\alpha = 0.05$

$$\text{Lower bound: } \hat{\mu}_n - t_{n-1,1-\alpha/2} \cdot \frac{\hat{\sigma}_n}{\sqrt{n}} = 3.522 - 2.045 \cdot \frac{1.74}{\sqrt{30}} = 2.872$$

$$\text{Upper bound: } \hat{\mu}_n + t_{n-1,1-\alpha/2} \cdot \frac{\hat{\sigma}_n}{\sqrt{n}} = 3.522 + 2.045 \cdot \frac{1.74}{\sqrt{30}} = 4.171$$

The corresponding bounds for the standard deviation are calculated from Equation (4.34),

$$\text{Lower bound: } \left\{ \frac{(n-1)\hat{\sigma}_n^2}{\chi_{n-1,1-\alpha/2}^2} \right\}^{1/2} = \left(\frac{29 \cdot 3.028}{45.722} \right)^{1/2} = 1.386$$

$$\text{Upper bound: } \left\{ \frac{(n-1)\hat{\sigma}_n^2}{\chi_{n-1,\alpha/2}^2} \right\}^{1/2} = \left(\frac{29 \cdot 3.028}{16.047} \right)^{1/2} = 2.339$$

There is significant uncertainty in the mean value and standard deviation of the force as evidenced by their wide confidence bounds.

The lower and upper bounds of the 95% confidence intervals for the parameters of the distribution are not calculated because the joint probability distribution of the mean and standard deviation is unknown.

c.   The probability that the impact force could exceed 8 MN is equal to one minus the CDF at this value,

$$P(X > 8) = 1 - F_X(8) = 0.023$$

$$\text{where } F_X(x) = \int_0^x \frac{1}{\sqrt{2\pi}\sigma x'} e^{-\frac{[\ln(x') - \mu]^2}{2\sigma^2}} dx'$$

We quantify the uncertainty in the above exceedance probability by varying the mean value of the force between its bounds, while keeping the standard deviation constant. When the mean value of the force becomes equal to its lower and upper bounds, while the standard deviation is fixed at 1.74, then the parameters of the distribution become $\mu = 0.9$, $\sigma = 0.559$ for the lower bound of the mean, and $\mu = 1.348$, $\sigma = 0.401$, for the upper bound. Figure 4.22 shows the PDFs of the ice impact force for $\mu = 1.15$ (gray curve), and $\mu = 1.348$ and 0.9 (dashed curves). The standard deviation of the force is 0.467, for all three PDFs in this figure. Then, the probability of exceedance ranges between 0.017 and 0.034.

We also vary the standard deviation of the force between its upper and lower bounds, while fixing the mean value to 3.522. Then the parameters of the distribution become and $\mu = 1.187$, $\sigma = 0.379$ and $\mu = 1.076$, $\sigma = 0.607$, respectively. The corresponding bounds of the probability of failure are 0.009 and 0.049. Uncertainty in the parameters of the distribution significantly affects the probability of exceedance of the ice force.

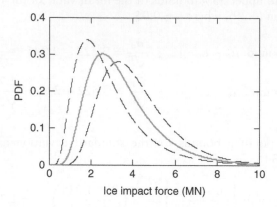

Figure 4.22 Effect of uncertainty in the mean value on PDF of the impact force. The dashed curves show the PDF when the mean value of the impact force equals to its lower and upper 95%-bounds.

Also, uncertainty in the standard deviation is more important than uncertainty in the mean value.

In general, we should use estimates of the distribution parameters together with their confidence bounds.

4.4.2.2 *The maximum likelihood method*

The maximum likelihood method for estimating the parameters of a distribution has become popular recently because it has a strong intuitive justification. Many commercial programs, such as @Risk, ExpertFit, and Minitab use this method. Loosely speaking, given observations $x_1, \ldots, x_n$, and a hypothesized probability distribution, this method finds the values of the distribution parameters that most likely produce these observations.

Suppose that the probability density function of random variable X is a function of parameter θ. We write this function as $f_X(x/\theta)$, to show the dependence of this function on θ. If we view $f_X(x/\theta)$ as a function of θ, then the likelihood function will be $L(\theta/x) = f_X(x/\theta)$.

Consider independent observations $x_1, \ldots, x_n$ and a vector of parameters θ. The likelihood of these observations is,

$$L(x_1, \ldots, x_n/\boldsymbol{\theta}) = \prod_{i=1}^{n} f_X(x_i/\boldsymbol{\theta}) \tag{4.78}$$

The maximum likelihood estimate of the parameters is the solution of the following optimization problem,

Find $\boldsymbol{\theta}$

To maximize $L(x_1, \ldots, x_n/\boldsymbol{\theta}) = \prod_{i=1}^{n} f_X(x_i/\boldsymbol{\theta})$ $\tag{4.79}$

Maximizing the logarithm of the likelihood function yields the same estimates of the parameters as maximizing the likelihood function because logarithm is a monotonically increasing function. Often, decision makers maximize the logarithm of the likelihood function,

$$\ln L(x_1, \ldots, x_n/\boldsymbol{\theta}) = \sum_{i=1}^{n} \ln f_X(x_i/\boldsymbol{\theta}) \tag{4.80}$$

because this simplifies the derivation of the optimum. One reason is that many families of PDFs, such as the normal, involve exponential functions of the random variable.

The maximum likelihood estimates of the distribution parameters make the derivative of the log-likelihood function zero,

$$\frac{\partial \ln L(x_1, \ldots, x_n/\hat{\boldsymbol{\theta}})}{\hat{\theta}_i} = 0 \quad \text{for } i = 1, \ldots, n \tag{4.81}$$

In addition, the second derivative of the log-likelihood function is negative, making sure that the optimum is indeed the maximum.

Example 4.30: Find the maximum likelihood estimators of the parameters of the lognormal distribution,

$$f_X(x/\mu, \sigma) = \frac{1}{x\sigma\sqrt{2\pi}} e^{-\frac{1}{2}\left[\frac{\ln(x)-\mu}{\sigma}\right]^2} \quad \text{for } x > 0$$

Solution:
The logarithm of the likelihood function is,

$$\ln L(x_1, \ldots, x_n/\mu, \sigma) = \sum_{i=1}^{n} \ln f_X(x_i/\mu, \sigma) \Rightarrow$$

$$\ln L(x_1, \ldots, x_n/\mu, \sigma) = \sum_{i=1}^{n} \ln\left(\frac{1}{x_i\sigma\sqrt{2\pi}} e^{-\frac{1}{2}\left[\frac{\ln(x_i)-\mu}{\sigma}\right]^2}\right) \Rightarrow$$

$$\ln L(x_1, \ldots, x_n/\mu, \sigma) = -\sum_{i=1}^{n} \ln(x_i) - n\ln(\sigma) - \frac{n}{2}\ln(2\pi) + \sum_{i=1}^{n}\left\{-\frac{1}{2}\left(\frac{\ln(x_i)-\mu}{\sigma}\right)\right\}^2$$

The derivative of the log-likelihood function is zero for $\mu = \hat{\mu}$ and $\sigma = \hat{\sigma}$,

$$\frac{\partial \ln L(x_1, \ldots, x_n/\hat{\mu}, \hat{\sigma})}{\hat{\mu}} = 0 \Rightarrow$$

$$\sum_{i=1}^{n} \frac{\ln(x_i) - \hat{\mu}}{\hat{\sigma}} = 0 \Rightarrow$$

$$\hat{\mu} = \frac{1}{n}\sum_{i=1}^{n} \ln(x_i) \tag{4.82}$$

Similarly,

$$\frac{\partial \ln L(x_1, \ldots, x_n/\hat{\mu}, \hat{\sigma})}{\hat{\sigma}} = 0 \Rightarrow$$

$$\frac{n}{\hat{\sigma}} - \sum_{i=1}^{n} \frac{\{\ln(x_i) - \hat{\mu}\}^2}{\hat{\sigma}^3} = 0 \Rightarrow$$

$$\hat{\sigma} = \sqrt{\frac{\sum_{i=1}^{n} \{\ln(x_i) - \hat{\mu}\}^2}{n}} \qquad (4.83)$$

There is no closed form expression for the maximum likelihood estimators of the parameters of some probability distributions such as beta and Weibull. In this case, a decision maker solves optimization problem (4.79) or a set of algebraic equations (4.80) numerically to estimate the distribution parameters.

The amount of information in a sample $x_1, \ldots, x_n$ for estimating the parameters of a distribution is quantified by the following matrix, called expected information matrix, and denoted by $\mathbf{I}$. The rank of this matrix is equal to the number of distribution parameters. The elements of this matrix are,

$$I_{i,j} = -E \left(\frac{\partial^2 \ln L(x_1, \ldots, x_n/\theta)}{\partial \theta_i \partial \theta_j} \right) \qquad (4.84)$$

where θ is the vector of distribution parameters. In the above equation, the expectation is taken with respect to $x_1, \ldots, x_n$. An important result that enables a decision maker to quantify the uncertainty in the estimates of the parameters of a distribution is that the covariance matrix of the parameters is equal to the inverse of the information matrix,

$$\mathbf{V} = \mathbf{I}^{-1} \Big|_{\theta = \hat{\theta}} \qquad (4.85)$$

where the right hand side is evaluated at the maximum likelihood estimates of the parameters. The diagonal elements of this matrix are equal to the variances of the estimates of the distribution parameters from the maximum likelihood method. The variance of the estimate of the ith distribution parameter is,

$$\sigma_{\hat{\theta}_i}^2 = V_{i,i} \qquad (4.86)$$

Many statisticians calculate the local information matrix instead of the expectation of the right hand side of equation (4.84),

$$I_{i,j} = -\frac{\partial^2 \ln L(x_1, \ldots, x_n/\theta)}{\partial \theta_i \partial \theta_j} \Big|_{\theta = \hat{\theta}} \qquad (4.87)$$

For large samples, a maximum likelihood estimator has the following properties,

1. It is asymptotically unbiased; its mean value converges to the true value of the parameter with n.
2. It has minimum variance.
3. Its sampling distribution is normal. The maximum likelihood estimate of a distribution parameter $\hat{\theta}$ converges to a normal variable with zero mean and variance $\frac{V}{n}$ with n.
4. Maximum likelihood estimators are invariant; the maximum likelihood estimator of a function of a parameter $\phi = f(\theta)$ is $\hat{\phi} = f(\hat{\theta})$.
5. In Bayesian estimation (section 6.4), we assume a prior probability distribution of an unknown distribution parameter and update it when observations become available. We select a uniform prior distribution when we are highly uncertain about the parameters. The maximum likelihood estimate coincides with the mode of the posterior probability distribution, when the support of the prior distribution contains the maximum likelihood estimate.

Compared to the method of moments, the maximum likelihood method has the advantage that it considers the information about the type of the hypothesized probability distribution of the parameters. As a result, often, maximum likelihood estimates are closer to the true values of the parameters of a distribution than estimates from the method of moments. On the other hand, the maximum likelihood estimator is only asymptotically unbiased, while the estimates of the moments of a variable are unbiased.

For small samples (e.g., less than 30), the estimates of the maximum likelihood method become biased, and the probability distribution of the parameter estimates deviates significantly from the normal. Use of the normality assumption could result in overestimation of the coverage of the confidence intervals of the distribution parameters, in this case. The reader should remember the shortcomings of the confidence intervals presented in section 4.2.4 (see paragraph "Accuracy of confidence interval for mean value".)

Example 4.31: Estimate the parameters of the lognormal distribution of the ice impact force by using the 30 observations in Table 4.13.

Solution:
The maximum likelihood estimates of the parameters of the lognormal distribution are,

$$\hat{\mu} = \frac{1}{n}\sum_{i=1}^{30} \ln(x_i) = 1.157\,\text{MN} \quad \text{and}$$

$$\hat{\sigma} = \sqrt{\frac{\sum_{i=1}^{n}\{\ln(x_i) - \hat{\mu}\}^2}{n}} = 0.446\,\text{MN}$$

where $x_1, \ldots, x_{30}$ are the observations in Table 4.13. We verified these results by solving the optimization problem defined by in Equations (4.79 and 4.80) numerically, by using software program Mathcad. Remember that, in example 4.29, we found that these parameters are $\hat{\mu} = 1.15$ MN and $\hat{\sigma} = 0.467$ MN, respectively, by the method of moments.

The information matrix is,

$$
\mathbf{I} = \begin{bmatrix} I_{11} & I_{12} \\ I_{21} & I_{22} \end{bmatrix}
$$

where the elements of the matrix are,

$$
I_{11} = \left. \frac{\partial^2 \ln L(x_1, \ldots, x_{30}/\mu, \sigma)}{\partial \mu^2} \right|_{\mu = \hat{\mu}, \sigma = \hat{\sigma}}
$$

$$
I_{22} = \left. \frac{\partial^2 \ln L(x_1, \ldots, x_{30}/\mu, \sigma)}{\partial \sigma^2} \right|_{\mu = \hat{\mu}, \sigma = \hat{\sigma}}
$$

$$
I_{12} = I_{21} = \left. \frac{\partial^2 \ln L(x_1, \ldots, x_{30}/\mu, \sigma)}{\partial \mu \partial \sigma} \right|_{\mu = \hat{\mu}, \sigma = \hat{\sigma}}
$$

In the above equations, $\ln L(x_1, \ldots, x_{30})$ is the natural logarithm of likelihood function,

$$
\ln L(x_1, \ldots, x_{30}/\mu, \sigma) = \ln \left\{ \prod_{i=1}^{30} \frac{1}{x_i \sigma \sqrt{2\pi}} e^{-\frac{1}{2} \left[\frac{\ln(x_i) - \mu}{\sigma} \right]^2} \right\}
$$

The covariance matrix of the estimates of the parameters is the inverse of the information matrix,

$$
\mathbf{V} = \mathbf{I}^{-1} = \begin{bmatrix} 6.618 \cdot 10^{-3} & 0 \\ 0 & 3.309 \cdot 10^{-3} \end{bmatrix}
$$

Therefore, the variance of $\hat{\mu}$ is $\sigma_{\hat{\mu}}^2 = 6.618 \cdot 10^{-3}$ and its standard deviation is 0.081.

The estimator of parameter μ is normally distributed for large samples. If we make this assumption, then the 95% confidence interval is,

$$
\hat{\mu} - z_{1-\alpha/2} \cdot \sigma_{\hat{\mu}} \leq \mu \leq \hat{\mu} + z_{1-\alpha/2} \cdot \sigma_{\hat{\mu}}
$$

Plugging the values of $\hat{\mu}$ and $\sigma_{\hat{\mu}}$ in the above equation, we find that the 95% confidence interval is [0.997, 1.316].

The variance of $\hat{\sigma}$ is $3.309 \cdot 10^{-3}$ and the standard deviation is 0.058. The 95% confidence interval of parameter $\hat{\sigma}$ is [0.333, 0.558]. The above estimates are uncorrelated because their covariance matrix $\mathbf{V}$ is diagonal, in this example.

Table 4.14 Comparison of the estimates of the parameters of a lognormal distribution by using the methods of moments and maximum likelihood.

			Moments		Maximum likelihood					
Obs.	Mean (3.5)	Stand. Dev.(2.0)	μ (1.111)	σ (0.532)	μ (1.111)	μ_l	μ_u	σ (0.532)	σ_l	σ_u
5	1.992	0.478	0.661	0.237	0.662	0.449	0.875	0.243	0.093	0.394
10	3.442	2.439	1.033	0.638	1.072	0.736	1.407	0.541	0.304	0.778
15	3.693	2.125	1.163	0.535	1.17	0.907	1.433	0.519	0.334	0.705
30	3.522	1.74	1.15	0.467	1.157	0.997	1.316	0.446	0.333	0.558
100	3.220	1.952	1.013	0.560	1.032	0.930	1.133	0.519	0.447	0.591
1000	3.375	1.890	1.080	0.522	1.080	1.047	1.112	0.523	0.500	0.546
10,000	3.508	1.979	1.117	0.526	1.115	1.105	1.126	0.531	0.522	0.537

Example 4.32:   Comparison of the methods of moments and maximum likelihood.

This example compares the accuracy of the methods of moments and maximum likelihood for different amounts of data. We generate 10,000 sample values from a lognormal probability distribution with mean value equal to 3.5 and standard deviation 2. The corresponding values of the parameters of the distribution are $\mu = 1.111$ and $\sigma = 0.532$. Estimate the parameters of the lognormal distribution by using both methods and subsets containing from 5 to 10,000 observations.

Solution:
Table 4.14 and Figure 4.23 compare the estimates of the parameters of the distribution from the two methods. Both methods have comparable accuracy in this example. The maximum error is 41% for μ and 54% for σ, when we use only 5 observations. Both methods estimated parameter μ accurately with 10 or more observations. However, accurate estimation of parameter σ required at least 100 observations. The maximum likelihood method underestimated σ by 16% and 2% when using 30 observations and 100 observations, respectively. Note that we calculated 95% confidence intervals of the parameters only for the maximum likelihood method. The reason is that calculation of the confidence intervals by the method of moments requires the joint PDF of the sample mean value and standard deviation of the lognormal distribution, which is not available.

Estimation of the parameters of a probability distribution with censored data
In some applications, decision makers can only collect censored data. For example, suppose that a test engineer wants to estimate the probability distribution of the time to failure of a timing belt of a new gas engine. The engineer simultaneously tests 100 belts under conditions designed to replicate the real operating conditions. The engineer cannot wait until all timing belts fail. Instead, he/she performs the test over a fixed time period and records times to failure. Then, he/she estimates the probability distribution of the time to failure from this information by using a version of the maximum likelihood method for censored data that we present here.

There are time-censored (type I) and failure-censored (type II) data. The engineer collects the first type of data if he/she stops the test after a fixed time. This process is

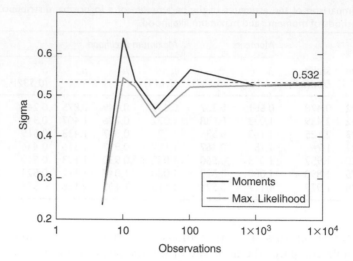

Figure 4.23 Estimates of parameter σ by using the methods of moments (black line) and maximum likelihood method (gray line).
The methods have same accuracy and require at least 100 observations to estimate the parameters accurately.

called type-I censoring. In the second case, he/she stops the test after a fixed number of belts fail, and this process is called type-II censoring. In the above example, the engineer followed a type-I censoring approach. Bury (1999, section 12), Cook et al. (2003) and the NIST Engineering Statistics handbook (2010) present methods for maximum likelihood estimation of probability distributions with both types of censored data.

Let $f_T(t/\boldsymbol{\theta})$ be the hypothesized PDF of the time to failure of a system. We collect observations of the time to failure up to a limit, t_{max}. The information from the test consists of the times to failure, $t_1, \ldots, t_r$, and the number of units that survived the test. We want to estimate the parameters of this PDF from this information.

Let n be the total number of units tested, and r the number of units that failed before time, t_{max}. The likelihood function is,

$$L(t_1, \ldots, t_r, \boldsymbol{\theta}) = \prod_{i=1}^{r} f_T(t_i/\boldsymbol{\theta}) \cdot \{1 - F_T(t_{max}/\boldsymbol{\theta})\}^{n-r} \tag{4.88}$$

The log-likelihood function is,

$$\ln L(t_1, \ldots, t_r, \boldsymbol{\theta}) = \sum_{i=1}^{r} \ln\{f_T(t_i/\boldsymbol{\theta})\} + (n - r)\ln\{1 - F_T(t_{max}/\boldsymbol{\theta})\} \tag{4.89}$$

For type-II censored data, we replace the time t_{max} by the end of test time. Note that the time that it takes for r units to fail is random.

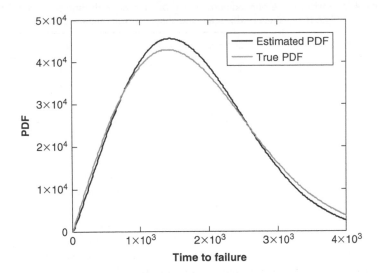

Figure 4.24 PDF of time to failure estimated by using only 44% of the observations.

We solve optimization problem (4.79) in order to find the distribution parameters. In this problem, we maximize the objective function in Equation (4.89).

Example 4.33: The Weibull distribution is known to represent well time to failure data. In order to demonstrate this method in the above example with the reliability engineer, we consider that he/she tested $n = 1,000$ timing belts for $t_{max} = 1,500$ hours. We assume that the time to failure follows a two-parameter Weibull distribution with scale parameter 2,000 hours and shape parameter 2, which is right-skewed. We generate $n = 1,000$ observations from this distribution that represent the times to failure of 1,000 belts. The minimum and maximum times are 52 and 5,414 hours and the median is 1,662 hours, in this sample. We keep only those times to failure that are less than or equal to 1,500 hours. Thus, we create a time-censored sample with 435 observed times to failure. Now, we pretend that we do not know the probability distribution of the time to failure and estimate it from the censored data.

The parameters of the probability distribution of the time are estimated by maximizing the likelihood function (4.89), where

$$f_T(t/\theta, \beta) = \frac{\beta}{\theta} \left(\frac{t}{\theta}\right)^{\beta-1} \exp\left\{-\left(\frac{t}{\theta}\right)^{\beta}\right\} \quad \text{for } t \geq 0$$

and

$$F_T(t) = 1 - \exp\left\{-\left(\frac{t}{\theta}\right)^{\beta}\right\}$$

We find that the scale parameter θ is 1,956 hours and shape parameter is 2.12. These estimates, which were found by using 44% of the observations, should be compared

Table 4.15 Performance of maximum likelihood method with time-censored data. The true values of the scale and shape parameters are 2,000 and 2.

Observations	Scale Parameter, β	Shape Parameter, θ
100	1,799	1.8
200	1,940	1.87
500	2.209	1.87
1,000	1,956	2.12

to the true values that are 2,000 and 2. Figure 4.24 compares the true and estimated PDFs of the time to failure. The probabilistic model is quite accurate considering that we obtained it by using only about half of the total number of observations.

Table 4.15 shows the effect of the number of observed failures in the censored sample on the accuracy of the estimation. The method yields reasonably accurate results. The accuracy of the estimates fluctuates with the number of tests, but the error in the estimates of the distribution parameters is less than 10%.

4.4.2.3   The least squares method

Some researchers estimate the parameters of a probability distribution by using the least squares method. Commercial program Minitab currently uses this method to fit a probability distribution to the data. In the least squares method, a decision maker constructs a histogram of the observed values of a random variable. Then, he/she minimizes the sum of the squares of the differences between the observed and expected data to estimate the distribution parameters. A variant of this method uses a cumulative histogram to estimate the parameters.

4.4.3   Step 3: Assess fit of selected distributions to observed data

After a decision maker selects few families of probability distributions and estimates their parameters, he/she performs a goodness-of-fit test to decide which of these families are consistent with the data. This statistical hypothesis test evaluates the hypothesis that a family of distributions conforms to the data. Statisticians call this the "null hypothesis." Example 4.26 in section 4.3 explained and demonstrated the Chi-square test for this purpose.

Commercial computer programs enable a decision maker to estimate the parameters of alternative families of probability distributions from data, and test if these families fit to the data. The user can perform one or more of the following tests: Chi-square, Kolmogorov-Smirnov and Anderson-Darling tests. We demonstrate how to select a probabilistic model to represent the measured values of the ice impact force on an offshore platform in Table 4.13 by using @Risk software.

A decision maker knows that the ice force has a low bound of zero, but does not know its upper bound. He/she considers the Lognormal, Gamma, Weibull and Triangular distributions. Software @Risk estimates the parameters of these distributions by using the maximum likelihood method. Figure 4.25 compares the fitted Lognormal

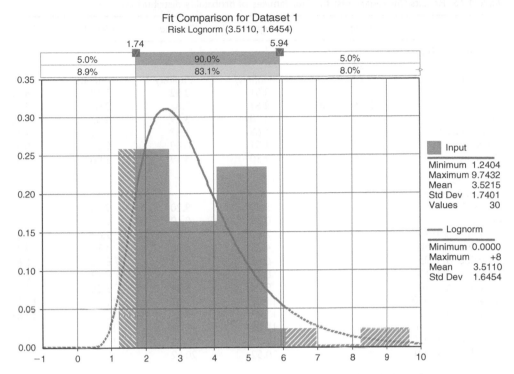

Figure 4.25 Histogram of ice force on offshore platform and fitted lognormal PDF.

distribution to a histogram of the data. The parameters of the fitted lognormal distribution are $\mu = 1.157$ and $\sigma = 0.446$, respectively. The mean value and standard deviation of the fitted distribution that correspond to these values are equal to 3.511 MN and 1.6454 MN, respectively. It is observed from Figure 4.25 that the ice force could exceed 5.94 MN with probability 0.08. The decision maker cannot decide from this figure if the Lognormal distribution fits the data.

The results of the goodness-of-fit test in Table 4.16 help the decision maker decide which of the above probability distributions fit the data. This table shows the values of the Chi-square statistic (Equation 4.77) and the p-value for each distribution. If the observations of the ice force were drawn from a hypothesized distribution, then the Chi-square statistic would rarely assume a high value located on the tail of the Chi-square distribution. The p-value is the probability that the Chi-square distribution could exceed the value that we observed in the test. Therefore, a high value of the Chi-square statistic and a very low value p-value (e.g., 0.05) would suggest that the ice forces do not follow the hypothesized distribution.

The observed Chi-square value for each distribution was calculated from Equation (4.77). For each distribution, the data is divided into six bins, each of which would contain 5 observations, on average, if the distribution perfectly explained the data. Here we demonstrate the calculation of the observed Chi-square value and the p-value for the lognormal distribution in the third column of Table 4.16. There are 4, 7, 6, 1,

Table 4.16 Results Chi-square test for four families of probability distributions.

	Input	Lognorm	Gamma	Weibull	Triang
Distribution Statistics					
Minimum	1.24	–	–	–	–
Maximum	9.74	+Infinity	+Infinity	+Infinity	10.19
Mean	3.52	3.51	3.52	3.53	4.10
Mode	2.0638 [est]	2.61	2.82	3.01	2.12
Median	3.08	3.18	3.29	3.37	3.78
Std. Deviation	1.74	1.65	1.57	1.71	2.19
Skewness	1.64	1.51	0.89	0.52	0.47
Kurtosis	7.25	7.30	4.19	3.06	2.40
Percentiles					
Chi–Squared Test					
Chi-Sq Statistic		5.20	9.20	9.20	11.60
P-Value		0.39	0.10	0.10	0.04
Cr. Value @ 0.750		2.67	2.67	2.67	2.67
Cr. Value @ 0.500		4.35	4.35	4.35	4.35
Cr. Value @ 0.250		6.63	6.63	6.63	6.63
Cr. Value @ 0.150		8.12	8.12	8.12	8.12
Cr. Value @ 0.100		9.24	9.24	9.24	9.24
Cr. Value @ 0.050		11.07	11.07	11.07	11.07
Cr. Value @ 0.025		12.83	12.83	12.83	12.83
Cr. Value @ 0.010		15.09	15.09	15.09	15.09
Cr. Value @ 0.005		16.75	16.75	16.75	16.75
Cr. Value @ 0.001		20.52	20.52	20.52	20.52
Chi-Sq Test (Binning Information)					
Bin #1 : Minimum		–	–	–	–
Bin #1 : Maximum		2.07	2.04	1.83	1.90
Bin #1 : Input		4.00	4.00	2.00	3.00
Bin #1 : Fit		5.00	5.00	5.00	5.00
Bin #2 : Minimum		2.07	2.04	1.83	1.90
Bin #2 : Maximum		2.62	2.69	2.64	2.78
Bin #2 : Input		7.00	8.00	9.00	10.00
Bin #2 : Fit		5.00	5.00	5.00	5.00
Bin #3 : Minimum		2.62	2.69	2.64	2.78
Bin #3 : Maximum		3.18	3.29	3.37	3.78
Bin #3 : Input		6.00	6.00	7.00	5.00
Bin #3 : Fit		5.00	5.00	5.00	5.00
Bin #4 : Minimum		3.18	3.29	3.37	3.78
Bin #4 : Maximum		3.85	3.98	4.17	4.95
Bin #4 : Input		1.00	–	2.00	8.00
Bin #4 : Fit		5.00	5.00	5.00	5.00
Bin #5 : Minimum		3.85	3.98	4.17	4.95
Bin #5 : Maximum		4.89	4.97	5.22	6.49
Bin #5 : Input		7.00	8.00	7.00	3.00
Bin #5 : Fit		5.00	5.00	5.00	5.00
Bin #6 : Minimum		4.89	4.97	5.22	6.49
Bin #6 : Maximum		+Infinity	+Infinity	+Infinity	10.19
Bin #6 : Input		5.00	4.00	3.00	1.00
Bin #6 : Fit		5.00	5.00	5.00	5.00

7 and 5 values in each bin for this distribution. Therefore, the Chi-square value of the test is,

$$\chi^2 = \frac{(4-5)^2}{5} + \frac{(7-5)^2}{5} + \frac{(6-5)^2}{5} + \frac{(1-5)^2}{5} + \frac{(7-5)^2}{5} + \frac{(5-5)^2}{5} = 5.20$$

There are six bins. The Lognormal distribution has two parameters. The number of degrees of freedom is in the range $[n-m-1, n-1]$ where n is the number of bins, and m the number of parameters. Therefore, the number of degrees of freedom ranges from 3 to 5. According to the suggestion in example 4.26 in section 4.3, we use the upper limit, in order to reduce the risk of rejecting the null hypothesis, while it is true. Then the p-value of the test statistic is one minus the value of the cumulative probability distribution of a chi-square variable with five degrees of freedom for the observed test statistic. This value is equal to 0.39.

In Table 4.16, the p-value of the Lognormal distribution is 0.39, while the p-values of the remaining three distributions are significantly lower, especially the triangular. Based on these results, the Lognormal distribution is selected because it conforms to the data better than the other distributions.

Conclusion

The conclusions of a probabilistic analysis are as good as the degree to which the probability distributions of the input variables are representative of the data. Therefore, it is critical to use all the information available to construct an accurate model of the inputs, that is, one that is consistent with the data. When we select an input model, we should be careful not to assume more than what we know and not to underestimate the probability of extreme catastrophic events.

The process of selecting input distributions involves three steps,

1. Hypothesize families of probability distributions that could represent the data.
2. Estimate the parameters of these distributions.
3. Assess the conformity of the model to the data.

There are well-established guidelines, methods and software for completing each step.

Uncertainty in a probabilistic model arises from uncertainty in the family of the probability distribution and its parameters. It is important perform sensitivity analysis, that is, investigate the effect of the uncertainty about the input on the final conclusion of a probabilistic study. For example, a decision maker should consider multiple input probability distributions and estimate the probability of failure of a system for each distribution.

Probabilities of events and mean values should be quoted together with confidence intervals. This information helps a decision maker select a course of action that reflects the amount of available information.

4.5 Modeling Dependent Variables

A complete probabilistic model of the random variables in a design problem consists of their joint probability distribution. Section 4.4 presented a method for modeling a single random variable. This method is adequate only for problems involving a single random variable or independent random variables. This section explains how to represent uncertainty in dependent variables. First, the section explains why it is important to model dependence accurately. Then it reviews and evaluates popular methods to model dependence. Finally, it presents copulas, which in many applications, are the most effective models for approximating a joint probability distribution.

4.5.1 *Overview of methods for modeling dependence*

It is challenging to estimate a joint probability distribution because this requires large amounts of data and because many experts are unaccustomed to making judgments about dependence. As a result, some decision makers estimate the marginal distributions of the random variables and make sweeping assumptions about their joint distribution that are not supported by the available evidence. These assumptions lead to significant errors in the estimated probability of failure of a system.

Many real-life problems involve dependent variables:

- The wind and wave loads on an offshore platform are dependent. Specifically, both loads become high in extreme weather and low in mild weather.
- Material and fatigue properties are dependent (Noh et al. 2009, Socie 2003).
- Consider a repair shop with two stations. Incoming parts are inspected and cleaned at the first station. The parts are repaired at the second station. On average, badly damaged parts require more time at both stations. Ignoring the positive dependence in the processing times of a part at the two stations can cause serious errors in the results of a simulation (Law, 2007, pp. 362–363, Mitchell et al. 1977).

Some common methods for modeling dependence are,

1. Assume independence.
2. Analyze the performance of a system in the extreme cases where the random variables have perfect and opposite dependence.
3. Transform dependent random variables into independent.
4. Represent dependence by a classical family of joint probability distributions, such as the bivariate normal and lognormal.
5. Represent dependence by copulas.

The following paragraphs discuss each method.

I. Assume independence

Often decision makers assume that the random variables are statistically independent. Some commercial computer programs for probabilistic analysis assume independent variables, by default. Then the joint PDF of the random variables reduces to the product of their margins. In most cases, decision makers do not justify the independence assumption or argue that it is reasonable to use a simple model when there is insufficient information to more accurately model the joint PDF of the random variables.

However, the independence assumption is dangerous because it can lead to underestimation of the probability of failure of a system. For example, if a civil engineer assumes that the wind and wave loads on an offshore platform are independent, then he will underestimate the probability of failure.

2. Analyze a system in both cases where the variables have perfect and opposite dependence

This method enables a decision maker to examine the sensitivity of a probabilistic assessment results to dependence. In this method, a decision maker estimates the probability of failure of a system or the consequences of a decision when the variables have perfect or opposite dependence. Dispersive sampling is useful for this purpose (Ferson et al. 2004). Sometimes, the decision maker would find that dependence is not significant and, therefore, it should be ignored. However, often, the decision maker cannot draw conclusions about a system and make a decision by considering extreme dependence cases, only.

3. Transform variables to independent

Sometimes, decision makers can transform a set of random variables into a second set of independent variables. In some applications, decision makers could transform a pair of dependent random variables X and Y to a pair of independent variables $Z = X + Y$ and $W = X - Y$. However, this method has limited applicability.

4. Use a classical family of joint probability distributions

Pair wise dependence is traditionally modeled by classical families of probability distributions, such as the normal and lognormal. Johnson, and Kotz, (1972) explain many of these families. This method is only applicable when the same family of univariate probability distributions can represent each random variable.

5. Represent dependence using a copula

The rest of this section explains and demonstrates how to model dependence by using copulas. One important advantage of this method is that it decouples the tasks of modeling the individual random variables and their dependence. Copulas are parametric functions approximating the joint cumulative distribution of a set of random variables in terms of their marginal distributions. Copulas have strong representation power; according to Sklar's representation theorem, for every set of continuous marginal distributions and their joint distribution, there is always a copula expressing the joint distribution in terms of the margins (Nelsen, 2006, section 2.3, Sklar, 1959).

There are copulas for multiple variables (Nelsen, 2006, section 2.10). This presentation is confined to models of two variables.

4.5.2 Copulas for modeling dependence

Copulas are functions approximating the joint CDF of multiple variables in terms of their margins.

The joint CDF of the two variables is expressed through the following parametric relation,

$$F_{X_1 X_2}(x_1, x_2) = C_\theta \{F_{X_1}(x_1), F_{X_2}(x_2)\} \tag{4.90}$$

where $F_{X_1 X_2}(x_1, x_2)$ is the unknown joint CDF of the variables, $F_{X_1}(x_1)$ and $F_{X_2}(x_2)$ the marginal CDFs, and θ is a vector of parameters. Copula $C_\theta(u, v)$ is a 2-increasing function that links the margins of two variables CDFs, $u = F_{X_1}(x_1)$ and $v = F_{X_2}(x_2)$, to their joint CDF. The domain of this function is $[0,1]^2$ and the range is $[0,1]$.

A copula fully and uniquely describes the dependence of two variables. According to Sklar's representation theorem, for any pair of marginal CDFs and a joint CDF consistent with the margins, there is always a unique copula linking the marginal CDFs to the joint CDF through Equation (4.90)

Consider three cases where the two variables are perfectly dependent, independent and oppositely dependent. The first row in Figure 4.26 shows the probability of the intersection of events $A = \{X_1 \leq x_1\}$ and $B = \{X_2 \leq x_2\}$ in each case. If the variables are perfectly dependent then one event contains the other and the probability of the intersection is equal to the smallest probability of these events. The probability of the intersection equals to the product of the probabilities of the two events if the events are independent. When the variables have opposite dependence, the two events are either disjoint, if their probabilities add up to less than or equal to one; otherwise they overlap. The joint probability is zero in the former case, and one minus the sum of the probabilities in the latter. The extreme values of the probability of the intersection of two events are called Frechet bounds.

The second row of Figure 4.26 shows the joint CDF of the two variables as a function of their marginal CDFs. Because the CDF is the intersection of events A and B this CDF is derived from the results in the first row. Copula $C_\theta(u, v)$ lies between the Frechet bounds, $\min(u, v)$ and $\max(0, u + v - 1)$, which represent the joint CDF of the random variables when these variables are perfectly dependent and oppositely dependent.

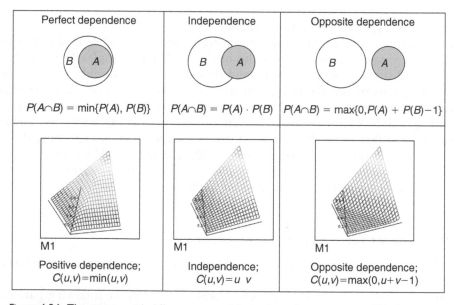

Figure 4.26 Three cases with different types of dependence between events "X_1 does not exceed a limit" and "X_2 does not exceed a limit" (upper panel). The lower panel shows the relation between the joint and the marginal CDFs of the above variables for these types of dependence.

Two variables are functionally dependent if and only if every value of one variable is associated with a single value of the other. Two random variables are functionally dependent whenever their underlying copula is equal to either one of the Frechet bounds. The left panel of Figure 4.27 shows sample values of two random variables with opposite dependence. Functional dependence is a weaker condition than linear dependence. Two variables are linearly dependent when one variable is a linear function of the other (Figure 4.27, right panel). Linear dependence implies functional dependence, but not vice versa.

Copulas enable a decision maker to construct the joint CDF into two steps: a) estimate the marginal probability distributions of the variables, and b) integrate them through a mathematical relation. Ferson et al. (2004) reviews methods for modeling probabilistic dependence between events and between random variables and the use of copulas for modeling dependence. Nelsen (2006) provides a comprehensive introduction to copulas. Genest and Favre (2007) present a practical introduction to copulas. They include guidelines for selecting a copula for a given problem, estimating its parameters and assessing the consistency of copula with the data, in this paper. This chapter walks the reader through the main steps for modeling dependence and demonstrates these steps on an engineering example with two random variables.

1.5.2.1 Monte-Carlo simulation with copulas

A primary application of copulas is in Monte Carlo simulation. Here we present a method for drawing a sample from a copula (Nelsen, 2006, section 2.6).

First, note that the marginal CDF, $u = F_X(x)$, of any continuous random variable X is uniform from 0 to 1. Indeed, since $u(x)$ is one-to-one function, its PDF is,

$$f_U(u) = \left. \frac{f_X(x)}{\frac{dF_X(x)}{dx}} \right|_{x=F_X^{-1}(u)} = 1 \quad \text{for } u \in [0, 1] \tag{4.91}$$

Equation (4.91) shows that the CDF, u, of a variable is uniform on [0, 1]. Therefore, bivariate function $C_\theta(u, v)$ is the CDF of two uniform random variables.

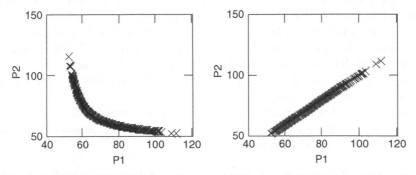

Figure 4.27 Sample values of two functionally dependent random variables (left panel) and two linearly dependent random variables (right panel). The variables in the left panel have opposite dependence.

One can consider CDFs, $u = F_{X_1}(x_1)$ and $v = F_{X_2}(x_2)$, as random variables. Then, copula $C_\theta(u, v)$ becomes the joint CDF of u and v. The second derivative of the copula with respect to u and v becomes the PDF of u and v,

$$c_\theta(u, v) = \frac{\partial^2 C_\theta(u, v)}{\partial u \partial v} \qquad (4.92)$$

We can draw a sample of a single random variable from a CDF $F_X(x)$ by using the inverse distribution function method (Melchers, 1999, pp. 67–68):

1. Generate a sample value u of a uniform random variable on [0,1].
2. Calculate $x = F_X^{-1}(u)$.

In order to generate a pair of values x_1 and x_2 from a copula $C_\theta(u, v)$, first we generate a pair of sample values u, v from the distribution $C_\theta(u, v)$ and then find x_1 and x_2 from the transformations $x_1 = F_{X_1}^{-1}(u)$ and $x_2 = F_{X_2}^{-1}(v)$. We generate u and v by the conditional distribution method; first we generate u, and then v from the conditional CDF of v given $U = u$, $F_V(v/u)$. This procedure is explained below,

1) Draw two sample values u and t from a uniform distribution from 0 to 1.
2) Calculate $v = F_V^{-1}(t/u)$.
3) The desired pair is u, and v.

4.5.2.2   Modeling the joint probability distribution of two random variables by using a copula

Here we explain a procedure to model dependence when data is available using an example. Table 4.17 shows the torque required to loosen 16 pairs of bolts. Each pair holds a plate. An engineer wants to construct a probabilistic model representing the data in this table.

Figure 4.28 summarizes a procedure that the engineer will follow to model the joint probability distribution of two random variables. The first step is to collect data for these variables. The engineer should be careful to record matching observations in this step. For example, each pair of measured torque values in Table 4.17 corresponds to the same plate. The engineer cannot model dependence using pairs of torque values that correspond to different plates. Next, the engineer models the marginal probability distributions of the random variables as explained in section 4.4. She evaluates the fit of the marginal distributions to the data and the confidence in the estimates of the parameters of these distributions. Then, she decides whether to collect more data and/or consider another family of distributions.

The engineer models dependence, after modeling the individual variables. First, she selects one or more parametric families of copulas. Then she estimates the parameters of each copula by using two methods:

a. Estimation of the parameters from measures of dependence such as Kendal's tau, and
b. The method of maximum pseudo-likelihood.

Table 4.17 Values of torque to loosen two bolts (ft lb)
Note: This table is identical to that in exercise 4.2.

Bolt 1	Bolt 2
15.70	16.42
14.72	15.24
16.24	16.08
17.28	16.31
17.20	16.82
17.73	16.75
13.82	15.16
15.77	15.82
17.10	16.53
14.91	15.35
15.31	15.10
14.31	14.64
14.15	14.40
15.02	15.20
15.23	15.50
13.88	14.97

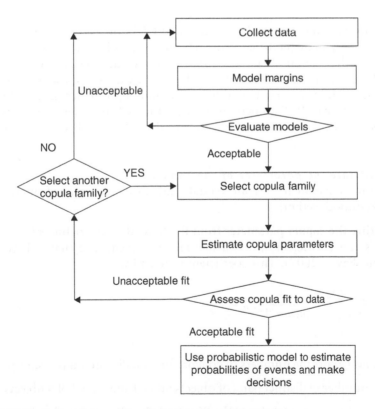

Figure 4.28 Procedure for modeling dependence by using copulas.

Both methods use measures of rank, only. If we change the values of the torque in Table 4.17 in a way that we do not change their ranks, then a measure of rank, such as Kendal's tau, will not change. The use of measures of rank is justified by the fact that the dependence captured by a copula has nothing to do with the behavior of the individual variables. The last step is to assess the fit of the copula to the data. We explain and demonstrate these steps below.

Step 2: Model the marginal distributions

The engineer determined that the normal probability distribution represents the torque of each bolt by using the methods in section 4.4. Let X_1 and X_2 be the torque values of the two bolts. Their mean values and standard deviations are, $\mu_{X_1} = 15.523$ ft lb, $\mu_{X_2} = 15.643$ ft lb, and $\sigma_{X_1} = 1.271$ ft lb, $\sigma_{X_2} = 0.76$ ft lb, respectively.

Step 3: Select a family of copulas to represent of the joint CDF

Many families of copulas have been proposed and studied. These include the Clayton, Frank, Farlie-Gumbel-Morgenstern and normal families (Nelsen, 2006, pp. 116–119). An empirical copula can also be used. In this example, the engineer models the dependence of the variables by using Frank's copula,

$$C_\theta(u, v) = \log_\theta \left\{ 1 + \frac{(\theta^u - 1)(\theta^v - 1)}{\theta - 1} \right\} \tag{4.93}$$

where u and v are the values of the individual CDFs of the two variables. In Frank's copula a single parameter, θ, controls dependence. This is a *comprehensive* copula, which means that the family in Equation (4.93) includes the cases of extreme dependence and independence. Indeed, perfect dependence arises for $\theta = 0$ (Figure 4.26, left panel), independence for $\theta = 1$ (middle panel) and opposite dependence for θ tending to infinity (right panel). Parameter θ can be estimated from measurements, or expert judgments. We demonstrate how to estimate the values of this parameter from data below.

Step 4: Estimate the parameter of the copula

We can estimate parameter θ from Kendall's tau coefficient, τ_n, or from the maximum pseudo-likelihood method.

a. Estimating the copula parameter from the Kendall's tau coefficient

Kendall's tau coefficient, τ_n, is a measure of dependence that is based on ranks (Genest and Favre, 2007). This coefficient is given by,

$$\tau_n = \frac{P_n - Q_n}{\binom{n}{2}} = \frac{4}{n(n - 1)} P_n - 1 \tag{4.94}$$

where P_n and Q_n are the numbers of concordant and discordant pairs, respectively and $\binom{n}{2}$ is the number of distinct pairs of objects taken from a pool of n objects, regardless of the order. Two pairs of values (X_{1_i}, X_{2_i}) and (X_{1_j}, X_{2_j}) are said to be concordant if

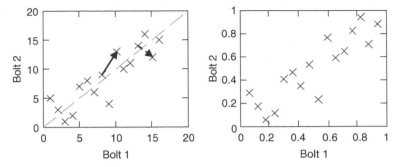

Figure 4.29 Ranks (left panel) and empirical CDFs (right panel) of the torque of bolts 1 and 2. The largest torque value has rank 16 and the smallest has rank 1. The line that connects two concordant pairs and starts from the point with the lowest x-value has positive slope (see continuous line on left panel). The line that connects discordant pairs and starts from the point with the lowest x-value has negative slope (see dashed line).

the order of the first two torque values X_{1_i}, X_{1_j} agrees with the order of the second, X_{2_i}, X_{2_j}. We mathematically express concordance with the following equation,

$$I_{X_{1_i}>X_{1_j}} \cdot I_{X_{2_i}>X_{2_j}} > 0 \tag{4.95}$$

where $I_{X_{1_i}>X_{1_j}}$ is a counter function that is one if the event denoted by the subscript is true and zero otherwise. For example, the first two pairs of values in Table 4.17 are concordant because both torque values in the first row are greater than their counterparts in the second row. Two pairs of values are discordant if the right hand side of Equation (4.95) is zero. For example, the fist and third pairs of values in Table 4.17 are discordant.

The left panel of Figure 4.29 shows the ranks of the 16 pairs of torque values. The largest torque value has rank 16 and the smallest 1. The right panel shows the values of the empirical CDFs, which are equal to the ranks normalized by $n+1$. Each pair is a point in this figure. The line that connects two concordant pairs and starts from the point with the lowest x-value has positive slope in this figure.

The number of concordant pairs is equal to,

$$P_n = \sum_{i=1}^{n}\sum_{i>j}^{n} (I_{X_{1_i}>X_{1_j}}) \cdot (I_{X_{2_i}>X_{2_j}}) + \sum_{i=1}^{n}\sum_{i>j}^{n} (I_{X_{1_i}<X_{1_j}}) \cdot (I_{X_{2_i}<X_{2_j}}) \tag{4.96}$$

From this equation, we find that there are 102 concordant pairs. Since there are 120 pairs, there are 18 discordant pairs. From Equation (4.94), $\tau_n = 0.7$.

Kendall's tau coefficient is more suitable than Pearson's correlation coefficient $\rho_{X_1 X_2}$ for characterizing dependence in copulas because tau expresses dependence of ranks. Tau varies from -1 to 1, like Pearson's correlation coefficient (Equation 3.51, Chapter 3). Functionally dependent variables have Kendall's tau equal to ± 1 (Figure 4.26, left panel). In contrast, only linearly dependent variables have Pearson's correlation coefficient equal to ± 1. Spearman's rho coefficient is another measure of functional dependence. This coefficient is equal to the correlation coefficient of the pairs of ranks of the data (Genest and Favre, 2007).

We estimate parameter θ of Frank's copula from the following one-to-one relation between Kendall's tau and θ. Parameter tau is zero for $\theta = 1$. Otherwise,

$$\tau(\theta) = 1 + \frac{4}{\ln(\theta)} - 4\frac{D\{-\ln(\theta)\}}{\ln(\theta)} \tag{4.97}$$

where $D(\theta) = \frac{1}{\theta}\int_0^\theta \frac{x}{e^x - 1}dx$ is the first Debye function.

Solving Equation (4.97) for θ we find that it is equal to $1.107 \cdot 10^{-5}$.

b. Maximum pseudo-likelihood method

The concept of the pseudo likelihood function is an extension of the likelihood function in Equation (4.78),

$$L(\theta) = \prod_{i=1}^{n} c_\theta(u, v) \tag{4.98}$$

where $c_\theta(u, v)$ is the joint probability density function, $c_\theta(u, v) = \frac{\partial^2 C_\theta(u,v)}{\partial u \partial v}$. This function is equal to 1 for $\theta = 1$, and

$$c_\theta(u, v) = \frac{\theta^{u+v} \cdot \ln(\theta) \cdot (\theta - 1)}{(\theta + \theta^{u+v} - \theta^u - \theta^v)^2} \tag{4.99}$$

otherwise.

The log-likelihood function is the logarithm of Equation (4.98).

We estimate parameter θ by maximizing the log-likelihood function. In this problem, we find that $\theta = 2.235 \cdot 10^{-5}$, numerically. The corresponding value of Kendall's tau is $\tau(\theta) = 0.684$. The latter value is close to that found directly from the data, which is 0.7.

Figure 4.30 shows copula $C_\theta(u, v)$ (left panel) and the probability density function $c_\theta(u, v)$ (right panel). The torque values of the two bolts are strongly dependent according to the figure, that is, when the torque in one bolt is high so is the torque in the other bolt.

Note that the margins of two variables and Kendall's tau do not completely define the joint PDF of two variables. Many different joint PDFs have identical margins and Kendall's tau. A decision maker should consider alternative families of copulas and fit the corresponding joint PDFs to the available observations or statistical summaries in order to select the most suitable one. Alternatively, the decision maker could calculate the probability of failure of a system for families that are consistent with the available evidence and determine bounds of the probability of failure.

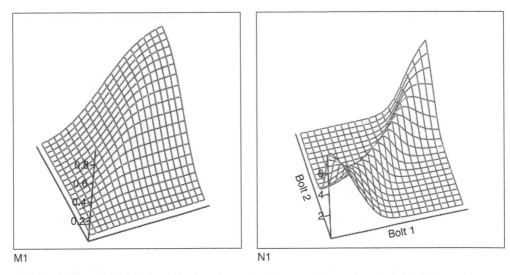

M1   N1

Figure 4.30 Copula and PDF from the results of the maximum pseudo likelihood method.

Step 5: Assess the fit of the copula to the data
The engineer should assess how well the probabilistic model conforms to the observed data, before using it. Genest et al. (2006) and Wang and Wells (2000) proposed methods for testing the goodness of fit of copulas to data. Here we evaluate the copula by comparing the observed torque values in Table 4.17 to simulated values drawn from the copula model.

We simulated 200 pairs of torque values by using the algorithm in the beginning of this section. In step 2 of this algorithm,

$$v = F_V^{-1}(t/u) = \frac{1}{\ln(\theta)} \ln\left(\frac{\theta \cdot t - \theta^u \cdot t + \theta^u}{t - \theta^u \cdot t + \theta^u}\right) \tag{4.100}$$

The left panel in Figure 4.31 shows 200 simulated (circles) and observed (crosses) CDF values. The right panel shows the corresponding torque values. The simulated and observed values look consistent.

Estimation of probabilities by using the copula model
The engineer calculates probabilities of critical events by using the fitted copula. Suppose that a system could fail when both bolt torque values become less than 14 ft lb. The probability of this event is,

$$P(X_1 < 14 \cap X_2 < 14) = C_\theta(u, v)$$
$$\text{where } u = F_{X_1}(14), v = F_{X_2}(14) \tag{4.101}$$

In the above equation $F_{X_1}(14)$ and $F_{X_2}(14)$ are the marginal CDFs of the torque of the two bolts evaluated at 14 ft lb. These CDF's are normal with mean values 15.523 and 15.643 ft lb and standard deviations 1.271 and 0.76 ft lb. Using this information, the

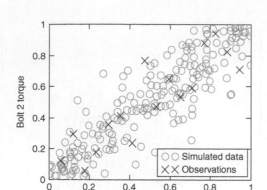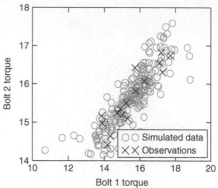

Figure 4.31   Simulated and observed CDF values (left panel) and the corresponding torque values (right panel).

engineer finds that the probability that both torque values could be less than 14 ft lb is 0.011, for both values of parameter θ found using Kendall's tau coefficient and the maximum pseudo-likelihood method. This probability is verified by Monte Carlo simulation with 40,000 replications.

We also calculate the probability that both torque values could be less than 14 ft lb in three cases where the torque values are perfectly dependent, independent and they have opposite dependence. Parameter θ is equal to 0, 1 and ∞ in these cases. From Equation (4.101), the probability of failure is 0.014, 0.0018 and 0 in these three cases. In this example, dependence significantly affects the results of probabilistic analysis. If system failure has minor consequences, then the engineer would decide that the system is safe based on the above three probabilities. Otherwise, the engineer would be unable to decide if the plate is safe.

Conclusions

A complete probabilistic model of a set of random variables is their joint distribution. There are several approaches for representing dependence and evaluating its effect on the results of a probabilistic assessment. These include simple approaches such as dispersive sampling and transformation of dependent variables into independent, and more rigorous methods such as those using parametric families of joint probability distributions and copulas. Copulas are practical, general tools for representing dependence. For any set of continuous variables and their joint CDF, there is always a copula linking together the margins to the joint CDF. Use of copulas simplifies the process of modeling dependence because it decouples the tasks of estimating the marginal distributions of the random variables and linking these distributions to a joint CDF.

There are many well–documented parametric families of copulas. This section presented and demonstrated methods for selecting a copula family, estimating the parameters of the family from data and assessing the conformity of the resulting copula to the data. In addition, the section presented an algorithm for Monte Carlo simulation by using a copula.

The following are the main conclusions of this section:

- Dependence significantly affects the results and conclusions of probabilistic analysis. Therefore, decision makers should quantify dependence accurately.
- The common practice of assuming independence often yields erroneous estimates of the probability of failure of a system.
- When collecting observations, it is important to record matching values of dependent quantities.
- Rank measures of dependence, such as Kendal's tau and Spearman's correlation coefficient are better than the traditional Pearson's correlation coefficient.
- When dependence cannot be quantified accurately, a decision maker should investigate its effect on the probabilities of critical events and quote an estimate of the probability of an event together with a confidence interval of this probability.
- Decision makers should seriously consider modeling dependence by using copulas, because they are practical and powerful tools for modeling dependence.

4.6   Conclusion of this Chapter

Statistical methods enable us to make inferences from observations. These tools are useful for constructing probabilistic models that decision makers use to make informed choices. This chapter presented methods to estimate the mean value of a variable, the probability of an event, and the probability distribution of random variables. In addition, it presented equations to quantify the accuracy of probabilistic models.

Then the chapter presented a systematic procedure to test a hypothesis, such as that the probability of an event is equal to some value. Finally, the chapter explained a process to select a probability distribution for a set of random variables.

Some important lessons learnt are as follows:

1. A decision maker rarely has enough data to represent uncertainty. Moreover, once she has collected a substantial amount of data (e.g., 100 observations), the law of diminishing returns kicks in: the decision maker needs a lot of additional data to improve even slightly the accuracy of her estimates. Therefore, in most decisions, besides data, she must use judgment to model uncertainty.
2. Dependence of variables must be considered. In order to represent uncertainty in a set of variables one needs their joint distribution. Moreover, positive dependence reduces the accuracy of the inferences from data.
3. A decision maker should study the effect of uncertainty in probabilities on the consequences of his decisions. He should quote estimates of a mean value or probability together with their confidence bounds.

   It is important to understand the interpretation of confidence intervals and the underlying assumptions of methods for calculating them. The true coverage of confidence intervals is often much smaller than the theoretical one.
4. Statistical hypothesis testing can only disprove a hypothesis. When little data is available, almost every hypothesis passes a hypothesis test. At the other extreme, the test rejects almost all hypotheses when there are massive amounts of data.

Questions and Exercises

Sample mean and variance

4.1 Present two-real life examples in which we need to estimate the properties of a population from a sample of it.

4.2 The table below shows the torques required to loosen two bolts holding a plate. Calculate the mean values of the torques of the two bolts. Calculate the sample means (from Equation 4.1) and the variances of the sample means of the torques of the two bolts from Equation (4.6).

4.3 Calculate the variance of the sample means in problem 2 by using Equation (4.8) and compare the results with those from Equation (4.6).

4.4 Calculate the sample variances of the torques by using both Equations (4.9) and (4.14).

4.5 How many additional values of the torque do we need in order to reduce the variances of the sample mean of the first column to one half of the values in problem 4.4?
How many additional values of the torque do we need in order to reduce the standard deviation of the sample means of the first column to one fourth of the standard deviation in problem 4.4?

4.6 Prove that the variance of a variable is equal to the mean value of the square of the variable minus the square of the mean value, $Var(X) = E(X^2) - \mu^2$. Hint: The variance is $Var(X) = E(X - \mu)^2$. Therefore, $Var(X) = E(X^2 + \mu^2 - 2X \cdot \mu)$. Use the fact that the mean is a linear operator.

4.7 In most practical problems, decision makers are uncertain about the values of the parameters of a probability distribution. For example, a bank manager knows that the interarrival time X of customers in a bank during lunch time is exponential, but she does not know the mean value of the interarrival time Θ. She believes that the interarrival time ranges from 5 min and 10 min. Chapters 2 and 6 explain that the value of a parameter can be quantified by using a subjective probability distribution even if there is no data about this variable. This probabilistic model represents the decision maker's belief about the uncertain quantity. The manager represents the mean value of the interarrival time, which she does not know, by a uniform random variable between 5 and 10 min. Then, both the variation of the interarrival time from customer-to-customer and the variation of its mean value contribute to the variance of the interarrival time.
Prove the law of total variance,

$$Var(X) = Var_\Theta\{E_X(X/\Theta = \theta)\} + E_\Theta\{Var_X(X/\Theta = \theta)\}$$

where $E_X(X/\Theta = \theta)$ and $Var_X(X/\Theta = \theta)$ are the mean value and variance of X conditioned upon the event $\Theta = \theta$ and $E_\Theta\{.\}$ and $Var_\Theta\{.\}$ are the mean value and variance of the variable in the bracket with respect to Θ.
Hint: Apply the equation for variance in the previous exercise to the variable $E(X^2/\Theta = \theta)$,
$Var(X) = E_\Theta\{E_X(X^2/\Theta = \theta)\} - E_\Theta\{E_X(X/\Theta = \theta)\}^2$. Then use the equation,

$$E_\Theta\{E_X(X^2/\Theta = \theta)\} = E_\Theta\{Var(X/\Theta = \theta)\} + E_\Theta\{E_X(X/\Theta = \theta)^2\}.$$

Table for Exercise 4.2: Torques of two bolts.

Bolt 1	Bolt 2
15.70	16.42
14.72	15.24
16.24	16.08
17.28	16.31
17.20	16.82
17.73	16.75
13.82	15.16
15.77	15.82
17.10	16.53
14.91	15.35
15.31	15.10
14.31	14.64
14.15	14.40
15.02	15.20
15.23	15.50
13.88	14.97

4.8   Explain why it takes fewer operations to calculate the variance of the sample mean using Equation (4.8) than (4.6).

Covariance and correlation

4.9   Make a scatter plot of the torque values of bolts 1 and 2 in problem 4.2.

4.10   Calculate the covariance and the correlation coefficient of the torques of the two bolts by using Equation (4.18).

4.11   Reproduce results in Example 4.6 for the calculation of the standard deviation of the sample in Figure 4.3 when the correlation coefficient is 0.75. See also Example 4.5.

4.12   Suppose that the correlation coefficient in Example 4.6 is equal to 1. What is the variance of the sample mean?

4.13   Explain in your own words why the accuracy of the sample mean deteriorates when observations are strongly correlated.

Confidence intervals

4.14   Write a computer program to calculate confidence bounds of the mean value in Equation (4.28) for a given coverage, $1 - \alpha$, and sample size n. Use a program for mathematical calculations such as Mathcad, Matlab or Mathematica, or use an Excel spreadsheet.

4.15   Write a computer program to calculate confidence bounds of the mean value in Equation (4.29) for a given coverage, $1 - \alpha$, and sample size n.

4.16   Use your code in the above problem to calculate the 90% confidence intervals for the mean values of the torques of the two bolts in table 4.17.

4.17 Repeat the calculations in Table 4.5 for an exponential probability distribution, with a mean value of 5 minutes, sample size 30 and expected coverage 90%. Perform 200 simulations.

4.18 Repeat the calculations in Example 4.8, this time assuming equal variances of the absorbance values of the control and cleaned plates.

4.19 Find the 95%-confidence interval for the variance of the interarrival time of the customers in a bank by using the sample values in Table 4.2.

4.20 In Example 4.9, estimate how many observations we need in order to reduce the width of the 95% interval by 50%.

4.21 Reproduce the results in Figures 4.10 and 4.11 by using Equations (4.45) and (4.46).

4.22 Derive the Wilson interval (Equations (4.49) and (4.50)).

4.23 Explain what approximations are involved in the calculation of the Wald confidence interval. Why is the coverage of this interval erratic? (see Figures 4.10 and 4.11)

4.24 Write a program to calculate the Wald, Wilson "Score" and Agresti-Coull intervals for a given coverage $1 - \alpha$, number of experiments, n, and number of occurrences of an event, k. Calculate the 95% confidence intervals for $n = 500$ and $k = 50$. Check your program by comparing its results to those from an online confidence interval calculator, such the one on the following link: www.measuringusability.com/wald.htm#score

4.25 Calculate the 95% Wilson confidence interval for $n = 500$ experiments and true probability equal to zero. Then, calculate the coverage of this interval and compare it to the value in Figure 4.13.

4.26 Justify the following definition of the confidence interval in objective probability: If we repeatedly calculate this interval using different samples drawn from the same population, the interval contains the true probability $1 - \alpha\%$ of the time.

Statistical hypothesis testing

4.27 a) Test the hypothesis that the probability of failure of an electronic circuit is 0.01. We tested 500 nominally identical circuits and observed one failure. Use a significance level $\alpha = 0.1$.

b) Is this an one-tailed or two-tailed test?

c) Find the probability that you could reject a hypothesis that is actually true in this test.

Hint: The test statistic for a sample proportion is,

$$\frac{\hat{P}(A) - P(A)}{\hat{\sigma}_{\hat{P}(A)}} \tag{4.102}$$

where $\hat{P}(A)$ is the observed proportion, and $\hat{\sigma}_{\hat{P}(A)} = \sqrt{\frac{\hat{P}(A)\{1 - \hat{P}(A)\}}{n_A}}$ is the standard deviation of the sample proportion. The statistic in Equation (4.102) follows the standard normal distribution.

4.28 We inspected two shipments consisting of $n_A = 334$ and $n_B = 116$ ball bearings from two factories. We found $x_A = 14$ defectives in the first shipment and

$x_B = 36$ in the second. Test the null hypothesis that the proportions of defective bearings from the two factories are equal. Use a level of significance equal to 0.05.

Hint: The test statistic for two sample proportions is,

$$\frac{\hat{P}(A) - \hat{P}(B)}{\sqrt{\hat{\sigma}^2_{\hat{P}(A)} + \hat{\sigma}^2_{\hat{P}(B)}}} \tag{4.103}$$

where $\hat{P}(A)$ and $\hat{P}(B)$ are the observed proportions, and $\hat{\sigma}_{\hat{P}(A)} = \sqrt{\frac{\hat{P}(A)\{1 - \hat{P}(A)\}}{n_A}}$

$\hat{\sigma}_{\hat{P}(B)} = \sqrt{\frac{\hat{P}(B)\{1 - \hat{P}(B)\}}{n_B}}$ are the standard deviations of these proportions. The statistic in Equation (4.103) follows the standard normal distribution.

4.29 Answer the following true-false questions:

a) The smaller the observed p-value the more likely it is to reject the null hypothesis. (T-F)
b) The probability of type I error is equal to the level of significance, α. (T-F)
c) Type I error occurs when we fail to reject a truly false hypothesis. (T-F)
d) When we increase the sample size we reduce the probability of Type II error. (T-F)
e) A hypothesis test is easy to pass when the sample size is small. (T-F)
f) If the alternative hypothesis is that the mean values of two populations are different, $\mu_1 \neq \mu_2$, then the test is called two-tailed test. (T-F)

Selecting input probability distributions

4.30 The following values of the times (in minutes) to drill a part in a factory have been recorded. What probability distributions could represent the data?

50 Observations of time to complete a task.

3.3349918	0.005779	0.0662127	0.4892728	0.2562124
0.8216964	1.063981	0.0225511	0.1949041	0.1539397
0.2680665	2.3595803	0.3087032	2.3655311	0.2789597
0.5244711	0.3158716	0.3860153	0.6438815	0.9433962
0.0974981	0.2539449	0.0741227	0.2655892	0.4276391
0.8739795	0.8971329	0.1244498	0.0886028	0.3297401
0.1708964	0.3983814	0.0016048	0.3618748	0.1428178
0.5953867	1.4318619	0.2459261	0.1480402	0.8889404
1.1961933	0.1221075	0.6617276	0.3904695	0.3547559
0.9575965	0.3270823	0.0871058	0.1475759	0.178514

4.31 Find the sample mean, sample variance, and sample standard deviation of the data in the previous problem. Also find the minimum and maximum values of the sample.

4.32 An industrial engineer believes that the Exponential, and Weibull distributions could fit the data (Section 3.2.1.2). The lower bound of the time to drill the part is zero. Estimate the parameters of these distributions by using the method of moments.

4.33 Same as problem 4.32 but estimate the parameters by using the maximum likelihood method.

4.34 Perform a goodness of fit test to test the conformity of the two hypothesized probability distributions to the data, problem 4.32. Use the estimates of the method of moments. Specifically, for each of the two distributions in problem 4.32, divide the data into eight bins, so that each bin is expected to contain approximately 6 observations.

4.35 Same as the previous problem but use the estimates from the maximum likelihood method.

Modeling dependent variables

4.36 Which of the statements below are true or false?

a. The marginal distributions and Pearson's coefficient of a pair of random variables completely define their joint probability distribution.

b. The marginal distributions and Pearson's coefficient of a pair of normal (Gaussian) random variables completely define their joint probability distribution.

c. If the Pearson's coefficient of a pair of random variables is zero then the random variables are statistically independent.

d. The copula of two random variables satisfies the following inequality,

$$C_\theta(u, v) \geq u \cdot v$$

where u and v are the marginal CDFs of the random variables.

e. Kendal's tau is equal to one for functionally dependent variables.

f. Pearson's correlation coefficient of two random variables is equal to one when the variables are functionally dependent.

g. The probability of an event depends on two random variables. We know the margins of these random variables but not their joint probability distribution. We can find bounds of the probability by considering only the two cases where the variables are perfectly and oppositely dependent.

4.37 The table below shows the thicknesses at two adjacent locations on 10 nominally identical plates. The thicknesses follow a normal probability distribution. Estimate the mean value and standard deviation of the thicknesses at the two locations.

Thicknesses of a plate at two locations in mm.

Location 1	Location 2
1.956	1.985
1.932	1.963
1.953	1.980
1.905	1.937
1.831	1.844
2.004	2.005
1.988	1.975
2.056	2.058
2.219	2.172
2.081	2.052
1.956	1.985

4.38   Make a scatter plot of the values in the table. Estimate Kendall's tau coefficient from the data. In addition, estimate Pearson's correlation and compare it with Kendall's tau.

4.39   Determine the joint cumulative probability distribution of the two quantities by using Frank's copula. Calculate the probability that the thicknesses at the two locations differ more than 0.1 mm.

4.40   Generate 500 random sample values of the load and the capacity by using the determined joint probability distribution.

4.41   Estimate the probability that the thicknesses at the two locations differ more than 0.1 mm from the sample in Problem 4.40.

Chapter 5

Probabilistic Analysis of Dynamic Systems

5.1 Introduction and Objectives

Many engineering systems are subjected to loads that are essentially random processes, e.g., random functions. Examples include the wings of an airplane under gust loads, the hull of a ship under wave loads or an automobile's suspension on a rough terrain. Moreover, the properties of many systems, such as their strength (stress at failure) are also random and vary in time. In random vibration analysis, one of the objectives is to determine the statistics of the system response and the probability of failure due to the response. Tools for time invariant reliability problems, such as first-order and second-order reliability methods (FORM and SORM), are not directly applicable to random vibration analysis.

Probabilistic analysis of a dynamic system under loads that are random processes involves the following steps:

1. Construct probabilistic models of the excitations using data obtained from measurements and experience. Using the stationarity assumption (described later in the chapter), the excitations are characterized using their means, autocorrelations and cross-correlations.
2. Calculate the statistics of the response.
3. Calculate the probability of failure. Failure may occur if the response exceeds a certain level (first excursion failure) or due to accumulation of fatigue damage (fatigue failure). An example of first excursion failure is the collapse of a structure because a stress exceeds a threshold.

This chapter presents methods for completing the above three steps, for a dynamic system with deterministic strength under time-varying random excitation. Emphasis is placed on how to characterize (or model) random processes in the time and frequency domains using analytical and simulation methods based on available data. The simulation methods include time series modeling and spectral representation methods.

5.1.1 Random Processes and Random Fields

In previous Chapters, we have characterized a random variable using various probability distributions. In real world however, we often deal with *random time functions*. These are random phenomena which depend on space *and* time. Such phenomena can

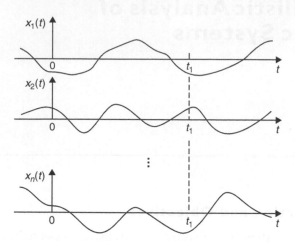

Figure 5.1 A continuous random process.

be for example, the sea surface, ship vibrations, and signals accompanied by undesired random waveform called noise. Here we will describe these random time functions probabilistically.

The concept of a random process generalizes the concept of a random variable to include time. A random variable X is a function of the possible outcomes e of an experiment; i.e. $X = X(e)$. If we assign a time function to every outcome e, we get a family of functions, denoted by $X(t, e)$, which is called a *random process*. We use x to denote a specific value (realization) of the random variable X and $x(t)$ to denote a specific time function of a random process. We can denote the random process by $X(t)$ in short. The parameter t usually represents time but it can also represent other quantities such as the distance from a reference point, for example.

The random process $X(t)$ consists of a family or ensemble of time functions $x_1(t)$, $x_2(t), \ldots, x_n(t)$, as shown in Figure 5.1. Each time function is called a *sample function, ensemble member* or *realization* of the random process. It shows the variation of the process with t for a particular realization of the process.

A random process degenerates to a random variable if t is fixed and e is a variable. The random variable $X(t_1) = X(t_1, e)$, often denoted by X_1 where the subscript 1 refers to time t_1, corresponds to a vertical slice through the ensemble at time t_1. The expected value of X_1 is called the *ensemble average* or the expected value of the random process at time t_1. In general, the expected value of a random process may not be constant through time.

There are four different types of random processes depending on the characteristics of time and space; namely, continuous random process, discrete random process, continuous random sequence and discrete random sequence. In this Chapter, we consider only continuous and discrete random processes.

Figure 5.2 shows four realizations of the elevation, $X(s)$, of a road covered with concrete slabs as a function of the distance, s, from a reference point. The slabs have fixed length and random height. This set of time histories belong to the *sample* or *ensemble* of time histories.

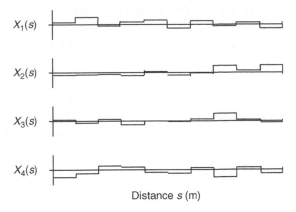

$X_1(s)$

$X_2(s)$

$X_3(s)$

$X_4(s)$

Distance s (m)

Figure 5.2 Four sample functions (realizations) of the elevation of a road covered with concrete slabs of fixed length.

The concept of *random fields* is a generalization of the concept of random processes. A random field is a random function of two or more parameters. The modulus of elasticity at a point in a solid, $E(x, y, z)$, varies with the location of the point (x, y, z) and is random. In this case, the modulus of elasticity is a random field because it is a random function of the three coordinates of the point of interest. The velocity field in a turbulent flow is also a random field. This is described by the velocity vector $V(x, y, z, t)$, where (x, y, z) are the coordinates of the point of interest, and t is time.

5.1.2 *Probabilistic Models of Random Processes*

A complete probabilistic model of a random process involves the joint probability distribution of the values of the random process for all possible values of the independent parameter, e.g. time. This information is impractical to obtain in most cases. We usually calculate only the first and second order statistics of the responses, that is, their means, autocorrelations and cross-correlations, because:

- It is impractical to collect sufficient data to determine the higher order statistics.
- In many problems, we assume that the random processes involved are jointly Gaussian. Then, the first and second order statistics completely define the processes.
- Even if the processes involved are not Gaussian, the first two moments contain important information. For example, we can determine an upper bound for the probability that the value of the process at an instant can be outside a given range centered about the mean value using Chebyshev inequality (Nigam, 1983, and Lin, 1967).

5.2 Modeling Random Processes in Time Domain

This section presents the basics on how to model or characterize (i.e. describe statistically) a random process in the time domain using correlation and covariance functions.

Details are also presented by Cooper and McGillem (1971), Melsa and Sage (1973), Newland (1993), and Peebles (1980), among many others.

5.2.1 Stationarity, Independence and Ergodicity

A random process $X(t)$ is called *stationary* if all its statistical properties (i.e. all moments) are time invariant, not changing under time shifts. In this case, the random variables $X(t_1)$ and $X(t_1 + \tau)$, where τ is a time shift, have therefore, the same moments for every τ. If the above is not true, the process is *non-stationary*.

The concept of stationarity in random processes corresponds to the steady-state concept in deterministic functions of time. For example, the wave elevation and the wind speed at a given point in the Atlantic Ocean can be considered stationary during short periods for which weather conditions do not change significantly. The force applied to the landing gear of an airplane during landing is a non-stationary process because this force changes rapidly in time.

A random process, $X(t)$, is called *strictly stationary* if the joint probability density function (PDF) of the process at all possible time instances is invariant under time shifts. Therefore,

$$f_{X(t+\tau)}(x) = f_{X(t)}(x)$$

$$f_{X(t_1+\tau),X(t_2+\tau)}(x_1,x_2) = f_{X(t_1),X(t_2)}(x_1,x_2)$$

$$\vdots$$

$$f_{X(t_1+\tau),X(t_2+\tau),\ldots,X(t_n+\tau)}(x_1,x_2,\ldots,x_n) = f_{X(t_1),X(t_2),\ldots,X(t_n)}(x_1,x_2,\ldots,x_n) \tag{5.1}$$

where $f_{X(t_1),X(t_2),\ldots,X(t_n)}(x_1,x_2,\ldots,x_n)$ is the joint PDF of the random process at $t_1,t_2,\ldots$, and t_n, respectively. In this case, the moments of $X(t)$ are also invariant under time shifts. For example, the elevation of the road consisting of concrete slabs in Figure 5.2 is a non-stationary random process because the joint PDF of the process at two locations, s_1 and s_2, changes under a distance shift. To define stationarity, we must therefore, define the distribution and density functions of the obtained random variables from the process.

The distribution function associated with the random variable $X_1 = X(t_1)$ is denoted by $F_X(x_1; t_1)$ and defined as

$$F_X(x_1; t_1) = P\{X(t_1) \leq x_1\} \tag{5.2}$$

where x_1 is a real number. For two random variables $X_1 = X(t_1)$ and $X_2 = X(t_2)$, we define the *second-order joint distribution function* as

$$F_X(x_1, x_2; t_1, t_2) = P\{X(t_1) \leq x_1, X(t_2) \leq x_2\} \tag{5.3}$$

The above definition can be easily extended to more than two random variables. The corresponding density functions are simply calculated by differentiating the distribution functions as

$$f_X(x_1; t_1) = \frac{\partial F_X(x_1; t_1)}{\partial x_1} \tag{5.4}$$

and

$$f_X(x_1, x_2; t_1, t_2) = \frac{\partial^2 F_X(x_1; t_1)}{\partial x_1 \partial x_2} \tag{5.5}$$

Two random processes $X(t)$ and $Y(t)$ are called *statistically independent* if the random variables $X(t_1), X(t_2), \ldots, X(t_N)$ are statistically independent from the random variables $Y(t_1^*), Y(t_2^*), \ldots, Y(t_N^*)$ for any choice of times $t_1, t_2, \ldots, t_N$ and $t_1^*, t_2^*, \ldots, t_N^*$. In this case, the following relationship holds

$$f_{X,Y}(x_1, x_2, \ldots, x_N, y_1, y_2, \ldots, y_M; t_1, t_2, \ldots, t_N, t_1^*, t_2^*, \ldots, t_M^*)$$
$$= f_X(x_1, x_2, \ldots, x_N; t_1, t_2, \ldots, t_N) \cdot f_Y(y_1, y_2, \ldots, y_M; t_1^*, t_2^*, \ldots, t_M^*) \tag{5.6}$$

First-Order, Second-Order, and Wide-Sense (Weak) Stationarity
A random process is called first-order stationary if its first-order density function does not change with a time shift; i.e.,

$$f_X(x_1; t_1) = f_X(x_1; t_1 + \tau) \tag{5.7}$$

for any time t_1 and any time shift τ. Because of *first-order stationarity*, the mean value $E[X(t)]$ of the process is constant; i.e.

$$E[X(t)] = \mu_X \tag{5.8}$$

where μ_X is a constant.

A random process is called *second-order stationary* if its second-order density function satisfies the following expression

$$f_X(x_1, x_2; t_1, t_2) = f_X(x_1, x_2; t_1 + \tau, t_2 + \tau) \tag{5.9}$$

for any x_1, x_2 and τ. The above expression is a function of the time difference $t_2 - t_1$, and not the absolute time. It should be noted that second-order stationarity implies first-order stationarity because the second-order density function determines the lower, first-order, density.

We can extend the above definitions to *N-order stationarity*. A process is *strict-sense stationary* if it is stationary to all orders as $N \to \infty$. The term stationary refers to the probability distributions and not to the sample functions themselves. All averages are therefore, independent of absolute time including the mean, mean square, variance and standard deviation. Because all engineering random processes have sample functions which are expressed for only a finite in time, they cannot be truly stationary. For practical purposes however, we often assume that they are stationary.

The autocorrelation function of a random process $X(t)$ denotes the correlation $E[X_1 X_2]$ of two random variables $X_1 = X(t_1)$ and $X_2 = X(t_2)$, and is expressed as

$$R_{XX}(t_1, t_2) = E[X(t_1)X(t_2)] \tag{5.10}$$

For a second-order stationary process R_{XX} is a function of the time difference $\tau = t_2 - t_1$, and not the absolute time, resulting in

$$R_{XX}(t_1, t_2) = R_{XX}(t_1, t_1 + \tau) = E[X(t_1)X(t_1 + \tau)] = R_{XX}(\tau) \tag{5.11}$$

A random process is called *wide-sense (weak) stationary* if the mean value of the process is constant, and auto-correlation function does not depend on absolute time; i.e.

$$E[X(t)] = \mu_X$$
$$E[X(t)X(x + \tau)] = R_{XX}(\tau) \tag{5.12}$$

The weak stationarity principle provides a practical way to characterize a random process using only its mean value and autocorrelation function.

Example 5.1: Show that the random process

$$X(t) = x \sin(\omega_0 t + \Theta)$$

is wide-sense stationary if x and ω_0 are constants and Θ is a uniformly distributed random variable on the interval $(0, 2\pi)$.

Solution:
The mean value of the process is equal to

$$E[X(t)] = \int\limits_0^{2\pi} x \sin(\omega_0 t + \theta) \underbrace{\frac{1}{2\pi}}_{f_\Theta(\theta)} d\theta = 0$$

and the autocorrelation function with $t_1 = t$ and $t_2 = t + \tau$ is

$$\begin{aligned}
R_{XX}(t, t + \tau) &= E[x \sin(\omega_0 t + \theta) x \sin(\omega_0 t + \omega_0 \tau + \theta)] \\
&= \frac{x^2}{2} E[\cos(\omega_0 \tau) - \cos(2\omega_0 t + \omega_0 \tau + 2\theta)] \\
&= \frac{x^2}{2} \cos(\omega_0 \tau) - \frac{x^2}{2} E[\cos(2\omega_0 t + \omega_0 \tau + 2\theta)] \\
&= \frac{x^2}{2} \cos(\omega_0 \tau)
\end{aligned}$$

Because the mean value is a constant, and the autocorrelation function depends only on the relative time τ, the random process $X(t)$ is wide-sense stationary.

Ergodicity
A stationary process is called *ergodic* if in addition to all ensemble averages being stationary with respect to a change in time, the averages taken along any sample function (time or temporal averages) are the same with the ensemble averages. This implies that the statistical moments, determined from generated data by traversing each

sample function horizontally are the same as the moments determined by traversing the ensemble vertically. In this case, any sample function can be used to characterize the random process statistically.

Note that if a process is ergodic, it is also stationary. Although ergodicity is a strict property, a variety of real-life random processes have been successfully characterized using the ergodicity assumption.

5.2.2 Correlation and Covariance Functions

This section presents the correlation (autocorrelation and cross-correlation) functions, covariance functions, and their properties for a random process $X(t)$.

Autocorrelation Function and its Properties

For a wide sense stationary process $X(t)$, the autocorrelation function $R_{XX}(\tau) = E[X(t)X(t+\tau)]$ depends only on the relative time τ, the mean value $E[X(t)]$ is constant, and the standard deviation $\sigma_{X(t)}$ is also a constant. If

$$E[X(t)] = E[X(t+\tau)] = \mu_X$$

and

$$\sigma_{X(t)} = \sigma_{X(t+\tau)} = \sigma_X$$

the correlation coefficient ρ for two random variables $X(t)$ and $X(t+\tau)$, is

$$
\begin{aligned}
\rho &= E\left[\frac{X(t) - \mu_X}{\sigma_X} \cdot \frac{X(t+\tau) - \mu_X}{\sigma_X}\right] \\
&= \frac{E[X(t)X(t+\tau)] - \mu_X E[X(t)] - \mu_X E[X(t+\tau)] + \mu_X^2}{\sigma_X^2} \\
&= \frac{R_{XX}(\tau) - \mu_X^2}{\sigma_X^2}
\end{aligned}
\tag{5.13a}
$$

or

$$R_{XX}(\tau) = \rho\sigma_X^2 + \mu_X^2 \tag{5.13b}$$

Because $-1 \le \rho \le 1$, the above expression yields

$$\underline{\text{Property 1}}: \quad -\sigma_X^2 + \mu_X^2 \le R_{XX}(\tau) \le \sigma_X^2 + \mu_X^2 \tag{5.14}$$

indicating that the autocorrelation function is bounded by $-\sigma_X^2 + \mu_X^2$ and $E[X^2(t)] = \sigma_X^2 + \mu_X^2$.

The definition of the autocorrelation function yields

$$\underline{\text{Property 2}}: \quad R_{XX}(0) = E[X^2(t)] \tag{5.15}$$

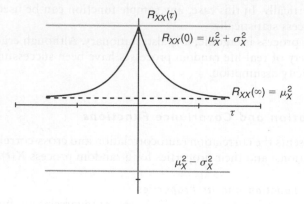

Figure 5.3 Typical autocorrelation function.

The random variables $X(t)$ and $X(t + \tau)$ become uncorrelated as $\tau \to \infty$ resulting in $\rho \to 0$. In this case,

$$\underline{\text{Property 3}}: \quad \lim_{|\tau| \to \infty} R_{XX}(\tau) = \mu_X^2 \tag{5.16}$$

From the definition of $R_{XX}(\tau)$ we also have

$$R_{XX}(\tau) = E[X(t)X(t + \tau)] = E[X(t - \tau)X(t)]$$

or

$$\underline{\text{Property 4}}: \quad R_{XX}(\tau) = R_{XX}(-\tau) \tag{5.17}$$

Figure 5.3 shows a typical autocorrelation function $R_{XX}(\tau)$ illustrating all properties graphically.

Example 5.2: A stationary process has the following autocorrelation function

$$R_{XX}(\tau) = \frac{1}{2 + \tau^2}$$

Calculate its mean value and variance.

Solution:
From property 3 of Equation (5.16), the square of the mean value of the process is

$$\mu_X^2 = \lim_{|\tau| \to \infty} R_{XX}(\tau) = 0$$

or

$$\mu_X = 0$$

The variance is equal to

$$\sigma_X^2 = E[(X(t) - \mu_X)^2] = E[X^2(t) - 2\mu_X X(t) + \mu_X^2]$$
$$= E[X^2(t)] - \mu_X^2$$
$$= R_{XX}(0) - \mu_X^2$$

resulting in

$$\sigma_X^2 = \frac{1}{2} - 0 = \frac{1}{2}$$

Example 5.3: An ergodic random process $X(t)$ is given. It has sample functions which are square waves of amplitude x_0 and period T, as shown in Figure 5.4. The phase of each sample function (time of first switching after $t = 0$) is a uniformly distributed random variable with a PDF $f(t_1) = \begin{cases} \frac{1}{T} & 0 \leq t_1 \leq T \\ 0 & \text{elsewhere} \end{cases}$ (Newland, 1993). Calculate the autocorrelation function.

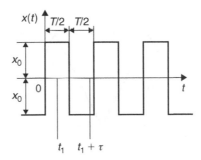

Figure 5.4 An ergodic random process with square wave sample function.

Solution:
Because of ergodicity, $R_{XX}(\tau)$ can be obtained by averaging along a single sample function considering the measuring time t_1 as a random variable which is uniformly distributed along the time axis; i.e.

$$R_{XX}(\tau) = E[X(t_1)X(t_1 + \tau)] = \int_{-\infty}^{\infty} x(t_1)x(t_1 + \tau)f(t_1)\, dt_1$$

or

$$R_{XX}(\tau) = \int_0^T x(t_1)x(t_1 + \tau)\frac{1}{T}\, dt_1$$

Special care is needed in calculating the above integral because all sample functions $x(t)$ are discontinuous. If $0 \leq \tau \leq T/2$, the integral is equal to

$$R_{XX}(\tau) = \frac{1}{T} \int_0^{T/2-\tau} x_0^2 \, dt_1 + \frac{1}{T} \int_{T/2-\tau}^{T/2} (-x_0^2) \, dt_1 + \frac{1}{T} \int_{T/2}^{T-\tau} x_0^2 \, dt_1 + \frac{1}{T} \int_{T-\tau}^{T} (-x_0^2) \, dt_1$$

$$= x_0^2 \left(1 - 4\frac{\tau}{T} \right)$$

and if $T/2 \leq \tau \leq T$, the integral is equal to

$$R_{XX}(\tau) = \frac{1}{T} \int_0^{T-\tau} (-x_0^2) \, dt_1 + \frac{1}{T} \int_{T-\tau}^{T/2} x_0^2 \, dt_1 + \frac{1}{T} \int_{T/2}^{3T/2-\tau} (-x_0^2) \, dt_1 + \frac{1}{T} \int_{3T/2-\tau}^{T} x_0^2 \, dt_1$$

$$= x_0^2 \left(-3 + 4\frac{\tau}{T} \right)$$

Figure 5.5 shows the calculated $R_{XX}(\tau)$ for the first period $(0 \leq \tau \leq T)$. It should be noted that $R_{XX}(\tau)$ is a triangular wave of constant amplitude x_0^2 extending to the left $(\tau < 0)$ and to the right $(\tau > T)$.

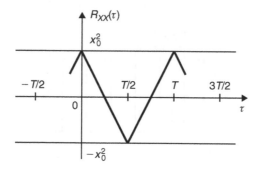

Figure 5.5 Autocorrelation function of the random process in Figure 5.4.

If $\tau = T, 2T, \ldots$ the sample values $X(t_1)$ and $X(t_1 + \tau)$ are in phase and perfectly correlated $(\rho = 1)$ with $R_{XX}(\tau) = E[X^2(t)] = x_0^2$. If $\tau = T/2, 3T/2, \ldots$ the sample values are in antiphase with perfect correlation and $R_{XX}(\tau) = -E[X^2(t)] = -x_0^2$. For other values of τ the sample functions are in phase or antiphase and the correlation is not perfect.

Cross-Correlation Function and its Properties
The cross-correlation function of two random processes $X(t)$ and $Y(t)$ is defined as

$$R_{XY}(t, t + \tau) = E[X(t)Y(t + \tau)] \tag{5.18}$$

If $X(t)$ and $Y(t)$ are at least wide-sense stationary, R_{XY} depends only on relative time τ, and is equal to

$$R_{XY}(\tau) = E[X(t)Y(t + \tau)] \tag{5.19}$$

The random processes X and Y are called *orthogonal processes* if

$$R_{XY}(t, t + \tau) = 0 \tag{5.20}$$

and are statistically independent if

$$R_{XY}(t, t + \tau) = E[X(t)] E[Y(t + \tau)] \tag{5.21}$$

If X and Y are statistically independent and at least wide-sense stationary,

$$R_{XY}(\tau) = \mu_X \mu_Y \tag{5.22}$$

where μ_X and μ_Y are the mean values of X and Y.

A few properties of the cross-correlation function are presented next. If $X(t)$ and $Y(t)$ are wide-sense stationary, $R_{XY}(\tau) = E[X(t)Y(t + \tau)] = E[X(t)Y(t - \tau)] = R_{XY}(-\tau)$ and $R_{YX}(\tau) = E[Y(t)X(t + \tau)] = E[Y(t)X(t - \tau)] = R_{YX}(-\tau)$. Therefore,

Property 1 : $\begin{aligned} R_{XY}(\tau) &= R_{XY}(-\tau) \\ R_{YX}(\tau) &= R_{YX}(-\tau) \end{aligned}$ $\tag{5.23}$

The correlation coefficient

$$\rho_{XY}(\tau) = E\left[\frac{X(t) - \mu_X}{\sigma_X} \cdot \frac{Y(t + \tau) - \mu_Y}{\sigma_Y} \right] \tag{5.24}$$

for two wide-sense stationary processes yields $R_{XY}(\tau) = \sigma_X \sigma_Y \rho_{XY}(\tau) + \mu_X \mu_Y$, and $R_{YX}(\tau) = \sigma_X \sigma_Y \rho_{YX}(\tau) + \mu_X \mu_Y$. Because $-1 \le \rho_{XY}, \rho_{YX} \le 1$, the limiting values of the cross-correlation functions R_{XY} and R_{YX} are $\pm \sigma_X \sigma_Y + \mu_X \mu_Y$ yielding

Property 2 : $-\sigma_X \sigma_Y + \mu_X \mu_Y \le R_{XY}(\tau), R_{YX}(\tau) \le \sigma_X \sigma_Y + \mu_X \mu_Y$ $\tag{5.25}$

For most random processes, there is no correlation between X and Y as the relative time τ becomes large; i.e. $\lim_{|\tau| \to \infty} \rho_{XY}(\tau) = 0$. Therefore,

Property 3 : $\lim_{|\tau| \to \infty} R_{XY}(\tau) = \lim_{|\tau| \to \infty} R_{YX}(\tau) = \mu_X \mu_Y$ $\tag{5.26}$

The cross-correlation function is also bounded as

Property 4 : $|R_{XY}(\tau)| \le \sqrt{R_{XX}(0)R_{YY}(0)} \le \frac{1}{2}[R_{XX}(0) + R_{YY}(0)]$ $\tag{5.27}$

It should be noted that the cross-correlation function is not necessarily an even function of the relative time τ exhibiting its maximum value at $\tau = 0$ as is the case with the autocorrelation function.

Example 5.4:   Consider the following two random processes

$$X(t) = x \cos(\omega t + \Theta)$$

and

$$Y(t) = y \cos(\omega t + \Theta + \phi)$$

where x, y, ω, and ϕ are constants while Θ is uniformly distributed random variable in the interval $(0, 2\pi)$, i.e.,

$$f_\Theta(\theta) = \begin{cases} 1/2\pi & \text{for } 0 \leq \theta \leq 2\pi \\ 0 & \text{otherwise} \end{cases}$$

Calculate the cross-correlation function.

Solution:
From the definition of $R_{XY}(\tau)$ in Equation (5.21) we have,

$$R_{XY}(\tau) = E[X(t)Y(t+\tau)] = E[xy \cos(\omega t + \theta) \cos(\omega t + \omega\tau + \theta + \phi)]$$

$$= \int_0^{2\pi} xy \cos(\omega t + \theta) \cos(\omega t + \omega\tau + \theta + \phi) \frac{1}{2\pi} d\theta$$

$$= \frac{1}{2} xy \cos(\omega\tau + \phi)$$

Covariance Functions
The concept of covariance of two random variables is extended to random processes. The autocovariance function is defined as

$$C_{XX}(t, t+\tau) = E[\{X(t) - E[X(t)]\} \cdot \{X(t+\tau) - E[X(t+\tau)]\}] \tag{5.28}$$

or equivalently,

$$C_{XX}(t, t+\tau) = R_{XX}(t, t+\tau) - E[X(t)]E[X(t+\tau)] \tag{5.29}$$

The cross-covariance function is similarly defined as

$$C_{XY}(t, t+\tau) = E[\{X(t) - E[X(t)]\} \cdot \{Y(t+\tau) - E[Y(t+\tau)]\}] \tag{5.30}$$

or

$$C_{XY}(t, t+\tau) = R_{XY}(t, t+\tau) - E[X(t)]E[Y(t+\tau)] \tag{5.31}$$

If $X(t)$ and $Y(t)$ are at least jointly wide-sense stationary, the above equations reduce to

$$C_{XX}(\tau) = R_{XX}(\tau) - \mu_X^2 \tag{5.32}$$

and

$$C_{XY}(\tau) = R_{XY}(\tau) - \mu_X \mu_Y \tag{5.33}$$

For a wide-sense stationary process, the variance does not depend on absolute time, and is given by

$$\sigma_X^2 = E[\{X(t) - E[X(t)]\}^2] = C_{XX}(0) = R_{XX}(0) - \mu_X^2 \tag{5.34}$$

Two random processes are *uncorrelated* if

$$C_{XY}(t, t + \tau) = 0 \tag{5.35}$$

or equivalently if

$$C_{XY}(t, t + \tau) = E[X(t)]E[Y(t + \tau)] \tag{5.36}$$

Similarly to random variables, if two random processes are statistically independent, they are also uncorrelated. The converse is *only* true for *jointly Gaussian processes*.

5.2.3   *Normal (Gaussian) Random Processes*

A random process is called normal, or Gaussian, if for any time t, the corresponding random variable $x(t)$ is normally distributed. The normal random process has the following three properties:

1.   Any random process obtained by a *linear* operation on the Gaussian random variables of a normal process is another normal process. For example, if $X(t)$ and $Y(t)$ are Gaussian, the processes $X(t) + Y(t)$, $\dot{X}(t)$, $\dot{Y}(t)$, and $\ddot{X}(t) + 3\ddot{Y}(t)$ are also Gaussian.
2.   The statistics of a random process are fully described by the expected value and the autocovariance function C_{XX} or equivalently, by the expected value and the autocorrelation function R_{XX}.
3.   If the process $X(t)$ is Gaussian and narrow-banded, the peak-to peak distances are Rayleigh distributed.

5.3   Fourier Analysis

Fourier analysis transforms a waveform (signal) from the time domain to the frequency domain and vice versa. It is commonly used to decompose a periodic function into its harmonic components. This section first introduces the concept of periodicity and the Fourier series, and then presents the basics of Fourier transform and the associated Fourier integrals.

5.3.1   *Fourier Series*

A function is *periodic* if it repeats after time T which is the period. This is expressed as

$$f(t) = f(t + T) = f(t + 2T) = \cdots = f(t + nT) \quad \text{for } n = 1, 2, 3, \ldots \tag{5.37}$$

where T (sec/cycle) is the period, and $1/T$ (cycles/sec $\equiv$ Hz) is the frequency.

Fourier series decomposes a periodic function $f(t)$ into a series of harmonic components as

$$f(t) = \frac{a_0}{2} + \sum_{n=1}^{\infty} \underbrace{[a_n \cos(n\omega t) + b_n \sin(n\omega t)]}_{n\text{-th harmonic component}} \tag{5.38}$$

where the $\omega = 2\pi/T$ is the fundamental frequency, $n\omega$ is the nth harmonic, and a_n and b_n are coefficients to be determined. Equation (5.38) can be also written as

$$f(t) = \frac{a_0}{2} + \sum_{n=1}^{\infty} d_n \cos(n\pi t + \phi_n) \tag{5.39a}$$

where

$$d_n = \sqrt{a_n^2 + b_n^2} \quad \text{and} \quad \phi_n = \tan^{-1}\left(-\frac{b_n}{a_n}\right) \tag{5.39b}$$

The above equations define the Fourier series.

The Fourier coefficients a_n and b_n are calculated using the orthogonality properties of the cosine and sine functions as

$$a_n = \frac{2}{T} \int_t^{t+T} f(t) \cos(n\omega t)\, dt \quad \text{for } n = 0, 1, 2, 3, \ldots \tag{5.40a}$$

and

$$b_n = \frac{2}{T} \int_t^{t+T} f(t) \sin(n\omega t)\, dt \quad \text{for } n = 0, 1, 2, 3, \ldots \tag{5.40b}$$

Note that α_0 from Equation (5.40a) is twice the mean value of the function $f(t)$. An infinite Fourier Series $(n \to \infty)$ is needed to represent a continuous periodic function exactly. However, in practice only a finite series is sufficient for most cases.

The Fourier series representation can be also expressed in the following complex form

$$f(t) = \sum_{n=-\infty}^{+\infty} c_n e^{jn\omega t} \tag{5.41a}$$

where the coefficients c_n are

$$c_n = \frac{1}{T} \int_t^{t+T} f(t) e^{-jn\omega t}\, dt \tag{5.41b}$$

The complex form provides the foundation for the forthcoming discussion on Fourier Transform and Fourier Integrals.

Example 5.5: Obtain the Fourier series of the following discontinuous periodic function

$$f(t) = \begin{cases} 0 & \text{for } -\pi < t \leq 0 \\ t & \text{for } 0 < t \leq \pi \end{cases}$$

which is shown in Figure 5.6.

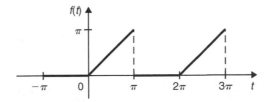

Figure 5.6 Periodic function for example 5.5.

Solution:
The period is $T = 2\pi$ and the rotational frequency is $\omega = \frac{2\pi}{2\pi} = 1$. The Fourier coefficients a_n are equal to

$$a_n = \frac{2}{T} \int_t^{t+T} f(t) \cos(n\omega t) \, dt = \frac{2}{2\pi} \int_{-\pi}^{0} 0 \cos(nt) \, dt + \frac{2}{2\pi} \int_0^{\pi} t \cos(nt) \, dt$$

$$= \frac{1}{\pi} \int_0^{\pi} t \cos(nt) \, dt$$

The above integral is evaluated using integration by parts as

$$a_n = \frac{1}{n\pi} \int_0^{\pi} t \frac{d \sin(nt)}{dt} \, dt = \frac{1}{n\pi} \left[t \sin(nt) - \int_0^{\pi} \sin(nt) \, dt \right]$$

$$= \frac{1}{n\pi} \left[t \sin(nt) + \frac{1}{n\pi} \cos(nt) \right]_0^{\pi}$$

$$= \frac{1}{n\pi} \left[t \sin(n\pi) + \frac{1}{n} \cos(n\pi) - t \sin 0 - \frac{1}{n} \cos 0 \right]$$

$$= \frac{1}{n\pi} \left[0 + \frac{1}{n}(-1)^n - 0 - \frac{1}{n} \right]$$

resulting in,

$$a_n = \frac{(-1)^n - 1}{n^2 \pi}$$

The coefficient a_0 is equal to

$$a_0 = \frac{2}{T} \int_t^{t+T} f(t)\, dt = \frac{2}{2\pi} \int_0^\pi t\, dt = \frac{1}{\pi} \frac{t^2}{2}\Big|_0^\pi = \frac{\pi}{2}$$

The b_n coefficients are equal to

$$b_n = \frac{2}{T} \int_t^{t+T} f(t) \sin(n\omega t)\, dt = \frac{2}{2\pi} \int_{-\pi}^0 0 \sin(nt)\, dt + \frac{2}{2\pi} \int_0^\pi t \sin(nt)\, dt$$

$$= \frac{1}{\pi} \int_0^\pi t \sin(nt)\, dt = \frac{1}{n\pi} \int_0^\pi (-1) t \frac{d\cos(nt)}{dt}\, dt = \frac{1}{n\pi} \left[-t\cos(nt) + \int_0^\pi \cos(nt)\, dt \right]$$

$$= \frac{1}{n\pi} \left[-t\cos(nt) + \frac{1}{n} \sin(nt) \right]_0^\pi = \frac{1}{n\pi} \left[-\pi \cos(n\pi) + \frac{1}{n} \sin(n\pi) + 0\cos 0 - \frac{1}{n} \sin 0 \right]$$

$$= -\frac{1}{n}(-1)^n$$

Therefore, the Fourier series of $f(t)$ is

$$f(t) = \frac{\pi}{4} + \sum_{n=1}^{\infty} \left[\frac{(-1)^n - 1}{n^2 \pi} \cos(nt) - \frac{1}{n}(-1)^n \sin(nt) \right]$$

Figure 5.7 shows $f(t)$ for $n = 5$, 10, and 20.

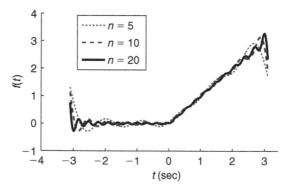

Figure 5.7 Fourier approximation of the periodic function in Figure 5.6, up to different values of the nth harmonic.

5.3.2 *Fourier Transform*

The coefficients a_n and b_n in the Fourier series representation of a periodic function are presented graphically in Figure 5.8. The frequency of the nth harmonic is

$$\omega_n = \frac{2\pi n}{T} = n\omega \tag{5.42}$$

and the spacing between adjacent harmonics is equal to

$$\Delta\omega_n = \omega_{n+1} - \omega_n = \omega = \frac{2\pi}{T} \tag{5.43}$$

The frequency spacing $\Delta\omega_n$ reduces with increasing T. At the limit, as $T \to \infty$, the discrete representation in Figure 5.8 becomes continuous. In this case, $f(t)$ is not a periodic function and the Fourier series becomes a *Fourier Integral* where the Fourier coefficients are continuous functions of frequency and are known as *Fourier Transforms*.

From equation (5.41), the Fourier series of a periodic function $f(t)$ can be written as

$$f(t) = \sum_{n=-\infty}^{+\infty} \left[\frac{1}{T} \int_{-T/2}^{T/2} f(t)e^{-j\omega_n t}\, dt \right] e^{j\omega_n t}$$

$$= \frac{1}{2\pi} \sum_{n=-\infty}^{+\infty} \left[\int_{-/2}^{T/2} f(t)e^{-j\omega_n t}\, dt \right] e^{j\omega_n t}\, \Delta\omega_n$$

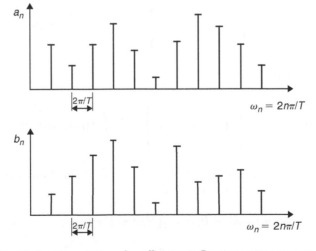

Figure 5.8 Graphical representation of coefficients in Fourier series representation.

As $T \to \infty$, $\Delta\omega_n \to d\omega_n$ and the infinite summation becomes an integral with the same infinite limits; i.e.,

$$f(t) = \frac{1}{2\pi} \int\limits_{-\infty}^{+\infty} \left[\int\limits_{-\infty}^{+\infty} f(t)e^{-j\omega_n t}\, dt \right] e^{j\omega_n t}\, d\omega_n$$

Because ω_n is a "dummy" variable, the above formula can be written as

$$f(t) = \frac{1}{2\pi} \int\limits_{-\infty}^{+\infty} F(\omega)e^{j\omega t}\, d\omega = \mathcal{F}^{-1}[F(\omega)] \tag{5.44}$$

where

$$F(\omega) = \int\limits_{-\infty}^{+\infty} f(t)e^{-j\omega t}\, dt = \mathcal{F}^{+1}[f(t)] \tag{5.45}$$

In Equations (5.44) and (5.45), $\mathcal{F}^{-1}$ denotes the inverse Fourier transform and $F(\omega)$ denotes the forward Fourier transform. The two equations form the Fourier Transform pair $f(t) \leftrightarrow F(\omega)$. A signal is transformed to the frequency domain using a forward Fourier transform $f(t) \to F(\omega)$, while the inverse Fourier transform $F(\omega) \to f(t)$ transforms the signal back to the time domain.

The forward Fourier transform $F(\omega)$ of a real-valued function $f(t)$ is a complex quantity

$$F(\omega) = P(\omega) + jQ(\omega) \tag{5.46}$$

where the real part is

$$P(\omega) = \mathrm{Re}\{F(\omega)\} = \mathrm{Re}\left\{ \int\limits_{-\infty}^{+\infty} f(t)e^{-j\omega t}\, dt \right\} = \int\limits_{-\infty}^{+\infty} f(t)\cos(\omega t)\, dt \tag{5.47}$$

And the imaginary part is

$$Q(\omega) = \mathrm{Im}\{F(\omega)\} = \mathrm{Im}\left\{ \int\limits_{-\infty}^{+\infty} f(t)e^{-j\omega t}\, dt \right\} = -\int\limits_{-\infty}^{+\infty} f(t)\sin(\omega t)\, dt \tag{5.48}$$

Because $P(\omega) = P(-\omega)$, the quantity $P(\omega)$ is an even function. Similarly, $Q(\omega)$ is an odd function with $Q(\omega) = -Q(-\omega)$. This yields,

$$F(-\omega) = P(-\omega) + jQ(-\omega) = P(\omega) - jQ(\omega) = F^*(\omega) \tag{5.49}$$

where $F^*(\omega)$ is the complex conjugate of $F(\omega)$.

It should be noted that the integral of an even function $f_e(t)$ times an odd function $f_o(t)$ over a symmetric distance is zero, resulting in

$$\int_{-\infty}^{+\infty} f_e(t) \sin(\omega t)\, dt = 0, \quad \int_{-\infty}^{+\infty} f_e(t) \cos(\omega t)\, dt = 2 \int_0^{+\infty} f_e(t) \cos(\omega t)\, dt \tag{5.50a}$$

$$\int_{-\infty}^{+\infty} f_o(t) \cos(\omega t)\, dt = 0, \quad \int_{-\infty}^{+\infty} f_o(t) \sin(\omega t)\, dt = 2 \int_0^{+\infty} f_o(t) \sin(\omega t)\, dt \tag{5.50b}$$

We can easily show that:

1. If $f(t)$ is a real-valued function,

$$f(t) = \frac{1}{2\pi} \int_{-\infty}^{+\infty} [P(\omega) \cos(\omega t) - Q(\omega) \sin(\omega t)]\, dt \tag{5.51}$$

2. If $f(t)$ is an even function,

$$Q(\omega) = 0 \tag{5.52a}$$

$$P(\omega) = 2 \int_0^{+\infty} f(t) \cos(\omega t)\, dt \tag{5.52b}$$

and

$$f(t) = \frac{1}{\pi} \int_0^{+\infty} [P(\omega) \cos(\omega t)]\, dt \tag{5.52c}$$

3. If $f(t)$ is an odd function,

$$P(\omega) = 0 \tag{5.53a}$$

$$Q(\omega) = -2 \int_0^{+\infty} f(t) \sin(\omega t)\, dt \tag{5.53b}$$

and

$$f(t) = -\frac{1}{\pi} \int_0^{+\infty} [Q(\omega) \sin(\omega t)]\, dt \tag{5.53c}$$

The Fourier transform of $f(t)$ exists if:

1. $f(t)$ is bounded with at most a finite number of maxima and minima and a finite number of discontinuities in any finite time interval (Dirichlet conditions), and
2. $f(t)$ is absolutely integrable; i.e.,

$$\int_{-\infty}^{+\infty} |f(t)|\, dt < \infty \tag{5.54}$$

Although the above two conditions are sufficient for $F(\omega)$ to exist, they are not necessary. Many functions of practical interest have transforms although they do not satisfy them. Two examples are the unit-impulse function $\delta(t)$ with a transform of $F(\omega) = 1$ and the unit-step function $u(t)$ with the transform of $F(\omega) = \pi\delta(\omega) + 1/j\omega$.

Example 5.6: Calculate the Fourier transform of the exponential time decay function

$$f(t) = \begin{cases} e^{-\alpha t} & \text{for } t \geq 0, \alpha > 0 \\ 0 & \text{otherwise} \end{cases}$$

Solution:
Using Equation (5.45),

$$F(\omega) = \mathcal{F}^{+1}[f(t)] = \int_{-\infty}^{+\infty} f(t)e^{-j\omega t}\, dt = \int_{0}^{+\infty} e^{-\alpha t}e^{-j\omega t}\, dt = \int_{0}^{+\infty} e^{-(\alpha+j\omega)t}\, dt$$

$$= -\frac{1}{\alpha + j\omega}e^{-(\alpha+j\omega)t}\Big|_{0}^{\infty} = \frac{1}{\alpha + j\omega} = \frac{\alpha - j\omega}{\alpha^2 + \omega^2}$$

Also from Equation (5.46),

$$F(\omega) = P(\omega) + jQ(\omega) = \frac{\alpha - j\omega}{\alpha^2 + \omega^2} = \frac{\alpha}{\alpha^2 + \omega^2} + j\frac{-\omega}{\alpha^2 + \omega^2}$$

yielding,

$$P(\omega) = \frac{\alpha}{\alpha^2 + \omega^2} \quad \text{and} \quad Q(\omega) = \frac{-\omega}{\alpha^2 + \omega^2}$$

The Fourier transform is shown in Figure 5.9.

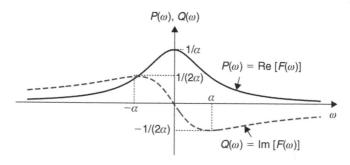

Figure 5.9 Fourier transform of the exponential decay function.

Example 5.7: Calculate the Fourier transform of the rectangular pulse

$$f(t) = \begin{cases} \alpha & \text{for } -b \leq t \leq b \\ 0 & \text{otherwise} \end{cases}$$

Solution:
Because $f(t)$ is an even function, its transform is real; i.e.

$$F(\omega) = P(\omega) = 2 \int_{0}^{+\infty} f(t) \cos(\omega t) dt = 2 \int_{0}^{+\infty} \alpha \cos(\omega t)\, dt = \frac{2\alpha}{\omega} \sin(\omega t)\Big|_{0}^{b}$$

$$= \frac{2\alpha}{\omega} \sin(\omega b)$$

with

$$\lim_{\omega \to 0} F(\omega) = \lim_{\omega \to 0} \frac{2\alpha}{\omega} \sin(\omega b) = 2\alpha b$$

Figure 5.10 shows $F(\omega)$.

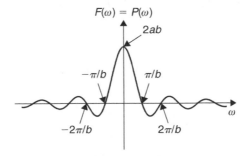

Figure 5.10 Fourier transform of the rectangular pulse function.

Note that $P(\omega) = 0$ if $\sin(\omega b) = 0$. This yields $\omega b = n\pi$ or

$$\omega = \frac{n\pi}{b} \quad n = \ldots, -2, -1, 1, 2, \ldots$$

Example 5.8: Calculate the Fourier transform of the following linear wave which symmetrically spans the vertical axis and is of finite duration

$$f(t) = \begin{cases} \alpha \cos(\omega_0 t) & \text{for } -b \leq t \leq b \\ 0 & \text{otherwise} \end{cases}$$

The function $f(t)$ is shown in Figure 5.11.

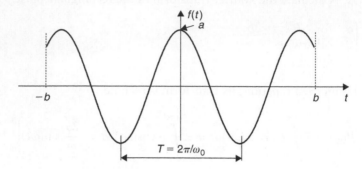

Figure 5.11   Linear wave for example 5.8.

Solution:
In this example, ω_0 is a constant frequency and should not be confused with ω. Because $f(t)$ is an even function, its Fourier transform is

$$F(\omega) = P(\omega) = 2 \int_0^{+\infty} f(t) \cos(\omega t)\, dt = 2\alpha \int_0^b \cos(\omega_0 t) \cos(\omega t)\, dt$$

$$= \alpha \int_0^b \{\cos[(\omega - \omega_0)t] + \cos[(\omega + \omega_0)t]\}\, dt$$

$$= \frac{\alpha \sin[(\omega - \omega_0)t]}{\omega - \omega_0}\Big|_0^b + \frac{\alpha \sin[(\omega + \omega_0)t]}{\omega + \omega_0}\Big|_0^b$$

$$= \frac{\alpha \sin[(\omega - \omega_0)b]}{\omega - \omega_0} + \frac{\alpha \sin[(\omega + \omega_0)b]}{\omega + \omega_0}$$

Figure 5.12 shows $F(\omega)$.

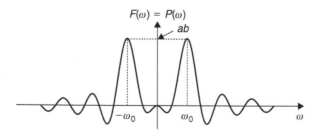

Figure 5.12   Fourier transform for function of Figure 5.11.

5.3.3 Delta (Dirac–Delta) Function

The delta function $\delta(t)$ is a rectangular pulse of infinitesimal width and infinite height so that its area is equal to one. It can be therefore, represented by the pulse of Example 5.7 if $b \to 0$ and $a = \frac{1}{2b} \to \infty$. Thus, the Fourier transform of $\delta(t)$ is

$$F(\omega) = \mathcal{F}^{+1}[\delta(t)] = \int_{-\infty}^{+\infty} \delta(t)e^{-j\omega t}\, dt = \lim_{b \to 0} \frac{2\frac{1}{2b}\sin(\omega b)}{\omega} = \lim_{b \to 0} \frac{\sin(\omega b)}{\omega b} = 1$$

or

$$\delta(t) \leftrightarrow 1 \tag{5.55}$$

Because

$$\int_{-\infty}^{+\infty} g(t)\delta(t - t_0)dt = g(t_0) \tag{5.56}$$

the Fourier transform of $\delta(t - t_0)$ where t_0 is a constant is

$$\mathcal{F}^{+1}[\delta(t - t_0)] = \int_{-\infty}^{+\infty} \delta(t - t_0)e^{-j\omega t}\, dt = e^{-j\omega t_0}$$

or

$$\delta(t - t_0) \leftrightarrow e^{-j\omega t_0} \tag{5.57}$$

Also,

$$\mathcal{F}^{+1}[\delta(t - t_0) + \delta(t + t_0)] = e^{-j\omega t_0} + e^{+j\omega t_0} = 2\cos(\omega t_0)$$

or

$$\delta(t - t_0) + \delta(t + t_0) \leftrightarrow 2\cos(\omega t_0) \tag{5.58}$$

and

$$\mathcal{F}^{-1}[\delta(\omega - \omega_0)] = \frac{1}{2\pi}\int_{-\infty}^{+\infty} \delta(\omega - \omega_0)e^{j\omega t}\, dt = \frac{1}{2\pi}e^{j\omega_0 t}$$

or

$$\delta(\omega - \omega_0) \leftrightarrow \frac{1}{2\pi}e^{j\omega_0 t} \tag{5.59}$$

yielding

$$\delta(\omega) \leftrightarrow \frac{1}{2\pi}$$

From Equation (5.59) we have

$$\mathcal{F}^{+1}\left[\frac{1}{2\pi}e^{j\omega_0 t}\right] = \delta(\omega - \omega_0)$$

or

$$\mathcal{F}^{+1}[e^{j\omega_0 t}] = 2\pi\delta(\omega - \omega_0) \tag{5.60a}$$

This result can be generalized as

$$\mathcal{F}^{+1}[ce^{\pm j\omega_0 t}] = 2\pi c\delta(\omega \mp \omega_0) \tag{5.60b}$$

where c is a constant.

Example 5.9:   Calculate the Fourier transform of

$$f(t) = \alpha\cos(\omega_0 t) \quad \text{for } -\infty < t < +\infty$$

and

$$f(t) = \alpha\sin(\omega_0 t) \quad \text{for } -\infty < t < +\infty$$

Solution:
From Equation (5.45),

$$F(\omega) = \mathcal{F}^{+1}[\alpha\cos(\omega_0 t)] = \int_{-\infty}^{+\infty} \alpha\cos(\omega_0 t)e^{-j\omega t}\,dt = \int_{-\infty}^{+\infty} \left(\frac{\alpha}{2}e^{j\omega_0 t} + \frac{\alpha}{2}e^{-j\omega_0 t}\right)e^{-j\omega t}\,dt$$

$$= \mathcal{F}^{+1}\left[\frac{\alpha}{2}e^{j\omega_0 t} + \frac{\alpha}{2}e^{-j\omega_0 t}\right]$$

Using Equation (5.59), the above transform becomes

$$\mathcal{F}^{+1}[\alpha\cos(\omega_0 t)] = \pi\alpha[\delta(\omega - \omega_0) + \delta(\omega + \omega_0)]$$

Similarly,

$$\mathcal{F}^{+1}[\alpha\sin s(\omega_0 t)] = \mathcal{F}^{+1}\left[-\frac{j\alpha}{2}e^{j\omega_0 t} + \frac{j\alpha}{2}e^{-j\omega_0 t}\right]$$

$$= j\pi\alpha[\delta(\omega + \omega_0) - \delta(\omega - \omega_0)]$$

Figures 5.13a and 5.13b show the Fourier transforms of $f(t) = \alpha\cos(\omega_0 t)$ and $f(t) = \alpha\sin(\omega_0 t)$, respectively.

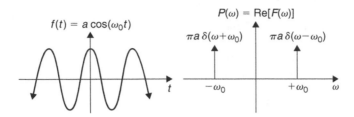

Figure 5.13a The function $a \cos \omega_0 t$ and its Fourier transform.

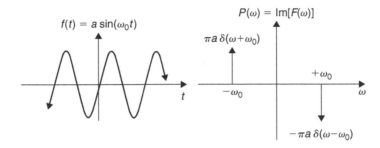

Figure 5.13b The function $a \sin \omega_0 t$ and its Fourier transform.

Comparison of the above transforms with the transform of the truncated cosine wave of Example 5.8 shows that the wave truncation results in the presence of all frequencies. However, most of the energy is concentrated around $-\omega_0$ and $+\omega_0$.

5.3.4 Properties of Fourier Transform

A number of very useful properties of the Fourier transform are presented in this section. We do not present the proof of these properties although some proofs are trivial. All proofs can be easily found in the literature. The Fourier transform of $f(t)$ is denoted by $F(\omega)$ and $F_n(\omega)$ is the transform of $f_n(\omega)$ with $n = 1, 2, 3, \ldots, N$.

1. *Linearity or Superposition*
   If a_n are constants (real or complex),

$$f(t) = \sum_{n=1}^{N} a_n f_n(t) \leftrightarrow \sum_{n=1}^{N} a_n F_n(\omega) = F(\omega) \tag{5.61}$$

2. *Scaling*
   If α is a real constant,

$$f(\alpha t) \leftrightarrow \frac{1}{|\alpha|} F\left(\frac{\omega}{\alpha}\right) \tag{5.62}$$

Note that if $\alpha = -1$,

$$f(-t) \leftrightarrow F(-\omega) = F^*(\omega)$$

3.   *Time and Frequency Shifting*
     If ω_0 and t_0 are real constants,

$$f(t - t_0) \leftrightarrow F(\omega)e^{-j\omega t_0} \tag{5.63a}$$

and

$$f(t - t_0)e^{-j\omega_0 t} \leftrightarrow F(\omega - \omega_0) \tag{5.63b}$$

4.   *Duality or Symmetry*

$$F(t) = 2\pi f(-\omega) \tag{5.64}$$

5.   *Modulation*

$$f(t)\cos(\omega_0 t) \leftrightarrow \frac{1}{2}[F(\omega - \omega_0) + F(\omega + \omega_0)] \tag{5.65a}$$

and

$$f(t)\sin(\omega_0 t) \leftrightarrow \frac{1}{2j}[F(\omega - \omega_0) - F(\omega + \omega_0)] \tag{5.65b}$$

where ω_0 is a real constant.

6.   *Differentiation*

$$\frac{d^n f(t)}{dt^n} = (j\omega)^n F(\omega) \tag{5.66a}$$

and

$$(-jt)^n f(t) = \frac{d^n F(\omega)}{d\omega^n} \tag{5.66b}$$

7.   *Integration*

$$\int_{-\infty}^{t} f(\tau)\,d\tau \leftrightarrow nF(0)\delta(\omega) + \frac{F(\omega)}{j\omega} \tag{5.67a}$$

and

$$\pi f(0)\delta(t) - \frac{f(t)}{jt} \leftrightarrow \int_{-\infty}^{+\infty} F(\xi)\,d\xi \tag{5.67b}$$

8. *Conjugation*

$$f^*(t) \leftrightarrow F^*(-\omega) \tag{5.68a}$$

and

$$f^*(-t) \leftrightarrow F^*(\omega) \tag{5.68b}$$

9. *Convolution*

$$f(t) = \int_{-\infty}^{+\infty} f_1(\tau)f_2(t - \tau)\, d\tau \leftrightarrow F_1(\omega)F_2(\omega) = F(\omega) \tag{5.69a}$$

and

$$f(t) = f_1(t)f_2(t) \leftrightarrow \frac{1}{2\pi} \int_{-\infty}^{+\infty} F_1(\xi)F_2(\omega - \xi)\, d\xi = F(\omega) \tag{5.69b}$$

10. *Correlation*

$$f(t) = \int_{-\infty}^{+\infty} f_1^*(\tau)f_2(t + \tau)\, d\tau \leftrightarrow F_1^*(\omega)F_2(\omega) = F(\omega) \tag{5.70a}$$

and

$$f(t) = f_1^*(t)f_2(t) \leftrightarrow \frac{1}{2\pi} \int_{-\infty}^{+\infty} F_1^*(\xi)F_2(\omega + \xi)\, d\xi = F(\omega) \tag{5.70b}$$

11. *Parseval's Theorem*

$$\int_{-\infty}^{+\infty} f_1^*(\tau)f_2(t + \tau)\, d\tau \leftrightarrow F_1^*(\omega)F_2(\omega) \tag{5.71}$$

or alternatively if $f_1(t) = f_2(t) = f(t)$,

$$\int_{-\infty}^{+\infty} [f_2(t)]^2\, dt \leftrightarrow \frac{1}{2\pi} \int_{-\infty}^{+\infty} |F_2(\omega)|^2\, d\omega \tag{5.72}$$

A table of useful Fourier transforms can be found in Peebles (1987).

5.4 Spectral Analysis

In Section 5.2, we have characterized a random process in the time domain using the autocorrelation, cross-correlation, and covariance functions. However, both time domain and frequency domain analysis methods exist for linear systems and deterministic signals. This section presents concepts to characterize a random process in the frequency domain.

The time history $x(t)$ of a sample function of a random process is usually not periodic and therefore, it can not be represented by a Fourier series. Also, for a stationary process, $x(t)$ extends from $t = -\infty$ to $t = +\infty$ and the condition of Equation (5.54) is not satisfied. Therefore, the Fourier transform of $x(t)$ may not exist. This difficulty is however, overcome by Fourier transforming the autocorrelation function $R_{XX}(\tau)$ of the process instead of its sample functions. This results in the *power spectral density* (PSD) function (or power density or power density spectrum) which provides information about the frequency content of the process indirectly, and indicates therefore, the distribution of energy of a random process over the frequency spectrum. If we properly adjust the random process so that its mean value is zero, then $\lim_{\tau \to \infty} R_{XX}(\tau) = 0$ and the condition $\int_{-\infty}^{+\infty} |R_{XX}(\tau)| \, d\tau < \infty$ is satisfied indicating that the Fourier transform of $R_{XX}(\tau)$ exists.

5.4.1   Power Spectral Density and Its Properties

In Section 5.3.2, we mentioned that the Fourier transform

$$X(\omega) = \int_{-\infty}^{+\infty} x(t) e^{-j\omega t} \, dt$$

of a deterministic signal $x(t)$ provides the spectral properties of $x(t)$. The function $X(\omega)$ is the *spectral density* of $x(t)$. Because the spectral density $X(\omega)$ of a sample function $x(t)$ may not exist, we cannot in general, obtain a spectral description of a random process by simply using the spectral density. However, the *power* of a random process as a function of frequency does exist and is called the *power spectral density* (PSD).

Power Spectral Density
Let $x_T(t)$ denote a truncated sample function of the random process $X(t)$ between $-T$ and T,

$$x_T(t) = \begin{cases} x(t) & \text{for } -T < t < T \\ 0 & \text{otherwise} \end{cases} \tag{5.73}$$

If T is finite, the Fourier transform

$$X_T(\omega) = \int_{-T}^{T} x_T(t) e^{-j\omega t} \, dt = \int_{-T}^{T} x(t) e^{-j\omega t} \, dt$$

of $x_T(t)$ exists because $\int_{-T}^{T} |x_T(t)| \, dt < \infty$.

A measure of the energy in $x(t)$ in the interval $(-T, T)$ is provided by

$$E(T) = \int_{-T}^{T} x^2(t) \, dt = \int_{-T}^{T} x_T^2(t) \, dt \tag{5.74}$$

which according to Parserval's theorem of Equation (5.72), is equal to

$$E(T) = \frac{1}{2\pi} \int_{-T}^{T} |X_T(\omega)|^2 \, d\omega \tag{5.75}$$

The average power $P(T)$ in the interval $(-T, T)$ is obtained by dividing the energy $E(T)$ by the time $2T$,

$$P(T) = \frac{E(T)}{2T} = \frac{1}{2\pi} \int_{-\infty}^{+\infty} \frac{|X_T(\omega)|^2}{2T} \, d\omega \tag{5.76}$$

where $|X_T(\omega)|^2/(2T)$ is the power spectral density of the truncated sample function of $x(t)$.

Because $P(T)$ is the average power of the truncated sample function, the average power of the entire sample function is obtained as $T \to \infty$. Also, because $P(T)$ is the average power of one sample function, each sample function of the process has its own average power. Therefore, $P(T)$ is a random variable. The expected value of this random variable provides a measure of the average power P_{XX} of the random process,

$$P_{XX} = \lim_{T \to \infty} \frac{1}{2T} \int_{-T}^{T} E[X^2(t)] \, dt = \frac{1}{2\pi} \int_{-\infty}^{+\infty} \lim_{T \to \infty} \frac{E[|X_T(\omega)|^2]}{2T} \, d\omega \tag{5.77}$$

Equation (5.77) indicates that the average power P_{XX} is given by the time average of the second moment of $X(t)$, and that it can be obtained by integration in the frequency domain. Note that if the random process is at least wide-sense stationary,

$$E[X^2(t)] = \mu_X^2 = \text{constant}.$$

If we define the *power spectral density* of the process as

$$S_{XX}(\omega) = \lim_{T \to \infty} \frac{E[|X_T(\omega)|^2]}{2T} \tag{5.78}$$

the average power P_{XX} is given by

$$P_{XX} = \frac{1}{2\pi} \int_{-\infty}^{+\infty} S_{XX}(\omega) \, d\omega \tag{5.79}$$

Some examples are presented next in order to illustrate the definition of the power spectral density and the derivation of the average power P_{XX}.

Example 5.10: If $x(t)$ is a real-valued function and $x(t) \leftrightarrow X(\omega)$ show the following form of Parseval's Theorem

$$\int\limits_{-\infty}^{+\infty} x^2(t)\, dt = \frac{1}{2\pi} \int\limits_{-\infty}^{+\infty} |X(\omega)|^2\, d\omega$$

Solution:
The proof is as follows:

$$\int\limits_{-\infty}^{+\infty} x^2(t)\, dt = \int\limits_{-\infty}^{+\infty} x(t)x(t)\, dt = \int\limits_{-\infty}^{+\infty} x(t)\left[\frac{1}{2\pi}\int\limits_{-\infty}^{+\infty} X(\omega)e^{j\omega t}\, d\omega\right] dt$$

$$= \frac{1}{2\pi}\int\limits_{-\infty}^{+\infty}\int\limits_{-\infty}^{+\infty} X(\omega)x(t)e^{j\omega t}\, d\omega\, dt = \frac{1}{2\pi}\int\limits_{-\infty}^{+\infty} X(\omega)\left[\underbrace{\int\limits_{-\infty}^{+\infty} x(\omega)e^{j\omega t}\, dt}_{X(-\omega)=X^*(\omega)}\right] d\omega$$

$$= \frac{1}{2\pi}\int\limits_{-\infty}^{+\infty} X(\omega)X^*(\omega)\, d\omega = \frac{1}{2\pi}\int\limits_{-\infty}^{+\infty} |X(\omega)|^2\, d\omega$$

Example 5.11: For the random process

$$X(t) = \cos(\omega_0 t + \Theta)$$

where ω_0 is a real constant and Θ is a random variable uniformly distributed between 0 and $\pi/2$, find the average power P_{XX}.

Solution:
The mean-squared value of $X(t)$ is

$$E[X^2(t)] = E[\cos^2(\omega_0 t + \theta)] = \frac{1}{2}E[1 + \cos(2\omega_0 t + 2\theta)]$$

$$= \frac{1}{2} + \int\limits_{0}^{\pi/2} \frac{1}{\pi}\cos(2\omega_0 t + 2\theta)\, d\theta = \frac{1}{2} - \frac{1}{\pi}\sin(2\omega_0 t)$$

indicating that the process is not even wide-sense stationary. According to Equation (5.77), the average power is

$$P_{XX} = \lim_{T \to \infty} \frac{1}{2T} \int_{-T}^{T} \left[\frac{1}{2} - \frac{1}{\pi} \sin(2\omega_0 t) \right] dt = \frac{1}{2}$$

Example 5.12:   For the random process of example 5.11, calculate the power spectral density $S_{XX}(\omega)$ and the average power P_{XX} using Equation (5.79).

Solution:
The spectrum of a truncated sample function is

$$X_T(\omega) = \int_{-T}^{+T} \cos(\omega_0 t + \theta) e^{-j\omega t} dt$$

$$= \frac{1}{2} e^{j\theta} \int_{-T}^{+T} e^{j(\omega_0 - \omega)t} dt + \frac{1}{2} e^{-j\theta} \int_{-T}^{+T} e^{j(\omega_0 + \omega)t} dt$$

$$= T e^{j\theta} \frac{\sin[(\omega - \omega_0)T]}{(\omega - \omega_0)T} + T e^{-j\theta} \frac{\sin[(\omega + \omega_0)T]}{(\omega + \omega_0)T}$$

which after some algebra yields,

$$\frac{E[|X_T(\omega)|^2]}{2T} = \frac{\pi}{2} \left\{ \frac{T}{\pi} \frac{\sin^2[(\omega - \omega_0)T]}{(\omega - \omega_0)^2 T^2} + \frac{T}{\pi} \frac{\sin^2[(\omega + \omega_0)T]}{(\omega + \omega_0)^2 T^2} \right\}$$

Considering that

$$\lim_{T \to \infty} \frac{T}{\pi} \left[\frac{\sin(cT)}{cT} \right]^2 = \delta(c)$$

where c is a constant, the power spectral density is

$$S_{XX}(\omega) = \frac{\pi}{2} [\delta(\omega - \omega_0) + \delta(\omega + \omega_0)]$$

and the average power is

$$P_{XX} = \frac{1}{2\pi} \int_{-\infty}^{+\infty} \frac{\pi}{2} [\delta(\omega - \omega_0) + \delta(\omega + \omega_0)] d\omega = \frac{1}{2}$$

Properties of Power Spectral Density

The power spectral density has the following properties:

1. $S_{XX}(\omega)$ is real with $S_{XX}(\omega) \geq 0$ $\hspace{4cm}$ (5.80)
2. $S_{XX}(\omega) = S_{XX}(-\omega)$; $\hspace{6cm}$ (5.81)

   The power spectral density is an even function. This is true if $X(t)$ is real.

3. $$\frac{1}{2\pi} \int_{-\infty}^{+\infty} S_{XX}(\omega)\, d\omega = AVR\{E[X^2(t)]\} \hspace{3cm} (5.82)$$

   where AVR denotes time average; i.e. $AVR(\bullet) = \lim_{T\to\infty} \frac{1}{2T} \int_{-T}^{T} (\bullet) dt$.

4. The power spectral density of the derivative $\dot{X}(t) = dX(t)/dt$ is

$$S_{\dot{X}\dot{X}}(\omega) = \omega^2 S_{XX}(\omega) \hspace{4cm} (5.83)$$

5. $\frac{1}{2\pi}\int_{-\infty}^{+\infty} S_{XX}(\omega)e^{j\omega t}\, d\omega = AVR[R_{XX}(t,t+\tau)]$ and $S_{XX}(\omega) = \int_{-\infty}^{+\infty} AVR[R_{XX}(t,t+\tau)]e^{-j\omega\tau}d\tau$.

   Therefore,

$$AVR[R_{XX}(t,t+\tau)] \leftrightarrow S_{XX}(\omega) \hspace{3cm} (5.84)$$

If $X(t)$ is at least wide-sense stationary, $AVR[R_{XX}(t,t+\tau)] = R_{XX}(\tau)$, and the above Fourier pair of Equation (5.84) yields

$$S_{XX}(\omega) = \int_{-\infty}^{+\infty} R_{XX}(\tau)e^{-j\omega\tau}\, d\tau \hspace{3cm} (5.85a)$$

and

$$R_{XX}(\tau) = \frac{1}{2\pi} \int_{-\infty}^{+\infty} S_{XX}(\omega)e^{j\omega\tau}\, d\omega \hspace{3cm} (5.85b)$$

or

$$R_{XX}(\tau) \leftrightarrow S_{XX}(\omega) \hspace{4cm} (5.86)$$

Equation (5.86) expresses the so-called *Wiener-Khintchine* theorem. A proof is provided in Section 5.4.4. Some texts divide the right hand side of Equation (5.85a) by 2π instead of the right hand side of Equation (5.85b).

The Wiener-Khintchine Theorem provides the basic relationship between the time domain description (correlation functions) and the frequency domain (power spectral density) description of random processes.

Because both $R_{XX}(\tau)$ and $S_{XX}(\omega)$ are real and even functions, the Wiener-Khintchine relations can be also expressed as

$$S_{XX}(\omega) = 2 \int_{0}^{+\infty} R_{XX}(\tau)\cos(\omega\tau)\, d\tau \hspace{3cm} (5.87)$$

and

$$R_{XX}(\tau) = \frac{1}{\pi} \int\limits_{0}^{+\infty} S_{XX}(\omega) \cos(\omega\tau) \, d\omega \qquad (5.88)$$

Example 5.13:   The elevation of a rough road, $X(s)$, is modeled as a stationary Gaussian random process with autocorrelation function $R_{XX}(d) = \sigma^2 e^{-\frac{|d|}{0.5}}$, where d is the distance between two locations, s and $s+d$, in meters (m), and σ is the standard deviation of the road elevation. The standard deviation σ is equal to 0.02 m. A car travels on the road with constant speed $V = 16$ m/sec. Find the PSD (power spectral density) the wheel of the car, assuming that it remains always in contact with the ground.

Solution:
The distance that a wheel travels is $s = Vt$. Therefore, the autocorrelation of the elevation of the wheel at two time instances, $t + \tau$ and t, is equal to the autocorrelation function of the road elevation at two locations separated by distance $V\tau$:

$$R_{XX}(\tau) = \sigma^2 e^{-\frac{|V\tau|}{0.5}} = \sigma^2 e^{-32|\tau|}$$

The units of the autocorrelation are m^2.

The PSD is the Fourier transform of the autocorrelation function. Using Equation (5.86a) we find it equal to

$$S_{XX}(\omega) = \sigma^2 \frac{64}{\omega^2 + 32^2}$$

Figure 5.14 shows the PSD for positive values of frequency.

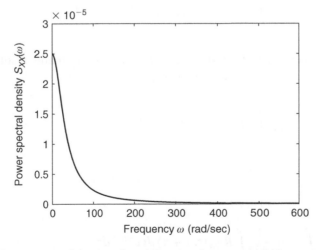

Figure 5.14   Power spectral density of road elevation for example 5.13.

Example 5.14:   Prove the relation in Equation (5.83).

Solution:
From Equation (5.78) we have

$$S_{\dot{X}\dot{X}}(\omega) = \lim_{T\to\infty} \frac{E[|\dot{X}_T(\omega)|^2]}{2T} = \lim_{T\to\infty} \frac{E[|j\omega X_T(\omega)|^2]}{2T}$$

$$= \omega^2 \lim_{T\to\infty} \frac{E[|X_T(\omega)|^2]}{2T} = \omega^2 S_{XX}(\omega)$$

Similarly, we can show that

$$S_{\ddot{X}\ddot{X}}(\omega) = \omega^4 S_{XX}(\omega)$$

5.4.2   Relationship Between Autocorrelation Function and Power Spectral Density

The *Wiener-Khintchine* theorem provides the relation between the autocorrelation function $R_{XX}(t)$ of process $X(t)$ and its power spectral density $S_{XX}(\omega)$. It states that the power spectral density and the time average of the autocorrelation function form the Fourier pair of Equation (5.84). This section provides a proof.

From the definition of $S_{XX}(\omega)$ in Equation (5.78), we have

$$S_{XX}(\omega) = \lim_{T\to\infty} E\left[\frac{1}{2T} \underbrace{\int_{-T}^{T} X(t_1)e^{j\omega t_1}\, dt_1}_{X_T(\omega)} \underbrace{\int_{-T}^{T} X(t_2)e^{j\omega t_2}\, dt_2}_{X_T^*(\omega)=X_T(\omega)} \right]$$

$$= \lim_{T\to\infty} \frac{1}{2T} \int_{-T}^{T}\int_{-T}^{T} \underbrace{E[X(t_1)X(t_2)]}_{R_{XX}(t_1,t_2)} e^{-j\omega(t_2-t_1)}\, dt_2 dt_1$$

If the variable changes $t = t_1$, and $\tau = t_2 - t_1 = t_2 - t$, are made, resulting in $dt = dt_1$ and $d\tau = dt_2$, the above equation becomes

$$S_{XX}(\omega) = \lim_{T\to\infty} \frac{1}{2T} \int_{-T}^{T}\int_{-T}^{T} R_{XX}(t_1, t_2)\, e^{-j\omega(t_2-t_1)}\, dt_2 dt_1$$

$$= \lim_{T\to\infty} \frac{1}{2T} \int_{-T-t}^{T-t}\int_{-T}^{T} R_{XX}(t, t+\tau)\, dt\, e^{-j\omega\tau}\, d\tau$$

$$= \int_{-\infty}^{+\infty} \lim_{T\to\infty} \frac{1}{2T} \int_{-T}^{T} R_{XX}(t, t+\tau)\, dt\, e^{-j\omega\tau}\, d\tau$$

indicating that

$$S_{XX}(\omega) = \int\limits_{-\infty}^{+\infty} AVR[R_{XX}(t, t+\tau)]\, e^{-j\omega\tau}\, d\tau$$

or

$$AVR[R_{XX}(t, t+\tau)] \leftrightarrow S_{XX}(\omega)$$

Example 5.15: Calculate the mean square value and autocorrelation function for a stationary random process $X(t)$ whose power spectral density is shown in Figure 5.15.

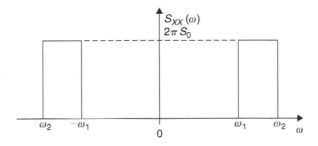

Figure 5.15 Power spectral density for the stationary process in example 5.15.

This power spectral density is characteristic of a *narrow band* random process.

Solution:
The mean square value and autocorrelation function are

$$\mu_X^2 = P_{XX} = \frac{1}{2\pi} \int\limits_{-\infty}^{+\infty} S_{XX}(\omega)\, d\omega = \frac{1}{2\pi}\, 2\pi\, S_0\, 2(\omega_2 - \omega_1) = 2S_0(\omega_2 - \omega_1)$$

and

$$R_{XX}(\tau) = \frac{1}{\pi} \int\limits_{0}^{+\infty} S_{XX}(\omega) \cos(\omega\tau)\, d\omega = \frac{1}{\pi} \int\limits_{0}^{+\infty} 2\pi S_0 \cos(\omega\tau) d\omega$$

$$= 2S_0 \left[\frac{1}{\tau} \sin(\omega\tau) \right]_{\omega_1}^{\omega_2} = \frac{2S_0}{\tau}[\sin(\omega_1\tau) - \sin(\omega_2\tau)]$$

$$= \frac{4S_0}{\tau} \cos\left[\frac{\omega_1 + \omega_2}{2}\tau \right] \sin\left[\frac{\omega_1 - \omega_2}{2}\tau \right]$$

Figure 5.16 shows the autocorrelation function.

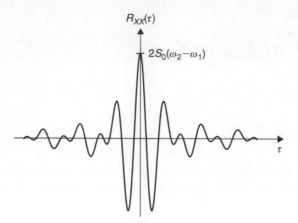

Figure 5.16   Autocorrelation function for the process in example 5.15.

Example 5.16:   Calculate the power spectral density of a random process with an autocorrelation function

$$R_{XX}(\tau) = \cos(\omega_0 \tau)$$

where ω_0 is a constant.

Solution:
Using Equation (5.85a), the power spectral density is

$$S_{XX}(\omega) = \int\limits_{-\infty}^{+\infty} R_{XX}(\tau)e^{-j\omega\tau}\,d\tau = \int\limits_{-\infty}^{+\infty} \cos(\omega_0\tau)e^{-j\omega\tau}\,d\tau$$

$$= \int\limits_{-\infty}^{+\infty} \frac{1}{2}(e^{j\omega_0\tau} + e^{-j\omega_0\tau})e^{-j\omega\tau}\,d\tau$$

$$= \frac{1}{2}\int\limits_{-\infty}^{+\infty} e^{j(\omega_0-\omega)\tau}\,d\tau + \frac{1}{2}\int\limits_{-\infty}^{+\infty} e^{-j(\omega_0+\omega)\tau}\,d\tau$$

If the two integrals in the above equation are calculated using Equation (5.60b), the power spectral density becomes

$$S_{XX}(\omega) = \pi[\delta(\omega - \omega_0) + \delta(\omega + \omega_0)]$$

Example 5.17: A wide-sense stationary process $X(t)$ has the following correlation function

$$R_{XX}(\tau) = e^{-|\tau|}$$

Calculate the power spectral density.

Solution:
Equation (5.85a) yields

$$S_{XX}(\omega) = \int\limits_{-\infty}^{+\infty} R_{XX}(\tau)e^{-j\omega\tau}\,d\tau = \int\limits_{-\infty}^{+\infty} e^{-|\tau|}e^{-j\omega\tau}\,d\tau$$

$$= \int\limits_{-\infty}^{0} e^{(1-j\omega)\tau}\,d\tau + \int\limits_{0}^{+\infty} e^{-(1+j\omega)\tau}\,d\tau = \frac{1}{1-j\omega} + \frac{1}{1+j\omega}$$

$$= \frac{2}{1+\omega^2}$$

5.4.3 Estimation of the power spectral density from data

This section explains how to estimate the power spectral density of a random process from records of observed values of this process. The section presents three methods:

1. Select an idealized parametric spectrum and estimate its parameters from the data.
2. First estimate the autocorrelation function and then calculate the power spectral density.
3. Directly estimate the power spectral density.

We consider stationary, ergodic processes with zero mean in this section.

5.4.3.1 Use an idealized parametric spectrum

Approximate power spectral density functions are used to characterize many random processes, such as the wave elevation and the wind speed at a certain location in an ocean. These functions involve few parameters that an analyst can estimate from data. The analyst should select this method when he is confident that a particular parametric spectrum represents the random process of interest.

For example, in design of ships and offshore platforms it is important to predict the applied wave and wind loads. Various approximate power spectral density functions are used to characterize the wave elevation at a location in the ocean. The Pierson-Moskowitz spectrum is the simplest

$$S_{XX}(\omega) = \pi\frac{0.0081 \cdot g^2}{\omega^5}e^{-0.74(\frac{g}{V\omega})^4} \quad \text{for } \omega \in [-\infty, +\infty] \tag{5.89}$$

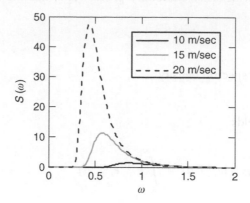

Figure 5.17 Wave spectra for different wind speeds.

This spectrum involves a single parameter, V, which is the wind speed at a height of 19.5 m above the still water level. Symbol g is the acceleration due to gravity, which is equal to 9.807 m/sec². Figure 5.17 shows the power spectral densities of waves for wind speeds of 10, 15 and 20 m/sec. It is observed that the wave energy is concentrated in a narrow frequency range, especially for high wind speeds. Both the bandwidth and the frequency of the peak decrease with the wind speed.

In order to determine this spectrum, an ocean engineer can measure the wind speed or the mean square of the waves. In the latter case, the engineer calculates the wind speed using Equation (5.90) below

$$V = \sqrt{\sigma_X / 0.00533} \tag{5.90}$$

where σ_X is the root mean square of the wave elevation. The modal frequency of the Pierson-Moskowitz spectrum is

$$\omega_m = 0.4 \sqrt{\frac{g}{H_s}} \tag{5.91}$$

where H_S is the significant wave height (the mean height of the 1/3 largest waves). This height is approximately equal to four times the root mean square of the wave elevation

$$H_S = 4\sigma_X \tag{5.92}$$

5.4.3.2 Estimate the spectrum from the autocorrelation function

The autocorrelation function is the mean value of the product of two values of the process at two points in time separated by time τ

$$R_{XX}(\tau) = E[X(t)X(x + \tau)]$$

Suppose that we have K records with N equally spaced values: $X^i(t_j)$, $i = 1, \ldots, K, j = 1, \ldots, N$. Then, the following is an unbiased estimator of the autocorrelation function

$$\overline{R}_{XX}(\tau) = \frac{1}{K \cdot N} \sum_{i=1}^{K} \sum_{j=1}^{N} X^i(t_j) \cdot X^i(t_j + \tau) \tag{5.93}$$

The power spectral density can be estimated by taking the Fourier transform of the right hand side of the above equation.

5.4.3.3   Directly estimate the spectrum from records

First, consider the case where we have one record of data $x(t)$, $0 \leq t \leq T$. By definition, (see Equation 5.78) the power spectral density is

$$S_{XX}(\omega) = \lim_{T \to \infty} \frac{E[|X(\omega)|^2]}{T}$$

where $X(\omega)$ is the Fourier transform of the process. Our intuition suggests the following estimator of the power spectral density

$$S_{XX}(\omega, T) = \frac{1}{T} \left| \int_{0}^{T} x(t) e^{-j\omega t} \, dt \right|^2 \tag{5.94}$$

This estimator is called periodogram. This is an unbiased estimator, that is, its mean value is equal to the true power spectral density function. However, the periodogram is a poor estimator, because its variance does not converge to zero with the record length T (Lutes and Sarkani, 2004, section 6.9). This means that, in an experiment, the estimate of the power spectral density will keep fluctuating as the record length increases.

The periodogram is a poor estimator because it fails to average fluctuations. The reason is that it tries to tune too sharply to the frequency of interest ω. We can correct this deficiency by averaging the value of the periodogram in Equation (5.94) over a spectral window with bandwidth δ that is centered on the frequency of interest. Then, the periodogram becomes

$$\overline{S}_{XX}(\omega, T) = \frac{1}{2\delta} \int_{\omega-\delta}^{\omega+\delta} S_{XX}(\omega, T) \, d\omega \tag{5.95}$$

The variance of the estimator in Equation (5.95) converges to zero with the record length. This averaging process introduces bias to the estimator, but the bias is not important if bandwidth 2δ is small.

When multiple records are available, we can estimate the power spectral density more accurately by averaging the estimates from each record

$$\bar{S}_{XX}(\omega, T) = \frac{1}{2K\delta T}\sum_{i=1}^{K}\left\{\int_{\omega-\delta}^{\omega+\delta}\left|\int_{0}^{T}x^i(t)e^{-j\omega t}\,dt\right|^2\,d\omega\right\} \tag{5.96}$$

where $x^i(t), i = 1, \ldots, k$ is the ith record.

Example 5.18: Estimation of the power spectral density of waves

An ocean engineer collected 30 independent records of the wave elevation under identical weather conditions. Each record has 200 values collected at a rate of one per second. Figure 5.18 shows the first 100 seconds of two of these records. We demonstrate how the engineer estimates the power spectral density using each of the above three methods.

Method 1:

The engineer believes that the wave spectrum is described by Equation (5.89). From the 30 records the engineer finds that the mean square of the wave elevation is 4.381 m². Based on this value, he finds that the wind speed V is equal to 19.817 m/sec using Equation (5.90). This information enables the engineer to completely define the wave spectrum. From Equations (5.91) and (5.92) the modal frequency is 0.433 rad/sec, which corresponds to a period of 14.514 sec.

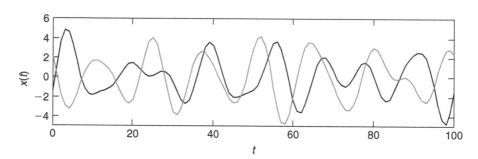

Figure 5.18 Two records of the wave elevation.

Method 2:

The engineer estimates the autocorrelation of the wave elevation from Equation (5.93). Figure 5.19 compares the autocorrelation function obtained from the 30 records using Equation (5.93) to the inverse Fourier transform of the spectrum found from method 1. The two autocorrelation functions are almost identical. Figure 5.20 compares the wave spectra corresponding to the autocorrelation functions in Figure 5.21. The two spectra agree well but the Pierson-Moskowitz spectrum is the most accurate. It is interesting that the autocorrelation function in Figure 5.19 looks more accurate than the power spectral density, even though they were both obtained from the same data.

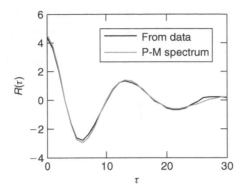

Figure 5.19 Autocorrelation function obtained from method 1 and method 2.

Method 3:

Finally, the wave spectrum was estimated directly from the Fourier transforms of the 30 records using Equation (5.96). The bandwidth δ was equal 1 sec in this equation. Figure 5.21 displays the results from a single record and from 30 records. The shape of the spectrum from 30 records is similar to that of the Pierson-Moskowitz spectrum. The spectrum from a single record is inaccurate. More data is needed in order to improve the estimates from the method 3.

Here are the main conclusions from this subsection.

- The analyst should try to impose some structure to the power spectral density function. If she believes that a particular family of spectra represents the data then she should use this family.

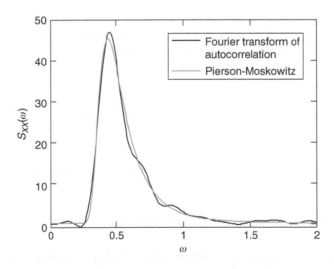

Figure 5.20 Power spectra estimated from method 1 and method 2.

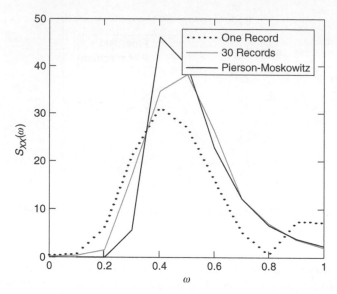

Figure 5.21 Power spectral density directly estimated from records using Equation (5.96).

- It is important to collect and use many independent records of observations in order to obtain accurate estimates of the power spectrum. Many independent short records could yield better results than a single long record.
- For the same random process, estimates of the autocorrelation function look more accurate than estimates of the power spectral density function.

5.4.4 Cross-Power Spectral Density and Its Properties

This section introduces the principle of cross-power spectral density by considering a random process $Z(t) = X(t) + Y(t)$ which is the sum of two other random processes $X(t)$ and $Y(t)$. The autocorrelation function of $Z(t)$ is

$$
\begin{aligned}
R_{ZZ}(t, t + \tau) &= E[Z(t)Z(t + \tau)] \\
&= E[[X(t) + Y(t)][X(t + \tau) + Y(t + \tau)]] \\
&= R_{XX}(t, t + \tau) + R_{YY}(t, t + \tau)
\end{aligned}
\tag{5.97}
$$

If the $X(t)$ and $Y(t)$ processes are at least wide-sense stationary, the Fourier transform of Equation (5.97) yields

$$
S_{ZZ}(\omega) = S_{XX}(\omega) + S_{YY}(\omega) + S_{XY}(\omega) + S_{YX}(\omega)
\tag{5.98}
$$

The last two terms of Equation (5.98) are the so-called *cross-power spectral densities* of the two processes $X(t)$ and $Y(t)$.

For two real random processes $X(t)$ and $Y(t)$, Fourier transforms of the the following truncated ensemble members

$$x_T(t) = \begin{cases} x(t) & \text{for } -T < t < T \\ 0 & \text{otherwise} \end{cases}$$

and

$$y_T(t) = \begin{cases} y(t) & \text{for } -T < t < T \\ 0 & \text{otherwise} \end{cases}$$

are denoted by $X_T(\omega)$ and $Y_T(\omega)$, respectively. The cross power $P_{XY}(T)$ of the two random processes in the time interval $(-T, T)$ is

$$P_{XY}(T) = \frac{1}{2T} \int_{-T}^{T} x_T(t) y_T(t) \, dt = \frac{1}{2T} \int_{-T}^{T} x(t) y(t) \, dt \tag{5.99}$$

or according to Parserval's theorem, the above equation becomes

$$P_{XY}(T) = \frac{1}{2T} \int_{-T}^{T} x(t) y(t) \, dt = \frac{1}{2\pi} \int_{-\infty}^{+\infty} \frac{X_T^*(\omega) Y_T(\omega)}{2T} \, d\omega$$

The cross power $P_{XY}(T)$ is a random variable because it corresponds to a particular ensemble member. The total average cross power P_{XY} is therefore, obtained by taking the expected value of the above expression and letting $T \to \infty$, as

$$P_{XY} = \lim_{T \to \infty} \frac{1}{2T} \int_{-T}^{T} \underbrace{E[X(t) Y(t)]}_{R_{XY}(t,t)} \, dt = \frac{1}{2\pi} \int_{-\infty}^{+\infty} \lim_{T \to \infty} \frac{E[X_T^*(\omega) Y_T(\omega)]}{2T} \, d\omega$$

If a *cross-power spectral density* is defined as

$$S_{XY}(\omega) = \lim_{T \to \infty} \frac{E[X_T^*(\omega) Y_T(\omega)]}{2T} \tag{5.100}$$

the total average cross power P_{XY} is equal to

$$P_{XY} = \frac{1}{2\pi} \int_{-\infty}^{+\infty} S_{XY}(\omega) \, d\omega \tag{5.101}$$

Similarly, we can define the cross-power spectral density

$$S_{YX}(\omega) = \lim_{T \to \infty} \frac{E[Y_T^*(\omega) X_T(\omega)]}{2T} \tag{5.102}$$

with the corresponding cross power of

$$P_{YX} = \frac{1}{2\pi} \int\limits_{-\infty}^{+\infty} S_{YX}(\omega)\, d\omega \qquad (5.103)$$

which is equal to P_{XY} (i.e. $P_{XY} = P_{YX}$).

The cross-power spectral densities $S_{XY}(\omega)$ and $S_{YX}(\omega)$ are complex functions of ω in contrast to the power spectral densities $S_{XX}(\omega)$ and $S_{YY}(\omega)$ which are real functions of ω. Physically, the total cross power $P_{XY} + P_{YX}$ represents the additional power the two random processes generate, because of their correlation, beyond their individual powers.

Properties of Cross-Power Spectral Density
This section provides the most important properties of the cross-power spectral densities of two random processes $X(t)$ and $Y(t)$. They are as follows:

1. $S_{XY}(\omega) = S_{YX}(-\omega) = S_{YX}^*(\omega)$ $\qquad\qquad$ (5.104)
2. $\mathrm{Re}[S_{XY}(\omega)]$ and $\mathrm{Re}[S_{YX}(\omega)]$ are even functions of ω while $\mathrm{Im}[S_{XY}(\omega)]$ and $\mathrm{Im}[S_{YX}(\omega)]$ are odd functions of ω $\qquad\qquad$ (5.105)

3. If $X(t)$ and $Y(t)$ are uncorrelated with constant mean values μ_X and μ_Y respectively,

$$S_{XY}(\omega) = S_{YX}(\omega) = 2\pi\mu_X\mu_Y\delta(\omega) \qquad (5.106)$$

4. $S_{XY}(\omega) = S_{YX}(\omega) = 0$ if $X(t)$ and $Y(t)$ are orthogonal $\qquad$ (5.107)
5. $AVR[R_{XY}(t, t + \tau)] \leftrightarrow S_{XY}(\omega)$ and $AVR[R_{YX}(t, t + \tau)] \leftrightarrow S_{YX}(\omega)$ $\qquad$ (5.108)

Equation (5.108) states that the time average of the cross-correlation function and the cross-power spectral density form a Fourier transform pair. We will prove this property in Section 5.4.4. If $X(t)$ and $Y(t)$ are jointly wide-sense stationary processes, the Fourier transform pair of Equation (5.108) yields

$$S_{XY}(\omega) = \int\limits_{-\infty}^{+\infty} R_{XY}(\tau)e^{-j\omega\tau}\, d\tau \qquad (5.109)$$

$$S_{YX}(\omega) = \int\limits_{-\infty}^{+\infty} R_{YX}(\tau)e^{-j\omega\tau}\, d\tau \qquad (5.110)$$

$$R_{XY}(\tau) = \frac{1}{2\pi} \int\limits_{-\infty}^{+\infty} S_{XY}(\omega)e^{j\omega\tau}\, d\omega \qquad (5.111)$$

and

$$R_{YX}(\tau) = \frac{1}{2\pi} \int\limits_{-\infty}^{+\infty} S_{YX}(\omega)e^{j\omega\tau}\, d\omega \tag{5.112}$$

Example 5.19: The cross-power spectral density for two jointly wide-sense stationary processes is

$$S_{XY}(\omega) = \begin{cases} 1 + j\omega/\omega_0 & \text{for } -\omega_0 < \omega < \omega_0 \\ 0 & \text{otherwise} \end{cases}$$

where $\omega_0 > 0$. Calculate the cross-correlation function.

Solution:
According to Equation (5.111),

$$R_{XY}(\tau) = \frac{1}{2\pi} \int\limits_{-\infty}^{+\infty} S_{XY}(\omega)e^{j\omega\tau}\, d\omega = \frac{1}{2\pi} \int\limits_{-\omega_0}^{+\omega_0} (1 + j\omega/\omega_0)e^{j\omega\tau}\, d\omega$$

$$= \frac{1}{2\pi} \int\limits_{-\omega_0}^{+\omega_0} e^{j\omega\tau}\, d\omega + j\frac{1}{2\pi\omega_0} \int\limits_{-\omega_0}^{+\omega_0} \omega e^{j\omega\tau}\, d\omega$$

$$= \frac{1}{2\pi} \left[\frac{1}{j\tau} e^{j\omega\tau} \right]_{-\omega_0}^{\omega_0} + j\frac{1}{2\pi\omega_0} \left\{ e^{j\omega\tau} \left[\frac{\omega}{j\tau} - \frac{1}{(j\tau)^2} \right]_{-\omega_0}^{\omega_0} \right\}$$

$$= \frac{1}{\pi\omega_0\tau^2} [(\omega_0\tau - 1)\sin(\omega\tau) + \omega_0\tau\cos(\omega\tau)]$$

Example 5.20: If the cross-power spectral densities are decomposed into real and imaginary parts as

$$S_{XY}(\omega) = R_{XY}(\omega) + jI_{XY}(\omega) \tag{5.113}$$

and

$$S_{YX}(\omega) = R_{YX}(\omega) + jI_{YX}(\omega) \tag{5.114}$$

show that

$$R_{XY}(\omega) = R_{YX}(-\omega) = R_{YX}(\omega) \tag{5.115}$$

and

$$I_{XY}(\omega) = I_{YX}(-\omega) = -I_{YX}(\omega) \tag{5.116}$$

Solution:
The real parts of the cross-power spectral densities are

$$R_{XY}(\omega) = \mathrm{Re}[S_{XY}(\omega)] = \frac{1}{2}[S_{XY}(\omega) + S^*_{XY}(\omega)]$$

$$= \frac{1}{2}\lim_{T\to\infty}\frac{1}{2T}E[X^*_T(\omega)Y_T(\omega) + X_T(\omega)Y^*_T(\omega)]$$

and

$$R_{YX}(\omega) = \mathrm{Re}[S_{YX}(\omega)] = \frac{1}{2}[S_{YX}(\omega) + S^*_{YX}(\omega)]$$

$$= \frac{1}{2}\lim_{T\to\infty}\frac{1}{2T}E[Y^*_T(\omega)X_T(\omega) + Y_T(\omega)X^*_T(\omega)]$$

indicating that

$$R_{XY}(\omega) = R_{YX}(\omega)$$

Similarly, the imaginary parts of the cross-power spectral densities are

$$I_{XY}(\omega) = \mathrm{Im}[S_{XY}(\omega)] = \frac{1}{2j}[S_{XY}(\omega) - S^*_{XY}(\omega)]$$

$$= \frac{1}{2j}\lim_{T\to\infty}\frac{1}{2T}E[X^*_T(\omega)Y_T(\omega) - X_T(\omega)Y^*_T(\omega)]$$

and

$$I_{YX}(\omega) = \mathrm{Im}[S_{YX}(\omega)] = \frac{1}{2j}[S_{YX}(\omega) - S^*_{YX}(\omega)]$$

$$= \frac{1}{2j}\lim_{T\to\infty}\frac{1}{2T}E[Y^*_T(\omega)X_T(\omega) - Y_T(\omega)X^*_T(\omega)]$$

indicating that

$$I_{XY}(\omega) = -I_{YX}(\omega)$$

Next, we will show that the relationship $R_{XY}(\omega) = R_{YX}(-\omega)$ between the cross-power spectral densities holds. We have,

$$R_{YX}(-\omega) = \frac{1}{2}\lim_{T\to\infty}\frac{1}{2T}E[Y^*_T(-\omega)X_T(-\omega) + Y_T(-\omega)X^*_T(-\omega)]$$

However, based on the Fourier transform $X_T(\omega) = \int_{-T}^{T} x(t)e^{-j\omega t}\,dt$ of a truncated sample function $x(t)$,

$$X^*_T(\omega) = X_T(-\omega)$$

and similarly,

$$Y_T^*(\omega) = Y_T(-\omega)$$

Because of the above two equations,

$$R_{YX}(-\omega) = \frac{1}{2}\lim_{T\to\infty}\frac{1}{2T}E[Y_T(\omega)X_T^*(\omega) + Y_T^*(\omega)X_X(\omega)] = R_{YX}(\omega)$$

and

$$I_{YX}(-\omega) = \frac{1}{2j}\lim_{T\to\infty}\frac{1}{2T}E[Y_T^*(-\omega)X_T(-\omega) - Y_T(-\omega)X_T^*(-\omega)]$$

$$= \frac{1}{2j}\lim_{T\to\infty}\frac{1}{2T}E[Y_Y(\omega)X_T^*(\omega) - Y_T^*(\omega)X_T(\omega)]$$

$$= I_{YX}(\omega)$$

Example 5.21:   Using the results from example 5.20, prove the relation of Equation (5.104); i.e.

$$S_{XY}(\omega) = S_{YX}(-\omega) = S_{YX}^*(\omega)$$

Solution:
The proof is as follows:

$$S_{XY}(\omega) = R_{XY}(\omega) + jI_{XY}(\omega) = R_{YX}(-\omega) + jI_{YX}(-\omega) = S_{YX}(-\omega)$$

and

$$S_{XY}(\omega) = R_{XY}(\omega) + jI_{XY}(\omega) = R_{YX}(\omega) - jI_{YX}(\omega) = S_{YX}^*(\omega)$$

5.4.5   Relationship Between Cross-correlation Function and Cross-Power Spectral Density

This section provides the proof of the following relation (see Equation 5.108)

$$S_{XY}(\omega) = \int_{-\infty}^{+\infty}\left\{\lim_{T\to\infty}\frac{1}{2T}\int_{-T}^{T}R_{XY}(t,\,t+\tau)\,dt\right\}e^{-j\omega\tau}\,d\tau$$

$$= \int_{-\infty}^{+\infty}AVR[R_{XY}(t,\,t+\tau)]e^{-j\omega\tau}\,d\tau$$

The product $X_T^*(\omega)Y_T(\omega)$ is calculated first, as

$$X_T^*(\omega)Y_T(\omega) = \int_{-T}^{T} x(t_1)e^{j\omega t_1}\,dt_1 \int_{-T}^{T} y(t_2)e^{-j\omega t_2}\,dt_2$$

$$= \int_{-T}^{T}\int_{-T}^{T} x(t_1)y(t_2)e^{-j\omega(t_2-t_1)}\,dt_1 dt_2$$

If the variable changes, $t = t_1$, and $\tau = t_2 - t_1 = t_2 - t$, are made, resulting in $dt = dt_1$ and $d\tau = dt_2$, the above product becomes

$$X_T^*(\omega)Y_T(\omega) = \int_{-T}^{T} x(t_1)e^{j\omega t_1}\,dt_1 \int_{-T}^{T} y(t_2)e^{-j\omega t_2}\,dt_2$$

$$= \int_{-T-t}^{T-t}\int_{-T}^{T} x(t)y(t+\tau)e^{-j\omega\tau}\,dt\,d\tau$$

From the definition of the cross-power spectral density $S_{XY}(\omega)$, we have,

$$S_{XY}(\omega) = \lim_{T\to\infty}\frac{E[X_T^*(\omega)Y_T(\omega)]}{2T}$$

$$= \lim_{T\to\infty}\int_{-T-t}^{T-t}\left[\frac{1}{2T}\int_{-T}^{T} R_{XY}(t,\,t+\tau)\,dt\right]e^{-j\omega\tau}\,d\tau$$

$$= \int_{-\infty}^{+\infty}\left[\underbrace{\lim_{T\to\infty}\frac{1}{2T}\int_{-T}^{T} R_{XY}(t,\,t+\tau)\,dt}_{AVR[R_{XY}(t,\,t+\tau)]}\right]e^{-j\omega\tau}\,d\tau$$

or,

$$S_{XY}(\omega) = \int_{-\infty}^{+\infty} AVR[R_{XY}(t,\,t+\tau)]e^{-j\omega\tau}\,d\tau$$

The inverse transform of the above equation yields

$$AVR[R_{XY}(t,\,t+\tau)] = \frac{1}{2\pi}\int_{-\infty}^{+\infty} S_{XY}(\omega)e^{j\omega\tau}\,d\omega$$

It should be noted that for jointly wide-sense stationary processes,

$$AVR[R_{XY}(t, \ t + \tau)] = R_{XY}(\tau)$$

In this case, the Fourier transform pair is simplified as

$$S_{XY}(\omega) = \int_{-\infty}^{+\infty} R_{XY}(\tau)e^{-j\omega\tau} \, d\tau \tag{5.117a}$$

and

$$R_{XY}(\tau) = \frac{1}{2\pi} \int_{-\infty}^{+\infty} S_{XY}(\omega)e^{j\omega\tau} \, d\omega \tag{5.117b}$$

or

$$R_{XY}(\tau) \leftrightarrow S_{XY}(\omega)$$

Example 5.22:   The cross-correlation function of two random processes $X(t)$ and $Y(t)$ is

$$R_{XY}(t, \ t + \tau) = \sin(\omega_0\tau) + \cos[\omega_0(t + \tau)]$$

where ω_0 is a constant. Calculate the cross-power spectral density $S_{XY}(\omega)$.

Solution:
The time average of R_{XY} is

$$AVR[R_{XY}(t, \ t + \tau)] = \lim_{T \to \infty} \frac{1}{2T} \int_{-T}^{T} R_{XY}(t, \ t + \tau) \, dt$$

$$= \lim_{T \to \infty} \frac{1}{2T} \int_{-T}^{T} \{\sin(\omega_0\tau) + \cos[\omega_0(t + \tau)]\} \, dt$$

$$= \sin(\omega_0\tau) + \lim_{T \to \infty} \frac{1}{2T} \int_{-T}^{T} \cos[\omega_0(t + \tau)] \, dt$$

$$= \sin(\omega_0\tau)$$

From Equation (5.108), we therefore, have

$$S_{XY}(\omega) = \int\limits_{-\infty}^{+\infty} \sin(\omega_0 \tau) e^{-j\omega\tau}\, d\tau$$

$$= -j\pi[\delta(\omega - \omega_0) - \delta(\omega + \omega_0)]$$

5.4.6 Coherence Function

The coherence function $\gamma(\omega)$, also denoted as $\gamma^2(\omega)$, is defined as

$$\gamma(\omega) = \frac{|S_{XY}(\omega)|^2}{S_{XX}(\omega)S_{YY}(\omega)} \tag{5.118}$$

It always varies between 0 and 1; i.e.

$$0 \le \gamma(\omega) \le 1 \tag{5.119}$$

If $\gamma(\omega) \to 0$, the random processes $X(t)$ and $Y(t)$ are linearly uncorrelated. If $\gamma(\omega) \to 1$, they are linearly correlated in the sense that $X(t)$ tends to increase (decrease) as $Y(t)$ increases (decreases). The coherence function $\gamma(\omega)$ is similar to the correlation coefficient

$$\rho = \frac{Cov(X, Y)}{\sqrt{Var(X)Var(Y)}}$$

5.5 Calculation of the Response

In the analysis of a dynamic system we need to construct a mathematical model that simulates the behavior of the actual system. It can be an assembly of discrete elements, such as masses, springs and dampers or continuous elements, such as beams and plates. By applying the laws of mechanics, we obtain a deterministic mathematical model of the system relating the responses to the excitations. The model is usually a set of differential equations relating the response (output) vector, $y(t)$, to the excitation (input) vector, $x(t)$.

A dynamic system can by thought as a multi-input, multi-output system, such as the one in Figure 5.22. Inputs and outputs are the elements of the excitation vector and response vector, respectively. The following equation relates the response to the excitation:

$$\mathbf{D}(\mathbf{p}, t)[\mathbf{y}(t)] = \mathbf{x}(t) \tag{5.120}$$

where $\mathbf{D}(\mathbf{p}, t)[\cdot]$ is a matrix of differential operators applied to the elements of the response. The size of $\mathbf{D}(\mathbf{p}, t)[\cdot]$ is $m \times n$, where m is the number of inputs and n is the number of outputs. $\mathbf{p}$ is a set of parameters associated with the system. We can solve Equation (5.120) analytically or numerically to compute the response.

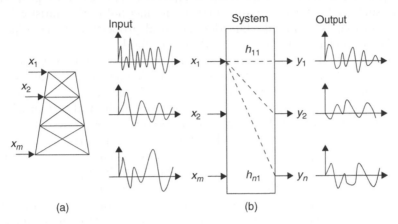

Input   System   Output

(a)   (b)

Figure 5.22 A structure subjected to excitations and its representation as a multi-input, multi-output system.

Systems can be categorized as linear or nonlinear depending on whether the operator $D(\mathbf{p}, t)$ is linear or nonlinear. The response of a linear, time invariant, single-input, single-output system can be calculated from the following convolution integral

$$y(t) = \int\limits_{-\infty}^{t} h(t - t_1)x(t_1)\, dt_1 \qquad (5.121)$$

where $h(t - t_1)$ is the *impulse response function*, or *Green's function*. This function represents the response of a system at time t due to a unit impulse that is applied at time t_1. The derivation of the convolution integral of Equation (5.121) is described in Section 5.5.1.

Dynamic systems can be also classified under the following categories:

- *Time invariant* systems whose parameters are deterministic. An example is a system that consists of a spring, a mass and a viscous damper where the values of the spring, mass and damping are deterministic constants. The response of a time invariant system at time t depends *only* on the elapsed time between the instant at which the excitation occurred and the current time t. If a time invariant system is also linear the response is provided by Equation (5.121).
- Time invariant systems whose parameters, $\mathbf{p}$, are random and time invariant. In this case parameters, $\mathbf{p}$, are random variables.
- *Time variant* systems whose parameters are deterministic and are functions of time. An example is a rocket whose mass decreases as fuel is burned.
- Time variant systems whose parameters are random. The parameters of these systems can be modeled as random processes.

Clearly, Equation (5.121) cannot be used directly to assess the safety of a dynamic system if the response is random because it is not enough to know the response due to a given output. We need to know the relationship between the statistical properties of the response (such as the mean and autocorrelation function) to the statistical properties of the excitation. We can calculate the statistical properties of the response using Monte-Carlo simulation methods or analytical methods.

In previous sections of this Chapter, we have modeled a random signal as a sample function of a random process. For that, either time domain analysis was used based on correlation functions or frequency domain analysis based on power spectral densities. This section describes the *response* characteristics of a physical system which is subjected to *random excitation*. The *output*, or response, $y(t)$ of the system is predicted if the input forcing, or excitation, $x(t)$ is known. These systems are common in engineering and science. Examples include vibration analysis, ship motions, electronic circuits, etc. Figure 5.22 provides a schematic representation of the input-output relationship of a general system with multiple inputs and multiple outputs.

We will restrict our discussion to single-input, single-output, linear systems for which the response $y(t)$ is related to the excitation $x(t)$ by a linear differential equation of the form

$$a_n \frac{d^n y}{dt^n} + a_{n-1} \frac{d^{n-1} y}{dt^{n-1}} + \cdots + a_0 y = b_m \frac{d^m x}{dt^m} + b_{m-1} \frac{d^{m-1} x}{dt^{m-1}} + \cdots + b_0 x \qquad (5.122)$$

where the coefficients a and b are *constant*, resulting in a *time invariant* linear system.

A linear system satisfies the *principle of linear superposition*

$$L[a x_1(t) + b x_2(t)] = a y_1(t) + b y_2(t) \qquad (5.123)$$

where a and b are constants and L is a linear operator. It states that if $y_1(t)$ is the response to excitation $x_1(t)$, and $y_2(t)$ is the response to excitation $x_2(t)$, the response to the linear combination $a x_1(t) + b x_2(t)$ of $x_1(t)$ and $x_2(t)$ is the same linear combination $a y_1(t) + b y_2(t)$ of $y_1(t)$ and $y_2(t)$. The principle of superposition greatly simplifies the analysis of a linear problem because it allows us to calculate the response to many simpler excitations individually and then superimpose the individual responses to obtain the system response to the original excitation.

Although most of physical phenomena are inherently nonlinear, the linearity assumption is reasonable for systems with small response such as vibratory displacement or small sea surface deformations. The response of a linear system occurs at the same frequency or frequencies with the input although the amplitude and phase of the response are different. For a nonlinear system however, the response *does not* necessarily occur at the input frequency or frequencies. It often contains frequencies which are not present in the input. In many cases, a nonlinear system can be linearized locally. It is recommended however, to calculate the coherence function in order to quantify if the system is linear.

5.5.1   *Characterization of a Linear System*

A linear system is characterized by determining the effect of a known input to the system output. The characterization can be performed either in the time domain or in the frequency domain. The *impulse response function* $h(t)$ characterizes the system in the time domain and the *frequency response function* (or *transfer function*) $H(\omega)$

characterizes the system in the frequency domain. If the governing differential equation is known, both these functions can be determined analytically. Otherwise, they can be estimated experimentally.

The next two sections describe how to calculate $h(t)$ and $H(\omega)$ for a *linear, time invariant* system, and also provide a physical interpretation of $h(t)$ and $H(\omega)$.

5.5.1.1 Impulse Response Function of a Linear, Time-invariant System

A single input-single output linear system is represented as

$$y(t) = L[x(t)] \tag{5.124}$$

where the linear operator L operates on the input $x(t)$ to produce the output $y(t)$. Using the definition of the unit impulse function,

$$x(t) = \int\limits_{-\infty}^{+\infty} x(t_1)\delta(t - t_1)\, dt_1 \tag{5.125}$$

in Equation (5.124), we get

$$y(t) = L[x(t)] = L\left[\int\limits_{-\infty}^{+\infty} x(t_1)\,\delta(t - t_1)dt_1 \right]$$

$$= \int\limits_{-\infty}^{+\infty} x(t_1)L[\delta(t - t_1)]\, dt_1 \tag{5.126}$$

because the linear operator operates on the time function $\delta(t - t_1)$ only. In Equation (5.126), $L[\delta(t - t_1)]$ is the *impulse response function $h(t - t_1)$* representing the response of the linear system to the unit impulse function $\delta(t - t_1)$; i.e.

$$h(t - t_1) = L[\delta(t - t_1)] \tag{5.127}$$

Because of Equation (5.127), Equation (5.125) becomes

$$y(t) = \int\limits_{-\infty}^{+\infty} x(t_1)h(t - t_1)\, dt_1 \tag{5.128}$$

This is known as the *convolution integral* of $x(t)$ and the unit impulse response function $h(t)$. It is also denoted as

$$y(t) = x(t) * h(t) \tag{5.129}$$

The convolution integral is a very important input-output relationship for a linear system. Provided that $\int_{-\infty}^{+\infty} |h(t)|\, dt < \infty$, Equation (5.128) is valid for any input $x(t)$ whose magnitude $|x(t)|$ is bounded.

Alternative forms to the convolution integral of Equation (5.128) are

$$y(t) = \int_{-\infty}^{t} x(t_1)h(t - t_1)\, dt_1 \qquad\qquad (5.130)$$

$$y(t) = \int_{0}^{+\infty} h(t_1)x(t - t_1)\, dt_1 \qquad\qquad (5.131)$$

and

$$y(t) = \int_{-\infty}^{+\infty} h(t_1)x(t - t_1)\, dt_1 \qquad\qquad (5.132)$$

The convolution integrals of Equations (5.128), (5.130), (5.131), and (5.132) can be used to calculate the response of a time-invariant, linear system to an arbitrary input.

Example 5.23:   Figure 5.23 represents a massless spring-damper system. Determine its impulse response function (Newland, 1993).

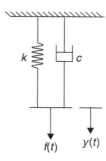

Figure 5.23 A spring-damper system.

Solution:
The governing differential equation (equation of motion) of the given mechanical system is

$$c\dot{y}(t) + ky(t) = f(t)$$

If the applied force is $f(t) = \delta(t)$, the impulse response function $h(t)$ will be the response of the system; i.e.

$$c\dot{h}(t) + kh(t) = \delta(t)$$

For $t > 0$, $\delta(t) = 0$, and the above equation becomes $c\dot{h} + kh = 0$ which has the solution

$$h(t) = h_0 e^{-kt/c}$$

with h_0 being a constant to be determined from the initial conditions at $t = 0$.
   Integration across $t = 0$ results in

$$c\int_{0^-}^{0^+} \dot{h}\, dt + k\int_{0^-}^{0^+} h\, dt = \int_{0^-}^{0^+} \delta(t)\, dt = 1$$

When the impulse is applied the response is a sudden movement with an instantaneously infinite velocity. Therefore, as $\dot{h} \to \infty$ and $dt \to 0$, the integral $\int_{0^-}^{0^+} \dot{h}\, dt$ has a finite value. Also, $h(0)$ is not infinite and therefore, its integral across zero is equal to zero; i.e. $\int_{0^-}^{0^+} h\, dt = 0$. Therefore, the above equation reduces to

$$c\int_{0^-}^{0^+} \dot{h}\, dt = 1$$

which results in $ch(0) = 1$ or $h(0) = \frac{1}{c}$. Thus,

$$h(0) = h_0 e^{-k0/c} = h_0 = \frac{1}{c}$$

and the solution becomes

$$h(t) = \begin{cases} \dfrac{1}{c} e^{-kt/c} & \text{for } t > 0 \\[2mm] 0 & \text{for } t < 0 \end{cases}$$

Example 5.24:   Calculate the response at time $t > 0$ of the spring-damper system of example 5.23 if the system is subjected to a step input $f(t) = f_0$ for $t \geq 0$.

Solution:
From the impulse response function of example 5.23,

$$h(t - t_1) = \begin{cases} \dfrac{1}{c} e^{-k(t-t_1)/c} & \text{for } t_1 < t \\[2mm] 0 & \text{for } t_1 > t \end{cases}$$

According to the convolution integral of Equation (5.130), the system response is equal to

$$
y(t) = \int_{-\infty}^{t} f(t_1) h(t - t_1) \, dt_1
$$

$$
= \int_{0}^{t} f_0 \frac{1}{c} e^{-k(t-t_1)/c} \, dt_1
$$

$$
= \frac{f_0}{c} (1 - e^{-kt/c}) \quad \text{for } t > 0
$$

The response $y(t)$ is obviously zero for $t < 0$ before the step input is applied.

Example 5.25: Rework example 5.24 using the convolution integral of Equation (5.131).

Solution:
From Equation (5.131), the response is

$$
y(t) = \int_{0}^{+\infty} \frac{1}{c} e^{-kt_1/c} f(t - t_1) \, dt_1
$$

To calculate the above integral, we must determine $f(t - t_1)$ as a function of t_1 for a constant t. It should be noted that the variable t_1 is measured backwards along the time axis. The system response $y(t)$ is therefore, equal to

$$
y(t_1) = \begin{cases} 0 & \text{for } t < 0 \\ \int_{0}^{t} \frac{1}{c} e^{-kt_1/c} f_0 \, dt_1 = \frac{f_0}{c}(1 - e^{-kt/c}) & \text{for } t > 0 \end{cases}
$$

which is the same as in example 5.24.

5.5.1.2 Frequency Response Function of a Linear, Time-invariant System

In the previous section, we characterized a linear, time-invariant system in the time domain using an impulse function. This section provides an equivalent characterization in the frequency domain by Fourier transforming the system response $y(t)$.

If $X(\omega), Y(\omega)$, and $H(\omega)$ are the Fourier transforms of $x(t)$, $y(t)$, and $h(t)$, respectively,

$$Y(\omega) = \int_{-\infty}^{+\infty} y(t)e^{-j\omega t}\,dt = \int_{-\infty}^{+\infty}\left[\int_{-\infty}^{+\infty} x(t_1)h(t-t_1)\,dt_1\right]e^{-j\omega t}\,dt$$

$$= \int_{-\infty}^{+\infty} x(t_1)\left[\int_{-\infty}^{+\infty} h(t-t_1)e^{-j\omega(t-t_1)}\,dt\right]e^{-j\omega t}\,dt_1$$

$$= \int_{-\infty}^{+\infty} x(t_1)H(\omega)e^{-j\omega t}\,dt_1 = H(\omega)X(\omega) \tag{5.133}$$

The function $H(\omega)$ is the *frequency response function* (FRF) or *transfer function* (TF) of the system. From Equation (5.184), the frequency response function is

$$H(\omega) = \frac{Y(\omega)}{X(\omega)} \tag{5.134}$$

Equation (5.133) shows that the Fourier transform of the response of a linear, time-invariant system is equal to the product of the Fourier transform of the input and the Fourier transform of the system impulse response function.

Example 5.26: Calculate the frequency response function $H(\omega)$ of the single degree-of-freedom mass-spring-damper mechanical system in Figure 5.24.

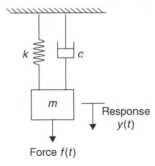

k c

m

Response $y(t)$

Force $f(t)$

Figure 5.24 A mass-spring-damper system.

Solution:
The governing differential equation is

$$m\ddot{y}(t) + c\dot{y}(t) + k y(t) = f(t)$$

For an exponential input force $f(t) = e^{j\omega t}$, the response is $y(t) = y_0 e^{j\omega t}$, where y_0 is a constant. Substitution of $x(t)$ and $y(t)$ in the differential equation yields

$$(-m\omega^2 + j\omega c + k)y_0 e^{j\omega t} = e^{j\omega t}$$

According to Equation (5.134),

$$H(\omega) = \frac{y_0 e^{j\omega t}}{e^{j\omega t}} = \frac{1}{-m\omega^2 + j\omega c + k}$$

Note that $H(\omega)$ is complex. The same result can be obtained by calculating the forward Fourier transform of the left and right hand sides of the governing differential equation, and then using Equation (5.133).

5.5.2   Response of a Linear System to Random Input

This section determines the characteristics of the response of a stable, linear, time-invariant system if the applied input is an ensemble member $x(t)$ of a random process $X(t)$. Figure 5.25 shows this conceptually.

The linear system is characterized by its impulse response function $h(t)$ or the corresponding frequency response function $H(\omega)$. This section provides expressions for the temporal characteristics of the response such as the mean value, mean-squared value, autocorrelation function, and cross-correlation functions. Section 5.5.3 provides the spectral characteristics of the response.

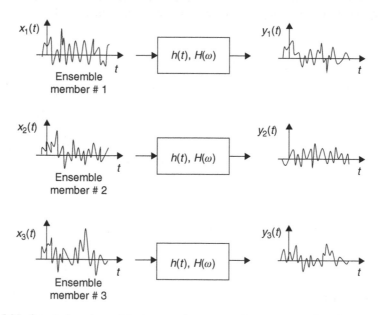

Figure 5.25   Sample functions of the input and output random processes for a linear, time-invariant system.

Mean and Mean-Squared Values of System Response

The convolution integral of Equation (5.132) provides the response $y(t)$ of the linear system. It can be generalized as

$$Y(t) = \int_{-\infty}^{+\infty} h(t_1)X(t - t_1)\, dt_1 \tag{5.135}$$

where the input random process $X(t)$ is operated upon to provide the output random process $Y(t)$. If $X(t)$ is a wide-sense stationary process, the mean value of the system response $Y(t)$ is

$$E[Y(t)] = E\left[\int_{-\infty}^{+\infty} h(t_1)X(t - t_1)\, dt_1\right]$$

$$= \int_{-\infty}^{+\infty} h(t_1)\underbrace{E[X(t - t_1)]}_{\mu_X}\, dt_1 = \mu_X \int_{-\infty}^{+\infty} h(t_1)\, dt_1 \tag{5.136}$$

indicating that the mean value of $Y(t)$ equals the mean value of $X(t)$ times the area under the impulse response function. Considering that the frequency response function $H(\omega) = \int_{-\infty}^{+\infty} h(t)e^{-j\omega t}\, dt$ is the Fourier transform of the impulse response function $h(t)$, the area under the impulse response function is equal to $H(0)$; i.e.

$$H(0) = \int_{-\infty}^{+\infty} h(t)\, dt \tag{5.137}$$

Because of Equation (5.137), Equation (5.136) becomes

$$E[Y(t)] = \mu_Y = H(0)\mu_X \tag{5.138}$$

indicating that $H(0)$ is the ratio of means of the output process and the input process.
The mean-squared response of $Y(t)$ for a wide-sense stationary input process $X(t)$ is

$$\overline{Y^2} = E[Y^2(t)] = E\left[\int_{-\infty}^{+\infty} h(t_1)X(t - t_1)\, dt_1 \int_{-\infty}^{+\infty} h(t_2)X(t - t_2)\, dt_2\right]$$

$$= \int_{-\infty}^{+\infty}\int_{-\infty}^{+\infty} \underbrace{E[X(t - t_1)X(t - t_2)]}_{R_{XX}(t_1 - t_2)} h(t_1)h(t_2)\, dt_1\, dt_2 \tag{5.139}$$

because $E[X(t - t_1)X(t - t_2)] = R_{XX}(t_1 - t_2)$ for a wide-sense stationary process $X(t)$.

Example 5.27: Calculate $\overline{Y^2}$ for a linear system which is excited by white noise with

$$R_{XX}(t_1 - t_2) = \delta(t_1 - t_2)$$

Solution:
Equation (5.139) yields

$$\overline{Y^2} = \int\limits_{-\infty}^{+\infty} \int\limits_{-\infty}^{+\infty} \delta(t_1 - t_2) h(t_1) h(t_2)\, dt_1\, dt_2$$

$$= \int\limits_{-\infty}^{+\infty} h^2(t)\, dt$$

which states that the output power is proportional to the area under the square of the impulse response function $h(t)$.

Autocorrelation Function of System Response
If $X(t)$ is a wide-sense stationary process, the autocorrelation function of $Y(t)$ is

$$R_{YY}(t, t + \tau) = E[Y(t) Y(t + \tau)]$$

$$= E\left[\int\limits_{-\infty}^{+\infty} h(t_1) X(t - t_1)\, dt_1 \int\limits_{-\infty}^{+\infty} h(t_2) X(t + \tau - t_2)\, dt_2 \right]$$

$$= \int\limits_{-\infty}^{+\infty} \int\limits_{-\infty}^{+\infty} E[X(t - t_1) X(t + \tau - t_2)]\, h(t_1) h(t_2)\, dt_1 dt_2$$

which reduces to

$$R_{YY}(\tau) = \int\limits_{-\infty}^{+\infty} \int\limits_{-\infty}^{+\infty} R_{XX}(\tau + t_1 - t_2)\, h(t_1) h(t_2)\, dt_1 dt_2 \tag{5.140}$$

if $X(t)$ is a wide-sense stationary process. Equation (5.140) shows that if $X(t)$ is a wide-sense stationary process, so is $Y(t)$.

Cross-Correlation Functions of System Input and Output

The cross-correlation function of the input $X(t)$ and the output $Y(t)$ is

$$R_{XY}(t, t + \tau) = E[X(t)Y(t + \tau)] = E\left[X(t) \int\limits_{-\infty}^{+\infty} h(t_1)X(t + \tau - t_1)\, dt_1 \right]$$

$$= \int\limits_{-\infty}^{+\infty} E[X(t)X(t + \tau - t_1)]h(t_1)\, dt_1$$

If the input process $X(t)$ is wide-sense stationary, the above expression becomes

$$R_{XY}(\tau) = \int\limits_{-\infty}^{+\infty} R_{XX}(\tau - t_1)h(t_1)\, dt_1 \qquad (5.141a)$$

indicating that the cross-correlation function $R_{XY}(\tau)$ is the convolution of $R_{XX}(\tau)$ and $h(\tau)$; i.e.

$$R_{XY}(\tau) = R_{XX}(\tau) * h(\tau) \qquad (5.141b)$$

Similarly, the following expression for the cross-correlation function $R_{YX}(\tau)$ holds

$$R_{YX}(\tau) = \int\limits_{-\infty}^{+\infty} R_{XX}(\tau - t_1)h(-t_1)\, dt_1 \qquad (5.142a)$$

or

$$R_{XY}(\tau) = R_{XX}(\tau) * h(-\tau) \qquad (5.142b)$$

Equations (5.141) and (5.142) show that the cross-correlation function does not depend on absolute time. Therefore, if $X(t)$ is wide-sense stationary, $X(t)$ and $Y(t)$ are *jointly* wide-sense stationary because $Y(t)$ is also wide-sense stationary according to Equation (5.140).

Using Equation (5.141a) into Equation (5.140), yields the following relationship between the autocorrelation function $R_{YY}(\tau)$ and the cross-correlation function $R_{XY}(\tau)$

$$R_{YY}(\tau) = \int\limits_{-\infty}^{+\infty} R_{XY}(\tau + t_1)h(t_1)\, dt_1 \qquad (5.143a)$$

or

$$R_{XY}(\tau) = R_{XY}(\tau) * h(-\tau) \qquad (5.143b)$$

Similarly,

$$R_{XY}(\tau) = R_{YX}(\tau) * h(\tau) \qquad (5.144)$$

Example 5.28: Calculate the cross-correlation functions $R_{XY}(\tau)$ and $R_{YX}(\tau)$ for example 5.27.

Solution:
From Equations (5.141a) and (5.142a),

$$R_{XY}(\tau) = \int\limits_{-\infty}^{+\infty} \delta(\tau - t_1)h(t_1)\, dt_1 = h(\tau)$$

and

$$R_{YX}(\tau) = \int\limits_{-\infty}^{+\infty} \delta(\tau - t_1)h(-t_1)\, dt_1 = h(-\tau) = R_{XY}(-\tau)$$

Note that the above results satisfy property #1 (Equation 5.23) of the cross-correlation function.

5.5.3 Spectral Characteristics of the Response of a Linear System

This section provides an expression for the power spectral density $S_{YY}(\omega)$ of the response, and the cross-power spectral densities $S_{XY}(\omega)$ and $S_{XY}(\omega)$ of the input and output, as a function of the input power spectral density and the frequency response function of the system. We assume that the input process $X(t)$ is wide-sense stationary so that $Y(t)$ and $X(t)$ are jointly wide-sense stationary processes.

We first show that the response power spectral density $S_{YY}(\omega)$ of a linear, time-invariant system with frequency response function $H(\omega)$ is given by $S_{YY}(\omega) = S_{XX}(\omega)|H(\omega)|^2$ where $S_{XX}(\omega)$ is the power spectral density of the input process $X(t)$. According to the Wiener-Khintchine theorem of Equation (5.85a), the response power spectral density is $S_{YY}(\omega) = \int_{-\infty}^{+\infty} R_{YY}(\tau)e^{-j\omega\tau}\, d\tau$. Substitution of Equation (5.140) yields

$$S_{YY}(\omega) = \int\limits_{-\infty}^{+\infty} h(t_1) \int\limits_{-\infty}^{+\infty} h(t_2) \int\limits_{-\infty}^{+\infty} R_{XX}(\tau + t_1 - t_2)\, e^{-j\omega\tau}\, d\tau\, dt_2\, dt_1$$

If $t = \tau + t_1 - t_2$, or $dt = d\tau$, the above relation becomes

$$S_{YY}(\omega) = \underbrace{\int\limits_{-\infty}^{+\infty} h(t_1)\, e^{-j\omega t_1}\, dt_1}_{H^*(\omega)} \underbrace{\int\limits_{-\infty}^{+\infty} h(t_2)\, e^{-j\omega t_2}\, dt_2}_{H(\omega)} \underbrace{\int\limits_{-\infty}^{+\infty} R_{XX}(\tau + t_1 - t_2)e^{-j\omega t}\, dt}_{S_{XX}(\omega)}$$

or

$$S_{YY}(\omega) = H^*(\omega)H(\omega)S_{XX}(\omega) = |H(\omega)|^2 S_{XX}(\omega) \tag{5.145}$$

This is an important relation between the input and output spectra of a linear, time-invariant system. It can be also proven using the definition of the power spectral density and Equation (5.78). Namely,

$$S_{YY}(\omega) = \lim_{T \to \infty} \frac{E[|Y_T(\omega)|^2]}{2T}$$

or because $Y_T(\omega) = H(\omega)X_T(\omega)$ (see Equation 5.133),

$$S_{YY}(\omega) = \lim_{T \to \infty} \frac{E[|H(\omega)|^2 |X_T(\omega)|^2]}{2T} = |H(\omega)|^2 \underbrace{\lim_{T \to \infty} \frac{E[|X_T(\omega)|^2]}{2T}}_{S_{XX}(\omega)}$$

yielding,

$$S_{YY}(\omega) = |H(\omega)|^2 S_{XX}(\omega)$$

If the power spectral density $S_{YY}(\omega)$ of the response is known, the average power P_{YY} in the system's response can be calculated by integrating $S_{YY}(\omega)$ throughout the frequency range as

$$P_{YY} = \frac{1}{2\pi} \int_{-\infty}^{+\infty} S_{YY}(\omega)\, d\omega = \frac{1}{2\pi} \int_{-\infty}^{+\infty} |H(\omega)|^2 S_{XX}(\omega)\, d\omega \qquad (5.146)$$

Example 5.29:   Determine the output power spectral density $S_{YY}(\omega)$ for the single degree-of-freedom system of Figure 5.24 for a forcing function $f(t)$ whose spectral density is $S_{FF}(\omega) = S_0$.

Solution:
The differential equation describing the motion of the system is

$$m\ddot{y}(t) + c\dot{y}(t) + ky(t) = f(t)$$

Taking the Fourier transform of both left hand and right hand sides yields, $(-m\omega^2 + jc\omega + k)Y(\omega) = F(\omega)$, and therefore, the frequency response function is

$$H(\omega) = \frac{Y(\omega)}{F(\omega)} = \frac{1}{-m\omega^2 + jc\omega + k}$$

From Equation (5.145), the output spectral density is

$$S_{YY}(\omega) = |H(\omega)|^2 S_{FF}(\omega) = \frac{S_0}{(k - m\omega^2)^2 + c^2\omega^2}$$

Cross-Power Spectral Densities of Input and Output of a Linear,
Time-Invariant System

This section proves that the cross-power spectral densities of input and output are given by

$$S_{XY}(\omega) = S_{XX}(\omega)H(\omega) \tag{5.147}$$

and

$$S_{YX}(\omega) = S_{XX}(\omega)H(-\omega) \tag{5.148}$$

For jointly wide-sense stationary processes $X(t)$ and $Y(t)$, the cross-power spectral density is $S_{XY}(\omega) = \int_{-\infty}^{+\infty} R_{XY}(\tau)e^{-j\omega\tau}\,d\tau$ (see Equation 5.109). Substitution of $R_{XY}(\tau)$ from Equation (5.141a) yields,

$$S_{XY}(\omega) = \int_{-\infty}^{+\infty}\int_{-\infty}^{+\infty} R_{XX}(\tau - t_1)h(t_1)\,dt_1 e^{-j\omega\tau}\,d\tau$$

If $t = \tau - t_1$ so that $dt = d\tau$, the above equation becomes

$$S_{XY}(\omega) = \underbrace{\int_{-\infty}^{+\infty} R_{XX}(t)e^{-j\omega t}\,dt}_{S_{XX}(\omega)}\ \underbrace{\int_{-\infty}^{+\infty} h(t_1)e^{-j\omega t_1}\,dt_1}_{H(\omega)}$$

or

$$S_{XY}(\omega) = S_{XX}(\omega)H(\omega)$$

Equation (5.148) is derived from Equation (5.147) based on the properties of the cross-power spectral densities and the power spectral density $S_{XX}(\omega)$; i.e.

$$S_{YX}(\omega) = S_{XY}(-\omega) = S_{XX}(-\omega)H(-\omega)$$

or

$$S_{YX}(\omega) = S_{XX}(\omega)H(-\omega)$$

because $S_{XX}(\omega)$ is an even function.

5.5.4 Response of a Linear System to Nonstationary Input

Sections 5.5.2 and 5.5.3 described how to calculate the stochastic response of a linear, time-invariant system due to stationary random inputs. However for some practical problems, the inputs are nonstationary. In this case, the responses are also nonstationary.

The *generalized spectral density* of a nonstationary process is defined as the generalized Fourier transform of the covariance, $C_{XX}(t_1, t_2)$, (Lin, 1967) as

$$S_{XX}(\omega_1, \omega_2) = \int\limits_{-\infty}^{+\infty} \int\limits_{-\infty}^{+\infty} C_{XX}(t_1, t_2) e^{-i(\omega_1 t_1 - \omega_2 t_2)} dt_1 dt_2 \tag{5.149}$$

The generalized spectral density of the output of a single-input, single-output linear system can be calculated from the generalized spectral density of the input as

$$S_{YY}(\omega_1, \omega_2) = S_{XX}(\omega_1, \omega_2) H(\omega_1) H^*(\omega_2) \tag{5.150}$$

Note that Equation (5.150) is analogous to Equation (5.145) for the stationary case. Following are some studies of the response of structures to nonstationary excitation:

- Caughey and Stumpf (1961) – study of the transient response of a single-degree-of-freedom system under white noise excitation
- Fung (1955) – analysis of stresses in the landing gear of an aircraft
- Amin (1966) – analysis of the response of buildings to earthquakes
- Nikolaidis, et al. (1987) – analysis of torsional vibratory stresses in diesel engine shafting systems under a special class of nonstationary processes, whose statistics vary periodically with shifts of time.

5.6   Random Process Characterization using Simulation

Sections 5.2 and 5.3 described how to model (or characterize) a random process in the time and frequency domains. We mentioned that a weakly stationary process is fully characterized in the time domain if we know its autocorrelation function. In the frequency domain, it is characterized by its power spectral density which is obtained from the autocorrelation function according to the Wiener-Khintchine theorem of Equation (5.86a).

It is common in practice however, to only have discretized information of only one sample function of the process during a finite time. In this case, the process can be characterized using time series modeling or spectral representation methods. This section describes the basics of time series modeling and two common spectral representation methods. After the process is characterized, sample functions can be easily generated and used in Monte Carlo simulation to determine the out-crossing probability, for example.

5.6.1   *Time Series Modeling*

Time-series modeling is a feedback-based approach which uses past data to estimate future data. It can characterize both stationary and non-stationary processes (Rupert, 2004, Singpurwalla, 1975 and 1978, Stavropoulos and Fassois, 2000). Time-series modeling techniques have been widely used in the fields of finance, econometrics, and bio-medical engineering, among others, where forecasting under stochastic conditions is needed.

A time series is a sequence of random observations taken over time such as a sequence of the daily price of a stock, or the yearly peak temperature. The time series models use past observations to develop statistical models that "closely" represent the phenomenon over time. Based on the ergodicity assumption, a time series model is a statistical model which is developed using data from only one sample function of the process.

It is generally recommended that the time series model use as few parameters as possible (Rupert, 2004). Too many parameters can improve the fitting accuracy but they increase the likelihood of the estimation error. On the other hand, too few parameters may result in an inadequate fitting. A statistical model with only few parameters is called "parsimonious," and is usually adequate for stationary processes. For a non-stationary process, the series is differenced in order to produce a new time series consisting of the changes in the original series. After differencing, the new time series usually represents a stationary process. If this is not true, a non-constant variance is assumed and additional parameters are needed to fit the model.

The most common time series models are the Auto-Regressive Integrated Moving Average (ARIMA), and the Generalized Auto-Regressive Conditional Heteroskedasticity (GARCH) models. ARIMA models are generally sufficient for stationary or non-stationary random processes with a constant standard deviation. For non-stationary processes with a time-dependent standard deviation, GARCH models are used.

5.6.1.1 ARIMA(p,d,q) Models

ARIMA models have been extensively used to characterize a random process (Samaras, et al. 1985, Gersh and Yonemoto, 1977). They combine Auto-Regressive (AR) models, Integrated (I) models, and Moving Average (MA) models (Rupert, 2004). The Integrated (I) model is used if the process is non-stationary. These models use a feedback mechanism based on either past observations (AR models), past standard errors (MA), or a combination of the two (ARMA) to determine the conditional expectation of future observations. AR models are the most commonly used. They provide a weighted average of past observations in addition to a white noise error term, to capture the correlation at instances t_1 and t_2 where $\tau = t_2 - t_1$ is small.

An ARIMA model is denoted by ARIMA(p, d, q), where the parameter p indicates the order of the autoregressive process which in turn, indicates the level of correlation to past observations. A higher p shows correlation over a larger span of past observations. The parameter p denotes the number of times the original non-stationary process must be differenced until it becomes stationary, and the parameter q denotes the order of the moving average process representing the level of dependency on random noise from past observations.

Consider a random process $X(t)$. A sample function $x(t)$ is discretized in the time interval $[0, T]$ using a uniform time step Δt so that $x_i = x(t_i)$, where $t_i = i \cdot \Delta t$. For an AR(p) model, the discretized sample function is represented as

$$x_i - \mu = \phi_1(x_{i-1} - \mu) + \phi_2(x_{i-2} - \mu) + \cdots + \phi_p(x_{i-p} - \mu) + \varepsilon_i \qquad (5.151)$$

where μ is the temporal mean of the process, $\varepsilon_i \equiv N(0, \sigma_e^2)$ is Gaussian white noise and $\phi_1, \phi_2, \ldots, \phi_p$, are feedback parameters to be estimated. Different order AR models can be created to determine the best fit. For an AR(2) model, the variance σ_e^2 of the Gaussian white noise is determined from

$$\gamma(0) = Var(X_i) = \frac{\sigma_e^2}{1 - \phi_1 \rho_1 - \phi_2 \rho_2 - \cdots - \phi_p \rho_p} \tag{5.152}$$

where $\gamma(0)$ is the variance of the random process, and ρ_p is the value of the autocorrelation function at the time lag $\tau = p \cdot \Delta t$. Similar expressions exist for higher order AR models. Details are provided in (Rupert, 2004).

After the feedback parameters are estimated, a residual series $E(t) = X(t) - \hat{X}(t)$ is formed as the difference between the actual $X(t)$ and the estimated $\hat{X}(t)$ processes and statistical tests are performed to make sure that the random variables E_t and $E_{t+\tau}$ are uncorrelated for every τ.

5.6.1.2 GARCH(p,q) Models

The ARIMA models assume that the process has a constant standard deviation. If this is not true, GARCH models are used to characterize a random process with a time-dependent standard deviation. In this case, a non-constant standard deviation a_t of the residual noise is calculated from

$$a_t = \sigma_t \varepsilon_t \tag{5.153}$$

where ε_t is Gaussian white noise with zero variance, and

$$\sigma_t = \sqrt{\alpha_0 + \sum_{i=1}^{q} \alpha_i a_{t-i}^2 + \sum_{i=1}^{p} \beta_i \sigma_{t-i}^2} \tag{5.154}$$

is a conditional standard deviation given the past values of $a_{t-1}, a_{t-2}, \ldots, a_{t-i}$ and $\sigma_{t-1}, \sigma_{t-2}, \ldots, \sigma_{t-i}$. The unknowns p, q, α_0, α_i and β_i must be estimated. The GARCH(p, q) model is a zero mean random process with a standard deviation provided by Equation (5.174). GARCH models can also be combined with AR(p) models as,

$$x_i - \mu = \phi_1(x_{i-1} - \mu) + \phi_2(x_{i-2} - \mu) + \cdots + \phi_p(x_{i-p} - \mu) + a_t \tag{5.155}$$

5.6.1.3 Model Identification and Selection

The model type can be identified by visually inspecting the plots of the autocorrelation function (ACF) and the partial sample autocorrelation function (PACF) for different lags (multiples of Δt). The autocorrelation function provides information on the statistical dependence between random variables $X(t)$ and $X(t + \tau)$ where $\tau = h \cdot \Delta t$ and h is an integer denoting the lag. For a stationary random process, the autocorrelation depends only on τ. For autoregressive models, the autocorrelation function dies out quickly with increasing τ. The sample autocorrelation function $\hat{\rho}(h)$ is defined as

$$\hat{\rho}(h) = \frac{n^{-1} \sum_{i=1}^{n-h} (x_{i+h} - \mu)(x_i - \mu)}{\hat{\sigma}^2} \tag{5.156}$$

where $\hat{\sigma}$ is the estimated standard deviation of the random process. Note that the above autocorrelation function is actually the correlation coefficient as defined in section 2.2.

The partial autocorrelation of lag h represents the autocorrelation between X_i and X_{i+h} with the linear dependence of X_{i+1} through X_{i+h-1} removed. This is equivalent to the autocorrelation between X_i and X_{i-h} that is not accounted for by lags 1 to $h - 1$, inclusive. The partial autocorrelation is useful in identifying the order of an autoregressive model. For an AR(p) model, it is zero for lags greater or equal to $p + 1$.

Figure 5.26 shows an example of the identification process for two different AR(p) models. The autocorrelation function (ACF) is plotted on the left, and the partial autocorrelation function (PACF) is plotted on the right. In Figure 5.26a, the PACF becomes zero (or statistically insignificant) after the first lag indicating that the model can be represented by an AR(1) model. Similarly, Figure 5.26b indicates an AR(2) model because of the two significant spikes in the PACF plot. An exponential decline of ACF indicates the presence of a positive feedback parameter ($\phi_1 > 0$). If the ACF alternates between positive and negative values and it simultaneously declines exponentially, there is at least one negative feedback parameter. Rupert (2004) provides detailed information on model identification.

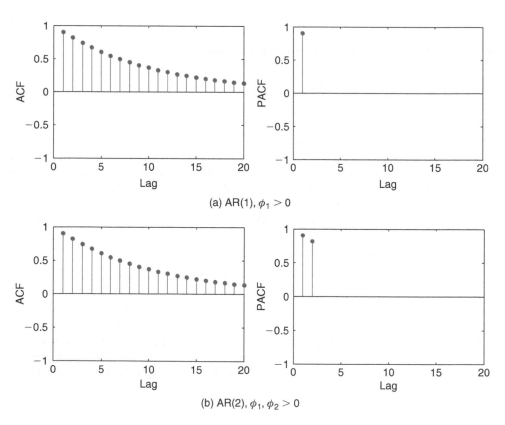

(a) AR(1), $\phi_1 > 0$

(b) AR(2), $\phi_1, \phi_2 > 0$

Figure 5.26 AR model identification.

After the order p of the model is identified, the ϕ's and the mean μ are estimated either by using the Yule-Walker equations (Yule, 1927, and Walker, 1931) or equivalently, by minimizing

$$\sum_{i=p+1}^{n} [\{x_i - \mu) - \phi_1(x_{i-1} - \mu) - \cdots - \phi_p(x_{i-p} - \mu)]^2 \tag{5.157}$$

Subsequently the residual series is tested to determine if it represents white noise. This is a necessary condition for the AR model to accurately represent the actual random process. Different statistical tests are available. Rupert (2004) presents the commonly used tests.

If different time series models are fit (or identified), the Akaike (AIC) and Bayesian (BIC) information criteria (Rupert, 2004) can be used to select the best alternative by penalizing models with increased number of parameters based on parsimony. The model with the lowest values of AIC and BIC is considered optimal. The AIC and BIC values can be calculated using statistical software such as SAS (SAS Institute, 1993).

Example 5.30:   Consider a Gaussian random process $F(t)$, representing a force, with a given autocorrelation of

$$\rho_F(t_1, t_2) = \exp\left(-\left(\frac{t_2 - t_1}{\psi}\right)^2\right) \tag{5.158}$$

where $\psi = 8.33 \times 10^{-2}$ years is a correlation length. The mean and standard deviation of $F(t)$ are $\mu = 3500\,\text{N}$ and $\sigma_F = 700\,\text{N}$. Fit an autoregressive model.

Solution:
The type and order of the time series model is first identified. Because the autocorrelation function is known, the order of an AR model can be easily determined by solving the following least-squares optimization problem

$$\underset{\phi_1,\ldots,\phi_p}{Minimize} \sum_{i=1}^{n} (\rho(h_i) - \hat{\rho}(h_i))^2 \tag{5.159}$$

where ρ is the given autocorrelation function and $\hat{\rho}$ is the estimated autocorrelation function from the Yule-Walker equations (Yule, 1927, and Walker, 1931), $\hat{\rho}(h_i) = \sum_{q=1}^{p} \phi_q \hat{\rho}(h_{i-q})$ where $i = 1, 2, \ldots, n$. It should be noted that if the autocorrelation function is not known, a sample function of the random process can be used to estimate the model parameters.

We fit an autoregressive model AR(p) represented by

$$f_i - \mu_F = \phi_1(f_{i-1} - \mu) + \phi_2(f_{i-2} - \mu) + \cdots + \phi_p(f_{i-p} - \mu) + \varepsilon_i \tag{5.160}$$

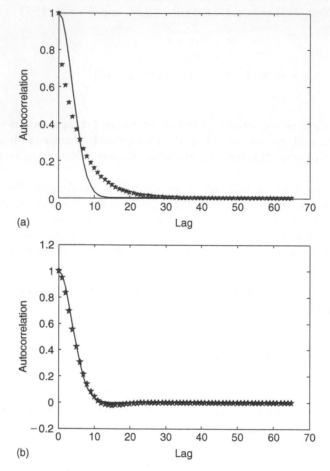

(a)

(b)

Figure 5.27 AR model fitting using a given autocorrelation function. (a) AR(1) model; Given (solid line) and AR(1) (dotted line) autocorrelation; (b) AR(2) Model; Given (solid line) and AR(2) (dotted line) autocorrelation.

where $f(t)$ is a sample function of the process $F(t)$, μ is the mean of the process, $\varepsilon_i \equiv N(0, \sigma_e^2)$ is Gaussian white noise, and $\phi_1, \phi_2, \ldots, \phi_p$, are feedback parameters to be estimated. Models of various orders are usually evaluated and the best fit is determined using the AIC and BIC criteria.

Two models are considered; the AR(1) model with one feedback parameter, and the AR(2) model with two feedback parameters. The order of the appropriate model is determined based on the optimal least-squares fit of the given autocorrelation function according to the optimization problem of Equation (5.160).

Figure 5.27a shows the AR(1) model fit using the calculated optimal value of $\phi_1 = 0.8461$ of the single feedback parameter. Figure 5.27b shows the AR(2) model fit using the calculated optimal values of $\phi_1 = 1.5128$ and $\phi_2 = -0.5979$ of the

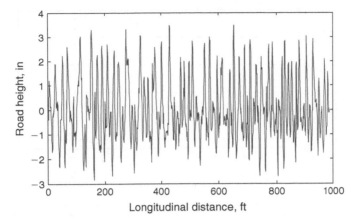

Figure 5.28 Sample function $u(t)$ of stochastic terrain in example 5.31.

two feedback parameters. The AR(2) model fits the given autocorrelation function better.

Example 5.31:   Figure 5.28 shows a sample function $u(t)$ of finite duration for a stochastic terrain. This represents the only information we have for the terrain. Identify an AR model.

Solution:
According to section 5.6.1.3, the model identification is performed by inspecting the plots of autocorrelation and partial autocorrelation functions (Figure 5.29) of the given road profile. The figure also shows the 95% confidence bands (two horizontal solid lines) defining the range within which the autocorrelation is treated as practically equal to zero. Figure 5.29a shows that the autocorrelation decays quickly with increasing lag from a value of 1 to practically zero. This indicates that an autoregressive type (AR) model with at least one positive feedback parameter may be a good fit. The fluctuation in the autocorrelation plot for a lag greater than 10 seconds, indicates the presence of at least one negative feedback parameter. Because the fluctuations eventually die out, the effect of the negative feedback parameter is less important compared with the effect of the positive feedback parameter. The partial autocorrelation plot of Figure 5.29b shows three noticeable spikes at small lags, suggesting that an AR(3) model may suffice. The Econometrics toolbox of MATLAB was used to estimate the three parameters of an AR(3) model as $\phi_1 = 1.2456, \phi_2 = -0.2976$, and $\phi_3 = -0.1954$. The toolbox also estimated the standard deviation $\sigma_\varepsilon = 0.5132$ of the zero-mean residual white noise process ε. Therefore, the random process of the terrain can be represented by the following AR(3) model

$$u_i = 1.2456u_{i-1} - 0.2976u_{i-2} - 0.1954u_{i-3} + \varepsilon_i(0, 0.5132^2) \tag{5.161}$$

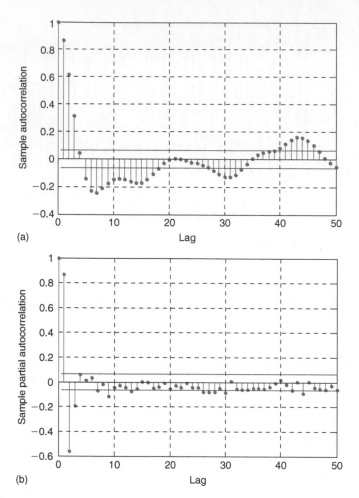

Figure 5.29 Autocorrelation and partial autocorrelation of stochastic terrain in example 5.31. a) Sample autocorrelation function (ACF); b) Sample partial autocorrelation function (PACF).

Figure 5.30 shows the sample autocorrelation function of the residual process indicating that it is practically white noise with no correlation at any lag. The sample partial autocorrelation function (not shown) also verified that the residual process is white noise. The AR(3) model is therefore, a good fit.

The Ljung-Box test (Kennedy, 2003) accepted the null hypothesis that the residuals are random and the Dickey-Fuller test (Kennedy, 2003) rejected the null hypothesis that the residuals are not random. Finally, the autocorrelation function of ε_i^2 was inspected to check for the presence of any arch effects, or for a non-constant standard deviation. Figure 5.31 shows that the sample autocorrelation of the residual squared process ε^2 is close to zero, indicating that the assumption of a constant standard deviation is sufficient.

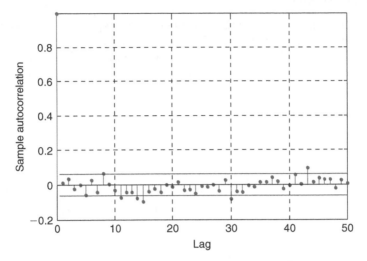

Figure 5.30 Sample autocorrelation function of residual process in example 5.31.

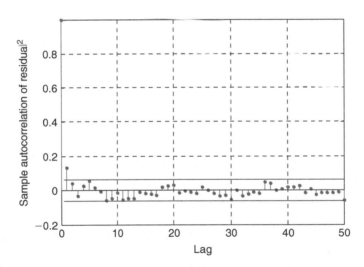

Figure 5.31 Sample autocorrelation function of residual squared process ε^2 in example 5.31.

5.6.2 *Spectral Representation Methods*

Besides time series modeling, spectral representation methods are also used to characterize a random process and generate sample functions. Multiple spectral decomposition based techniques are available (Schueller, 1997), including the Expansion Optimal Linear Estimation (EOLE) (Sudret and Der Kiureghian, 2000, and Li and Der Kiureghian, 1993), the Karhunen-Loeve (K-L) expansion (Loeve, 1978, and Choi, et al. 2007), and the Borgman spectral representation (Borgman, 1969).

5.6.2.1 EOLE Method

The EOLE method is a variant of the Karhunen-Loeve expansion method that involves decomposition of the covariance matrix of a random process to generate independent eigenfunctions. The number of eigenvalues can be optimally determined.

The random process, $X(t)$ can be approximated by $\hat{X}(t)$ using the correlated random variables, $\chi_i, i = 1, 2, \ldots, N$ as

$$\hat{X}(t) = \sum_{i=1}^{N} \chi_i \varphi_i(t) \tag{5.162}$$

where $\varphi_i(t)$ are shape or basis functions which depend on the chosen correlation structure. The vector of correlated random variables $\chi = \{\chi_i, i = 1, 2, \ldots, N\}$ are transformed into the vector of uncorrelated standard random variables $\xi = \{\xi_i, i = 1, \ldots, N\}$ using

$$\chi = \varphi \cdot \Lambda^{1/2} \cdot \xi \tag{5.163}$$

where Λ is a diagonal matrix containing the significant eigenvalues (first few eigenvalues with highest value) of the covariance matrix, and φ is the matrix of corresponding eigenvectors.

The random process in Equation (5.162) is then expressed (Sudret and Der Kiureghian, 2000) as,

$$\hat{X}(t) = \mu(t) + \sum_{i=1}^{\infty} \sqrt{\lambda_i} \xi_i \varphi_i(t) \tag{5.164}$$

where $\mu(t)$ is the mean of the random process, λ_i is an eigenvalue, and φ_i is the corresponding eigenvector of the covariance matrix, C whose terms are

$$C_{t,t_i} = \{\rho_X(t - t_i), i = 1, \ldots, N\} \tag{5.165}$$

In Equation (5.165), $\rho_X(t - t_i)$ is the autocorrelation of the random process as a function of relative time $(t - t_i)$.

For a weakly stationary process, if we use only M significant terms, Equation (5.164) becomes

$$\hat{X}(t) \cong \mu + \sum_{i=1}^{M} \frac{\xi_i}{\sqrt{\lambda_i}} \varphi_i^T C_{t,t_i} \tag{5.166}$$

where

$$C_{t,t_i} \varphi_i = \lambda_i \varphi_i \tag{5.167}$$

and $i = 1, 2, \ldots, M$.

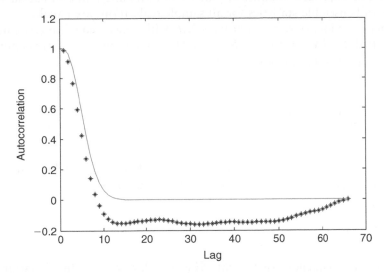

Figure 5.32 Autocorrelation in example 5.30; Given (solid line) and EOLE (dotted line).

To implement the EOLE method for a stationary process, we first estimate the covariance matrix of Equation (5.165) using sample data. Then, the eigenvalues $\lambda_i, i = 1, 2, \ldots, M$ with the largest value, and the corresponding eigenvectors φ_i are determined from the eigenvalue problem of Equation (5.167). The lowest M is determined so that the variance of the estimation error is minimized. The variance of the error (Sudret and Der Kiureghian, 2000) is calculated from

$$\text{Var}[X(t) - \hat{X}(t)] = \sigma_X^2 - \sum_{i=1}^{M} \frac{1}{\lambda_i} (\varphi_i^T C_{t,t_i}) \tag{5.168}$$

where σ_X is the standard deviation of the random process.

The EOLE method is demonstrated using the force random process $F(t)$ from the example 5.30. The covariance function is first calculated using the autocorrelation function of Equation (5.158). An eigenvalue analysis is subsequently performed according to Equation (5.167). Fifteen eigenvalues are needed in order to keep the error variance below 1%. Figure 5.32 shows that the autocorrelation function from the EOLE method is similar to the given autocorrelation function of Equation (5.158).

It should be noted that for broad-band processes the number of required eigenvalues may be large. This can compromise the efficiency of the EOLE method significantly.

5.6.2.2 Borgman Spectral Representation

The Borgman spectral representation method (Borgman, 1969) is based on the Fourier transform. The phase angles of all harmonics are treated as independent random variables. The approach assumes ergodicity in mean and autocorrelation so that the temporal and ensemble statistics for all sample realizations are equal.

Consider a stationary random process $X(t)$ with a mean equal to zero and a single-sided spectral density function of $S_{XX}(\omega)$. The random process $X(t)$ is approximated by $\hat{X}(t)$ using the following series as $N \to \infty$

$$\hat{X}(t) = \sum_{n=1}^{N} A_n \cos(\omega_n t + \phi_n) \tag{5.169}$$

where $\phi_1, \phi_2, \ldots, \phi_N$ are uncorrelated random phase angles which are uniformly distributed between 0 and 2π. The amplitudes are given by,

$$A_n = \sqrt{S_{XX}(\omega_n)\Delta\omega_n/\pi} \quad n = 1, 2, \ldots, N \tag{5.170}$$

Let ω_N be a frequency such that $S_{XX}(\omega) = 0, \forall \omega \geq \omega_N$. In Equation (5.170), $\Delta\omega_n = \omega_n - \omega_{n-1}$ where the discrete frequencies ω_n are chosen such that the area under the spectrum in the interval $\omega_{n-1} < \omega < \omega_n$ is equal to n/N of the total area under the spectrum in the interval $0 < \omega < \omega_N$. As $N \to \infty$, the approximated random process $\hat{X}(t)$ asymptotically approaches a Gaussian process because of the central limit theorem.

Shinozuka (1972) proposed an alternative approach according to which the frequencies ω_n in Equation (5.169) are randomly selected using the cumulative distribution function (CDF),

$$F(\omega) = \int_{0}^{\omega} f(\omega_1)\,d\omega_1 \tag{5.171}$$

where

$$f(\omega) = \frac{S_{XX}(\omega)}{\int_{0}^{\omega} S_{XX}(\omega_1)\,d\omega_1} \tag{5.172}$$

is the probability density function obtained from the spectrum.

The Shinozuka (1972) spectral representation method is demonstrated using an example of a stochastic terrain with the height $u(t)$ (Figure 5.33) representing one sample function realization of the random process $U(t)$. The power spectral density $S_{UU}(\omega)$ of $u(t)$ is calculated using Welch's method (Welch, 1967) (Figure 5.34). The probability density function $f(\omega)$ and the cumulative distribution function $F(\omega)$ are computed from Equations (5.171) and (5.172). A random set of $N = 25$ frequencies from $F(\omega)$, and uniformly distributed phase angles, are used in Equation (5.169). N is chosen so that the autocorrelation function of the approximated random process is similar to the given autocorrelation function (Figure 5.35).

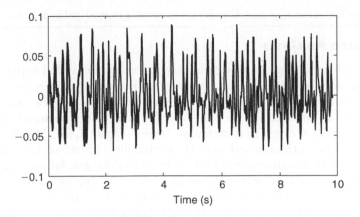

Figure 5.33 Sample function $u(t)$ of experimental road height process $U(t)$.

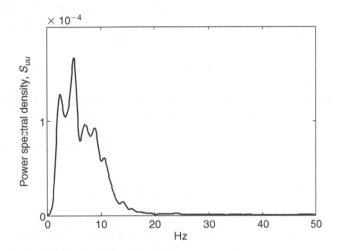

Figure 5.34 Power spectral density of height random process.

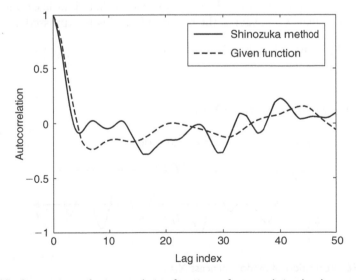

Figure 5.35 Comparison of autocorrelation functions of true and simulated process.

5.7   Failure Analysis

A structure subjected to a random, time-varying excitation can fail in two ways; a) the response exceeds a certain level, e.g., the wing of an aircraft flying in gusty weather can break if the stress exceeds a maximum allowable value, and b) the excitation causes fatigue failure. In this section, we only consider failure due to exceedance of a level. This type of failure is called *first excursion failure* (Figure 5.36). The objective of this section is to review methods to calculate the probability of first excursion failure during a given period and also discuss briefly how to analyze fatigue failure (section 5.7.5).

In general, the problem of finding the probability of first excursion failure during a period is equivalent to the problem of finding the probability distribution of the maximum of a random vector that contains infinite random variables. These variables are the values of the process at all time instants within the period. No closed form expression exists for the first excursion probability for a general random process. Approximate results are available for some special cases. In this section, we will review methods to estimate the following quantities for a random process:

- the number of crossings of a level α over a period,
- the PDF of the peaks,
- the PDF of the first passage time, T_f, which is the time at which the process up-crosses a certain level for the first time (Figure 5.36).

Nigam (1983) presents an overview of the methods to estimate the probability of first excursion failure.

5.7.1   Number of Outcrossings during a Period

It is often desirable to calculate the reliability of components or systems subject to loading which can be represented by a random process. For that, the probability that the component fails during a specified time interval must be determined. We assume that failure occurs at time T_f when a threshold α is exceeded (Figure 5.36). The probability of failure $P_f(T)$ is the probability that there is at least one time t in the interval [0, T] so that $X(t)$ exceeds the threshold α. This is expressed as

$$P_f(T) = P[\exists t \in [0, T] \text{ such that } X(t) \geq \alpha] \tag{5.173a}$$

or

$$P_f(t) = 1 - P[X(t) < \alpha, \forall t \in [0, T]] \tag{5.173b}$$

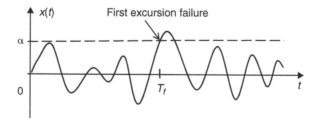

Figure 5.36   Realization of random process $X(t)$.

We present a number of different methods to estimate the probability in Equations (5.173) in the following sections.

5.7.1.1 Rice's Series

If p_k is the probability of exactly k out-crossings in the interval $[0, T]$, the probability of no out-crossings is

$$p_0 = 1 - P_f(T) = 1 - \sum_{k=1}^{\infty} p_k = 1 + \sum_{k=1}^{\infty} p_k \sum_{i=1}^{\infty} (-1)^i \binom{k}{i}$$

$$= 1 + \sum_{i=1}^{\infty} \frac{(-1)^i}{i!} \sum_{k=1}^{\infty} i! \binom{k}{i} p_k$$

$$= 1 + \sum_{i=1}^{\infty} \frac{(-1)^i}{i!} \sum_{k=1}^{\infty} k(k-1)\cdots(k-i+1) p_k \tag{5.174}$$

$$= 1 + \sum_{i=1}^{\infty} \frac{(-1)^i}{i!} m_i$$

$$= m_0 - m_1 + \frac{1}{2} m_2 - \frac{1}{6} m_3 + \cdots$$

where m_i is the ith factorial moment of the number of out-crossing with

$$m_0 = 1 \tag{5.175}$$

and

$$m_i = \sum_{k=1}^{\infty} k(k-1)\cdots(k-i+1) p_k \quad \text{for } i \geq 1 \tag{5.176}$$

where $\binom{k}{i} = 0$ if $i > k$. Note that m_1 is the mean number of out-crossings.

Equation (5.174) provides an exact solution to the level crossing problem and is known as the Rice's series (Rice, 1944, 1945, 1954). The computational effort to evaluate $P_f(T)$ from Equation (5.174) is however, high. Normally the series is truncated after the first term providing the upper limit

$$P_f(T) \leq m_1 \tag{5.177}$$

for the failure probability. It should be noted that $P_f(T)$ can only be approximated by m_1 if the out-crossing probability is very small; i.e. $P_f(T) << 1$.

5.7.1.2 The Poisson Assumption

Let $N^+(t, \alpha)$ be a counting process with ensemble members whose value increases by one each time the ensemble member $x(t)$ of the process $X(t)$ exceeds the threshold α. Also, assume that $N^+(0, \alpha) = 0$.

The $N^+(t, \alpha)$ is a Poisson process if 1) the probability of having two or more out-crossings in $[t, t + \Delta t]$ is negligible compared to the probability of having exactly one out-crossing, 2) Δt is sufficiently small, and 3) the out-crossings in $[t, t + \Delta t]$ are statistically independent of the previous out-crossings in $[0, t]$. The probability that the number of out-crossings $N^+(t, \alpha)$ is equal to n is then provided by

$$P(N^+(t, \alpha) = n) = \frac{1}{n!}(\lambda(t, \alpha))^n e^{-\lambda(t,\alpha)} \tag{5.178}$$

where $\lambda(t, \alpha)$ is the mean value of $N^+(t, \alpha)$ in the interval $[0, t]$; i.e.

$$\lambda(t, \alpha) = E[N^+(t, \alpha)] = m_1 \tag{5.179}$$

Thus, the probability of failure is equal to

$$P_f(T) = 1 - P(N^+(T, \alpha) = 0) = 1 - e^{-m_1} \tag{5.180}$$

For broad-band processes the maxima between adjacent zero up-crossings are practically uncorrelated. The out-crossings from the safe domain will therefore, be independent and Equation (5.180) is valid. For narrow-band processes, the out-crossings for low to medium threshold levels tend to occur in clumps (Figure 5.37) implying that the crossing events are correlated and Equation (5.180) is not valid. However, at higher threshold levels only the highest peak in a clump is likely to imply an out-crossing. Therefore, the out-crossings tend to become independent as $\alpha \to \infty$. This hypothesis can be formally proved for Gaussian processes (Crammer and Leadbetter, 1967).

Equations (5.174) and (5.180) do not consider that the process may start in the failure region; i.e. $X(0) > \alpha$. Accounting for this initial condition, the failure probability is defined as

$$P_f(T) = 1 - (1 - P_f(0))P(X(t) < \alpha \; \forall t \in [0, T] \,|\, X(0) < \alpha) \tag{5.181}$$

where $P_f(0) = P(X(0) < \alpha)$ is a time-invariant reliability problem.

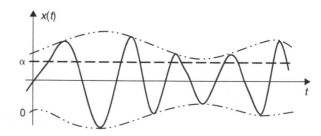

Figure 5.37 Out-crossings of a narrow-band process.

5.7.1.3 Average Number of Out-crossings

In order to determine the mean number of out-crossings of process $X(t)$ for level α, we consider the random process

$$Y(t) = H(X(t) - \alpha) \tag{5.182}$$

where $H(\cdot)$ denotes the Heaviside step function. Differentiation of Equation (5.182) yields the derivative process

$$\dot{Y}(t) = \dot{X}(t)\delta(X(t) - \alpha) \tag{5.183}$$

where $\delta(\cdot)$ is the Dirac delta function, and $X(t)$ is assumed to be a differentiable process. For a realization $x(t)$ of $X(t)$, Figure 5.38 shows the corresponding realizations $y(t)$ of $Y(t)$ and $\dot{y}(t)$ of $\dot{Y}(t)$. The process $\dot{Y}(t)$ consists of a series of unit pulses which occur each time $X(t)$ has an out-crossing. The number of out-crossings $N(T, \alpha)$ in the time interval $[0, T]$ is determined by integrating the absolute value of $\dot{Y}(t)$, as

$$N(T, \alpha) = \int_0^T \left| \dot{Y}(\tau) \right| d\tau = \int_0^T \left| \dot{X}(\tau) \right| \delta(X(\tau) - \alpha)\, d\tau \tag{5.184}$$

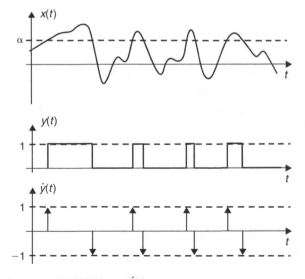

Figure 5.38 Realizations of $X(t), Y(t)$, and $\dot{Y}(t)$.

The mean number of out-crossings is then equal to

$$E[N(T,\alpha)] = \int_0^T E[|\dot{X}(\tau)|\delta(X(\tau) - \alpha)]\,d\tau$$

$$= \int_0^T \int_{-\infty}^{+\infty} \int_{-\infty}^{+\infty} |\dot{x}(\tau)|\delta(x(t) - \alpha)f_{X\dot{X}}(x, \dot{x}, \tau)\,dx\,d\dot{x}\,d\tau \qquad (5.185)$$

$$= \int_0^T \int_{-\infty}^{+\infty} |\dot{x}(\tau)|f_{X\dot{X}}(\alpha, \dot{x}, \tau)\,d\dot{x}\,d\tau$$

where $f_{X\dot{X}}$ is the joint density function of $X(t)$ and $\dot{X}(t)$. The above equation accounts for both the up-crossings and down-crossings. It is reasonable to assume that any positive crossing is followed by a negative crossing. Thus,

$$E[N^+(T,\alpha)] = E[N^-(T,\alpha)] = \frac{1}{2}E[N(T,\alpha)] \qquad (5.186)$$

where $N^-(T,\alpha)$ counts the number of down-crossings of $X(t)$ from level α. Due to Equation (5.186), Equation (5.187) becomes

$$E[N^+(T,\alpha)] = \int_0^T \int_0^\infty |\dot{x}(\tau)|f_{X\dot{X}}(\alpha, \dot{x}, \tau)\,d\dot{x}\,d\tau = \int_0^T v^+(\tau, \alpha)\,d\tau = m_1 \qquad (5.187)$$

where the number of out-crossings per unit time, or out-crossing rate, $v^+(t, \xi)$ is defined by

$$v^+(t,\alpha) = \int_0^\infty |\dot{x}(t)|f_{X\dot{X}}(\alpha, \dot{x}, t)\,d\dot{x} \qquad (5.188)$$

Equation (5.188) is the *Rice's formula* (Rice, 1954). For stationary processes the out-crossing rate does not depend on time; i.e. $v^+(t, \alpha) = v^+(\alpha)$.

Equation (5.176) provides the following upper bound to the probability of failure in the time interval $[0, T]$

$$P_f(T) \leq m_1 = \int_0^T v^+(t, \xi)\,dt \qquad (5.189)$$

which becomes

$$P_f(T) \leq v^+(t, \xi)T \qquad (5.190)$$

if the process is stationary.

If the probability of $X(0) < \alpha$ is accounted for, the failure probability in Equation (5.180) becomes

$$P_f(T) = 1 - (1 - P_f(0))P(N^+(T, \alpha) = 0 \mid X(0) < \alpha)$$

Because of Equation (5.179), the above equation yields

$$P_f(T) = 1 - (1 - P_f(0)) \exp\left(-E(N^+(T, \alpha)) \mid X(0) < \alpha\right) \tag{5.191}$$

which can be approximated by

$$P_f(T) = 1 - (1 - P_f(0)) \exp\left(-E(N^+(T, \alpha))\right) \tag{5.192}$$

It has been observed however, that the mean number of out-crossings given $X(0) < \alpha$, is approximated more accurately by

$$E[N^+(T, \alpha) \mid X(0) < \alpha] \approx \frac{E[N^+(T, \alpha)]}{1 - P(0)} \tag{5.193}$$

This provides a better approximation of the probability of failure as

$$P_f(T) = 1 - (1 - P_f(0)) e^{-\frac{E[N^+(T, \alpha)]}{1 - P(0)}} \tag{5.194}$$

Equation (5.194) usually yields accurate results even for relatively low threshold levels, where the out-crossings are not statistically independent.

5.7.2 *Probability Density Function of Peaks*

A random process can have several peaks in a time interval $[0, T]$ (Figure 5.39). The expected number of peaks $M(\alpha, t)$ above level α per unit time is

$$M(\alpha, t) = \int_{\alpha}^{\infty} \int_{-\infty}^{0} |\ddot{x}| f_{X, \dot{X}, \ddot{X}}(\alpha, 0, \ddot{x}) \, d\ddot{x} \, dx \tag{5.195}$$

where $f_{X, \dot{X}, \ddot{X}}(x, \dot{x}, \ddot{x})$ is the joint PDF of $X(t)$ and its first two derivatives at t.

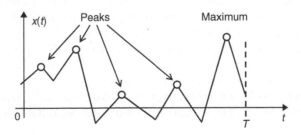

Figure 5.39 Local peaks and maximum value of process $X(t)$ in time interval $[0, T]$.

The expected number of peaks per unit time, $MT(t)$, regardless of the level is obtained by integrating Equation (5.195) with respect to x from $-\infty$ to $+\infty$, as

$$MT(t) = \int_{-\infty}^{0} |\ddot{x}| f_{X,\ddot{X}}(0,\ddot{x})\, d\ddot{x} \tag{5.196}$$

We can calculate the cumulative probability distribution of a peak at a level by normalizing the expected number of peaks below that level using the expected number of peaks regardless of the level. The PDF of a peak is the derivative of the cumulative distribution of a peak with respect to the level. It can be shown that this PDF of a peak at level α is

$$f_{peak}(\alpha, t) = \frac{1}{MT(t)} \int_{-\infty}^{0} |\ddot{x}| f_{X,\dot{X},\ddot{X}}(\alpha,\ 0,\ddot{x})\, d\ddot{x} \tag{5.197}$$

5.7.3   First Passage Time

This section describes how to calculate the probability distribution of the first passage time T_f (Figure 5.36) under the Poisson assumption for a stationary or non-stationary process.

Consider a stationary process first. If the level is high, e.g., larger than the mean value plus six standard deviations of the process, the up-crossing process can be assumed to be a Poisson process with stationary increments and with interarrival rate $v^+(\alpha)$. Then according to Equations (5.178) and (5.179), the first passage time, T_f, follows the exponential distribution

$$F_{T_f}(t) = 1 - e^{-v^+(\alpha)t} \tag{5.198}$$

where $v^+(\alpha)$ can be calculated from Rice's formula in Equation (5.188). Equation (5.198) assumes that $P_f(0) = 0$. The probability of first excursion failure, $P(F)$, is the probability that the process will exceed α during the time interval $[0, T]$. It is given by

$$P(F) = P(t \le T) = F_{T_f}(T) = 1 - e^{-v^+(\alpha)T} \tag{5.199}$$

It should be emphasized that Equation (5.199) assumes that the peaks of the process are statistically independent. However, subsequent peaks are usually positively correlated, especially if the process is narrowband (e.g., the energy of the process is confined to a narrow frequency range). Yang and Shinozuka (1971, 1972) proposed a method to account for correlation between subsequent peaks of a process assuming that these peaks are subjected to a Markov chain condition.

If the random process is nonstationary, the out-crossing rate $v^+(t, \alpha)$ is a function of time. Then the probability of first excursion failure during and interval $[0, T]$ becomes

$$P(F) = 1 - e^{-\int_{0}^{T} v^+(t,\alpha)\, dt} \tag{5.200}$$

The value of the cumulative probability distribution of the first passage time, T_f, is equal to the probability of failure during the period $[0, t]$; i.e.

$$F_{T_f}(t) = P(T_f \leq t) = P(F) = 1 - e^{-\int_0^t v^+(\tau, \alpha)d\tau} \qquad (5.201)$$

5.7.4 First Excursion Failure

Consider a stationary Gaussian process $X(t)$ with zero mean. If the mean is not zero, we can easily create another process by subtracting the mean from $X(t)$. The density function of a zero mean Gaussian process $X(t)$ is

$$f_{X\dot{X}}(x, \dot{x}) = \frac{1}{2\pi\sigma_X\sigma_{\dot{X}}} \exp\left[-\frac{1}{2}\left(\left(\frac{x}{\sigma_X}\right)^2 + \left(\frac{\dot{x}}{\sigma_{\dot{X}}}\right)^2\right)\right] \qquad (5.202)$$

For a given threshold α, the out-crossing rate can be determined from Equation (5.188) as

$$
\begin{aligned}
v^+(\alpha) &= \int_0^\infty \dot{x} f_{X\dot{X}}(\alpha, \dot{x})\, d\dot{x} \\
&= \int_0^\infty \dot{x} \frac{1}{2\pi\sigma_X\sigma_{\dot{X}}} e^{-\frac{1}{2}\left(\left(\frac{\alpha}{\sigma_X}\right)^2 + \left(\frac{\dot{x}}{\sigma_{\dot{X}}}\right)^2\right)}\, d\dot{x} \\
&= \frac{1}{2\pi\sigma_X\sigma_{\dot{X}}} e^{-\frac{1}{2}\left(\left(\frac{\alpha}{\sigma_X}\right)^2\right)} \int_0^\infty \dot{x} e^{-\frac{1}{2}\left(\frac{\dot{x}}{\sigma_{\dot{X}}}\right)^2}\, d\dot{x} \\
&= \frac{\sigma_{\dot{X}}}{2\pi\sigma_X} e^{-\frac{1}{2}\left(\frac{\alpha}{\sigma_X}\right)^2}
\end{aligned}
\qquad (5.203)
$$

where σ_X^2 is the variance of the process and $\sigma_{\dot{X}}^2$ is the variance of the derivative of the process; i.e. $\sigma_{\dot{X}}^2 = \frac{1}{2\pi}\int_{-\infty}^{+\infty} S_{\dot{X}\dot{X}}(\omega)d\omega = \frac{1}{2\pi}\int_{-\infty}^{+\infty} \omega^2 S_{XX}(\omega)d\omega$. For $\alpha = 0$, the zero-crossing rate is

$$v^+(0) = \frac{1}{2\pi} \frac{\sigma_{\dot{X}}}{\sigma_X} \qquad (5.204)$$

After we have calculated the out-crossing rate, we can estimate the probability of first excursion failure from Equation (5.199).

The peaks (both positive and negative) of a stationary Gaussian process with zero mean follow the *Rice PDF* which is a weighted average of a Gaussian PDF and a Rayleigh PDF as

$$f_{peak}(\alpha) = (1 - \xi^2)^{1/2} \frac{1}{(2\pi)^{1/2}\sigma_X} e^{-\frac{\alpha^2}{2\sigma_X^2(1-\xi^2)}} + \xi\Phi\left[\frac{\xi\alpha}{\sigma_X(1-\xi^2)^{1/2}}\right] \frac{\alpha}{\sigma_X^2} e^{-\frac{\alpha^2}{2\sigma_X^2}} \qquad (5.205)$$

where $\Phi(\cdot)$ is the cumulative probability distribution function of a standard Gaussian random variable (zero mean, unit standard deviation). The parameter ξ is the ratio of the expected zero upcrossing rate to the expected number of peaks; i.e. $\xi = \frac{v^+(0)}{MT} = \frac{\sigma_{\dot{X}}^2}{\sigma_X \sigma_{\ddot{X}}}$. This is the *bandwidth parameter*. It expresses the degree to which the energy of the random process is confined to a narrow frequency range. The derivation of the Rice PDF can be found in the book by Lutes and Sarkani (2004, pp. 491–492). $\sigma_{\ddot{X}}^2$ is the variance of the acceleration provided by $\sigma_{\ddot{X}}^2 = \frac{1}{2\pi} \int_{-\infty}^{+\infty} S_{\ddot{X}\ddot{X}}(\omega)d\omega = \frac{1}{2\pi} \int_{-\infty}^{+\infty} \omega^4 S_{XX}(\omega)d\omega$. The PDF of the normalized level, $\eta = \frac{\alpha}{\sigma_X}$, is

$$f_H(\eta) = f_{peak}(\eta \sigma_X)\sigma_X \tag{5.206}$$

Distribution of Local Extremes for a Narrow-band Stationary Process

A narrow-band process is a process whose energy is confined to a narrow frequency range (Figure 5.37). The number of peaks of a narrow-band process is almost equal to the number of zero up-crossings. Therefore, the bandwidth parameter is close to one ($\xi \cong 1$). If a stationary Gaussian process is also narrow-band then the peaks follow the Rayleigh PDF

$$f_{peak}(\alpha) = \frac{\alpha}{\sigma_X^2} e^{-\frac{\alpha^2}{2\sigma_X^2}} \cdot 1(\alpha) \tag{5.207}$$

where $1(\alpha)$ is a unit step function (e.g., a function that is one for positive values of the independent variable and zero otherwise).

Distribution of Local Extremes for a Broad-band Stationary Process

At the other extreme, a broad-band process has bandwidth parameter equal to zero ($\xi = 0$). In this case, the PDF of the peaks reduces to a Gaussian density

$$f_{peak}(\alpha) = \frac{1}{(2\pi)^{1/2}\sigma_X} e^{-\frac{\alpha^2}{2\sigma_X^2}} \tag{5.208}$$

Shinozuka and Yang (1971), and Yang and Liu (1980) estimated the probability distribution of the peaks of a nonstationary process. If the time period is much longer than the expected period of the process, then the cumulative probability distribution of the peaks at a level α can be approximated using the following equation

$$F_{peak}(t, \alpha) \cong 1 - \frac{\int_0^t v^+(\tau, \alpha)\, d\tau}{\int_0^t v^+(\tau, 0)\, d\tau} \tag{5.209}$$

The ratio on the right hand side of the above equation is the expected number of up-crossings of level α divided by the number of zero up-crossings. Shinozuka and

Yang (1971) showed that the distribution of the peaks fits the Weibull distribution very well

$$F_{peak}(T, \alpha) = 1 - e^{-(\frac{\alpha}{\sigma})^{\beta}} \qquad (5.210)$$

where the parameters β and σ are the shape and scale parameters of the distribution, respectively. These parameters depend on the characteristics of the process and the duration T.

5.7.5 Fatigue Failure

Fatigue failure is another important failure mode for a vibrating system. It is due to accumulation of damage that is inflicted to the system by the oscillating stresses, which individually are not enough to cause failure. For example, a spot weld in a car body may fail due to accumulation of damage caused by vibrating stresses. There are two approaches to the problem of fatigue failure; a cumulative damage approach that is usually based on Miner's rule or a fracture mechanics approach. The first approach is popular because it is simple and also yields reasonably accurate estimates of the lifetime of a component. Fracture mechanics based approaches are however, more complex. In the following, we will use Miner's rule (Miner, 1945, Wirsching, et al. 1995) to analyze fatigue failure of a dynamic system under a random, time varying excitation.

Traditionally, the S-N curve is used to characterize the behavior of materials in fatigue. It relates the number of stress cycles, which a material can withstand before failing due to fatigue, to the vibration amplitude of these stress cycles

$$NS^b = c \qquad (5.211)$$

where N is the number of stress cycles, S is the amplitude and b and c are constants that depend on the material type. The exponent b ranges from 3 to 6. Note that the mean stress is assumed zero and the amplitude is assumed constant. Experimental results have shown that the above equation and the values of b and c are independent of the exact shape of the stress time history.

Suppose that a structure is subjected to n cycles of amplitude S, where n is less than N. Then we can define the damage D inflicted to the structure as the ratio of n over N; i.e.

$$D = \frac{n}{N} \qquad (5.212)$$

If the structure is subjected to η_1 cycles of amplitude S_1, η_2 cycles of amplitude $S_2, \ldots, \eta_n$, cycles of amplitude S_n, then we can calculate the damage inflicted by each set of stress cycles and then add them to get the total damage using Miner's rule, as

$$D = \sum_i \frac{n_i}{N_i} \qquad (5.213)$$

where N_i is the number of cycles of amplitude S_i to failure. According to Miner's rule, failure occurs when the damage exceeds 1. Also Miner's rule states that D is independent of the order of cycles. Although experimental evidence has shown that the order is important, and several rules that account for the order have been proposed, none of these rules correlates with experimental results better than Miner's rule.

If the stress is a random variable, D will also be a random variable. Then the problem of calculating the probability of fatigue failure is equivalent to that of calculating the probability that D exceeds one. Miles (1954) estimated the mean value of the damage for a stationary, narrow-band process over a period T, as

$$E(D(T)) = v^+(0)T \int_0^\infty \frac{f_{peak}(\alpha)}{N(\alpha)} d\alpha \tag{5.214}$$

where $N(\alpha)$ is the number of stress cycles of amplitude α to failure. If the random process is also Gaussian

$$E(D(T)) = \frac{v^+(\alpha)T}{c} (\sqrt{2}\sigma_X)^b \Gamma\left(1 + \frac{b}{2}\right) \tag{5.215}$$

Mark (1961) derived an expression for the variance of the damage for a second-order mechanical system subjected to a stationary Gaussian process.

If the applied stress is a wide-band process, we can scale the damage obtained from Equation (5.215) by a correction factor, λ so that

$$E(D(t)) = \lambda \frac{v^+(\alpha)T}{c} (\sqrt{2}\sigma_X)^b \Gamma\left(1 + \frac{b}{2}\right) \tag{5.216}$$

Wirsching and Light (1980) provide estimates of the correction factor λ.

5.7.6 Estimation of Damage due to Cyclic Loading and Probability of Fatigue Failure

Most engineering structures, such as airplanes, ships, and offshore platforms, are subjected to nonstationary excitations which are due to changes in the loading, operating or weather conditions. To calculate the fatigue damage of these structures, we can divide the spectrum of loading and operating conditions into cells. In each cell, these conditions are assumed constant. The time of exposure to the conditions in each cell is estimated using data from previous designs. Then we calculate the fatigue damage corresponding to operation in each cell and add up the damage over all cells. An alternative approach finds an equivalent stationary load that causes damage equal to the damage due to operations in all cells and the damage due to the equivalent load. Wirsching and Chen (1988) reviewed methods for estimating damage in ocean structures.

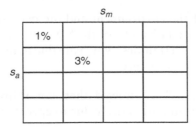

Figure 5.40 Load spectrum for alternating and mean stresses.

Consider that the load spectrum (percentage of lifetime during which a structure is subjected to a given mean and alternating load) is given (Figure 5.40). From this spectrum, we can estimate the joint PDF $f_{S_a S_m}(s_a, s_m)$ of the mean stress S_m and the alternating stress S_a.

For every cell in Figure 5.40, we can find an equivalent alternating stress that causes the same damage as the alternating and mean stresses when they are applied together. This equivalent stress is $\frac{s_{a_i}}{(1 - s_{m_j}/S_Y)}$ where s_{a_i} and s_{m_j} are the alternating and mean stresses corresponding to the (i, j) cell, and S_Y is the yield stress. The above expression is obtained using the Goodman diagram. Suppose n_{ij} is the number of cycles corresponding to that cell. Then, the damage due to the portion of the load spectrum corresponding to the (i, j) cell is $D_{ij} = \frac{n_{ij}}{N_{ij}}$. N_{ij} is the number of cycles at failure when s_{a_i} and s_{m_j} are applied. From the S-N curve we can find the number of cycles as

$$N_{ij} \left[\frac{s_{a_i}}{1 - \frac{s_{m_j}}{S_Y}} \right]^b = c \Rightarrow N_{ij} = \frac{c \left(1 - \frac{s_{m_j}}{S_Y} \right)^b}{s_{a_i}^b} \tag{5.217}$$

Therefore, the damage corresponding to the (i, j) cell is

$$D_{ij} = \frac{n_{ij} s_{a_i}^b}{c (1 - \frac{s_{m_j}}{S_Y})^b} \tag{5.218}$$

The total damage is

$$D = \sum_i \sum_j D_{ij} = \sum_i \sum_j \frac{n_{ij} s_{a_i}^b}{c \left(1 - \frac{s_{m_j}}{S_Y} \right)^b} = \sum_i \sum_j \frac{p_{ij} s_{a_i}^b}{c \left(1 - \frac{s_{m_j}}{S_Y} \right)^b} \tag{5.219}$$

where p_{ij} is the percentage of the number of cycles corresponding to the (i, j)th load cell. This percentage is obtained from information on the load spectrum.

Sources of uncertainty in damage calculations include:

- the random nature of the environment and the material properties (for example uncertainties in parameters b and c of the S-N curve).
- epistemic (modeling) uncertainties, which are due to idealizations in the models of the environment and the structure.

Nikolaidis and Kaplan (1992) showed that, for ships and offshore platforms, uncertainty in fatigue life is almost entirely due to epistemic (modeling) uncertainties.

5.8 Evaluation of the Stochastic Response of Nonlinear Systems

Analysis of nonlinear systems is considerably harder than that of linear systems. A principal reason is that the superposition principle does not hold for nonlinear systems. Closed form solutions have been found for few simple nonlinear systems.

Methods for analysis of nonlinear systems include,

- methods using the Fokker-Plank equation, which shows how the PDF of the response evolves in time
- statistical linearization methods (Roberts and Spanos, 1990)
- methods using state-space and cumulant equations.

The above methods are presented in the book by Lutes and Sarkani (2004).

5.9 Concluding Remarks

Many systems are subjected to excitations that can be modeled as random processes (random functions) or random fields because these excitations vary in time and/or in space. Probabilistic analysis of dynamic systems involves three main steps; modeling of the excitations, calculation of the statistics of the response and calculation of the probability of failure. To develop a probabilistic model of a random process we need to specify the joint probability distribution of the values of a random process for all possible values of the index parameter (e.g., the time). This is often impractical. Therefore, often we determine the second-order statistics of the random process, that is, the mean value, and the autocorrelation function. We have analytical tools, in the form of closed form expressions, for the second-order statistics of linear systems. However, few analytical solutions are available for nonlinear systems. Simulation methods are also used in developing a probabilistic model of a random process. This chapter provided the basics of the commonly used time series modeling and spectral representation simulation methods.

A system may fail due to first excursion failure, in which the response exceeds a level, and fatigue failure due to damage accumulation. Tools are available for estimating the probability of failure under these two modes. For first excursion failure we have equations for approximating the average upcrossing rate of a level, the PDF of the local peaks of the response and the PDF of the time to failure. For fatigue failure, we have equations for the cumulative damage. This chapter introduced methods for dealing with these kinds of failure when a system is subjected to loads that follow a stochastic process. In the next chapter, we talk about subjective assessment of probabilities and probability distributions. Good understanding of subjective probability helps us make decisions when data is limited and also helps reduce the epistemic uncertainty.

Questions and Exercises

Modeling Random Processes in Time Domain

5.1   A binary process is defined by

$$X(t) = x_0 \text{ or } -x_0 \text{ for } (n-1)T < t < nT$$

where the levels x_0 and $-x_0$ occur with equal probability, T is a positive constant and $n = 0, \pm 1, \pm 2, \ldots$ Sketch a sample function.

5.2   Using the following relationship,

$$f_X(x_1, x_2; t_1, t_2) = f_X(x_1, x_2; t_1 + \tau, t_2 + \tau)$$

prove that

$$R_{XX}(t_1, t_1 + \tau) = E[X(t_1) X(t_1 + \tau)] = R_{XX}(\tau)$$

is true.

5.3   For the random process of Problem 5.1, calculate

a) The mean value $E[X(t)]$
b) $R_{XX}(t_1 = 0.1T, t_2 = 0.8T)$

5.4   A random process $X(t)$ consists of three sample functions $x_1(t) = 2$, $x_2(t) = 2\cos t$, and $x_3(t) = \sin t$ each occurring with equal probability. Is the process stationary or wide-sense stationary?

5.5   Two statistically independent, zero-mean random processes $X(t)$ and $Y(t)$ have autocorrelation functions

$$R_{XX}(\tau) = e^{-|\tau|} \quad \text{and} \quad R_{YY}(\tau) = \sin(\pi\tau)$$

a) Find the autocorrelation function of the random process $Z_1(t) = X(t) + Y(t)$.
b) Find the autocorrelation function of the random process $Z_2(t) = X(t) - Y(t)$.
c) Find the cross-correlation function of Z_1 and Z_2.

5.6   Two random processes are defined as

$$X(t) = f_1(t + \varepsilon) \quad \text{and} \quad Y(t) = f_2(t + \varepsilon)$$

where $f_1(t)$ and $f_2(t)$ are periodic waveforms with period T and ε is a random variable uniformly distributed in $(0, T)$. Find an expression for the cross-correlation function $E[X(t) Y(t + \tau)]$.

5.7 A stationary ergodic random process is given with the following auto-correlation function.

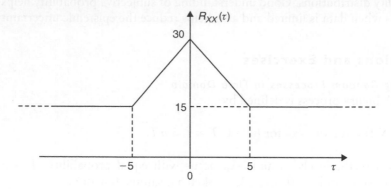

Calculate a) $E[X(t)]$, b) $E[X^2(t)]$, and c) σ_X^2.

5.8 For an at least wide-sense stationary process $X(t)$, a new process $Y(t) = X(t) - X(t+\tau)$ is defined.

a) Show that the mean value of $Y(t)$ is zero even if $X(t)$ has a non-zero mean.
b) Show that $\sigma_Y^2 = 2[R_{XX}(0) - R_{XX}(\tau)]$
c) If $Y(t) = X(t) + X(t+\tau)$, find $E[Y(t)]$ and σ_Y^2. How do these results compare with the results from parts a) and b)?

5.9 Two random processes $X(t)$ and $Y(t)$ are given. Find expressions for the autocorrelation function of $Z(t) = X(t) + Y(t)$ if

a) $X(t)$ and $Y(t)$ are correlated.
b) $X(t)$ and $Y(t)$ are uncorrelated.
c) $X(t)$ and $Y(t)$ are uncorrelated with zero means.

5.10 Assume that a random process $X(t)$ is ergodic with an autocorrelation function of $R_{XX}(\tau) = 12 + \frac{1}{\tau^2}[1 + \cos(5\tau)]$

a) Find μ_X
b) What is the average power in $X(t)$?

5.11 For a random process $X(t)$ with $R_{XX}(t, t+\tau) = 16 + 5e^{-|\tau|}$, calculate the total average power in $X(t)$. Is $X(t)$ first-order stationary, ergodic or wide-sense stationary?

5.12 The random processes $X(t)$ and $Y(t)$ are defined as $X(t) = x_0 \cos(\omega_0 t + \Theta)$ and $Y(t) = y_0 \sin(\omega_0 t + \Theta)$ where x_0, y_0, and ω_0 are constants and Θ is a random variable uniformly distributed in $(0, 2\pi)$.

a) Show that $X(t)$ and $Y(t)$ are zero-mean, wide-sense stationary processes with autocorrelation functions $R_{XX}(\tau) = \frac{x_0^2}{2}\cos(\omega_0\tau)$ and $R_{YY}(\tau) = \frac{y_0^2}{2}\cos(\omega_0\tau)$.
b) Calculate the cross-correlation function $R_{XY}(t, t+\tau)$, and show that $X(t)$ and $Y(t)$ are jointly wide-sense stationary.

5.13   Consider the random processes $X(t)=x_0 \cos(\omega_0 t+\Theta)$ and $Y(t)=y_0 \cos(\omega_1 t+\Phi)$ where x_0, y_0, ω_1 and ω_0 are constants and Θ and Φ are statistically independent random variables each uniformly distributed in $(0, 2\pi)$. Show that $X(t)$ and $Y(t)$ are jointly wide-sense stationary.

Fourier analysis

5.14   Calculate the Fourier transform of

$$x(t) = \begin{cases} x_0 & \text{for } -\tau/2 < t < \tau/2 \\ 0 & \text{otherwise} \end{cases}$$

where $\tau > 0$ and x_0 are real constants.

5.15   Calculate the Fourier transform of $y(t) = x(t) \cos(\omega_0 t + \theta_0)$ where $x(t)$ is given in Problem 5.14 and ω_0 and θ_0 are real constants.

5.16   Calculate the Fourier transform of

$$x(t) = \begin{cases} \dfrac{x_0 |t|}{\tau} & \text{for } |t| \leq \tau \\ 0 & \text{for } |t| > \tau \end{cases}$$

where $\tau > 0$ and x_0 are real constants.

5.17   Calculate the Fourier transform of $x(t) = u(t)e^{j\omega_0 t}$ where $u(t)$ is the unit-step function and ω_0 is real constant.

5.18   Calculate the Fourier transform of the signal

$$x(t) = \begin{cases} x_0 t^2 & \text{for } 0 < t < \tau \\ 0 & \text{otherwise} \end{cases}$$

if $\tau > 0$ and x_0 are real constants.

5.19   Let $X(\omega)$ be the Fourier transform of $x(t)$. Calculate the transforms of the following functions in terms of $X(\omega)$:

a) $x(t-4)e^{j\omega_0 t}$

b) $\dfrac{dx(t)}{dt} e^{j\omega_0(t+1)}$

c) $x(t+3) = x(2t)$

where ω_0 is a real constant.

5.20   If $x(t) \leftrightarrow X(\omega)$, calculate the inverse transforms of the following functions in terms of $x(t)$:

a) $X(\omega)X^*(\omega + \omega_0)$

b) $X(\omega - \omega_0)\dfrac{dX(\omega)}{d\omega}$

where ω_0 is a real constant.

5.21   Using the Fourier transform properties, show that the following relations are true:

a) $c \leftrightarrow c(2\pi)\delta(\omega)$ where c is a constant.

b) $\cos(\omega_0 t) \leftrightarrow \pi[\delta(\omega - \omega_0) + \delta(\omega + \omega_0)]$ where ω_0 is a real constant.

5.22   Show that a frequency-domain impulse can be represented by

$$\delta(\omega) = \frac{1}{2\pi} \int\limits_{-\infty}^{\infty} e^{-j\omega t} dt$$

5.23   Show that a time-domain impulse can be represented by

$$\delta(t) = \frac{1}{2\pi} \int\limits_{-\infty}^{\infty} e^{j\omega t} d\omega$$

Spectral analysis

5.24   A random process $x(t) = u(t)x_0 \cos(\omega_0 t + \Theta)$ is given, where x_0 and ω_0 are constants, Θ is a uniformly distributed random variable in $(0, \omega)$, and $u(t)$ is the unit-step function.

   a) Is $X(t)$ wide-sense stationary?
   b) Calculate the power in $X(t)$.

5.25   A random process $X(t) = x_0 \cos(\omega_0 t) + y_0 \sin(\omega_0 t)$ is given where x_0 and y_0 are random variables and ω_0 is a real constant. Calculate the power spectral density.

5.26   Determine which of the following functions represent a valid power spectral density. For those that cannot be, explain why.

   a) $\frac{\omega^2}{\omega^4 + 1}$,   b) $e^{-\omega}$,   c) $\omega^2 - \delta(\omega)$,   d) $\frac{\omega}{1 + j\omega^2}$.

5.27   A wide-sense stationary random process $X(t)$ with a mean value of 1.0 cm has the following power spectral density

$$S_{XX}(\omega) = \begin{cases} 0.004 \, \text{cm}^2 \cdot \text{sec} & \text{for } 20 \, \text{rad/sec} \leq |\omega| \leq 1200 \, \text{rad/sec} \\ 0 & \text{otherwise} \end{cases}$$

   Calculate:

   a) the RMS value of $X(t)$, and
   b) the standard deviation of $X(t)$.

5.28   The autocorrelation function of the random process $X(t)$ is $R_{XX}(\tau) = e^{-|\tau|} \cos(\omega_0 \tau)$ where ω_0 is a real constant. Calculate the power spectral density of $X(t)$.

5.29   A random process $X(t)$ with a power spectral density of $S_{XX}(\omega) = \frac{2\omega^2}{(1 + \omega^2)}$ is applied to an ideal differentiator. Calculate the power spectral density of the differentiator's output.

5.30   A deterministic signal $\cos(\omega_0 t)$ where ω_0 is a real constant, is added to a noise process $N(t)$ with $S_{NN}(\omega) = \frac{\omega^2}{\omega_0^2 + \omega^2}$ where $\omega_0 > 0$ is a constant. Calculate the ratio of average signal power to average noise power.

5.31   Show that the following two properties of the cross-power spectral density are valid:

a) If $X(t)$ and $Y(t)$ are orthogonal then $S_{XY}(\omega) = S_{YX}(\omega) = 0$.
b) If $X(t)$ and $Y(t)$ are uncorrelated with constant means μ_X and μ_Y, then
   $S_{XY}(\omega) = S_{YX}(\omega) = 2\pi\mu_X\mu_Y\delta(\omega)$.

5.32   The autocorrelation function of a random process $X(t)$ is $R_{XX}(\tau) = 1 + e^{-\tau^2}$.

a) Calculate the power spectral density of $X(t)$.
b) What is the average power in $X(t)$?
c) What fraction of the power is in the $-1 \le \omega \le +1$ frequency band?

5.33   If $X(t)$ and $Y(t)$ are real random processes, determine which of the following are valid. For those what are not, state at least one reason why.

a) $R_{XX}(\tau) = e^{-|\tau|}$,   b) $|R_{XY}(\tau)| \le j\sqrt{R_{XX}(0)R_{YY}(0)}$,
c) $R_{XX}(\tau) = \sin(\tau)$,   d) $S_{XX}(\omega) = \frac{2}{2+\omega^3}$,
e) $S_{XX}(\omega) = \frac{e^{-|\tau|}}{1+\omega^2}$,   f) $S_{XY}(\omega) = 1 + j\omega^2$, and
g) $S_{XY}(\omega) = 2\delta(\omega)$.

5.34   The cross-correlation of the jointly wide-sense stationary processes $X(t)$ and $Y(t)$ is equal to $R_{XY}(\tau) = u(\tau)e^{-\omega_0\tau}$ where $\omega_0 > 0$ is a constant. Calculate

a) $R_{YX}(\tau)$, and
b) $S_{XY}(\omega)$ and $S_{YX}(\omega)$.

Calculation of the Response

5.35   An input $x(t) = e^{-t}$ is applied to a linear system which has an impulse response $h(t) = \omega_0 e^{-\omega_0 t}$ where ω_0 is a real positive constant. Calculate the system's response using Equation (5.132).

5.36   For the linear system of Problem 5.35, calculate the power spectral density $Y(\omega)$ of the response using Equation (5.133).

5.37   The input

$$x(t) = \begin{cases} x_0 & \text{for } 0 < t < T \\ 0 & \text{otherwise} \end{cases}$$

where x_0 and T are positive real constants, is applied to the system of Problem 5.35. Calculate the time response $y(t)$.

5.38   Two systems with transfer functions $H_1(\omega)$ and $H_2(\omega)$ are connected in series. Show that the transfer function $H(\omega)$ of the combined system is $H(\omega) = H_1(\omega)H_2(\omega)$.

5.39   A random process $X(t) = x_0 \sin(\omega_0 t + \Theta)$ where x_0 and ω_0 are real positive constants and Θ is a random variable uniformly distributed in $(-\pi, \pi)$, is applied to the linear system of Problem 5.35. Calculate the system's response process using Equation (5.135).

5.40   A random process $X(t)$ is applied to a linear time-invariant system. If the resulting response is $Y(t) = X(t) + X(t - \tau)$ where τ is a real constant, calculate the system's transfer function.

5.41   The autocorrelation function of a random process $X(t)$ is $R_{XX}(\tau) = 1 + e^{-|\tau|}$.
       Calculate the mean value of the response of a system having an impulse response function

$$h(t) = \begin{cases} e^{-\omega_1 t} \sin(\omega_0 t) & \text{for } 0 < t \\ 0 & \text{for } t < 0 \end{cases}$$

where ω_1 and ω_0 are real positive constants.

5.42   A random process $X(t)$ with the autocorrelation function of $R_{XX}(\tau) = e^{-|\tau|}$, is input to a system with the impulse response function

$$h(t) = \begin{cases} \omega_0 e^{-\omega_0 t} & \text{for } 0 < t \\ 0 & \text{for } t < 0 \end{cases}$$

where ω_0 is a real positive constant. Calculate the autocorrelation function of the response $Y(t)$.

5.43   The transfer function of a linear system is $H(\omega) = \frac{e^{j\omega/10}}{(10 + j\omega)^2}$. Calculate and sketch its impulse response.

5.44   An input $x(t) = e^{-t}$ is applied to a linear system with an impulse response function of $h(t) = te^{-\omega_0 t}$, where $\omega_0 > 0$ is a real constant. Using Equation (5.132), calculate the response $y(t)$.

5.45   A stationary random process $X(t)$ is applied to a linear system with an impulse response function $h(t) = t^2 e^{-4t}$. If $E[X(t)] = 2$, what is the mean value of the system's response $Y(t)$?

5.46   A random process $X(t)$ is applied to a linear system with an impulse response function of $h(t) = te^{-t}$. The cross-correlation of $X(t)$ with the output $Y(t)$ is $R_{XY}(\tau) = \tau e^{-\tau}$.

a) Calculate the autocorrelation of $Y(t)$.
b) What is the average power in $Y(t)$?

5.47   Two identical systems are connected in series. The impulse response function of each system is $h(t) = te^{-5t}$. A wide-sense stationary process $X(t)$ is applied to the combined system.

a) Calculate the response $Y(t)$ of the series system.
b) If $E[X(t)] = 4$, calculate $E[Y(t)]$.

Chapter 6

Subjective (Bayesian) Probability

6.1 Introduction

Most people think that probability is the long-term relative frequency of an outcome in a repeatable random experiment. This definition is easy to understand, but it is not applicable to decision problems involving unique events, as we discussed in Chapter 2. For example, scientists cannot estimate the probability that the earth's climate changes because of carbon emissions and deforestation by observing as they would, the outcomes of a repeatable experiment. Similarly, one cannot estimate the probabilities that the following propositions are true from repeated observations:

1. The shortest route from Portland, Oregon to Los Angeles is less than 2,000 miles.
2. By 2020, humans will land in Mars.
3. An engineer predicted the stress at a given hot point in a joint of an offshore platform by using a commercial finite element code. The true value of the stress exceeds by 10% the calculated value.
4. A new flu virus will cause more than 200,000 deaths in the US in 2015.

Sometimes, it is impractical to estimate probabilities even for outcomes for which many observations of a repeated experiment seem to be available. For example, suppose that a driver in the US wants to estimate the probability that he/she could die in a car accident in 2015. Approximately, 40,000 people die each year in car accidents in the US out of a population of 300,000,000. Therefore, the driver may think that this probability is 1.3 per 10,000. However, this estimation ignores that each person drives a different car, lives at a different location, and drives differently.

The concept of subjective probability is essential in decision making because it enables a decision maker to quantify the likelihood of the outcomes that affect the consequences of alternative options. Chapter 2 defined subjective probability to be a person's degree of belief that an outcome will occur or that a proposition is true. We assess this probability by observing how the person is inclined to bet on the outcome of an uncertain event or on the truth of a proposition. Specifically, the probability of an outcome is the fair price of a lottery ticket that is worth $1 if the outcome materializes and zero otherwise. This definition is cyclical because it uses the concept of risk neutral attitude, which is ultimately connected to the concept of subjective probability.

This chapter will present the following alternative definition that circumvents this difficulty:

> Subjective probability of an event is the long-term frequency for which the decision maker is indifferent between a lottery ticket that pays some prize (for example, $100) with this frequency and a second ticket that pays the same prize if and only if the event occurs.

This definition of subjective probability uses the long-term relative frequency as a yardstick. A decision maker can estimate the probability of a one-time event using this definition. By following common sense rules (axioms), we will show that subjective probability calculations follow the same rules as the long-term frequency does.

Some authors maintain that the concept of objective probability is meaningless. Howard (2004) claims that all probabilities are subjective because estimation of probabilities from observed data always requires judgment. He says that the term "objective probability" is an oxymoron to him. French (1986, p. 233) claims that the view of probability as a long-term relative frequency is inappropriate to our needs in decision analysis because we cannot conduct an infinitely repeatable experiment in real life.

Strictly speaking, the above concerns are valid. However, relative frequency is very useful to quantify uncertainty because in practice, we can obtain sufficient data to estimate the long-term frequency of many events accurately enough to decide on what outcome to bet. For example, we can calculate the reliability (probability of no failure) of many consumer products from warranty claims records or from consumer magazines. Similarly, we can calculate probabilities of catastrophic airline accidents from publicly available records such as those of the Aircraft Crashes Record Office (http://www.baaa-acro.com/). Finally, objective probability serves as a yardstick to define the concept of subjective probability.

Types of uncertainty

There are two types of uncertainty: random and epistemic. Random (or aleatory) uncertainty is due to inherent randomness in physical phenomena or processes. Epistemic uncertainty is due to incomplete understanding about a phenomenon or process that is not in itself random and, in principle, is knowable. Table 6.1 presents examples of events in which there is random and epistemic uncertainty. One can reduce or eliminate epistemic uncertainty by acquiring knowledge. However, one cannot reduce random uncertainty because this uncertainty is due to inherent randomness.

Some events involve both types of uncertainty. Suppose that we flip a thumbtack. There are two sources of uncertainty about the outcome, a) uncertainty in the long-term

Table 6.1 Events or propositions involving random and epistemic uncertainty.

Random	Epistemic
• Heads-up in a flip of a fair coin • Six in a roll of a fair die • Pick up a king of spades card from a deck of 52 cards	• The distance from New York City to Paris is less than 5,000 miles • The diameter of Mars is less than 10,000 km • Mathematical constant e is less than 3.

frequency of a tip-up flip, and b) inherent variability in the outcome of a flip. The first type of uncertainty is epistemic and the second random. Suppose that we flip the thumbtack many times and record the outcomes. This information reduces the epistemic component of uncertainty because it enables us to estimate the long-term frequency of a tip-up flip. However, the random component remains unchanged; no matter how much we know about the probability of a tip-up flip, we cannot predict the outcome of a next flip because of inherent variability.

Epistemic uncertainty is systemic; it tends to affect an entire population of objects in the same way. Suppose that an automotive engineer made a wrong assumption that led to overestimation of the strength of a suspension control arm. As a result of this error the manufacturer may get many warranty claims around the same time. On the other hand, manufacturing variability and variability in driving conditions induce variability in the life of a control arm from one car to another.

Outline of this chapter

This chapter studies in depth the concept of subjective probability. Section 6.2 presents an alternative definition of subjective probability to that in Chapter 2 that is not cyclical. This definition compares a lottery ticket whose payoff depends on the outcome of interest to a reference ticket whose payoff depends on an outcome with known long-term relative frequency. The event that the arrow of a well oiled, balanced probability wheel will stop in a given sector is an example.

The definition of subjective probability in section 6.2 is based on common sense rules (axioms) that describe how a rational decision maker chooses a risky option among alternatives. The same section proves that calculations of subjective probabilities follow the same rules as those of objective probability in Chapter 3. For example, the subjective probability of the union of disjoint events is equal to the sum of the probabilities of these events.

Section 6.3 describes procedures for eliciting an expert's subjective probability of an outcome and the probability distribution of a variable. Section 6.4 explains Bayesian analysis, which is a method for updating initial subjective estimates of probabilities, by using observations from a repeatable experiment or expert judgments. Section 6.5 explains that humans do not always act rationally when assessing probabilities and identifies some root causes of this irrational behavior. This discussion should help decision makers avoid situations where they could act irrationally.

6.2 Definition of Subjective Probability

6.2.1 Overview

This section defines subjective probability of an event by using a reference experiment as a benchmark. The objective probabilities (long-term relative frequencies) of the outcomes of the reference experiment are obvious. This experiment is a spin of a well oiled probability wheel with a balanced needle (Figure 6.1). We define subjective probability in two steps;

1. We describe six common sense rules (axioms) that govern a rational person's estimates of the relative likelihoods of events.

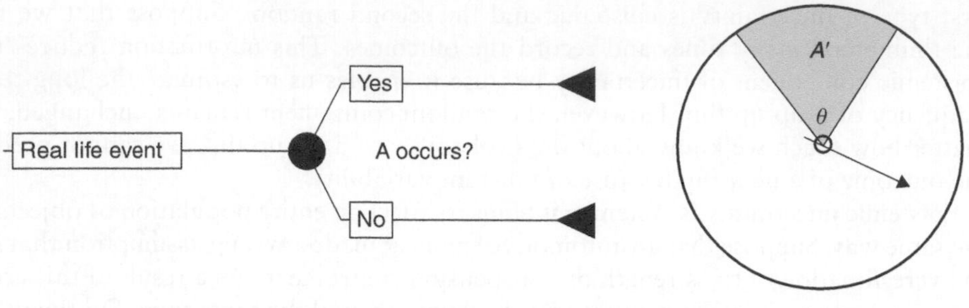

Figure 6.1 Using a probability wheel as a ruler to measure the likelihood of a real life event.

Ticket A is worth $1 if event A occurs and zero otherwise.	Ticket B is worth $1 if event B occurs and zero otherwise.

Case 1: Decision maker strictly prefers ticket A. ⇒ Decision maker believes that A is more likely than B.
Case 2: Decision maker strictly prefers ticket B. ⇒ Decision maker believes that B is more likely than A.
Case 3: Decision maker selects a ticket randomly but indicates that will happily exchange it for the other. ⇒ Decision maker believes that A and B are equally likely.

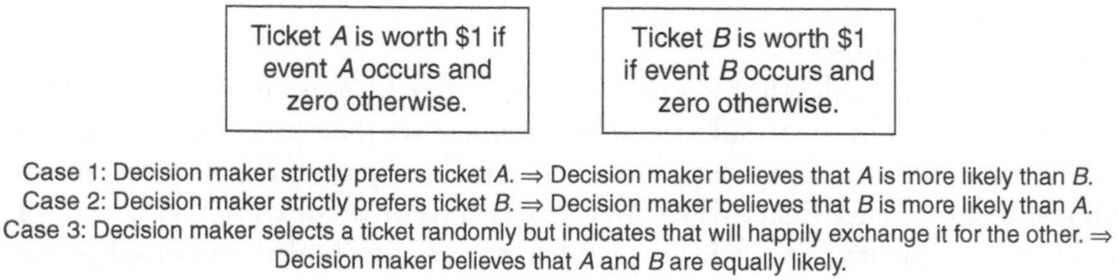

Figure 6.2 Determining a decision maker's belief about the relative likelihood of events A and B by observing the decision maker's inclination to bet on each event.

2. We introduce the reference experiment and use it as a ruler to measure the likelihood of a real life event. We assume that for each real life event, the decision maker can determine an equally likely event on the probability wheel.

The presentation in this subsection is largely based on the approach by French (1986).

The first step is to explain the meaning of the statement "the decision maker believes that event A is more likely than B." For this purpose, consider a facilitator who asks the decision maker to choose between two lottery tickets that are both worth $1 if events A and B occur, respectively (Figure 6.2). Then the facilitator asks the decision maker if he/she is willing to exchange the selected ticket for the other. If the decision maker selects one ticket (say ticket A) and does not want to exchange it for the other, then this means that the decision maker believes that A is more likely than B. If the decision maker agrees to exchange each ticket for the other, it means that he/she believes that A and B are equally likely.

6.2.2 Axiomatic definition of probability

Step 1: A decision maker compares the likelihoods of events and ranks them according to his/her beliefs about their relative likelihood. Different decision makers would rank the same events differently, but the ranking should follow some common sense rules (axioms). For example, if a decision maker believes that event A is more likely than B and B is more likely than C then A must be more likely than C.

We present six axioms (common sense rules) that must govern the relation of the likelihoods of events. The following notation is used:

Relation $A \geq_l B$ means that the decision maker believes that outcome A is at least as likely as B. This is called weak preference. Relation $A >_l B$, which means that the decision maker believes that outcome A is strictly more likely than B, is called strong preference.

Axiom 1 maintains that a rational decision maker can always order events in terms of their likelihood; a rational decision maker does not vacillate when comparing the likelihood of events.

Axiom 1: Comparability

The decision maker can always compare two outcomes A and B and decide that,

- A is at least as likely as B, $A \geq_l B$, or
- B is at least as likely as A, $B \geq_l A$, or
- A is equally likely as B, $A \sim_l B$.

This axiom is equivalent to the statement "there are no outcomes A and B that the decision maker cannot compare." This is probably the most contentious axiom as Section 2.2.4 stated. Often, people vacillate when they compare disparate or rare events. The developers of imprecise probability were motivated by the need of a structured approach to decision making under uncertainty that relaxes the comparability axiom (Walley, 1991, Ferson, 2004, Aughenbaugh and Paredis 2006). The authors have the following comments on the comparability axiom:

- If a decision maker is motivated to seriously consider the consequences of a decision, then he/she will be able compare the likelihood of the outcomes (e.g., French, 1986, p. 226 and p. 237). When people cannot decide which outcome is more likely, they tend to select the default option (e.g., they do nothing and resign to their fate). This is irrational behavior and often has bad consequences.
- In many decisions, we can compare the likelihood of two outcomes by comparing them to outcomes of a reference experiment, such as, the roll of a die or the spin of a probability wheel (see axiom 7 below). The following example demonstrates this point.

Compare the probabilities of the outcomes of the following two experiments. In both experiments, the decision maker draws a ball at random from two urns.

> **Experiment 1:** The decision maker knows that the first urn contains equal numbers of red and black balls.
> **Experiment 2:** The second urn contains red and black balls but their ratio is unknown.

Is the decision maker more likely to draw a red ball in experiment 1 or in experiment 2?

In experiment 1, the relative frequency of outcome "red ball" is $P_1(red) = 0.5$. In the second experiment, the relative frequency is unknown. Many people think that it is impossible to compare the likelihoods of drawing a red ball in the two experiments. Suppose that the decision maker compares two tickets each of which is worth \$100 only if he/she correctly predicts the color of a ball drawn in each experiment. Most people prefer the first ticket because they know the probability of winning $100 in the first experiment.

However, both tickets are equivalent (Raiffa, 1961). In order to show this, Raiffa modified experiment 2 as follows.

> **Modified Experiment 2**: Flip a fair coin. If you get heads bet on red; otherwise bet on black.

When asked to compare experiment 2 to the modified experiment, people typically say that they have the same chance to win $100 in both experiments. This is reasonable because when you have no reason to believe that one color is more likely than the other, you can flip a fair coin to decide how to bet. Consequently, the likelihood of picking up a red ball in experiment 2 is equal to the probability to win $100 in the modified experiment,

red ball in experiment 2 $\sim_l$ win $100 in modified experiment 2

The probability of winning $100 in the modified experiment, $P_{2'}(win)$, is equal to one half of the sum of the fractions of red and black balls in the second urn,

$$P_{2'}(win) = p_{red} \cdot P(heads) + p_{black} \cdot P(tails)$$
$$= 0.5 \cdot (p_{red} + p_{black})$$
$$= 0.5$$

where p_{red} and p_{black} are the fractions of red and black balls. But the probability of picking up a ball from the first urn is also 0.5. Therefore, it is equally likely to pick a red ball from the first and the second urn.

The above example demonstrates that we can compare the likelihoods of the outcomes in a decision by comparing these outcomes to those of a reference experiment in which we know the relative frequencies of the outcomes.

Axiom 2: Transitivity
Transitivity axiom does not require justification: if $A \geq_l B$ and $B \geq_l C$ then $A \geq_l C$.

Axiom 3: Consistency of indifference and weak preference
If the decision maker believes that outcome A is at least as likely as B, and B is at least as likely as A then the decision maker believes that the two outcomes are equally likely, and vice versa.

If $A \geq_l B$ and $B \geq_l A$ then $A \sim_l B$

Axiom 4: Consistency of strict and weak preference
If $A >_l B$ then relation $B \geq_l A$ is false.

Axiom 5: Independence of common events
Addition of an extra common possibility to two events does not change their relative likelihood. Suppose that "rain at noon tomorrow" is at least as likely as "snow at noon

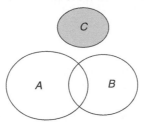

Figure 6.3 Illustration of axiom of independence of common outcomes: Adding an extra common possibility, *C*, to both events *A* and *B* does not change the relative likelihood.

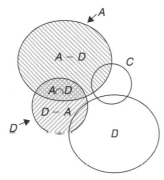

Figure 6.4 Decomposition of events *A* and *D* into three disjoint events in order to prove Lemma 1.

tomorrow." Consider that sunshine and rain cannot occur simultaneously and that the same is true for sunshine and snow. Then "rain or sunshine at noon tomorrow" is as least as likely as "snow or sunshine at noon tomorrow."

Suppose that outcomes A and C are disjoint and so are outcomes B and C. If $A \geq_l B$ then, $A \cup C \geq_l B \cup C$ (Figure 6.3).

We can derive useful facts from this axiom.

Lemma 1: If $A \geq_l C$ and $B \geq_l D$, and A and B and C and D are disjoint pairs, then $A \cup B \geq_l C \cup D$.

Proof: The first step is to decompose the union of events A and D into three disjoint events as Figure 6.4 shows,

$$A \cup D = (A - D) \cup (D - A) \cup (A \cap D)$$

We observe that, based on the independence of common events axiom,

$$(A - D) \cup B \geq_l (A - D) \cup D \quad \text{and}$$
$$A \cup (D - A) \geq_l C \cup (D - A)$$

Since both the right hand side of the first inequality and the left hand side of the second are equal to the union of A and D,

$$(A - D) \cup B \geq_l C \cup (D - A)$$

Now we add the possibility of the intersection of events A and D to both sides. Then using the independence of common events axiom, we conclude that,

$$(A \cap D) \cup (A - D) \cup B \geq_l C \cup (D - A) \cup (A \cap D) \Leftrightarrow$$
$$A \cup B \geq_l C \cup D \quad \text{Q. E. D.}$$

Lemma 2: From Lemma 1, we will show that if $A \sim_l C$ and $B \sim_l D$ and A and B, and C and D are disjoint pairs of events, then $A \cup B \sim_l C \cup D$.

Proof: The condition where the pairs of events A and C, and B and D are equally likely is a special case of the condition where A is at least as likely as C, and B is at least as likely as D. Therefore, from Lemma 1, $A \cup B \geq_l C \cup D$.

Similarly, the condition where the pairs of events A and C, and B and D are equally likely is a special case of the condition where C is at least as likely as A, and D is at least as likely as C. Therefore, $C \cup D \geq_l A \cup B$.

In view of these results, $A \cup B \sim_l C \cup D$. Q. E. D.

We have shown that if each member of a pair of disjoint events is as likely as a corresponding member of another pair of disjoint events, then the union of the events of the first pair is equally likely as the union of the second pair.

This lemma applies to finite sets of disjoint events: If $A_1, \ldots, A_n$ are disjoint events, $B_1, \ldots, B_n$ are also disjoint, and $A_i \sim_l B_i$, for $i = 1, \ldots, n$ then $A_1 \cup \ldots \cup A_n \sim_l B_1 \cup \ldots \cup B_n$.

Lemma 2, plays a pivotal role in constructive an axiomatic foundation of subjective probability because it enables us to prove that subjective probabilities must be calculated by using the same rules as objective probabilities.

Axiom 6: Non-triviality
The certain event S is at least as likely as the empty event $S \geq_l \emptyset$ and the same is true for all non empty events.

Step 2: At this stage we have established a set of rules that enable a decision maker to compare the events in a decision. Now we assume that the decision maker can map any real life event to an equally likely event of a reference experiment whose objective probability (long-term frequency) is obvious. Then we will define the probability of the real life event to be equal to the objective probability of its counterpart on the reference experiment. Finally, we will prove that calculations of subjective probabilities must obey the rules of objective probability (Chapter 3).

Axiom 7: Benchmarking
This axiom enables us to estimate the probability of a real life event. Consider the reference experiment where the decision maker spins the probability wheel in Figure 6.5. A rational person should agree that if the needle is well balanced and the wheel is well

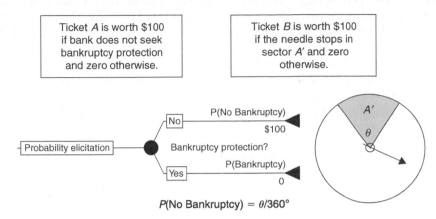

Figure 6.5 Estimation of the probability of a real life event.

oiled, then the objective probability of the needle stopping in a sector will be equal to the angle of that sector normalized by 360°.

The decision maker can compare the likelihood of an event with that of another event on the probability wheel if he/she thinks seriously about these events. For a real life event, the decision maker can determine a sector on the probability wheel for which the likelihood that the needle will stop in this sector is equal to the likelihood of the real life event. The angle of this sector is unique. Moreover, we define the probability of the event to be equal to the objective probability (relative frequency) of the needle stopping in a sector, which is equal to the angle of that sector normalized by 360°,

$$P(event) = \frac{\theta}{360°} \tag{6.1}$$

Consider that the decision maker wants to estimate the likelihood that a particular bank will not file for bankruptcy protection during the next year (event A). Then, the decision maker can determine a sector on the probability wheel on which the needle is equally likely to stop as the event that the bank will not seek bankruptcy protection. A facilitator can help the decision maker determine this sector as follows. First the facilitator selects a very large angle (e.g., 359°) and asks the decision maker which ticket in Figure 6.5 he/she prefers. Suppose that the decision maker selects ticket B (otherwise he/she would not worry about the bankruptcy). The facilitator decreases angle θ incrementally until the decision maker becomes indifferent between the two tickets. This shows that the decision maker's probability that the bank will not be bankrupt in the next year is $P(A) = \theta/360°$.

In reality, the elicitation procedure is more complex. The facilitator should elicit the same probability by using different elicitation procedures, check for consistency and give feedback to the expert. If the expert's answers lead to different probability estimates, the facilitator should point this out to the decision maker and ask the expert to correct his/her responses. For example, the facilitator could start from a very small angle θ and increase it incrementally until the decision maker becomes indifferent between the two bets in Figure 6.5. If starting from a small angle yields a significantly different probability estimate than starting from a very large one, then the elicitation process must be repeated. Section 6.3 will present processes for subjective probability elicitation.

We have presented an axiomatic definition of subjective probability and a procedure for estimating this probability that is consistent with this definition. We also need to show that subjective probability calculations follow the same rules as those for objective probability. For this purpose we will show that subjective probability satisfies Kolmogorov's axioms (section 3.1.1.2),

1.   The probability of any event is nonnegative.
2.   The probability of the sure event S is one.
3.   The probability of the union of disjoint events is equal to the sum of the probabilities of these events.

The first axiom is satisfied by the definition in Equation (6.1). The second axiom also follows from Equation (6.1). A decision maker will exchange a ticket that pays \$1 for sure with a ticket that pays the same amount if the needle of the probability wheel will stop anywhere on this wheel (that is in the sector with $\theta = 360°$), and vice-versa. Therefore, according to the definition in Equation (6.1), $P(S) = 1$.

For finite sums, we only need to show that subjective probability obeys the third axiom for two disjoint events A and B,

$$P(A \cup B) = P(A) + P(B) \tag{6.2}$$

Let C and D be two disjoint events on the probability wheel that are equally likely as A and B,

$$\begin{aligned} P(A) &= P(C) \\ P(B) &= P(D) \end{aligned} \tag{6.3}$$

According to the axioms of objective probability in section 3.1.1.2,

$$P(C \cup D) = P(C) + P(D)$$

According to Lemma 2,

$$\begin{aligned} A \cup B &\sim C \cup D \Leftrightarrow \\ P(A \cup B) &= P(C \cup D) \Leftrightarrow \\ P(A \cup B) &= P(C) + P(D) \end{aligned} \tag{6.4}$$

Finally, we derive Equation (6.2) by substituting $P(A)$ for $P(C)$ and $P(B)$ for $P(D)$ in the last row in Equation (6.4). Q.E.D.

6.2.3   Conditional probability

A decision maker's beliefs are based on his/her knowledge. A decision maker revises his/her beliefs with new information. Consider that a driver complains to a mechanic that her car makes a loud noise that seems to be coming from a wheel. Based on the customer's complaint, the mechanic suspects that the source can be either: a) a loose wheel (proposition A), b) a faulty wheel bearing (proposition B), c) a damaged tire (proposition C), d) a sticky brake caliper (proposition D). The mechanic inspects the tires and the wheel nuts and finds that they are both in good condition. In view of this information, the mechanic revises his initial beliefs, and now he suspects that the source of noise is either a faulty wheel bearing or a sticky caliper. The mechanic has no reason to believe that one possibility is more likely than the other. The mechanic test drives the car and hears a whining noise. In view of this evidence, the mechanic revises his belief and now he feels that a faulty bearing is more likely than a stuck caliper.

In this subsection, we will study how a decision maker should update his/her probabilities when new information is obtained. Suppose that before inspecting the car, the mechanic believes that propositions A to D are equally likely to be true,

$$A \sim B \sim C \sim D \tag{6.5}$$

After the inspection, the mechanic determines that all tires are in good condition (proposition A is false) and the wheel bolts are properly tightened (proposition C is false). In view of this information, the mechanic believes that the problem is equally likely to be a faulty bearing or a sticky caliper. Mathematically, this statement is represented as follows,

$$[B/(A^C \cap C^C)] \sim [D/(A^C \cap C^C)] \tag{6.6}$$

where the above expression means that, after learning that that propositions A and C are false, the mechanic believes that propositions B and D are equally likely to be true.

After the test drive, the mechanic updates his beliefs again. Now he believes that a faulty bearing is more likely than a sticky caliper,

$$[B/(A^C \cap C^C \cap T)] \geq_l [D/(A^C \cap C^C \cap T)] \tag{6.7}$$

where T represents the evidence from the test drive.

In general, if a decision maker knows that a proposition E is true, then the decision maker will compare the likelihood of other propositions based on the likelihood of each of these propositions and proposition E being true simultaneously. Thus, propositions are compared based on their intersection with the proposition that is known to be true.

Axiom 8: Comparison of propositions conditioned on the available information
For any propositions G and H, G is at least as likely as H given evidence E, if and only if the intersection of G and E is at least as likely as the intersection of H and E,

$$(G/E) \geq_l (H/E) \Leftrightarrow (G \cap E) \geq_l (H \cap E) \tag{6.8}$$

We need to assess probabilities of events conditioned upon another event E. The conditional probability of an event H must be proportional to the probability of the intersection of this event and the conditioning event according to Axiom 8. We normalize this probability of the intersection by the probability of the conditioning event so that the conditional probabilities of all events that partition the sample space add up to one. Therefore,

$$P(H/E) = \frac{P(H \cap E)}{P(E)} \tag{6.9}$$

This definition is identical to the definition of the conditional probability in section 3.1.1.3.

6.2.4 *Principle of insufficient reason*

How do you estimate the probability distribution of a variable if you only know that it varies in a specific range? Laplace and Bernoulli introduced the principle of insufficient reason or principle of indifference. This principle asserts that a decision maker should consider that two or more outcomes are equally likely if there is no reason that one outcome is more likely than the others (or that there is no lack of symmetry). For example, suppose that a decision maker has no reason to believe that one side of a die is more likely than the others. Then, the decision maker concludes that all sides have 1/6th probability.

In the authors' opinion, if the comparability axiom in section 6.1.2 is true, then so is the insufficient reason principle. According to the comparability axiom, a decision maker can always compare two outcomes A and B and decide that either one is more likely than the other or that they are equally likely. Stated differently, this principle asserts that a rational decision maker cannot be indecisive. Therefore, if the decision maker has no reason to believe that one event is more likely than the other, then the decision maker should consider the events equally likely.

Although this principle looks reasonable, many authors have tried to disprove it through counterexamples (for example French, 1986, pp. 218–219). Suppose that a decision maker only knows that a variable is in the range from zero to one. Then, many analysts use a uniform PDF by invoking the principle of insufficient reason. However, there are serious issues with this approach. Using probability calculus, one can find that that the square of this variable is more likely to assume values less than 0.5 than higher values. This result is inconsistent with the decision maker's belief because, often, the decision maker is not only ignorant about the variable but also about its square.

The authors believe that this example does not disprove the principle of insufficient reason. First, a decision maker may know that there is a cause-effect relation between a variable and its nth power. For example, if the variable is the cyclic stress amplitude at a notch in a structure, then the resulting fatigue damage could be proportional to the 5th power of the stress. Since stress causes fatigue damage, it is reasonable to assume that the stress is uniform and to derive the probability distribution of the fatigue damage by using probability calculus.

Second, in the authors' opinion, the comparability axiom – not the principle of insufficient reason could be the cause of this paradox. If the decision maker has no reason to believe that outcome A is more likely than B but he has limited information, then the decision maker could be unable to decide which outcome is more likely or that the two outcomes are equally likely.

Sinn (1980) presented an approach for distributing a probability over events for which there is no lack of symmetry that addresses the above counterexamples. This approach is based on three axioms: the first two are the comparability and transitive order axioms in Section 6.2.2. The third axiom states the following; if there is a choice between two lotteries, both of which yield the same consequence with probability $1 - p$ but different consequences with probability p then the ordering of the lotteries should be the same as that of the latter two consequences. For example, a rational person prefers a lottery that yields $10 and $20 both with probability 0.5 to a lottery that yields $10 and $15 both with probability 0.5. Sinn's approach uses all the information available, and only this information, in order to construct a second order probabilistic model (a model whose parameter(s) are random). We will present Sinn's approach by using two examples.

Example 6.1:   Principle of insufficient reason (1)
An urn contains two balls. A label on the urn says that each ball could be black or red. Without looking, one has to answer the following question: How many red balls are in the urn?

Solution:
There could be 0, 1 or 2 red balls in the urn, and there is no reason that one number is more likely than the others. Therefore, the probabilities of 0, 1 or 2 red balls are all equal to 1/3 according the principle of insufficient reason.

Many authors argue that this approach does not correctly count the possibilities regarding the contents of the urn. These authors consider how we select balls of each type and put them in the urn. We pick up one ball at a time from a large box containing practically infinite numbers of red and black balls and put it in the urn without looking. Then, for two balls, there are four possibilities:

1)   Pick up two black balls and put them in the urn (event BB).
2)   Pick up a black ball first and then a red one and put them in the urn (event BR).
3)   Pick up a red ball first and then a black one and put them in the urn (event RB).
4)   Pick up two red balls and put them in the urn (event RR).

According to this view, there are four equally likely possibilities: BB, BR, RB and RR. Then the probability of no red balls is ¼, the probability one red ball, ½ and the probability of two red balls ¼.

Therefore, there are two possible PMFs of the number of red balls in the urn (Figure 6.6)

$$
f_R(r) = \begin{cases} \dfrac{1}{3}\delta(r) + \dfrac{1}{3}\delta(r-1) + \dfrac{1}{3}\delta(r-2) & \text{or} \\[2mm] \dfrac{1}{4}\delta(r) + \dfrac{1}{2}\delta(r-1) + \dfrac{1}{4}\delta(r-2) \end{cases} \tag{6.10}
$$

where R is the number of red balls, $f_R(r)$ the PMF of the number of red balls and $\delta(\cdot)$ the unit impulse function. We assume that these PMFs are equally likely, by virtue of the principle of insufficient reason.

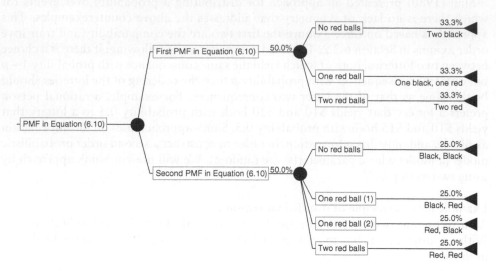

Figure 6.6 Two equally likely PMFs of the number of red balls in a box.

In compact form, the PMF of the number of red balls is,

$$f_R(r) = \frac{1}{2}(1-p)\delta(r) + p\delta(r-1) + \frac{1}{2}(1-p)\delta(r-2) \tag{6.11}$$

where p is the probability of one red and one black ball. This parameter could be equal to one half or one third with probability 0.5. This type PMF whose parameter(s) is random is called second order probabilistic model and it is presented in section 6.3.5.

Example 6.2: Principle of insufficient reason (2)

A cube is hidden in a box and a label on the box says that its length is in the range [0.8, 1.2] cm. What is the PDF of the length?

Solution:

One may assume that, according to the principle of insufficient reason, the length is uniform in [0.8, 1.2] cm. However, from the information on the box label, one could conclude that the area of one side of the cube or its volume is uniform. Therefore, there are three possibilities:

1. The length of the side of the cube could be uniform in the range [0.8, 1.2] cm.
2. The area of one side of the cube could be uniform in the range [0.8^2, 1.2^2] cm².
3. The volume could be uniform in the range [0.8^3, 1.2^3] cm³.

In the latter two cases, from Equation (3.116), the PDF of the length of the side of the cube is,

$$
\begin{aligned}
f_L(l) &= 2l \cdot U(l^2, 0.8^{1/2}, 1.2^{1/2}) && \text{if area is uniform in } [0.8^2, 1.2^2] \\
f_L(l) &= 3l^2 \cdot U(l^3, 0.8^{1/3}, 1.2^{1/3}) && \text{if volume is uniform in } [0.8^3, 1.2^3]
\end{aligned}
\tag{6.12}
$$

where L denotes the side length and l its value. Symbol $U(x, a, b)$ is the uniform PDF of a variable x and a and b the lower and upper bounds of this variable. Figure 6.7 shows the three possible PDFs of the length.

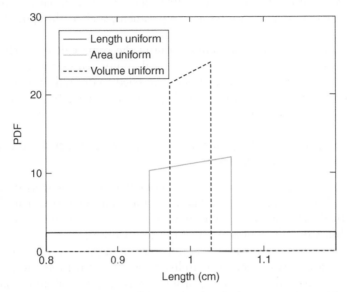

Figure 6.7 Three PDFs of the length of the cube side, which are consistent with the information on the label on the box. The legend shows the assumptions upon which each PDF was derived.

There is no reason that one of the above three possibilities is more likely than the others. Therefore, each possibility could be true with probability 1/3. In conclusion, the PDF of the length of the cube is,

$$f_L(l) = \begin{cases} U(l, 0.8, 1.2) & \text{with probability 1/3} \\ 2l \cdot U(l^2, 0.8^{1/2}, 1.2^{1/2}) & \text{with probability 1/3} \\ 3l^2 \cdot U(l^3, 0.8^{1/3}, 1.2^{1/3}) & \text{with probability 1/3} \end{cases} \qquad (6.13)$$

More compactly, the PDF of the length of the cube side is,

$$f_L(l) = nl^{n-1} \cdot U(l^n, 0.8^{1/n}, 1.2^{1/n})$$

where n is a hyperparameter (a distribution parameter that is random) that could be equal to 1, 2 or 3 with probability 1/3. This type of probabilistic model whose parameter is a random variable is presented in section 6.3.5.

6.2.4.1 Summary of the section on the definition of subjective probability

One-time events affect the consequences of most decision problems. It is necessary to define the concept of subjective probability in order to represent a decision maker's belief about the likelihood of these events. This section defined the concept of subjective probability. It presented eight axioms (common sense rules) that describe how a rational person makes decisions under uncertainty. The most controversial axiom is the one that stipulates that a decision maker can always compare the likelihood of uncertain events. A decision maker could resolve this issue by comparing one-time events indirectly to the outcomes of a reference experiment, such as the spin of a probability wheel.

Subjective and objective probabilities follow the same calculus. The principle of insufficient reason is useful for estimating probabilities of events when there is very limited or no information. Some authors have tried to discredit this principle. These authors must explain why the comparability axiom is wrong because this principle is a consequence of the comparability axiom.

6.3 Eliciting Expert's Judgments in Order to Construct Models of Uncertainty

Elicitation, in the context of uncertainty modeling, is the process of capturing the knowledge and beliefs of an expert about an uncertain quantity. The objective is to obtain statistical summaries that will be used to estimate the probability of an event or the probability distribution of a set of random variables.

First, this subsection describes a general, structured process for eliciting judgments of experts and for constructing probabilistic models from these judgments. Then it explains how to elicit information in order to estimate probabilities and probability distributions. In probabilistic analysis of high consequence systems, such as a nuclear plant, we have to estimate probabilities of rare events. This subsection reviews methods for estimating small probabilities (less than 0.01). Finally, an approach called hyperparameterization for representing uncertainty in the estimated probability distributions is presented. The subjective probabilistic models estimated from this approach can be updated when observations from experiments or additional evidence from experts become available by using Bayesian analysis (section 6.4).

6.3.1 Elicitation process

In order to make an informed decision, a decision maker should accurately estimate the probabilities of all events that affect the consequences of the decision. Often, the decision maker has to rely on judgment because he/she has limited data. Subjective estimates should be elicited carefully so that they correctly represent the decision maker's beliefs.

Assessment of probabilities is challenging when there is limited data. Often a decision maker lacks the knowledge and skills required to estimate these probabilities. Therefore, the decision maker should seek help from experts. For example, a patient who wants to choose among two alternative medical treatments needs to know the possible outcomes of these treatments and their likelihood. Physicians can provide some of this information. A friend may help the patient ask the physician the right questions, compile the answers and estimate the likelihood of the outcomes of these treatments.

Extensive research has been conducted and documented on elicitation of expert judgments and estimation of probabilities from these judgments. Structured procedures and standards for the elicitation process are available in the literature (Winkler, 1967, Clemen 1997, pp. 285–291, 291–294, Cooke and Goosens, 2004, Booker and McNamara, 2005, and O'Hagan et al. 2006, pp. 28–31).

Participants in the elicitation process must play four roles:

a) The decision maker who wants to use the results of elicitation in order to make a decision
b) Subject matter experts who are knowledgeable about the uncertain quantities in the particular decision
c) A statistician who trains the expert, interprets and assesses answers, and provides the expert with feedback
d) A facilitator who manages the elicitation process and interviews the expert

Usually, one person plays multiple roles. For example, a statistician manages the interview of the subject matter expert and interprets the answers. A probability elicitation process consists of the following steps.

I Background

The first step is to understand the objectives of the elicitation and to identify those events and variables whose probabilistic models need to be elicited. The decision maker and the statistician should list those probabilities and probability distributions that they want to estimate. They should search the literature for data or previously estimated models.

The decision maker should plan how to use the results of elicitation. If he/she plans to collect data later in order to update the probability distribution of a variable, then he/she should structure the elicitation process so that he/she can estimate parametric probability distributions of this variable. The parameters of this probabilistic model are random. In this case, a decision maker should estimate the probability distributions of both the random variable and its parameters.

The results of this step will help the decision maker and the statistician identify and recruit subject matter experts. This step also includes planning the entire process. The statistician/facilitator should prepare a set of questions in order to elicit expert judgments for each variable or event. Questions must be clear, easy to understand and should be formatted in a way that they do not bias the experts. A graphic format for the answers should also be prepared. Figure 6.5 shows an example of a question for estimating the probability of an event and the format in which the expert will express the answer. The facilitator asks the expert to specify a sector on a probability wheel for which the expert is indifferent between a lottery ticket that pays $100 if an event of interest occurs and a reference ticket that pays the same amount if the arrow stops in the sector.

It is important to estimate the same quantity in different ways or estimate dependent quantities (such as the probabilities of an event and its complement or the probabilities of disjoint events and their union) and check the estimates for consistency. The facilitator should bring possible inconsistencies to the expert's attention and ask the expert to revise his/her responses.

2   Structuring and decomposition of the elicitation

Decomposition is the process of breaking down the probability elicitation process into smaller, easier tasks. Decomposition is useful because, in many problems, it is easier to decompose an event to its constituents and estimate their probabilities than to estimate directly the probability of this event.

Examples of such events include:

- Failure of a high consequence system, such as a nuclear power plant
- A serious medical error such as a wrong blood transfusion
- Unauthorized disclosure of confidential information of a company (Alemi, 2006).

In the last example, the following procedure is easier than estimating directly the probability of unauthorized disclosure,

a) Identify scenarios in which confidential information of a company could be accessed and released to outsiders.
b) Estimate the probabilities of these scenarios.
c) Estimate the probability of unauthorized disclosure from this information.

This procedure could be better than direct estimation of the probability of unauthorized disclosure because plenty of data from previous unsuccessful attempts to access confidential data could be available. An additional advantage of decomposition is that some of the elicited probabilities in one problem could be useful in other problems.

The example below demonstrates the decomposition process in detail.

Example 6.3:   Estimating the probability of a medical error
Wrong blood transfusion is rare, fortunately. A surgeon estimates that, on average, four wrong transfusions occur per 1,000 operations when the operating room is under-staffed. When there is adequate staff in the operating room, one wrong transfusion occurs per 10,000 operations. There is staff shortage in one out of five operations. Estimate the probability of a wrong blood transfusion in an operation

Solution:
We calculate the probability of a wrong blood transfusion by using the total probability theorem (Section 3.1.1.4),

$$
\begin{aligned}
P(\text{wrong transfussion}) = &\, P(\text{wrong transfussion/no staff shortage}) \\
&\cdot P(\text{no staff shortage}) \\
&+ P(\text{wrong transfussion/staff shortage}) \\
&\cdot P(\text{staff shortage})
\end{aligned}
\tag{6.14}
$$

From the above data,

$$
\begin{aligned}
P(\text{wrong transfussion/no staff shortage}) &= 10^{-4} \\
P(\text{wrong transfussion/staff shortage}) &= 4 \cdot 10^{-3} \\
P(\text{staff shortage}) &= 0.2 \quad \text{and} \\
P(\text{no staff shortage}) &= 0.8
\end{aligned}
\tag{6.15}
$$

We calculate the probability of a wrong transfusion by plugging the results in Equation (6.15) to Equation (6.14),

$$P(\text{wrong transfusion}) = 0.8 \cdot 10^{-4} + 0.2 \cdot 4 \cdot 10^{-3} = 0.88 \cdot 10^{-3}$$

In conclusion, the probability of a wrong transfusion in an operation is approximately one per thousand. Of all wrong transfusions, ten times more occur while the operating room is understaffed compared to when it is not[1].

3   Identification and recruiting of experts

The selection of experts can be obvious or difficult. Experts could be available within the decision maker's organization. For example, suppose that the CEO of company A (Smith) will negotiate the terms of a multi-million contract with the CEO of company B (Johnson). Smith wants to estimate the likelihood that Johnson will accept a particular offer. Lawyers in company A who have already negotiated contracts with Johnson and they have seen how he responds to various offers could serve as subject matter experts in this negotiation. In addition, Smith should check if any executives in her company have worked for Johnson before and are familiar with the way he negotiates and decides.

A facilitator should select experts by considering the following criteria (Hora and von Winterfeldt, 1997)

a)   Tangible evidence of expertise
b)   Reputation
c)   Availability and willingness to participate in the elicitation
d)   Understanding of the decision problem
e)   Honesty and impartiality
f)   No evidence of conflict of interest

4   Motivating and training experts

The facilitator should explain to the experts the objectives of the elicitation and how their judgments will be used to estimate probabilities. Many experts are scientists or engineers and they feel uncomfortable making subjective assessments, especially when they feel that these judgments could be proven wrong. It is important to explain the experts the following in order to address their concerns:

a)   Important decisions usually involve uncertainty about one-time events.
b)   In order to make informed decisions it is important to characterize this uncertainty. For one-time events, this can only be done by using expert judgment.
c)   The objective of the elicitation is to capture the expert's knowledge and convert it into estimates of the probabilities of these events.

---

[1] This is different from saying that when an operating room is understaffed 10 times more wrong transfusions happen than when it is not (this number is actually 40). The information that the operating room is understaffed only 20% of the time is the key here.

The facilitator/statistician should perform a dry run in order to train the experts to estimate probabilities. The experts should also be trained to recognize the most important heuristics and biases in human judgment (section 6.5). The facilitator/statistician must be prepared to change the format of the questions based on the results of this dry run in order to avoid biasing the experts.

Probability calibration is a powerful tool for training and motivating experts. In a calibration procedure, the facilitator asks experts to estimate probabilities of events and then the facilitator compares these estimates to observed relative frequencies. For example, during 2009, a meteorologist predicted that it will snow in Southeast Michigan with 75% probability in 20 days. Records show that it actually snowed in 14 of these 20 days, or 70% of the time. This is evidence of a well-calibrated expert. On the other hand, if a second meteorologist predicted that it would snow with probability 80% in 30 days, but it actually snowed in 15 of these days (i.e. 50% of the time) then the second meteorologist is biased toward overestimating the probability of snow. Hubbard (2007, pp. 65–69) reported that experts who took 3-5 probability calibration tests and received feedback, became better at quantifying real-life uncertainties. Generally, monitoring the correlation between elicited and observed probabilities and providing incentives to experts to make unbiased estimates improves the expert's skills at assessing probabilities. For example, meteorologists are well calibrated because the accuracy of their assessments affects their careers.

5   Probability elicitation and feedback

This step involves iterations. First, the facilitator/statistician helps the expert elicit statistical summaries, such as a fractile, and the mean value or the standard deviation of a variable. For example, the facilitator shows the expert a lottery ticket that pays $100 with objective probability (long-term frequency) p and asks the expert for what value of p the ticket is equivalent to another ticket that pays the same amount if and only if the event of interest occurs (Figure 6.5). Then the statistician checks the elicited summaries for consistency, and if the summaries are inconsistent the statistician brings this to the expert's attention. Finally, the statistician fits a probability distribution to these summaries, checks if this distribution is reasonable and gives feedback to the expert.

6.3.2   Eliciting probabilities

This subsection presents procedures for estimating subjective probabilities. First, the subsection explains two approaches for testing the accuracy of estimated probabilities. The first is a calibration procedure for testing and improving a person's ability to assess subjective probabilities of one-time events. The second approach may only falsify an estimated probability by observing if an event, which is thought to be unlikely, occurs.

Then the subsection presents procedures for estimation of a) probabilities of discrete events, b) probability distributions of one or more variables and c) parametric probability distributions.

6.3.2.1   Probability calibration

As mentioned earlier, calibration of subjective probability assessments is the process of comparing these assessments with observed frequencies (O'Hagan et al. 2006, section 4.2, Hubbard, 2007, Chapter 5, pp. 53–59). The objectives of this process are to:

a) Determine how good an expert is at assessing subjective probabilities, and
b) Help the expert improve his/her ability to assess probabilities by providing the expert with feedback about the accuracy of his/her estimates.

In a calibration test, a person estimates the probabilities of many events, groups these estimates into subsets by magnitude of probability and then compares these probabilities with observed relative frequencies. For example, suppose that an expert identified 20 propositions that he/she believes that they are true with probability 0.8. Then a facilitator counts how many of these propositions are actually true and compares the relative frequency of true propositions to the expert's probability. A relative frequency that is close to 0.8 suggests that the expert is unbiased. Two calibration tests are presented here on which readers can practice on eliciting probabilities and obtain feedback about the accuracy of their estimates.

The first test involves True/False questions, while the second involves estimation of intervals that contain some value with a given high probability (e.g., 90%). In the first test, a facilitator asks a decision maker to answer True/False questions and estimate the probability that he/she is right for each question. The facilitator explains to the decision maker that the estimated probability that he/she is right should be close to the observed relative frequency that he/she is actually right. After the expert takes the test, the facilitator calculates the relative frequency that the decision maker was actually right and compares it to the decision maker's estimated probability of being right. The estimates of a well-calibrated decision maker are close to the corresponding relative frequencies.

Figure 6.8 shows the results of a calibration exercise with 50 true/false test questions that the first author took. This test is available on following web site; http://calibratedprobabilityassessment.org/. Marks on the horizontal axis show the author's confidence levels in this test. The vertical axis shows the percentage of questions that the author actually answered correctly, for each confidence level.

The straight diagonal line represents perfect calibration (that is, answering correctly 70% of the questions with 70% confidence). Points above the perfect calibration line indicate lack of confidence (e.g., the author answered 70% of questions right about which he was only 50% confident), and points below the perfect calibration line indicate overconfidence (the author answered only 50% of questions right about which he was 90% confident).

Figure 6.8 shows that, generally, the first author is well calibrated. When the author is very confident that he is right (his probability of being right exceeds 85%) he is actually right with higher relative frequency. When the author is less confident than 85%, then his accuracy is less consistent.

Test 1: True/False Questions
The questionnaire on the next page has 20 propositions that can be true or false. Most readers do no not know for sure which propositions are true. However, for most propositions, they should feel that a particular proposition is more likely to be true than false. For example, the author thinks that proposition "Mars is always further

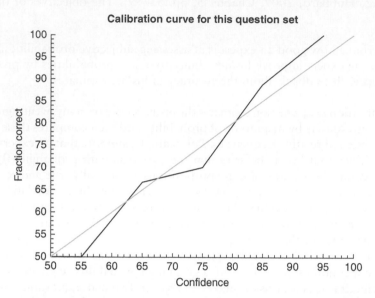

Figure 6.8   Results of calibration test with 50 true/false questions. The straight line represents a perfectly calibrated expert. Points below the straight line show that the expert was overconfident (he/she was right less frequently than what he/she thought). Points above the straight line show lack of confidence.

from the earth than Venus" is more likely to be false than true because he has heard repeatedly that NASA plans to send astronauts to Mars but not to Venus.

Answer whether each proposition is true or false. Then circle the probability that represents your confidence in the answer. If you are sure about your answer then circle probability 100%. If you are inclined to say that your answer is correct but you are not sure about it then circle probability 75%. If you have no reason to believe that a proposition is more likely to be true than false, this means that your subjective probabilities of "true" and "false" are both 50%. Then circle one answer at random, just like you must do in a true/false exam in which only the right answers count and no points are deducted for wrong answers. In addition, circle probability 50% to indicate your complete lack of confidence.

When you estimate the probability that your answer to a question is correct, ask yourself if you would put your money where your mouth is. For example, if you think that your probability to be correct is 75%, would you bet $7.5 for the opportunity to win $10 if you are correct and zero if you are wrong? If not, then reconsider your estimates and revise them.

After answering all questions, count the number of propositions for which you indicated that you are 50% confident that you are correct, and do the same for the other two confidence levels. Then look at the answers on page 31 and mark correct and wrong answers. Count and record the number of correct and wrong answers for each probability level. Finally, calculate the percent of questions that you answered correctly for each confidence level.

Finally, assess your ability to estimate probabilities. For each confidence level on the left column, compare the percent of answers that you answered correctly to the corresponding confidence levels. If for those questions that you said that you are 75% confident that you are correct, you were actually correct only 20% of the time, the test results indicate that you are overconfident. On the other hand, if you were correct 95% of the time, then this means that you may lack confidence in your assessment.

Questionnaire

No	Proposition	True/False (Circle one)	Confidence that you are correct (Circle one)
1	Traffic accidents cause less than 10% of deaths in US	T F	50%, 75%, 100%
2	Automotive company Hyundai is from Japan.	T F	50%, 75%, 100%
3	Soviet Union was the second country to develop nuclear weapons.	T F	50%, 75%, 100%
4	In Excel symbol "**" means to raise a number into a power.	T F	50%, 75%, 100%
5	More American homes have telephones than microwave ovens.	T F	50%, 75%, 100%
6	The Old Testament has less than 10 books.	T F	50%, 75%, 100%
7	The probability of the union of two events is equal to the sum of their probabilities.	T F	50%, 75%, 100%
8	The conditional probability of an event A conditioned upon another event B can be less than the unconditional probability of A.	T F	50%, 75%, 100%
9	The Normal (Gaussian) probability distribution is symmetric.	T F	50%, 75%, 100%
10	The minimum value of a variable that follows the exponential probability distribution is zero. Then, the mean value of this variable is grater than its standard deviation.	T F	50%, 75%, 100%
11	The median of a random variable whose PDF is symmetric is equal to the mean value of this variable.	T F	50%, 75%, 100%
12	A symmetric PDF has positive skewness.	T F	50%, 75%, 100%
13	The wingspan of an Airbus A330-200 is smaller than its overall length.	T F	50%, 75%, 100%
14	Johann Carl Friedrich Gauss was born in Germany.	T F	50%, 75%, 100%
15	Wolfgang Amadeus Mozart died after 1800.	T F	50%, 75%, 100%
16	A basket ball is larger than a volley ball.	T F	50%, 75%, 100%
17	The specific weight of motor oil is lower than that of water.	T F	50%, 75%, 100%
18	The seating capacity of a Boeing 737-400 is more than 250.	T F	50%, 75%, 100%
19	The distance from Toledo, OH, to New York is smaller that the distance from Toledo, OH to Washington DC.	T F	50%, 75%, 100%
20	Pierre-Simon Laplace was born before 1600.	T F	50%, 75%, 100%

Assessment:

Estimated probability correct	No. of propositions	No. Correct	No. Wrong	True probability correct
50%				
75%				
100%				
Total				

Answers
1.   Traffic accidents cause less than 10% of deaths in the US.   T
2.   Automotive company Hyundai is from Japan.   F
3.   Soviet Union was the second country to develop nuclear weapons.   T
4.   In Excel symbol "**" means to raise a number into a power.   F
5.   More American homes have telephones than microwave ovens.   T
6.   The Old Testament has less than 10 books.   F
7.   The probability of the union of two events is equal to the sum of their probabilities.   F
8.   The conditional probability of an event A conditioned upon another event B can be less than the unconditional probability of A.   T
9.   The Normal (Gaussian) probability distribution is symmetric.   T
10.   The minimum value of a variable that follows the exponential probability distribution is zero. Then, the mean value of this variable is grater than its standard deviation.   F
11.   The median of a random variable whose PDF is symmetric is equal to the mean value of this variable.   T
12.   A symmetric PDF has positive skewness.   F
13.   The wingspan of an Airbus A330-200 is smaller than its overall length.   T
14.   Johann Carl Friedrich Gauss was born in Germany.   T
15.   Wolfgang Amadeus Mozart died after 1800.   F
16.   A basket ball is larger than a volley ball.   T
17.   The specific weight of motor oil is lower than that of water.   T
18.   The seating capacity of a Boeing 737-400 is more than 250.   F
19.   The distance from Toledo, OH, to New York is smaller that the distance from Toledo, OH to Washington DC.   F
20.   Laplace was born before 1600.   F

Results of first calibration test

Figure 6.9 shows the results of the calibration test taken by 8 industrial engineering students in a senior level Operations Research class at the University of Toledo. There are three categories of questions, each corresponding to an assessed probability of 50%, 75% and 100% that an answer was correct. The horizontal axis shows the probabilities that the students estimated that they were right, and the vertical axis shows the observed relative frequency of being right. The solid line marked with diamonds shows the observed relative frequency that the students correctly answered the questions, while the dashed line corresponds to the results of a perfectly calibrated person whose estimate of the probability of being right is equal to the observed relative frequency. For example, students estimated that they answered correctly with a probability of 70% 62 questions. Actually, 39 of the answers to these questions were correct (63%) and the remaining 23 answers (37%) were wrong. According to these results, the students were overconfident in this test, especially when they thought that they were right with 100% probability.

Test 2: Confidence Intervals
The second questionnaire involves 20 questions about the value taken by some quantity. The objective is to estimate upper and lower bounds for the quantity in question.

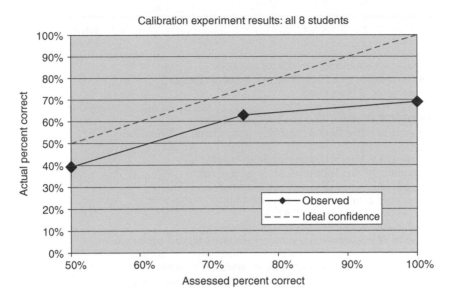

Figure 6.9 Results of calibration experiment of 8 students.

The range of these bounds should be sufficiently wide so as to contain the true value of each quantity with 90% probability, but not wider than that. The level of difficulty of each question varies, but you can always answer a question correctly by sufficiently broadening the range of the quantity that you are supposed to estimate.

In order to avoid bias, for each question, think about what you know first and then consider why you could be wrong. After answering the questions, consider the following experiment involving the choice of two lottery tickets. The first ticket is worth $100 if your interval contains the true value of the quantity that you want to estimate, and zero otherwise. The second ticket is also worth $100 with objective probability 90%. For example, if you spin the arrow of the probability wheel in Figure 6.5 you win $100 if and only if the arrow stops in a sector that subtends an angle of $360° \times 0.9 = 324°$. Your range should be such that you are indifferent between the two tickets. Widen the range if you prefer the ticket for the probability wheel, and narrow it if you prefer the ticket whose value depends on the correctness of the interval you estimate. Iterate until you become indifferent between the two tickets.

Questionnaire: Estimate the quantities below:

No	Question	Lower 90% bound	Upper 90% bound
1	The wing span of a Boeing 737-400 in ft		
2	The seating capacity of a Boeing 737-400, 1-class standard		
3	The shortest distance from Toledo OH to New York City, New York in miles		
4	The shortest distance from Toledo OH to Ann Arbor, MI in miles		

No	Question	Lower 90% bound	Upper 90% bound
5	The year when Alexander the Great was born		
6	The Curb weight of a 2010 Chevrolet Impala in pounds		
7	The diameter of Mars in km		
8	The year when Laplace was born		
9	The number of countries in the European Union		
10	The value of mathematical constant e (estimate up to 3 significant digits)		
11	The age of Dr. Martin Luther King when he died		
12	The number of days for the moon to orbit earth		
13	The year when Chernobyl accident occurred		
14	The number of millimeters in one inch		
15	The number of countries in NAFTA		
16	The diameter of a circle with circumference 2 m		
17	The value of mathematical constant π (estimate up to 3 significant digits)		
18	The probability that a normal random variable is in within an interval plus and minus two standard deviations from the mean value		
19	The year that the first World War ended		
20	The age of John F. Kennedy when he was inaugurated as president of the US		

Answers

No	Question	Answer
1	The wing span of a Boeing 737-400 in ft	119.5 ft
2	The seating capacity of a Boeing 737-400, 1-class standard	159
3	The shortest distance from Toledo OH to New York City, New York	502 m
4	The shortest distance from Toledo OH to Ann Arbor, MI	43 m
5	The year when Alexander the Great was born.	356 BC
6	The Curb weight of a 2010 Chevrolet Impala	3555 lb
7	The diameter of Mars in km.	6792
8	The year when Laplace was born	1749
9	The number of countries in the European Union	27
10	The value of mathematical constant e (estimate up to 3 significant digits)	2.72
11	The age of Dr. Martin Luther King when he died	39
12	The number of days for the moon to orbit earth	27
13	The year when Chernobyl accident occurred	1986
14	The number of millimeters in one inch	25.4
15	The number of countries in NAFTA	3
16	The diameter of a circle with circumference 2 m.	0.637
17	The value of mathematical constant π (estimate up to 3 significant digits)	3.14
18	The probability that a normal random variable is in within an interval plus and minus two standard deviations from the mean value	0.954
19	The year that the first World War ended	1918
20	The age of John F. Kennedy when he was as inaugurated president of the US	44

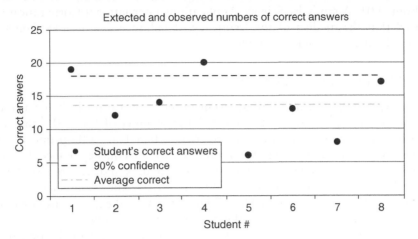

Figure 6.10 Results of second calibration test.

Results of the second calibration test

Figure 6.10 compares the frequency with which each student in the Operations Research class was right to the probability that he/she thought that was right. The figure also compares the average frequency of being right for the eight students. Students 1, 4 and 8 accurately estimated the probability that they were right. The remaining students were overconfident.

The overall performance of the students in this second test was slightly better than that in the first test. Specifically, although students exhibited overconfidence in both tests, their degree of overconfidence in the second test was lower than the first. The reason could be that they may have learnt to assess their confidence better after taking the first test and receiving feedback.

6.3.2.2 Testing the estimated probability of an event that is thought to be rare

Matheron (1989) argued that a probabilistic model is falsified if an event of zero probability in the model actually occurs. We can extend this proposition to assess the estimated probabilities of rare events. If an event that is thought to have very low probability (e.g., 1 in 100,000) occurs in the first few trials, people will be surprised and question the accuracy of this estimate. Below is an example.

In the early 1980's, NASA administrators estimated that the probability of a catastrophic failure of the space shuttle was 10^{-5} (Cooke, 1991). When the space shuttle exploded in 1986 after 25 flights, people questioned this low estimate. Some experts revised this estimate to approximately one failure in less than one hundred flights. A second catastrophic failure occurred in the 113th shuttle launch, when space shuttle Columbia exploded in space in 2003. This incident provided additional evidence that the original estimate of NASA administrators was incorrect.

Since the first catastrophic failure, there have been many formal assessments of the probability of this type of failure. Dr. Richard Feynman, who was a member of the committee that investigated the accident, estimated that the probability of failure was about 0.01. A study by Science Applications International Corporation in 1995 estimated that the median of the frequency of failure is one in 145 flights, which corresponds to a probability of 0.0069.

6.3.2.3   Elicitation of probabilities

One can express the likelihood of an event in terms of a probability or a frequency. For example, an expert can say that there is a probability of 1.4×10^{-4} that a person in the US will die in a car accident in one year, or that 1 in 7,000 people die in a car accident in one year. Expressing likelihood in terms of frequency reduces bias because relative frequencies are unambiguous and easy to understand (Gigerenzer, 2002). The likelihood of an event could be expressed using verbal expressions such as "the event is probable" or "the event is very likely." However, verbal expressions introduce bias and interpersonal variation because these expressions are vague.

A facilitator could use direct or indirect methods to elicit probabilities. It is challenging to estimate probabilities of rare events, i.e. events with probability less than 1/100. Howard (1989) suggested expressing probabilities of rare events on a scale of microprobabilities (1 occurrence per million). For example, the risk of dying in a vehicle crash per year in the US is about 140 microprobabilities per year. Woloshin et al. (2000) reported that expressing probabilities on a magnified logarithmic scale (i.e. 1 in 100, 1 in 1000, 1 in 10,000 trials etc.) yielded more accurate estimates than expressing these probabilities on a direct analogue scale (i.e. 1 in 100, 2 in 100 etc.). One reason is that experts are more comfortable expressing small probabilities on a logarithmic scale than on a direct scale.

People are often uncomfortable directly estimating the probabilities of events. This is particularly true for subject matter experts who do not understand the concept of subjective probability. We will present three indirect methods to estimate subjective probabilities. In the first method, a facilitator estimates an expert's probability of an event by observing how the expert is inclined to bet on the event. For example, suppose that the facilitator wants to estimate the probability that University of Michigan will beat Ohio State University in a football game. One approach is to ask the expert for his/her maximum buying price for a ticket that pays $100 if University of Michigan wins and zero otherwise (Figure 6.11). The more confident the expert is that Michigan will win the more money he/she will pay for the ticket. If the decision maker is risk neutral then his/her subjective probability is equal to the maximum buying price of the ticket normalized by the prize amount of $100. We explained this method in section 2.2.4.

This method assumes that the expert is risk neutral. The facilitator should repeat this process for different prizes and compare the estimated probabilities in order to check if the expert is risk neutral for the amounts involved in these bets. In addition, the facilitator should elicit the expert's minimum selling price of the same ticket. This is the minimum amount of money for which the expert is willing to assume the risk to pay $100 if University of Michigan beats Ohio State. The expert's probability that University of Michigan will win is also equal to the maximum selling price of the ticket. The facilitator should check if the minimum selling price is equal to the maximum buying price.

> This ticket will be worth $100
> if University of Michigan
> beats Ohio State University;
> otherwise it will be worthless.

Figure 6.11 First method for estimating probabilities: The probability that University of Michigan will beat Ohio State is equal to the maximum buying price of the ticket normalized by the prize amount of $100.

Bet for Michigan: Win $A if Michigan wins
Lose $B if Michigan loses
Bet against Michigan: Lose $A if Michigan wins
Win $B if Michigan loses

Figure 6.12 Second method for estimating probabilities. An expert's probability that Michigan could win is equal to $B/(A + B)$, where A and B are the amounts for which the expert is indifferent between the bet for Michigan and against it.

The second method also assumes a risk neutral decision maker. The facilitator asks the expert to adjust the amounts for which two bets for and against an event become equivalent (Clemen, 1997, pp. 269–273). For example, consider the two bets in Figure 6.12.

The more confident the expert is that Michigan will win, the lower the amount A is relative to B for which the expert thinks that the two bets are equivalent. If the expert is indifferent between the two bets, then the expected monetary profits of the two tickets are equal:

$$p(\text{Michigan wins}) \cdot A - [1 - p(\text{Michigan wins})] \cdot B$$
$$= -p(\text{Michigan wins}) \cdot A + [1 - p(\text{Michigan wins})] \cdot B \tag{6.16}$$

Therefore, the expert's probability that Michigan wins is,

$$p(\text{Michigan wins}) = \frac{B}{A + B} \tag{6.17}$$

The third method for eliciting the probability of an event is to ask the expert to compare the likelihood of this event to that of a reference event whose relative frequency is obvious. The outcomes of the flip of a coin, the roll of die, or the spin of the arrow of a probability wheel, are examples of such events.

For example, the expert compares two bets: the first bet pays $100 if and only if the University of Michigan beats Ohio State in the next football game; the second bet also pays $100 if the arrow of a probability wheel will stop in a sector A'. The facilitator adjusts the angle that sector A' subtends until the expert becomes indifferent between the two bets. Then, the expert's probability that Michigan will win is equal to the probability that the arrow will stop in sector A' (Figure 6.13).

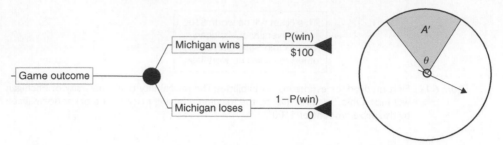

Figure 6.13 Third method: Expert believes that bet on Michigan and bet on arrow stopping in sector A' are equivalent $\Rightarrow p$(Michigan win) $= \theta/360°$.

Compared to the first two methods, the third method has the advantage that it does not assume a risk neutral decision maker. Theoretically, the estimated probability of an event from this method should be independent of the prize of the two bets.

It is important to estimate more probabilities than needed or to estimate the probability of the same event differently. For example, if we want to estimate the probabilities of two disjoint events, we should also estimate the probability of their union. Then we should check if the sum of the probabilities is equal to the probability of the union. If the estimates are inconsistent, then we should review the elicitation process and repeat some steps in order to correct the inconsistencies.

6.3.3 *Estimation of probabilities of rare events*

In risk analysis of high consequence systems (e.g., aircraft, buildings, offshore platforms and nuclear power plants) analysts have to estimate probabilities of rare events. Examples of such events are:

- An earthquake with magnitude 8 or greater on the Richter scale could occur over the next 50 years in California.
- Confidential information of a company could be disclosed without authorization to another party.
- A cyber attack could disrupt the operation of power generators or aircraft control systems with devastating consequences on the US economy and national security.

This book explained how to estimate probabilities by using observations in chapter 4, and expert judgment in this chapter. It is challenging to estimate probabilities of rare events by using these methods because a) there is limited data, and b) people have difficulty estimating low probabilities. For example, in 1980s, NASA administrators estimated that the probability of a catastrophic failure of the space shuttle was about 10^{-5}. Unfortunately, this estimate was lower than the true failure probability by three orders of magnitude as evidenced by the Challenger and Columbia disasters in 1986 and 2003.

This section will propose two methods for estimating probabilities of rare events: decomposition and statistics of extremes. Section 6.3.1 explained the decomposition approach. Failures of high consequence systems are the result from a sequence of failures of components or subsystems or human errors. We can estimate the probabilities of these contributing events from test data or from expert judgment. In decomposition, we break down the probability of a catastrophic event into events whose probabilities we can estimate, and use this information to estimate the probability of failure of the system.

6.3.3.1 Decomposition

We demonstrate the decomposition approach on two examples, one involving the failure of a pressure tank of a power plant and the second involving unauthorized release of confidential information.

Example 6.4: Finding the failure probability of a pressure tank in a nuclear plant using decomposition (Vesely et al. 1981)
Consider a pressure tank system of a nuclear power plant consisting of a tank T that contains compressed air. A control system regulates a pump that compresses air into the tank. The control system consists of five components listed in Table 6.2. An engineer wants to estimate the probability that the pressure tank could rupture. It is difficult to estimate directly the probability of this rare event. The following approach for estimation of the probability of failure is easier;

1. Identify those critical components whose failure leads to tank rupture.
2. Estimate the probabilities of failure of these components.
3. Develop a logical model for the relation between the failure of these components and the tank rupture.
4. Estimate the probability of the latter event by using probability calculus.

Table 6.2 presents the probabilities of failure of the system components. These could be estimated from tests performed by manufacturers and expert judgment.

Table 6.2 Critical components and probability of failure.

Component	Failure probability
Pressure Tank T	5×10^{-6}
Relay K2	3×10^{-5}
Switch S	1×10^{-4}
Relay K1	3×10^{-5}
Relay R	1×10^{-4}
Switch S1	3×10^{-5}

Figure 6.14 is a fault tree of the system. This is a logical model that shows the possible ways that the tank could rupture. According to this diagram the following three sequences of events could cause rupture of the tank (RT):

- Switch S fails AND relay R fails OR relay K1 fails OR switch S1 fails
- Relay K2 fails
- Pressure tank T fails.

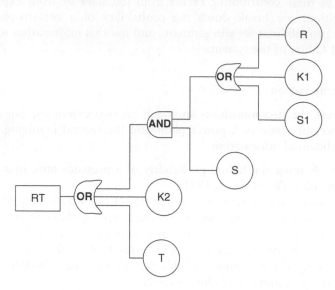

Figure 6.14 Failure tree of pressure tank.

Solution:
The fault tree is equivalent to the following logical expression,

$$RT = T \cup K2 \cup \{S \cap (R \cup K1 \cup S1)\} \tag{6.18}$$

The engineer believes that components fail independently, that is, failure of one component does not change the likelihood of failure of another component.
   The probability of rupture of the tank is,

$$P(RT) = 1 - \{1 - P(T)\} \cdot \{1 - P(K2)\} \cdot \{1 - P(S) \cdot P(R \cup K1 \cup S1)\} \tag{6.19}$$

$$\text{where } P(R \cup K1 \cup S1) = 1 - \{1 - P(R)\} \cdot \{1 - P(K1)\} \cdot \{1 - P(S1)\} \tag{6.20}$$

   Using the estimated probabilities in Table 6.2 we find that the probability of tank rupture is equal to 3.5×10^{-5}. It would be impractical to directly estimate this probability. In addition, we can calculate the sensitivity of the probability of tank rupture from Equations (6.19) and (6.20) to the probabilities of failure of the components. For example, the reliability of the tank affects the reliability of the system much more than the reliability of switch S. If we double the probability of failure of the tank, the probability of failure of the system will increase to $4 \cdot 10^{-5}$ whereas if we double the probability of failure of switch S the system failure probability will not change appreciably.
   An additional advantage of the decomposition approach is that there is carryover of information from one application to another. We can use the estimated failure probabilities of the components of the system in Table 6.2 in reliability analysis of other systems involving these components.

Example 6.5:   Estimating probability of a leak of a company's confidential data to a competitor

The example presented here has been adapted from Alemi (2006). The management of a pharmaceutical company wants to estimate the likelihood that a competitor could obtain confidential information of this company. The management cannot estimate directly the relative frequency of this catastrophic event because the event has never occurred. Therefore, the management asks security experts to determine the ways in which a competitor could eventually gain access to the data. Security experts think that this could happen if the competitor could either buy the data from a company's employee or from a hacker.

Figure 6.15 explains these two scenarios. For an employee to sell confidential data, three events must occur: First there must be a disgruntled employee willing to steal and sell the data, second this employee must have access to the data and third the employee must have access to a buyer. The second way is for a hacker to steal and sell the data. In order for this to happen, first, there must be a vulnerability in the company's computer security, second there must be a hacker determined to get the data, and third the hacker must have access to a buyer.

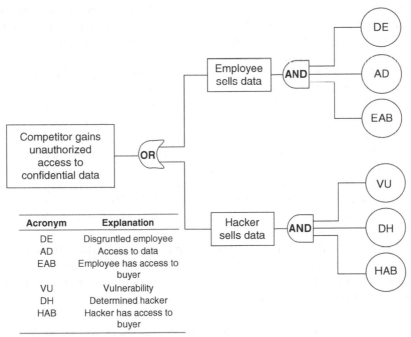

Figure 6.15 How a competitor could gain unauthorized access to a company's confidential data.

Equation (6.21) shows how the event that the company's data are sold to a competitor is broken down to its constituents,

$$\text{Competitor buys data} = \text{Employee sells data} \cup \text{Hacker sells data}$$
$$= (DE \cap AD \cap EAB) \cup (VU \cap DH \cup HAB) \tag{6.21}$$

Figure 6.15 explains the symbols in the above equation.

Security experts believe that it is extremely unlikely that the competitor will buy the data from both an employee and a hacker. This means that the corresponding events on the right hand side of the first line in Equation (6.21) are disjoint. Also the events in the parentheses in Equation (6.21) are statistically independent. For example, the likelihood that there is a disgruntled employee does not affect the likelihood that this employee has access to the data.

Therefore, the probability that the competitor could buy the data either from an employee or a hacker is,

$$P(\text{competitor buys data}) = P(DE)P(AD)P(EAB) + P(VU)P(DH)P(HAB) \quad (6.22)$$

The next step is to estimate the probabilities of the events in Equation (6.22). In order for an employee to attempt to steal and sell data, the employee must be dissatisfied, greedy and have little or no concern about risk. After talking to the human resources department and examining records of incidents involving disgruntled employees the security experts estimated that the probability that an employee in the company could try to steal and sell confidential data is 0.1. They also estimated that only 10% of employees have access to confidential data. Finally, the experts estimated that the probability that an employee could find a buyer willing to buy the data is 0.05.

The security experts also consulted with the information technology department in order to assess the vulnerability of the company's computers. They estimated that the probability that a determined hacker could exploit some vulnerability and gain access to the data was 1%. The experts also estimated that there is a 5% probability that there exists a hacker who is determined to try to break into the company's network. Finally, they estimated that a hacker has a probability of 0.01 to find a willing buyer. Table 6.3 summarizes the probabilities of these events.

Plugging the probabilities in Table 6.3 into Equation (6.22), the management finds that the probability that a competitor could be able to buy the company's confidential data is 5.05×10^{-4}. The probabilities that the competitor could buy the data from an employee or from a hacker are 5.0×10^{-4} and 5.0×10^{-6}, respectively. Therefore, in order to protect confidential information, the security department should watch problematic employees rather than hackers.

Note that the probability of privacy violation is very small, though the probabilities of the events contributing to this outcome are relatively high. This example demonstrates how we can estimate the probability of a rare outcome by finding the different ways in which this outcome could occur.

Table 6.3 Events that must occur so that an employee or a hacker would sell the company's data to a competitor.

Acronym	Explanation	Probability
DE	Disgruntled employee	0.1
AD	Access to data	0.1
EAB	Employee has access to buyer	0.05
VU	Weak access control	0.01
DH	Determined hacker	0.05
HAB	Hacker has access to buyer	0.01

6.3.3.2 *Calculating probabilities of rare events by using statistics of extremes*

Extreme distributions deal with the maximum and the minimum values of observations of a random quantity (for example the 50-year maximum wind speed, the maximum ground acceleration during an earthquake and the 500-year volume flood). Section 3.2.1.2 explained that we can estimate the probability of a rare event by estimating the parameters of the extreme probability distribution that represents this event. For example, if the 50-year maximum wind speed at some location in the North Sea follows the Gumbel probability distribution, then we can estimate the probability that the speed will exceed some limit by finding the shape and location parameters of this distribution.

Example 6.6: Using statistics of extremes and tail approximation to estimate the probability of a severe earthquake in Southern California

The risk of a disaster from a large earthquake has greatly increased because of the growth of big cities. Many cities, such as Los Angeles and San Francisco are located in seismically active areas. Five earthquakes in the 20th century caused more than 100,000 deaths each.

A magnitude 8 or 9 earthquake (in Richter scale) in California would be devastating. Although there is plenty of data about earthquakes in California, it is not possible to estimate the frequency of earthquakes with magnitudes higher than 8 directly from the data because such earthquakes are extremely rare. In order to get a perspective of the magnitude of the strongest earthquakes in the world consider the following facts:

- The biggest earthquake since 1900 that was recorded with modern equipment occurred on May 22, 1960, in Chile and measured 9.5 on the Richter scale.
- The 2004 Indian Ocean Earthquake is the next biggest earthquake ever recorded, measuring 9.1 on the Richter scale. It generated a formidable tsunami, which damaged much of the coastline of Southeast Asia.
- The biggest earthquake to hit the United States occurred in Prince William Sound, Alaska, on March 28, 1964 and had a magnitude of 9.2. The destruction was massive, but only 125 lives were lost as a result of the earthquake (15) and ensuing tsunami (110).
- The strongest earthquake that has been documented in California since 1900 occurred in 1906. Its magnitude is estimated to be between 7.7 and 8.3 on the Richter scale.

The following example shows how to estimate the likelihood that an earthquake of magnitude higher than 8 and 9 could occur in Southern California over the next 30 years and 60 years by using two methods: a) extreme value statistics and b) tail approximation.

a) Extreme value statistics method

The proposed approach is based on two assumptions,

a) The magnitude of the strongest earthquake that occurs in Southern California each year follows the Gumbel probability distribution, and
b) The maximum annual earthquake magnitudes are mutually independent.

Table 6.4 Annual Maximum Earthquake Magnitudes (source: Southern California Earthquake Data Center, http://www.data.scec.org/catalog_search/evid.php).

Magnitude (in Richter scale)				
7.5	6	5.37	5.19	4.71
7.3	6	5.37	5.19	4.71
7.1	5.95	5.3	5.19	4.69
6.9	5.8	5.3	5.17	4.65
6.7	5.77	5.29	5.14	4.64
6.6	5.75	5.28	5.13	4.61
6.6	5.75	5.28	5.12	4.56
6.6	5.75	5.27	5.1	4.55
6.6	5.7	5.26	5.08	4.48
6.5	5.69	5.26	5.03	4.47
6.42	5.65	5.25	4.99	4.37
6.4	5.51	5.24	4.99	4.37
6.4	5.47	5.23	4.96	4.36
6.4	5.46	5.23	4.73	4.29
6.3	5.42	5.22	4.73	
6.3	5.39	5.2	4.72	

The CDF of the magnitude of the strongest earthquake that could occur in Southern California in one year is,

$$F_X(x) = \exp\{-e^{-\alpha_1(x-w_1)}\} \tag{6.23}$$

where α_1 and w_1 are the shape and location parameters of the one-year maximum earthquake magnitude. The location parameter is also the most likely value of the magnitude. Table 6.4 lists the 78 annual strongest earthquake magnitudes observed in Southern California from 1932 to 2009. The strongest earthquake occurred in 1952 and it had magnitude 7.5.

First, we will fit a Gumbel distribution representing the annual maximum earthquake magnitude to the data. Then, we will derive the probability distributions of the 30-year and 60-year maximum earthquake magnitudes.

The mean value of the earthquake magnitude in Table 6.4 is 5.461 and the standard deviation is 0.741. The scale parameter is calculated from Equation (3.92),

$$\alpha_1 = \frac{\pi}{\sqrt{6}\sigma_X} = 1.73 \tag{6.24}$$

The location parameter is calculated from Equation (3.91),

$$w_1 = E(X) - \frac{\gamma}{\alpha_1} = 5.127 \tag{6.25}$$

The same parameters were found to be $\alpha_1 = 1.687$ and $w_1 = 5.315$ by using the maximum likelihood method (Chapter 4). In our calculation of the 30- and 60-year maximum earthquake magnitudes, we will use the estimates from the method of moments.

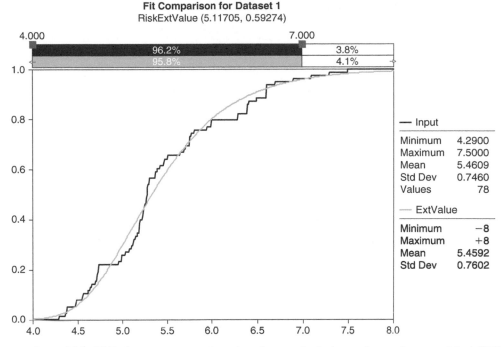

Figure 6.16 CDF of maximum annual earthquake magnitude (smooth curve) vs. empirical CDF (This CDF was calculated by using software @RISK5.5). The heading specifies the location parameter and the inverse of the shape parameter.

Figure 6.16 compares the fitted CDF of the annual earthquake magnitude and the empirical CDF. It is observed that, according to the fitted CDF, an earthquake with magnitude 7 or higher could occur in one year with probability 0.041. This probability is close to the relative frequency of the same event (3/78) in Table 6.4.

Now we will find the CDF of the maximum earthquake magnitude in 30 years. This magnitude is equal to the maximum of the magnitudes of thirty, one-year earthquakes. According to the reproductive property of the Gumbel distribution (Equation 3.93), the CDF of the 30-year maximum has same scale parameter as the one-year earthquake and location parameter, $w_{30} = w_1 + \ln(30)/\alpha_1 = 7.093$. Therefore the CDF of the 30-year earthquake is,

$$F_{Y_{30}}(y) = \exp\{-e^{-1.73(y-7.093)}\} \tag{6.26}$$

Similarly, the probability distribution of the 60-year earthquake is,

$$F_{Y_{60}}(y) = \exp\{-e^{-1.73(y-7.493)}\} \tag{6.27}$$

Figure 6.17 shows the PDF and CDF of the 30-year earthquake magnitude. From Equation (6.26), we find that the probabilities that no earthquake with magnitude greater than 8 and 9 could occur in 30 years are 0.812 and 0.964, respectively. This

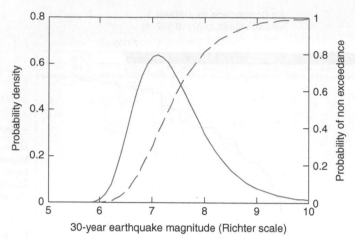

Figure 6.17 Probability density and distribution of 30-year earthquake in Southern California.

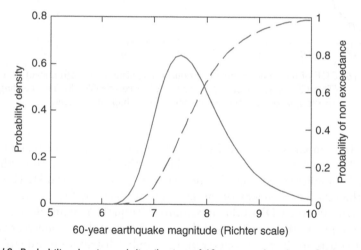

Figure 6.18 Probability density and distribution of 60-year earthquake in California.

means that there is a 0.19 probability that an earthquake with magnitude greater than 8 will occur in the next 30 years. The corresponding probability for an earthquake with magnitude greater than 9 is 0.04.

Figure 6.18 shows the distribution of the 60-year earthquake. The probabilities that no earthquake with magnitude greater than 8 and 9 could occur in 60 years are equal to 0.66 and 0.929, respectively. This means that there is a 0.44 probability that an earthquake with magnitude greater than 8 will occur over the next 60 years. The corresponding probability for an earthquake with magnitude greater than 9 is 0.07.

Ang and Tang (1984) (pp. 236–237) used data about the annual earthquake magnitudes in California from 1932 to 1962 to estimate the parameters of the Gumbel

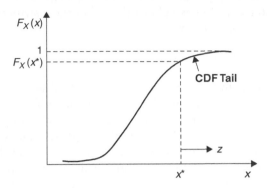

Figure 6.19 Notation for tail modeling.

distribution, representing the annual earthquake magnitude. They found that the shape and location parameters are $\alpha_1 = 2$ and $w_1 = 5.7$. From these parameter values we find that the annual probabilities of earthquakes exceeding magnitudes of 8 and 9 are 0.01 and 0.0014. The probabilities of earthquakes exceeding magnitudes of 8 and 9 in a thirty year period are 0.26 and 0.04. In a sixty year period, the probabilities of the same events are 0.45 and 0.08.

b) Tail approximation method

The tail modeling method approximates the right tail of the CDF of a random variable, which corresponds to rare events, with a standard distribution. We estimate the parameters of this distribution from observations of frequent events. For example, we estimate the probability distribution of the maximum flood to occur in 100 years by observing floods that occur each year.

The fundamental concept of the tail modeling is based on the property of tail equivalence. According to Maes and Breitung, (1993), and Caers, and Maes, (1998), two distribution functions $F_X(x)$ and $G_X(x)$ are called *tail equivalent* if,

$$\lim_{x \to \infty} \frac{1 - F_X(x)}{1 - G_X(x)} = 1 \tag{6.28}$$

In this case, the tail model of $F_X(x)$ approximates that of CDF $G_X(x)$.

If x^* is a large threshold of the maximum earthquake magnitude in one year x (see Figure 6.19), then the generalized Pareto distribution (GPD) provides a general approximation of the conditional distribution $F_Z(z)$ of the exceedance of $z = x - x^*$ for the region $x \geq x^*$. For a large threshold x^*, distribution $F_Z(z)$, conditioned upon $x \geq x^*$, is (Castillo, 1988),

$$F_Z(z) = \begin{cases} 1 - \left(1 + \frac{\xi}{\sigma} z\right)_+^{-\frac{1}{\xi}} & \text{if } \xi \neq 0 \\ 1 - \exp\left(-\frac{z}{\sigma}\right) & \text{if } \xi = 0 \end{cases} \tag{6.29}$$

where $\langle A \rangle_+ = \max(0, A)$, $z \geq 0$, and ξ and σ are the shape and scale parameters. The shape parameter determines the weight of the tail of the distribution. The cumulative distribution of interest $F_X(x)$ for $x \geq x^*$ is,

$$F_X(x) = [1 - F_X(x^*)]F_Z(z) + F_X(x^*) \tag{6.30}$$

Tail modeling involves three steps. First we obtain N observations of the earthquake magnitude. Then we select a threshold value x^* and identify those N_{x^*} sample points (out of N) for which $x \geq x^*$. The selection of threshold x^* is important and has been the subject of extensive research. Hasofer (1996) suggests that $N_{x^*} \approx 1.5\sqrt{N}$. Finally, we estimate the shape and scale parameters ξ and σ by fitting the tail model to the empirical CDF using the maximum likelihood or the least square methods. Only the tail part of the data is used in estimating the parameters. The following unconstrained minimization problem is solved in the latter case,

$$\min_{\xi, \sigma} \sum_{i=N-N_{x^*}}^{N} [p_i - F_X(x_i)]^2 \tag{6.31a}$$

where the empirical CDF is,

$$p_i = \frac{i - 0.5}{N} \quad i = 1, \ldots, N \tag{6.31b}$$

We select the 14 strongest earthquakes in Table 6.4. These earthquakes have magnitudes from 6.4 to 7.5, and CDF values from 0.827 to 0.994. We estimate the shape and scale parameters of the Pareto CDF by minimizing the sum of the squares of the differences between the values of the approximating Pareto CDF and the empirical CDF from the data (Equation 6.31).

Figure 6.20 compares the fitted CDF to the empirical CDFs. The fitted CDF overestimates the probability of very strong earthquakes. For example a 8-Richter and a 9-Richter earthquakes could occur with probabilities 0.01 and 0.003, respectively, in a year. If annual earthquake magnitudes are mutually independent, then, according to these results, a 9-Richter could occur with probability 0.1 in 30 years. This probability is too high. We overestimated probabilities of strong earthquakes because we overemphasized the part of the CDF for magnitudes less than 6.5 at the expense of higher values, by minimizing the sum of the differences of the CDFs.

Consequently, we modified the objective function that we minimize in order to fit the tail model. The sum of the square differences between the safety indices of the Pareto CDF and empirical CDF became the new objective function[2]. This function evenly penalizes deviations of two CDFs in the entire region of displacements because the safety index varies more benignly than the CDF. Figure 6.21 compares the fitted to the Pareto CDF. The latter CDF has scale parameter 0.306 and shape parameter 0.146. From the Pareto CDF, we calculated that the probabilities that 8-Richter and 9-Richter earthquakes could occur in one year are 0.004 and 0.001, respectively.

---

[2] Safety index (or Reliability index) of a system is the R-quantile of the standard normal distribution, where R is the reliability of the system.

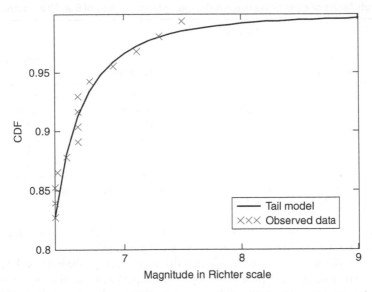

Figure 6.20 Tail approximation of CDF of the magnitude obtained by minimizing the sum of the square differences of the empirical and Pareto CDFs.

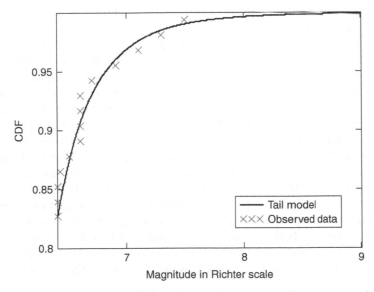

Figure 6.21 Tail approximation of CDF of the magnitude obtained by minimizing the sum of the square differences of the safety indices of the empirical and Pareto CDFs.

Assume that the annual earthquake magnitudes are independent. Then the CDF of the largest earthquake magnitude in n years is,

$$F_{Y_{30}}(y) = F_Y(y)^n \qquad (6.32)$$

Table 6.5 Probabilities of earthquake magnitudes exceeding the values of 8 and 9 in southern California.

Period (years)	Magnitude (Richter scale)	Method		
		Extreme value	Tail approximation	Extreme value (Ang and Tang, 1984)*
I	8	0.007	0.0035	0.0I
	9	0.0012	6.8×10^{-4}	0.0014
30	8	0.19	0.1	0.26
	9	0.04	0.02	0.04
60	8	0.44	0.19	0.45
	9	0.07	0.04	0.08

*The authors of this book calculated these probabilities by using statistical extrapolation and the parameters of the annual earthquake distribution in (Ang and Tang, 1984, pp. 236–237).

From the above equation, the probability that earthquakes with magnitudes 8 and 9 on the Richter scale could occur in 30 years are 0.1 and 0.02, respectively. In 60 years, earthquakes with the above magnitudes could occur with probabilities 0.19 and 0.04.

Table 6.5 summarizes the probabilities of earthquakes from the extreme value and the tail approximation methods. There is a considerable probability that an earthquake with magnitude greater than 8 could occur within the next 30 years in southern California. Such an earthquake could be devastating. The extreme value method predicts that earthquakes of magnitude 8 or 9 on the Richter scale are twice as likely as the tail approximation method does.

The extreme value method is overly conservative in our opinion, because it assumes that the annual earthquake magnitudes are statistically independent. Although earthquakes that occur very rarely (e.g., once every five years) are mutually independent, earthquakes that occur within a one-year period are generally dependent. Indeed, data from the United States Geological Survey (USGS) suggests that earthquakes occur in clusters within a one year period. Therefore, we believe that the results of the tail approximation method are more accurate than those of the extreme value method.

Finally, using the earthquake probability mapping of the USGS we found that the probability that an earthquake with magnitude 8 or higher could occur in the Los Angeles area (zip code 92401) in 30 years is between 0.02 and 0.03 (http://geohazards. usgs.gov/eqprob/2009/index.php). The probability of the same earthquake in 60 years is between 0.04 and 0.06. These probabilities are significantly lower than those in Table 6.5. The authors believe that when making decisions that could result in severe consequences, we should consider the most conservative estimates that are consistent with the available information, because it is better to err on the safe side.

6.3.4 Eliciting probability distributions

Sections 6.3.2 and 6.3.3 explained how to elicit an expert's beliefs about the probability of an event. This section extends the methods presented to probability distributions.

In principle, estimation of a cumulative probability distribution $F_X(x)$ of random variable X requires elicitation of infinite probabilities of the form, $P(X \leq x) = F_X(x)$. Often, three to five probabilities adequately represent an expert's beliefs about the distribution. Besides these probabilities, we can elicit location parameters (e.g., the mean value and median of a distribution), scale parameters (e.g., the standard deviation) and the weight of the tail (e.g., the shape parameter of the generalized Pareto distribution in Equation 6.29).

To facilitate the representation a decision maker's belief about one or more variables we should impose some structure on a probabilistic model. For this purpose, we represent the decision maker's belief by a parametric distribution, that is, a distribution that is fully defined once we specify the values of its parameters. Many parametric probability distributions are available for modeling a single random variable (Bury, 1999) and many variables (Kotz et al. 2000). Most of these probability distributions are suitable for specific physical quantities. For example, the Weibull distribution represents well the strength of a component, and the Gumbel distribution represents the lifetime load on a system. If a decision maker knows the bounds of a random variable but has limited data, then he/she should consider the uniform, triangular or beta distributions. Sections 3.2, 4.4 and 4.5 presented some of these families of distributions and guidelines about what distribution is suitable for a given problem.

This section presents and demonstrates methods for estimation of both univariate and multivariate probability distributions. These methods elicit probabilities that one or more variables assume values in certain ranges and fit a standard probability distribution to these estimates. In addition, the section addresses the following issues:

- What and how many statistical summaries should we elicit?
- How should we quantify imprecision in the elicited summaries?

6.3.4.1 *Eliciting univariate distributions*

For a random variable X, a facilitator could elicit probabilities $P(X \leq x)$ for several values of limit x, or quantiles for given probability values $P(X \leq x)$. In addition, he/she could elicit the median and the upper and lower quartiles of X, and credible intervals. Credible intervals contain the value of a variable with a given coverage probability.

In many decisions, a facilitator wants to elicit probabilities of one-time events. An example is the probability that the DJIA index will close above 15,000 on December 31, 2012. The facilitator should always consider that people understand and estimate relative frequencies more accurately than probabilities of one-time events. When training an expert to elicit probabilities, the facilitator should draw analogies between probabilities of one-time events and long-term frequencies. He/she should use gambling devices, such as a fair coin, a die and a probability wheel to train the expert to assess subjective probabilities of one-time events.

In the above probability elicitation example, the facilitator/statistician must explain to the expert the meaning of subjective probability and formulate questions in a way that the expert understands them easily and feels comfortable answering them. For example, the facilitator could ask the following question in order to estimate the probability that the DJIA index could close above 15,000 on December 31, 2012:

Consider a lottery ticket that pays \$100 with some long-term relative frequency value, p. What long-term frequency p makes this lottery ticket equivalent to a second ticket that also pays \$100 if the index closes above 15,000 on December 31, 2012?

Table 6.6 Demand for new car model in the US and in Canada.

Probability that demand does not exceed the values in next two columns	Cars, US	Cars, Canada
Minimum value	1,000	300
0.05	5,000	2,000
0.25	16,000	10,000
0.5	23,000	14,000
0.75	30,000	22,000
0.95	45,000	28,000

Example 6.7: Quantifying the demand for a new car model

An automaker wants to estimate the demand for a new Diesel-hybrid car in the US and in Canada in 2012. The automaker's CEO asks the vice president of global marketing to estimate the median of the demand in each country. Then the CEO asks for the first and third quartiles. Finally, the vice president estimates the 90% central credible intervals and lower bounds for the demand.

Table 6.6 summarizes the elicitation results. According to these results, the expert projects that the median numbers of cars that will be sold in the US and Canada are 23,000 and 14,000 respectively. The numbers in the second column indicate that the demand in the US ranges between 5,000 and 45,000 cars with 90% probability.

Once the statistical summaries in Table 6.6 have been elicited, a statistician should fit a model to the data to help the CEO make decisions about this model. In this example, a statistician fits two beta probability distributions to the elicited summaries in Table 6.6. The CDF of the demand is the four-parameter beta distribution (Equation 3.76),

$$F_X(x, x_{min}, x_{max}, \beta_1, \beta_2) = \frac{B_{x'}(\beta_1 + \beta_2)}{B(\beta_1, \beta_2)} \quad \text{for } 0 < x' < 1 \qquad (6.33)$$

where x is the projected demand, $x' = \frac{x - x_{min}}{x_{max} - x_{min}}$ is the normalized projected demand, x_{min} and x_{max} are the minimum and maximum values of the demand, $B(\beta_1, \beta_2)$ is the beta function, $B_x(\beta_1, \beta_2)$ is the incomplete beta function, and $\Gamma(\cdot)$ is the gamma function.

The statistician determines the parameters of the distributions of the projected demand by minimizing the sum of the square differences of the safety indices of the fitted and the estimated CDFs. Table 6.7 shows the values of the parameters of the fitted distributions. Figure 6.22 compares these distributions with the data in Table 6.6.

Overfitting and *feedback* are powerful tools for assessing and improving elicitation results. In the context of elicitation, overfitting is the process of eliciting more summaries than needed in order to fit a distribution. The statistician/facilitator should compare the fitted probability distribution to the data and determine if discrepancies are within the bounds of imprecision in the expert's judgments. Figure 6.22 shows a good agreement between the fitted beta distributions and the data.

Feedback is the process of checking the adequacy of a distribution that has been fitted to an expert's stated summaries, by telling the expert some implications of the

Table 6.7 Parameters of fitted distributions to elicited data.

Location/Parameter	US	Canada
Minimum (cars)	1,000	300
Maximum (cars)	57,370	30,150
Shape, β_1	0.625	0.844
Shape, β_2	0.414	0.831

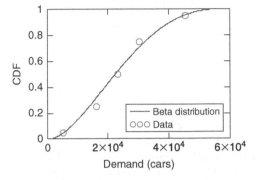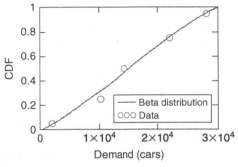

Figure 6.22   CDFs of the demand of a new diesel-hybrid car in the US (left panel) and Canada (right panel).

fitted distribution and asking the expert if these implications represent his/her beliefs. From the fitted CDF of the demand in US, in Figure 6.22, the statistician calculates that demand could be less than 20,000 and 10,000 cars with probabilities 0.42 cars and 0.11, respectively. The marketing CEO says that the probabilities are consistent with her beliefs.

People have difficulty comparing the likelihoods of a rare event and the outcome of a spin of a probability wheel. For example, an expert would be very imprecise about the 0.01 quantile of the demand in the US, in the above example. Therefore, a facilitator should avoid very low or high probabilities (e.g., less than 0.05 or greater than 0.95). If it is necessary to estimate small probabilities the expert should be allowed to estimate probability bounds or credible intervals for a probability.

Much research has investigated people's ability to assess rare events. For example, Alpert and Raiffa (1982) elicited 98% credible intervals of the number of foreign cars imported in US in 1968. They found that 43% percent of the assessments fell outside these intervals, instead of 2%. They found that feedback tempered, but did not eliminate, overconfidence. Motivating experts by providing them with feedback and incentives reduces bias. For example, meteorologists are well calibrated probability estimators because their performance is evaluated on the basis of the accuracy of their estimates.

Experiments show that people underestimate the width of credible intervals. For example, elicited 90%, 95% and 98% were found to contain the true answer only 40% to 70% of the time (Alpert and Raiffa, 1982). The authors observed the same trends in a probability calibration test that they performed at the University of Toledo (section 6.3.2, Figure 6.10).

There is conflicting evidence as to whether people are better at eliciting probabilities, quantiles or credible intervals. Use of an adaptive fixed interval method (Winman et al. 2004) or including other assessments such as the median (Soll and Klayman, 2004), improves calibration.

Saaty (1977) proposed estimating probability ratios of events instead of probabilities. Specifically, instead of eliciting the five probabilities of the demand for a new car model in the US in Table 6.6, an expert could estimate their ratios. For example, the expert could estimate the ratio of the probabilities that the demand would not exceed 5,000 and 25,000. The expert could estimate a maximum of $\binom{5}{2} = \frac{5 \cdot 4}{2} = 10$ probability ratios, but four probability ratios plus an estimate of one probability would suffice in order to determine the probabilities all of the events in Table 6.6.

Quantifying uncertainty in elicited probability distributions

There are two sources uncertainty in elicited probabilities or probability distributions:

a) Imprecision in the values of the elicited summaries,
b) Uncertainty about how representative the fitted probability distribution is of the expert's beliefs. For example, many distributions fit well to the data about the projected demand for the Diesel-hybrid car but only one could represent the marketing expert's beliefs. This uncertainty is epistemic.

One approach to quantify uncertainty in an elicited distribution is to determine a class of admissible probability distributions that are consistent with the decision maker's beliefs. Since we elicit probabilistic models in order to make decisions, we should investigate how the choice of the probability distribution from the class of admissible distributions affects the optimum course of action and its consequences. This approach is called *sensitivity analysis*.

Usually, we determine the class of admissible probability distributions by allowing the distribution parameters to vary from their estimated values. In example 6.7 we estimated the demand of a Diesel/hybrid model. Now we allow the estimates of the shape parameters of the CDF of the demand in the US in Table 6.7 to vary by ± 0.05 from the estimated values. Figure 6.23 shows four admissible CDFs obtained by varying their shape parameters plus the CDF for the nominal values of the parameters. The region defined by these CDFs contains the elicitation data. Note the model in Figure 6.23 accounts only for imprecision in the elicited values of the CDF's of the demand. It does not account for the uncertainty about how representative the beta distribution is of expert's beliefs about the demand.

6.3.4.2 Eliciting multivariate distributions

A complete probabilistic model of the random variables in probabilistic analysis and design problems consists of the joint probability distribution of these variables. It is challenging to estimate the joint probability distribution because of lack of data and because most experts are unaccustomed to making judgments about the dependence of variables. As a result, most researchers and practitioners estimate the marginal distributions of the random variables and make sweeping assumptions about the dependence

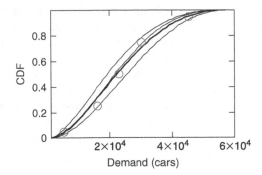

Figure 6.23 Expressing imprecision in the estimated CDF of the demand of the new car model by varying the values of the distribution parameters.

or association of the variables that are not supported by the available evidence, such as assuming stochastic independence. These assumptions often lead to significant errors in the expected utility of a course of action or the estimated reliability of a system.

There are three ways to model dependence:

a)   Restructure the problem by expressing it in terms of independent variables. For example, the yield stress and Young's modulus of a material, S_Y and E are dependent. It might be reasonable to assume that S_Y and E/S_Y are independent.
b)   Fit a standard multivariate probability distribution to the elicited summaries (Kotz et al. 2000).
c)   First, estimate the marginal distributions of the random variables and then use a copula to approximate their joint cumulative distribution in terms of the marginal distributions.

First, a facilitator could ask an expert to estimate the summaries of marginal distributions. Then the facilitator could ask the expert to estimate the probabilities of intersections of events such as $P(X > X_{0.5} \cap Y > Y_{0.5})$ where $X_{0.5}$ and $Y_{0.5}$ are the 0.5-quantiles of X and Y. If these variables were independent, then, this probability would be equal to the product of the probabilities of each of these variables exceeding their 0.5-quantiles, which is $0.5^2 = 0.25$. Alternatively, the expert could estimate the conditional probability $P(X > X_{0.5}/Y > Y_{0.5})$. If this conditional probability is greater than 0.5 then there is a positive association between the two variables. The expert should estimate many probabilities of intersections of events or conditional probabilities and examine if these probabilities are consistent. For example, the expert could estimate the probability $P(X > X_{0.5}/Y > Y_{0.75})$.

If there is positive association between X and Y, then,

$$P(X > X_{0.5}/Y > Y_{0.75}) > P(X > X_{0.5}/Y > Y_{0.5})$$

In general, it is better to elicit the conditional probability $P(X > X_{0.5}/Y > Y_{0.75})$ than the joint probability $P(X > X_{0.5} \cap Y > Y_{0.75})$ because experts understand relations between conditional probabilities better than joint probabilities. For example, it is easier to understand that $P(X > X_{0.5}/Y > Y_{0.75}) > P(X > X_{0.5}/Y > Y_{0.5})$ than that $P(X > X_{0.5} \cap Y > Y_{0.75}) > P(X > X_{0.5} \cap Y_{0.75} \geq Y > Y_{0.5})$, although the two inequalities are equivalent (O'Hagan et al. 2006, p. 109) (see also problem 6.21).

Experiments have shown that experts overestimate probabilities of intersections of events (e.g., the probability of ten heads flips of a coin in a row) and they underestimate probabilities of unions of events (e.g., the probability of at least one heads flip in a series of five flips). This is explained by anchoring bias; an expert starts with an estimate of the probability of one heads flip and fails to adjust it appropriately for the union of intersection of events. Therefore, it is better to elicit the probability of elementary events (a heads flip in one trial) rather than composite events.

Example 6.8:   Estimation of the joint CDF of the demand of a Hybrid/Diesel car in US and Canada

Copulas are functions approximating the joint cumulative distribution function (CDF) of multiple variables in terms of their margins. This presentation is confined to models of two variables.

The joint CDF of the projected demands in US and Canada, X and Y, is expressed through the following parametric relation,

$$F_{XY}(x, y) = C_\theta\{F_X(x), F_Y(y)\} \tag{6.34}$$

where $F_{XY}(x, y)$ is the unknown joint CDF of the demands in the two countries, $F_X(x)$ and $F_Y(y)$ the marginal CDFs of these demands and θ is a vector of parameters. As mentioned in Chapter 4, copula $C_\theta(u, v)$ is a 2-increasing function that links the marginal CDFs of two variables, $u = F_X(x)$ and $v = F_Y(y)$, to their joint CDF. The domain of this function is $[0,1]^2$ and the range is $[0,1]$.

Many families of copulas have been proposed and studied. These include the Clayton, Frank, Farlie-Gumbel-Morgenstern and normal families. An empirical copula can also be used. In this example, the dependence of the two demands was modeled using Frank's copula,

$$C_\theta(u, v) = \log_\theta\left\{1 + \frac{(\theta^u - 1)(\theta^v - 1)}{\theta - 1}\right\} \tag{6.35}$$

There is a single parameter θ that controls the dependence of the two demands. This is a *comprehensive* copula which means that the family in Equation (6.35) includes the cases of extreme dependence and independence. Indeed, perfect dependence arises for $\theta = 0$, independence for $\theta = 1$ and opposite dependence for θ tending to infinity. Parameter θ can be estimated from measurements, or expert judgments. For example, an expert may estimate Kendall's tau (Genest and Favre, 2007). There is a one-to-one relation between Kendall's tau and parameter θ. Tau is zero for $\theta = 1$. Otherwise,

$$\tau(\theta) = 1 + \frac{4}{\ln(\theta)} - 4\frac{D\{-\ln(\theta)\}}{\ln(\theta)} \tag{6.36}$$

where $D(s) = \int_0^s \frac{x/\theta}{e^x - 1} dx$ is the first Debye function.

The expert estimates that the probability that the demand in Canada could not exceed the median given that the demand in US does not exceed the median is 0.8. He/she also estimates that the probability that demand in Canada could not exceed the 0.75 quantile given that demand in US does not exceed the median is 0.9. The above estimates mean that the joint CDF of the demand in both counties are, $F_{XY}(23000, 14000) = 0.4$ and $F_{XY}(23000, 22000) = 0.45$. These values point to a strong association between demands in the two countries. Indeed, if demands were perfectly dependent, then both CDF values would be equal to 0.5,

$$F_{XY}(23000, 14000) = \min\{F_X(23000), F_Y(14000)\} = \min(0.5, 0.5) = 0.5$$

$$F_{XY}(23000, 22000) = \min\{F_X(23000), F_Y(22000)\} = \min(0.5, 0.75) = 0.5$$

On the other hand, if demands were stochastically independent, then

$$F_{XY}(23000, 14000) = F_X(23000) \cdot F_Y(14000) = 0.25 \quad \text{and}$$

$$F_{XY}(23000, 22000) = F_X(23000) \cdot F_Y(22000) = 0.375$$

We estimated parameter θ in Equation (6.35) by minimizing the sum of the square differences of the values of the joint CDF of the demand in both countries from Equation (6.35) and the two elicited values of this joint CDF with respect to θ,

Find θ
to minimize,
$$[C_\theta\{F_X(23,000), F_Y(14,000)\} - 0.4]^2 + [C_\theta\{F_X(23,000), F_Y(22,000)\} - 0.45]^2$$
$$= \{C_\theta(0.5, 0.5) - 0.4\}^2 + \{C_\theta(0.5, 0.75) - 0.45\}^2$$

We found that $\theta = 3.544 \cdot 10^{-3}$. The corresponding value of Kendall's correlation coefficient is $\tau = 0.495$. Therefore, from the copula approximation, there is a 0.372 probability that the demand in US and Canada could be less than or equal to 23,000 and 14,000 cars. In addition, the demand in the two countries could be less than or equal to 23,000 and 22,000 cars with probability 0.484. These probabilities are close to the elicited values of 0.4 and 0.45, respectively.

The left panel in Figure 6.24 shows the joint CDF of the demand. For comparison, the right panel shows the joint CDF in the hypothetical case where demands were independent. It is observed that values of demand in the two countries are clustered more near the diagonal on the left panel than on the right.

6.3.5 Representing uncertainty about an elicited distribution by a second-order probabilistic model

In many elicitations, the facilitator/statistician is uncertain about how accurately the elicited probability distribution represents the experts' beliefs. This subsection explains and demonstrates how to quantify this uncertainty by a second-order model.

A common approach to express uncertainty about an elicited probabilistic model is to consider its parameters as random variables with their own distributions. We call random distribution parameters *hyperparameters*, and the corresponding distribution *second-order model* (O'Hagan 2006, pp. 253–254). For example, when a

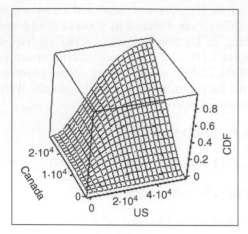 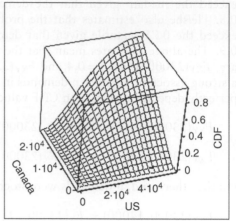

Figure 6.24 Joint CDF of demand in US and Canada. The left panel shows the fitted CDF to the expert estimates. The right panel shows the CDF if the demands in the two countries were independent.

decision maker is uncertain about the probability of a successful trial of an experiment, p, the decision maker quantifies the likelihood of the number of successful trials by a binomial distribution (Section 3.2.1.1) in which probability p is a random variable.

It is sometimes easier to think of distributions over subjective probabilities in terms of the following examples.

- Multiple experts assess the probability of an event. Usually, each expert provides a different value of the probability.
- Each time a decision maker views the same problem from a different perspective she estimates a different probability for the same event

If we have no reason to believe that one probability estimate is correct then we can think of the probability as following a distribution.

Example 6.9: Representing uncertainty about the probability distribution of the number of business class reservations in an airline
An airline manager wants to estimate the probability distribution of the reserved seats in the business class section for a flight. The manager knows that the Poisson distribution (Equation (3.68)) represents the number of reservations, but she only knows that the expected value of reservations ranges from 25 to 35. The manager expresses her uncertainty about the expected number of reservations Θ by assuming that it is random and that it follows a uniform probability distribution between 25 and 35.

Figure 6.25 shows the PDF (left panel) and CDF (right panel) of the number of reservations for expected values of 25, 30 and 35. The same figures show the expected

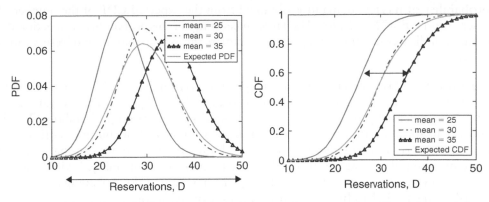

Figure 6.25 PDF and CDF of the number of business class reservations. The right panel shows the lower and upper bounds of the CDF for expected values of reservations of 25 and 35 passengers, respectively. The arrows on the left and right panels show the random and epistemic components of uncertainty in the number of reservations.

PDF and CDF of the number of reservations, which are calculated as follows,

$$\overline{f}_D(d) = \int_{25}^{35} f_D(d/\theta) f_\Theta(\theta)\, d\theta$$

$$\overline{F}_D(d) = \int_{25}^{35} F_D(d/\theta) f_\Theta(\theta)\, d\theta$$

(6.37)

In the above equations, D is the number of reservations, Θ, is the unknown expected value of D, $\overline{f}_D(d)$ and $\overline{F}_D(d)$ are the expected PDF and CDF of the number of reservations, and $f_D(d/\theta)$ and $F_D(d/\theta)$ are the conditional PDF and CDF of D conditioned upon $\Theta = \theta$. Note that, $\overline{f}_D(d) \neq f_D(d/\overline{\Theta})$, where $\overline{\Theta} = \int_{25}^{35} \theta f_\Theta(\theta) d\theta = 30$.

There is great uncertainty about the number of reservations according to Figure 6.25. This uncertainty has two components:

a) Uncertainty about the expected number of reservations, and
b) Uncertainty about the number of reservations given its expectation.

The first component is epistemic and the second is random uncertainty. The manager can reduce epistemic uncertainty by calculating the expected number of reservations from flight records, but she cannot reduce random uncertainty.

The number of reservations could vary from 10 to 50. The business class section has 32 seats. The manager calculates bounds of the probability of meeting the demand by using the bounds of the CDF on the right panel of Figure 6.25. The probability of meeting the demand could be as low as $F_D(32/\theta = 35) = 0.345$ and as high as $F_D(32/\theta = 25) = 0.929$. If the expectation of the number of reservations is 30, then the probability of meeting the demand is $F_D(32/\theta = 30) = 0.685$. Another estimate of the probability of meeting demand is obtained from the expected CDF of the number of reservations, $\overline{F}_D(32) = 0.668$. The large uncertainty in the probability of meeting the demand indicates that the manager must estimate the expected number of reservations accurately.

Howard (2004) claims that it is meaningless to assign a distribution to a probability in order to represent uncertainty about this probability. Specifically, Howard says: "Some people are tempted to assign probability distributions on probabilities. This is equivalent to saying they are uncertain about uncertainty. Probability is sufficient to describe uncertainty. One thing a clairvoyant knows nothing about is probability. If it were meaningful to assign a probability distribution to a probability, I could ask the clairvoyant to tell us the actual probability, which he cannot do. Assigning probability distributions to probabilities is creating a source of confusion."

The authors believe that thinking of a probability as a random variable with its own probability distribution is both a valid and useful approach to quantify uncertainty about a probabilistic model because,

a)   Any observable quantity has a probability distribution. Objective probabilities have probability distributions because they are observable.
b)   Subjective probabilities are also observable from a suitable elicitation procedure. Therefore, it is also valid to think of a subjective probability as a random variable with its own probability distribution.
c)   An expert can estimate the probability distribution of a subjective probability by following the procedure in Figure 6.1.

We explain the concept of modeling a distribution parameter by a random variable by an example. Suppose that we want to predict the number of tip-up flips, k, of a thumbtack in n independent flips. Since we flip a thumbtack, we do not know the probability of a tip-up flip. Imagine a very large box containing many thumbtacks with different sizes and shapes. The long-term frequency of a tip-up flip for different thumbtacks lies in the range [0.3,0.6]. Consider the sequence of the two experiments in Figure 6.26. First, we pick up a thumbtack randomly from the box, flip it n times and count the number of tip-up flips. We repeat this sequence of experiments. The number of tip-up flips in this sequence follows a binomial distribution with parameter p (the probability of a tip-up flip) uniformly distributed from 0.3 to 0.6. This distribution is a second-order model.

Next we justify why a second-order model is meaningful for modeling uncertainty in both an objective probability (long-term frequency) and a subjective probability.

We can only asses the probability of an observable event or the probability distribution of an observable quantity. An experiment whose outcome is observable passes the following clarity test: When all uncertain events have been resolved, we should be able to determine unequivocally the outcome of the experiment. No interpretation or judgment should be required (Howard, 1988). We cannot assess the probability of an event for which we must use judgment in order to determine if it has occurred. Because a cumulative probability distribution is a set of probabilities we need to investigate if these probabilities are observable. Long-term relative frequencies are certainty observable. Therefore, if a decision maker can repeat the same experiment many times,

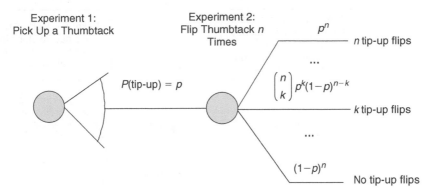

Figure 6.26 An experiment whose outcomes follow a second-order probability distribution. The number of tip-up flips k follows a binomial distribution with random parameter p (probability of a tip-up flip).

he/she can assess a probability distribution of a long-term frequency by the procedure depicted in Figure 6.26.

Consider again example 6.9 where the airline manager wants to estimate the probability of the event: {long-term frequency of more than 32 business class reservations $\leq p$}. First, the facilitator tells the manager that he/she will ask her to compare gambles on the above event. Eventually, the facilitator will collect sufficient data to find the true long-term relative frequency and determine what gambles the manager won. The facilitator asks the manager to compare lottery ticket A in Figure 6.27 to ticket B. The facilitator changes the probability that ticket B could win \$100 until the manager becomes indifferent between the two tickets. Then, the value of the cumulative distribution function of the long tem frequency is,

$$P(\text{relative frequency of more than 32 reservations} \leq p) = \frac{\theta}{360°} \qquad (6.38)$$

where angle θ is the angle subtended by sector A' in Figure 6.27 for which the manager is indifferent between the two tickets.

For example, suppose that the facilitator wants to elicit the manager's probability that the long-term relative frequency of more than 32 reservations does not exceed 0.5. The decision maker says that he/she is indifferent between a ticket that pays \$100 if the long-term frequency does not exceed 0.5 and a reference ticket that is worth the same amount if the needle of the probability wheel stops in a sector that subtends an angle $\theta = 80°$. Then the probability that the long-term relative frequency of more than 32 reservations does not exceed 0.5 is $80°/360° = 0.22$.

Now we explain how we define the probability distribution of a subjective probability. There are two views about the observability of subjective probabilities. Winkler (1967, p. 778) claims that a person has no built-in probability distribution in his/her mind. Rather the person has some knowledge, which can be elicited by a carefully designed process. There is no true subjective probability distribution according to this

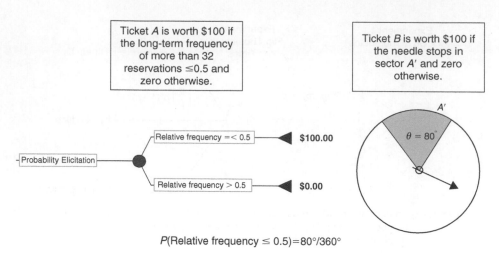

P(Relative frequency ≤ 0.5)=80°/360°

Figure 6.27 Estimation of the probability that the long-term frequency of more that 32 reservations is less than or equal to 0.5.

view, but a set of distributions that are equally valid in the sense that they are consistent with a person's beliefs. According to Winkler's view, subjective probability is not observable and consequently the concept of a probability distribution over a subjective probability is meaningless.

The other view is that there is a true subjective probability that would result if the person could perfectly assess his/her own beliefs (O'Hagan, 1988, and O'Hagan et al. 2006, pp. 22–23). According to this view, which we adopt in this book, a person's true probability is observable from a suitable elicitation procedure. On this basis, we can also assess the probability distribution of a subjective probability, $F_P(p)$, by following the procedure in Figure 6.1 where A is the event $\{P \leq p\}$.

In conclusion, any observable uncertain quantity has a probability distribution function. A person's subjective probabilities are observable from a suitable elicitation procedure. Therefore, it makes sense to consider a subjective probability as a random variable. The same is true for the parameters of a distribution because parameters are functions of probabilities. Finally, a facilitator can elicit an expert's belief about the probability distribution of a parameter by following the procedure in Figure 6.1, by using a standard gambling device, such as a probability wheel, as a ruler.

6.3.5.1   *Other representations of uncertainty in a probabilistic model*

Oakley and O'Hagan (2007) presented a method for modeling uncertainty in a probability distribution. The key idea is to consider that the true PDF of an uncertain parameter (e.g., the proportion of patients who have bleeding after taking a new anti-inflammatory drug), $f_\Theta(\theta)$ could deviate from a baseline, standard parametric PDF, $f_{\Theta_0}(\theta)$, in a way that the ratio $f_\Theta(\theta)/f_{\Theta_0}(\theta)$ is a stationary Gaussian random process. That means that the true PDF of the proportion of patients is equal to the baseline PDF scaled by the Gaussian random process. A facilitator determines this PDF ratio by estimating the parameters of the autocorrelation function of this random process. In the same paper, Oakley and O'Hagan presented a method for updating the PDF when additional information becomes available. This model is flexible, and unlike other second order models, it can represent multimodal probability distributions of a distribution parameter.

However, the assumption that $f_\Theta(\theta)/f_{\Theta_0}(\theta)$ is a stationary Gaussian process could be restrictive. Moreover, the user has to assume some form of the autocorrelation function of this process, which can be challenging.

An alternative way to model uncertainty in an elicited distribution is to use the *epsilon-contamination* class (Berger, 1985, Sections 3.5.3 and 4.7). This is a model of the class of admissible probability distributions of a distribution parameter that are close to an elicited probability distribution. Let $f_{\Theta_0}(\theta)$ be the elicited PDF of a parameter, ε a small number, which is less than one, and $q_\Theta(\theta)$ another PDF that belongs to a class L. Then the epsilon-contamination class is defined as follows,

$$\Gamma = \{f_\Theta(\theta) : (1 - \varepsilon) \cdot f_{\Theta_0}(\theta) + \varepsilon \cdot q_\Theta(\theta), \; q_\Theta(\theta) \in L\} \tag{6.39}$$

This family includes all PDF's that differ from the elicited PDF by a small value.

6.4 Bayesian Analysis

6.4.1 Motivation

In most design decisions, designers rely on judgment to estimate a probability distribution because they have limited or no data. For example, in the initial design stage of a system, the design is defined only vaguely and few or no tests have been performed. Therefore, designers rely primarily on expert judgment to estimate a performance attribute of the design.

Later, designers obtain data from tests, warranty records or customer surveys, and they want to update their initial probability distribution in view of the data. Often, they use judgment or ad-hoc rules for this purpose. Most designers when using this approach either overemphasize the initial distribution or the new data.

For example, in the early 1980s NASA management estimated that a disaster would occur in about 1 per 100,000 missions of the space shuttle. When Challenger was lost in the 25th mission, many people thought that this estimate was too optimistic. On the other hand, a stubborn person could argue that the original estimate of the probability of failure was correct; the Challenger failure was one of those rare failures that occur once in every 100,000 missions, on average. Although this example is extreme, it demonstrates the need from a formal method to combine expert judgment and data.

This section presents Bayesian analysis, which is a structured approach for updating probability distributions. This method uses Bayes' rule (Section 3.1.1.3), which is derived from the basic laws of probability. A decision maker who uses this method will quantify uncertainty more accurately than someone who uses ad-hoc methods, because the former decision maker will avoid biases of human judgment, such as stubbornness and indecision.

First, this section derives an equation for updating an initial estimate of the probability distribution of an unknown quantity, such as the probability of failure of a system or the occurrence rate of an earthquake, in light of new evidence. Then it explains how to use this equation when the new evidence consists of observed data or of expert judgments. The section demonstrates the use of Bayesian analysis to optimize the parameters of a test for reliability validation in order to maximize the accuracy of the estimated reliability.

Researchers have used Bayes' rule to assess the safety of a system when limited data is available. Wu et al. (1990) combined the opinions of experts about the probability of failure of a system. Coles and Tawn (1996) estimated the probability distribution of the annual rainfall in the U.K. by combining observed data and expert judgments. Gunawan and Papalambros (2006) and Zhu et al. (2008) used Bayes' rule for estimation of a system's failure probability in reliability based design optimization.

6.4.2 How to update a probability distribution using observations or expert judgment

Section 6.3.5 explained that decision makers can represent uncertainty about a probabilistic model by using a second-order model, that is, a distribution with random parameters. For example, the PDF of a random variable X could involve one uncertain parameter Θ. This PDF is denoted by $f_X(x/\theta)$, where X is the basic random variable and θ is a value of distribution parameter Θ.

The probability distributions of the parameters are estimated from observed values of the basic random variable X or from expert judgment. This section explains how to update the probability distributions of the parameters in both cases.

Consider the case where there is only one distribution parameter. Let $f_{\Theta_0}(\theta)$ be the prior PDF of the parameter. This PDF represents the decision maker's beliefs before he/she gets evidence E. According to Bayes' rule in chapter 3, the probability that the value of parameter Θ is in a small interval conditioned upon the evidence E is proportional to the unconditional probability of the same event, scaled by the likelihood of the evidence,

$$P(\theta_i < \Theta \leq \theta_i + \delta\theta/E) = \frac{P(E/\theta_i < \Theta \leq \theta_i + \delta\theta) \cdot P(\theta_i < \Theta \leq \theta_i + \delta\theta)}{k} \quad (6.40)$$

where θ_i is some value of Θ, $\delta\theta$ is a small interval, $P(E/\theta_i < \Theta \leq \theta_i + \delta\theta)$ is the likelihood of the evidence and k is a normalizing constant,

$$k = \sum_{i=1}^{n} P(E/\theta_i < \Theta \leq \theta_i + \delta\theta) \cdot P(\theta_i < \Theta \leq \theta_i + \delta\theta) \quad (6.41)$$

The likelihood of the evidence is the probability of obtaining evidence E given that the value of parameter Θ is in the range $[\theta_i, \theta_i + \delta\theta]$. In the limit, Equation (6.40) yields,

$$f_\Theta(\theta) \, d\theta = \frac{P(E/\theta) \cdot f_{\Theta_0}(\theta) \, d\theta}{k} \quad (6.42)$$

The last equation can be simplified to,

$$f_\Theta(\theta) = \frac{P(E/\theta) \cdot f_{\Theta_0}(\theta)}{k} \qquad (6.43)$$

where $k = \int_{-\infty}^{\infty} P(E/\theta) \cdot f_{\Theta_0}(\theta) \, d\theta$. Functions $f_{\Theta_0}(\theta)$ and $f_\Theta(\theta)$ are called prior and posterior PDFs, respectively. In plain English, Equation (6.4.2) says:

> The posterior PDF of parameter Θ is proportional to the likelihood of the evidence $P(E/\theta)$, times the prior PDF of Θ.

Equation (6.43) is the counterpart of Bayes' rule for events in section 3.1.1.3.

The conditional expectation of parameter Θ given evidence E is called Bayesian estimator of this parameter,

$$E(\Theta/E) = \int_{-\infty}^{\infty} \theta \cdot f_\Theta(\theta) \, d\theta \qquad (6.44)$$

If parameter Θ is discrete, then its PMF has the form, $f_\Theta(\theta) = \sum_{i=1}^{n} \delta(\theta - \theta_i) P(\Theta = \theta_i)$, where $\theta_i, i = 1, \ldots, n$ are the possible values of Θ, and $\delta(\cdot)$ is the unit impulse function.

Then, Equation (6.43) reduces to the following equation,

$$f_\Theta(\theta) = \frac{\sum_{i=1}^{n} \delta(\theta - \theta_i) P(E/\theta_i) \cdot f_{\Theta_0}(\theta_i)}{k} \qquad (6.45)$$

where the normalizing constant is, $k = \sum_{i=1}^{n} P(E/\theta_i) \cdot f_{\Theta_0}(\theta_i)$.

When there are multiple continuous parameters, the posterior PDF of the vector of parameters $\boldsymbol{\Theta}$ is,

$$f_{\boldsymbol{\Theta}}(\boldsymbol{\theta}) = \frac{P(E/\boldsymbol{\theta}) \cdot f_{\boldsymbol{\Theta_0}}(\boldsymbol{\theta})}{k} \qquad (6.46)$$

The above is an extension of Equation (6.43) from a single parameter to a vector of parameters.

The posterior PDF depends on both the information contained in the prior PDF and the evidence. The influence of these functions on the posterior PDF depends on the amount of information that these functions contain. A flat prior PDF or likelihood of some parameter, such as a probability, contain little information about this probability and indicate that the decision maker is imprecise about it. On the other hand, a PDF or likelihood function with a spike shows high confidence.

For example, if a decision maker is highly uncertain about the probability of failure of a product, then he/she will construct a flat prior. Suppose that the expert performs hundreds or thousands of tests. Then, the likelihood function will consist of a spike, and it will influence the posterior more than the prior. On the other hand, if the decision maker is confident about the probability of failure before performing the tests he/she will construct a prior PDF to reflect that, which will strongly influence the posterior.

There are two types of evidence;

1. Observed values of the basic variable X, $x_1, \ldots, x_n$.
2. Estimates of the values of the distribution parameter(s) obtained from experts.

We consider each case below.

6.4.2.1 Evidence consists of observed values of basic random variable

If we view the PDF of the basic variable $f_X(x/\theta)$ as a function of θ, then the likelihood function in Equation (6.43) will be $P(E/\theta) = f_X(x/\theta)$. Then, the posterior PDF of parameter θ becomes,

$$f_\Theta(\theta) = \frac{f_X(x/\theta) \cdot f_{\Theta_0}(\theta)}{k} \qquad (6.47)$$

If evidence consists of independent observations, then the likelihood is equal to the probability of observing values $x_1, \ldots, x_n$,

$$P(E/\theta) = \prod_{i=1,\ldots,n} f_X(x_i/\theta) \qquad (6.48)$$

Then, Equation (6.43) for the posterior PDF of Θ becomes,

$$f_\Theta(\theta) = \frac{\prod_{i=1}^{n} f_X(x_i/\theta) \cdot f_{\Theta_0}(\theta)}{k} \qquad (6.49)$$

where $k = \int_{-\infty}^{\infty} \prod_{i=1}^{n} f_X(x_i/\theta) \cdot f_{\Theta_0}(\theta) \, d\theta$.

An analyst can avoid the integration in Equation (6.49) by selecting a prior distribution of the parameter Θ that is *conjugate* to the underlying random variable X. Then the posterior distribution belongs to the same parametric family as the prior and its parameters are calculated very efficiently using algebraic equations. For example, the beta and binomial distributions are conjugate pairs. If the prior distribution of the probability of an event is beta, so is its posterior distribution.

In our opinion, a decision maker should focus primarily on selecting a prior distribution that represents her belief about the value of the unknown distribution parameter. Reducing computational cost is nice but less important than selecting a representative prior.

Example 6.10: Probability of a Space Shuttle disaster

At the time of this writing, the space shuttle has completed 129 missions. Two disasters occurred in 1986 and in 2003. We will demonstrate how to estimate the probability of failure of the Shuttle by combining judgment about this probability and the above data. Assume that failures occur independently and that the probability of failure does not change with time.

Assume that the prior PDF of the probability of failure is uniform between 0 and an upper limit,

$$
f_{P_0}(p) = \begin{cases} \dfrac{1}{p_u} & \text{if } 0 \le p \le p_u \\[2mm] 0 & \text{otherwise} \end{cases}
$$

where $f_{P_0}(p)$ is the prior PDF and p_u is the upper limit of the probability of failure.

Solution:

The number of catastrophic failures x that could occur in n flights follows the binomial distribution (Equation 3.66),

$$
f_X(x/p) = \binom{n}{x} p^x (1-p)^{n-x} = \frac{n!}{(n-x)!x!} p^x (1-p)^{n-x}
$$

This distribution represents the number of times an event occurs in n independent trials. We have evidence that $x=2$ failures occurred in $n=129$ missions. Therefore, the likelihood of this evidence is,

$$
P(E/p) = f_X(x/p) = \frac{129!}{127!2!} p^2 (1-p)^{127}
$$

The updated PDF of the probability of failure given that x failures occurred in n flights is,

$$
f_P(p) = \begin{cases} \dfrac{1}{k\,p_u} \cdot \dfrac{129!}{127!\,2!} \cdot p^2 \cdot (1-p)^{127} & \text{for } 0 \le p \le p_u \\[2mm] 0 & \text{otherwise} \end{cases}
$$

where the normalizing constants k is,

$$
k = \frac{1}{p_u} \int_0^{p_u} \frac{129!}{127!\,2!} p^2 (1-p)^{127} \, dp
$$

Note that the posterior PDF is a beta distribution truncated in the range from 0 to p_u.

A NASA scientist estimates that the probability of failure must be less than 0.1. Figure 6.28 shows the prior and posterior PDFs of the probability of failure in one flight. There is significant uncertainty in the probability of failure. For example, the failure probability could be in the range from [0.013, 0.033] with only 0.57 probability.

The scientist calculates that the expected value (Bayesian estimator) of the probability of failure is 0.023 by using the posterior PDF in this figure. If the scientist estimated the same probability only from the observed number of failures, the scientist would find that the probability would be $2/129 = 0.016$.

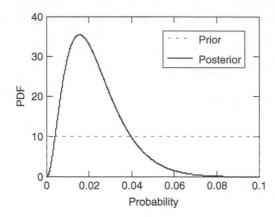

Figure 6.28 PDF of probability of a disaster in one Shuttle flight assuming that this probability cannot exceed 0.1.

The scientist investigates the effect of the prior PDF of the failure probability by changing the upper bound to 0.05 and to 0.15. The posterior PDF for the upper bound of 0.15 is practically identical to the one for the upper bound of 0.1. Figure 6.29 shows the posterior PDF when the upper bound is 0.05. This bound significantly affected the PDF by truncating it in the range [0, 0.05]. The mean value of the probability of failure is 0.021 based on this PDF. Unless there is strong evidence that the probability of failure could not exceed 0.05, this PDF should not be used.

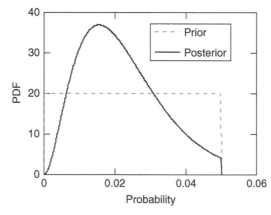

Figure 6.29 PDF of probability of a disaster in one Shuttle flight assuming that this probability cannot exceed 0.05.

Example 6.11: Assessment of the risk of a future catastrophic failure of the remaining Space Shuttles

There are three remaining Shuttles that have flown a total of 76 times. Assume that it is current practice to retire a Shuttle after 50 flights. Then, there are 74 remaining flights.

Find the probability of that no catastrophic failure will occur in these flights.

Solution:

The probability that there will be no disasters in the remaining 74 flights is, $P(\text{no distasters}) = (1 - p)^{74}$. We estimate this probability in three different ways:

Approach 1: Find the average of the probability of no disasters by using the posterior PDF of the probability of a disaster in one flight,

$$P(\text{no disaster}) = \int_0^{0.1} (1 - p)^{74} f_P(p)\, dp = 0.257$$

Approach 2: Calculate the probability of no disaster by using the Bayesian estimator of the probability of a disaster,

$$P(\text{no disaster}) = (1 - 0.023)^{74} = 0.18$$

Approach 3: Use the rate from the observed data,

$$P(\text{no disaster}) = (1 - 0.016)^{74} = 0.32$$

The estimate of the first approach is more accurate than those of the other two approaches for two reasons. Firstly, the first approach uses the correct equation from probability calculus to find the expected value of a function of a random variable (Equation 3.119), and secondly it considers both expert judgment and observations. The second approach approximates the expectation of a function of a random variable with the value of that function for the expectation of the variable. This approximation is inaccurate in this problem. The third approach ignores expert judgment.

Example 6.12: Estimation of the probability of a severe earthquake in southern California using Bayesian analysis

A seismologist wants to estimate the probability that an earthquake of magnitude greater than 7 in Richter scale could occur in the next 30 years in Southern California. Earthquakes of such intensity are rare. For example, three earthquakes with magnitude greater than 7 occurred in the last 78 years (Table 6.4). The seismologist assumes that the number of earthquakes follows the Poisson distribution (Equation 3.68). This is a reasonable assumption for rare events. The seismologist believes that the rate of earthquakes with magnitude greater than 7 could vary from 0 to 0.2 per year. The later rate corresponds to one earthquake in 5 years, on average.

Find a) the posterior PDF of the rate of earthquakes with magnitude greater than 7 and b) the probability of occurrence of such an earthquake during the next 30 years.

Solution:
The PMF of the number of earthquakes in T years is,

$$f_X(x/v) = \frac{(vT)^x e^{-vT}}{x!}$$

where X is the number of earthquakes, $f_X(x)$ is the PMF of the number of earthquakes, and v is the rate at which earthquakes above 7 in Richter scale occur (earthquakes per year).

The prior PDF of the rate of earthquakes is assumed uniform from 0 to 0.2,

$$f_{N_0}(v) = \begin{cases} 5 & \text{for } 0 < v \le 0.2 \\ 0 & \text{otherwise} \end{cases}$$

In this problem, the number of earthquakes X is the basic variable. Rate v is the parameter that the seismologist wants to estimate. The evidence is that three earthquakes occurred in the last 78 years. The likelihood of the evidence is $P(E/v) = f_X(x/v)$.

The seismologist uses Bayes' rule to update the prior PMF of the earthquake rate,

$$f_N(v) = \frac{1}{k} f_X(x/v) f_{N_0}(v)$$

where the normalization constant is $k = \int_0^{0.2} f_X(x/v) f_{N_0}(v)\, dv$. The value of this constant is 0.064.

Figure 6.30 shows the posterior PDF of the earthquake rate.

From the posterior PDF, the seismologist finds the Bayesian estimator of the earthquake rate,

$$E(N) = \int_0^{0.2} v f_N(v)\, dv = 0.051 \text{ earthquakes per year}$$

This means that one earthquake stronger than 7 in Richter scale occurs in 20 years, on average. If the seismologist estimated the same rate only from the observed number of 3 earthquakes in 78 years, he/she would find that the rate is 0.038, which corresponds to one earthquake every 26 years, on average.

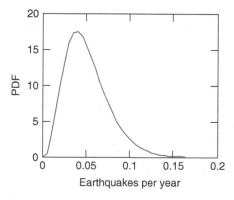

Figure 6.30 Posterior PDF of earthquake rate.

The probability of no earthquakes of magnitude 7 or greater in the next 30 years ($x = 0$) can be estimated in three different ways:

Approach 1: Find the average of this probability by using the posterior PDF of the earthquake rate,

$$P(\text{no earthquake}) = \int_0^{0.2} f_X(x/v) f_N(v)\, dv = \int_0^{0.2} \frac{(vT)^x e^{-vT}}{x!} f_N(v)\, dv = 0.272$$

Approach 2: Calculate the probability of no earthquake by using the Bayesian estimator of the earthquake rate,

$$P(\text{no earthquake}) = \frac{\{E(N) \cdot T\}^x e^{-E(N)T}}{x!} = \frac{(0.051 \cdot 30)^0 e^{-0.051 \cdot 30}}{0!} = 0.215$$

Approach 3: Use the rate from the observed data,

$$P(\text{no earthquake}) = \frac{(0.038 \cdot 30)^0 e^{-0.038 \cdot 30}}{0!} = 0.32$$

There is a considerable difference between the three probabilities. The probability from the first approach should be the most accurate as we explained in the end of example 6.11.

Example 6.13: Effect of number of observations on posterior probability of failure of a pump
The estimate of a parameter should become more accurate with the number of observations. In the limit, the posterior PDF of the parameter should converge to a unit impulse centered at the long-term relative frequency of this parameter. In addition, the effect of the prior PDF on the posterior should diminish with number of observations, provided that the support of the prior includes the long-term relative frequency.

An engineer estimates that one pump per ten fails, on average. Assume that the prior PDF of the probability of failure of a pump is beta, with parameters $\beta_1 = 1$ and $\beta_2 = 9$,

$$f_P(p) = \frac{\Gamma(\beta_1 + \beta_2)}{\Gamma(\beta_1)\Gamma(\beta_2)} p^{\beta_1 - 1} \cdot (1 - p)^{\beta_2 - 1} \quad \text{for } 0 < p < 1$$

a) Find the posterior PDF of the probability of failure for three cases where the engineer tested 20, 200 and 2000 systems and observed 1, 10 and 100 failures, respectively.

b) Find the 95% credible interval of the probability of failure.

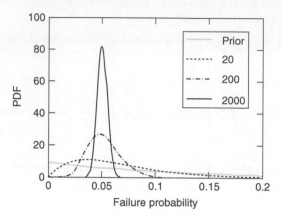

Figure 6.31 Effect of number of tests on estimate of probability of failure. Confidence about the estimated probability of failure increases with the number of tests.

Solution:
The number of failures x in n tests, follows the binomial distribution (Equation 3.66), Then, the likelihood of the evidence from the tests is,

$$P(E/p) = \frac{n!}{(n-x)!x!}p^x(1-p)^{n-x}$$

where n is the number of tests, and x the number of failures.

The engineer calculates the posterior PDF of the probability of failure from Equation (6.43). Figure 6.31 shows the prior and posterior PDFs of the probability of failure. The engineer is highly uncertain about the probability of failure, before performing tests, as evidenced by the large dispersion of the prior PDF. The engineer becomes increasingly confident as he/she observes the tests results. When the engineer sees the results of the 2000 tests, he/she becomes almost certain that the probability of failure is 0.05.

The engineer assesses the accuracy of the estimate of the failure probability by calculating the $1 - \alpha\%$ credible interval. This interval contains the probability of failure with probability $1 - \alpha$, so as it is equally likely that the true probability could be lower than the low bound and higher than the upper bound. We can find central credible intervals by solving numerically each of the following equations,

$$\int_0^{p_{low}} f_P(p)\,dp = 0.025 \quad \text{and}$$

$$\int_{p_{high}}^1 f_P(p)\,dp = 0.975$$

Table 6.8 shows the lower and upper bounds of the 95% credible intervals for the three posterior PDFs in Figure 6.31.

Table 6.8 95% credible intervals for probability of failure.

n	x	p_{low}	p_{high}
20	1	0.0085	0.178
200	10	0.027	0.086
2,000	100	0.041	0.06

6.4.2.2 Evidence consists of expert judgment about the values of the distribution parameters

If a facilitator has no observations of the basic variable then he/she can elicit an expert's estimate of a distribution parameter. In this case, the facilitator should quantify his confidence about this estimate. For example, the facilitator could elicit the expert's estimates of the 50% and 90% credible intervals of the elicited probability of failure, or estimate these intervals himself.

The likelihood of the evidence is the PDF of the elicited value of the parameter $\hat{\theta}$ viewed as a function of the true value of this parameter. In order to estimate the likelihood function of the evidence, the facilitator needs to assume some relation between the expert's estimate and the true value of the parameter. For example, the expert's estimate of the parameter is equal to the true value plus an error term,

$$\hat{\theta} = \theta + \varepsilon \tag{6.50}$$

where θ is the true value of the parameter, which is unknown, $\hat{\theta}$ is the estimate of this parameter and ε is the error. Let $f_E(\varepsilon)$ be the PDF of the error, E. Note that Greek letter E denotes the error; this should be distinguished from Latin letter E, which represents the evidence. The type of this PDF could be selected by either the facilitator or the expert. Then, the PDF of the estimate viewed as a function of the true value of the parameter is,

$$f_{\hat{\Theta}}(\hat{\theta}/\theta) = f_E(\hat{\theta} - \theta) \tag{6.51}$$

The above equation was derived from Equation (3.116), for the special case where $x = \varepsilon$, and $y = \hat{\theta}$. Then, the likelihood of the evidence is,

$$P(Evidence/\theta) = f_E(\hat{\theta} - \theta) \tag{6.52}$$

The posterior PDF of the parameter is,

$$f_\Theta(\theta) = \frac{f_E(\hat{\theta} - \theta)f_{\Theta_0}(\theta)}{k} \tag{6.53}$$

In the above equation, normalizing constant k is the integral of the numerator of the right hand side for all possible values of parameter θ.

Example 6.14: **Estimating the probability of failure of a system using expert judgment**
In the early design stage of a system, a designer (who is also a statistician) estimates that the probability of failure, p, of the system has median 0.1, and 99% percentile 0.3. He assumes that the PDF of this probability is lognormal.

The designer calculates the parameters of the lognormal distribution from the above estimates. Let σ be the standard deviation of the logarithm of the probability of failure Then the logarithm of the probability, $\ln(p)$, is normal with mean value $\mu = \ln(0.1) = -2.303$. From the designer's percentiles,

$$P\{\ln(p) \leq \ln(0.3)\} = 0.99 \Leftrightarrow$$

$$P\left\{\frac{\ln(p) - \mu}{\sigma} \leq \frac{\ln(0.3) - \mu}{\sigma}\right\} = 0.99 \Leftrightarrow$$

$$\frac{\ln(0.3) - \mu}{\sigma} = \Phi^{-1}(0.99)$$

and

$$P\{\ln(p) \leq \ln(0.1)\} = 0.5 \Leftrightarrow$$

$$\frac{\ln(0.1) - \mu}{\sigma} = \Phi^{-1}(0.5)$$

Therefore, the standard deviation of the logarithm of the probability of failure is,

$$\sigma = \frac{\ln(0.3) - \ln(0.1)}{\Phi^{-1}(0.99) - \Phi^{-1}(0.5)} = 0.472$$

Figure 6.32 shows the PDF and CDF of the probability of failure.

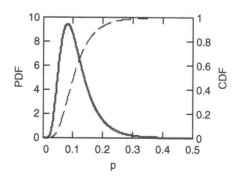

Figure 6.32 Prior PDF and CDF of probability of failure.

Later, when the design has evolved, the designer asks an expert to examine the system and estimate its probability of failure. The expert estimates that this probability is $p' = 0.1$. The designer believes that the expert's estimate is significantly more accurate than the initial estimate. Then the designer asks the expert how confident she is about this probability. She replies that the true probability could be in the ranges [0.075, 0.125] and [0.05, 0.15] with probabilities 0.5 and 0.95, respectively. The designer assumes that the expert's estimated probability is equal to the true probability scaled by a factor ε, $p' = p \cdot \varepsilon$. If ε is greater than one, this means that the expert overestimated the probability of failure.

Find the posterior PDF of the probability of failure.

Solution:
The designer needs to calculate the likelihood function of the expert's estimate. The designer assumes that factor ε is lognormally distributed. Therefore, $\ln(\varepsilon)$ is normally distributed. Let $\ln(b)$ and σ' be the mean value standard deviation of $\ln(\varepsilon)$. The values of b and σ' are estimated from the expert's credible intervals as follows,

$$P(0.075 < p \le 0.125) = 0.5 \Leftrightarrow$$

$$P\left(0.075 < \frac{p'}{\varepsilon} \le 0.125\right) = 0.5 \Leftrightarrow$$

$$P\left(\frac{0.075}{p'} < \frac{1}{\varepsilon} \le \frac{0.125}{p'}\right) = 0.5 \Leftrightarrow$$

$$P\left(\frac{p'}{0.125} < \varepsilon \le \frac{p'}{0.075}\right) = 0.5 \Leftrightarrow$$

$$P\{\ln(p') - \ln(0.125) < \ln(\varepsilon) \le \ln(p') - \ln(0.075)\} = 0.5$$

Since $\ln(\varepsilon)$ is normal with mean value $\ln(b)$ and standard deviation σ' the above equation becomes,

$$\Phi\left\{\frac{\ln(p') - \ln(0.075) - \ln(b)}{\sigma'}\right\} - \Phi\left\{\frac{\ln(p') - \ln(0.125) - \ln(b)}{\sigma'}\right\} = 0.5 \qquad (6.54)$$

Similarly, using the information about the 95% credible interval we obtain a second equation that the values of b and σ' should satisfy,

$$\Phi\left\{\frac{\ln(p') - \ln(0.05) - \ln(b)}{\sigma'}\right\} - \Phi\left\{\frac{\ln(p') - \ln(0.15) - \ln(b)}{\sigma'}\right\} = 0.95 \qquad (6.55)$$

The designer finds the values of the above parameters by solving Equations (6.54) and (6.55),

$$b = 1.313 \quad \text{and} \quad \sigma' = 0.25$$

The likelihood of the evidence is the PDF of the expert's estimate of the probability of failure conditioned upon the true value of the probability, viewed as a function of the true failure probability,

$$P(E/p) = f_{p'}(p'/p)$$

which is normal with mean value of $\ln(p \cdot b)$ and standard deviation σ'. Figure 6.33 shows the likelihood function,

Finally, the posterior PDF of the probability of failure is the product of the prior PDF times the likelihood divided by a normalizing constant. Figure 6.34 shows the posterior PDF together with the prior PDF and the likelihood. It is observed that use of the expert's estimates reduced uncertainty in the probability of failure dramatically.

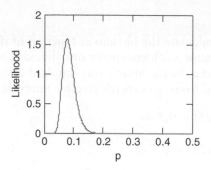

Figure 6.33   Likelihood function of probability of failure.

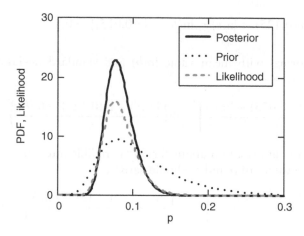

Figure 6.34   Posterior PDF of failure probability. The likelihood has been scaled by 10.

Example 6.15:   Estimation of PDF of a function of two variables

Oberkampf et al. (2004) challenged researchers on uncertainty quantification to solve a set of problems where there was limited information from expert judgment. We will solve one of the problems here. Consider function $Y = (A + B)^A$, where A and B are independent variables, which means that knowledge about one variable does not alter our belief about the other. There is no uncertainty in the functional relation between A, B and Y. There is only uncertainty in the values of A and B.

Two experts said that variable A is in the following intervals: [0.1, 0.4] and [0.3, 0.6]. Similarly they stated that the variable B is in the following intervals: [0.2, 0.5] and [0.4, 0.7]. The objective is to quantify the uncertainty in Y. The analyst assumes that the experts are unbiased and their errors are independent.

Assume that an analyst only knows that variable A is between 0.1 and 1, and variable B is between 0 and 1. Since the analyst has no other information regarding the prior the analyst assumes uniform prior probability distributions for A and B.

When the evidence given by experts about a random variable X, consists of intervals, the analyst needs a method to interpret the evidence and bring it into a form so that the analyst can use it in the Bayesian framework. To apply Bayes' theorem to this problem, the analyst needs to estimate the likelihood $P(E/X=x)$. The analyst makes the following assumptions to combine evidence from the experts:

a)   The analyst converts the interval provided by each expert about a variable into a point estimate. This can be the midpoint of the interval or another point obtained based on the analyst's judgment.

b)   The point estimate of the expert is equal to the true value of the variable plus an error. The analyst assumes a joint probability distribution of the errors of the experts.

Suppose that there is evidence in the form of intervals from n experts, $[x_i^{\min}, x_i^{\max}]$ for $i = 1, \ldots, n$ (Figure 6.35). The analyst assumes that the ith expert gives a point estimate $\hat{x}_i$, which is the midpoint of the ith interval:

$$\hat{x}_i = \frac{x_i^{\max} + x_i^{\min}}{2} \quad \text{for } i = 1, \ldots, n \tag{6.56}$$

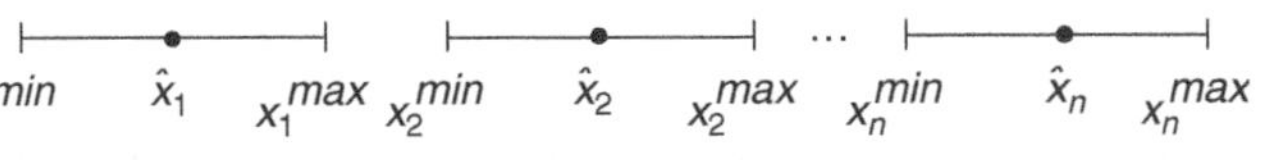

Figure 6.35 Evidence from n experts about variable X.

Let x be the true value of variable X. If the error in the ith expert's estimate is E_i, then:

Point Estimates $=$ Random Variables $+$ Errors of Experts, or

$$\hat{\mathbf{X}} = \mathbf{X} + \mathbf{E} \tag{6.57}$$

where $\hat{\mathbf{X}}$ is vector of the point estimates of the experts (size n), $\mathbf{X}$ is a vector whose elements are all equal to the true value denoted by x, and $\mathbf{E}$ is the vector of the errors.

We need the PDF of $\mathbf{E}$ to estimate the likelihood of the evidence. As an example, we assume that random vector $\mathbf{E}$ is normal with mean $\mathbf{b}$ and covariance matrix $\mathbf{C}$.

$$\mathbf{C} = \begin{bmatrix} \sigma_{E_1}^2 & \rho_{1,2}\sigma_{E_1}\sigma_{E_2} & \cdots \\ \vdots & \ddots & \vdots \\ \rho_{n,1}\sigma_{E_n}\sigma_{E_1} & \cdots & \sigma_{E_n}^2 \end{bmatrix} \tag{6.58}$$

where σ_{E_i} is the standard deviation in the estimate of the ith expert and ρ_{ij} is the correlation coefficient of the estimates of two experts. The ith element of vector b, b_i, is the bias of the ith expert. The analyst should estimate the above quantities or elicit them from the expert.

As an example, an analyst could assume that:

a) Bias b_i is zero,
b) The endpoints of the interval provided by ith expert are equal to the midpoint ±3 standard deviations of the error, respectively.

On the basis of the above assumptions, the analyst can calculate the standard deviation of the error in each expert's estimate:

$$\sigma_{E_i} = \frac{x_i^{max} - x_i^{min}}{6} \tag{6.59}$$

The next step is to determine the experts' error given the evidence from experts. We assume that the bias is zero for all the experts (that is the errors of the experts have zero mean). Figure 6.36 displays the experts' errors for variables A and B. Then the likelihood of the evidence is:

$$P(E/\mathbf{X} = \mathbf{x}) = f_E(\hat{\mathbf{X}} - \mathbf{X}/\mathbf{X} = \mathbf{x}) = \frac{1}{2 \cdot \pi \cdot |\mathbf{C}|^{1/2}} e^{-\frac{1}{2}(\hat{\mathbf{X}} - \mathbf{x} - \mathbf{b})^T \mathbf{C}^{-1}(\hat{\mathbf{X}} - \mathbf{x} - \mathbf{b})} \tag{6.60}$$

The posterior PDF of variable $\mathbf{X}$ can be calculated from the following equation,

$$f_\mathbf{X}(\mathbf{x}) = \frac{1}{k} P(E/\mathbf{X} = \mathbf{x}) f_{\mathbf{X}_0}(\mathbf{x}) \tag{6.61}$$

For variable A the likelihood function is,

$$P(E/\mathbf{A} = \mathbf{a}) = f_E(\hat{\mathbf{A}} - \mathbf{A}/\mathbf{A} = \mathbf{a}) = \frac{1}{2 \cdot \pi \cdot |\mathbf{C}|^{1/2}} e^{-\frac{1}{2}(\hat{\mathbf{A}} - \mathbf{a} - \mathbf{b}_A)^T \mathbf{C}^{-1}(\hat{\mathbf{A}} - \mathbf{a} - \mathbf{b}_A)} \tag{6.62}$$

The covariance matrix is diagonal because the errors of the experts are independent. The vector of the bias values is zero because experts are unbiased. From the expert's estimates and Equations (6.56) and (6.59),

$$\hat{\mathbf{A}} = \begin{pmatrix} 0.25 \\ 0.45 \end{pmatrix}, \quad \mathbf{C} = \begin{bmatrix} 0.05^2 & 0 \\ 0 & 0.05^2 \end{bmatrix}, \quad \mathbf{b}_A = \begin{pmatrix} 0 \\ 0 \end{pmatrix} \tag{6.63}$$

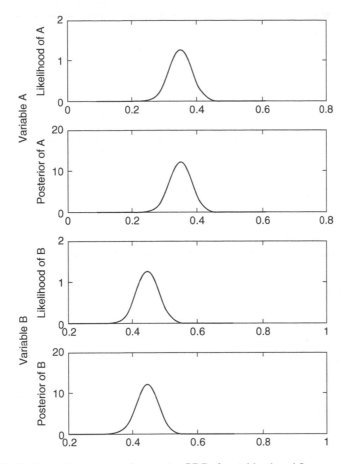

Figure 6.36 Likelihood functions and posterior PDF of variables A and B.

The likelihood function for variable B is,

$$P(E/\mathbf{B} = \mathbf{b}) = f_E(\hat{\mathbf{B}} - \mathbf{B}/\mathbf{B} = \mathbf{b}) = \frac{1}{2 \cdot \pi \cdot |\mathbf{C}|^{1/2}} e^{-\frac{1}{2}(\hat{\mathbf{B}} - \mathbf{b} - \mathbf{b}_B)^{\mathsf{T}} \mathbf{C}^{-1} (\hat{\mathbf{B}} - \mathbf{b} - \mathbf{b}_B)} \qquad (6.64)$$

$$\hat{\mathbf{B}} = \begin{pmatrix} 0.35 \\ 0.55 \end{pmatrix}, \quad \mathbf{C} = \begin{bmatrix} 0.05^2 & 0 \\ 0 & 0.05^2 \end{bmatrix}, \quad \mathbf{b}_B = \begin{pmatrix} 0 \\ 0 \end{pmatrix} \qquad (6.65)$$

Figure 6.36 shows the likelihood functions of variables A and B obtained from Equations (6.64) and (6.65). The same figure shows the posterior PDFs of the same variables. Both the likelihood of A and its posterior are centered on 0.35. The reason is that the prior PDF of this variable is uniform from 0.1 to 1 and the likelihood is centered on 0.35. Note that according to the two experts, A could vary from 0.1 to 0.6. When the analyst combines the two experts' intervals he finds that variable A could vary from 0.25 to 0.45.

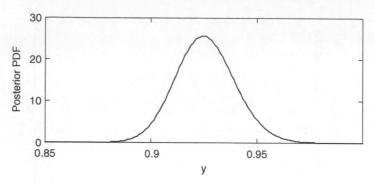

Figure 6.37 Posterior PDF of depended variable.

Finally, the analyst calculates the CDF of dependent variable,

$$F_Y(y) = P(Y \leq y) = P\{(a+b)^a \leq y\} \Rightarrow$$
$$F_Y(y) = P\{a+b \leq y^{1/a}\} \Rightarrow$$
$$F_Y(y) = \int_0^1 P(b \leq y^{1/a} - a) \cdot f_A(a)da \Rightarrow \tag{6.66}$$
$$F_Y(y) = \int_0^1 F_B(y^{1/a} - a) \cdot f_A(a)da$$

The PDF of Y is,

$$f_Y(y) = \int_0^1 \frac{1}{a} f_B(y^{1/a} - a) \cdot f_A(a) \cdot y^{(1-a)/a} da \tag{6.67}$$

Figure 6.37 shows the PDF of Y. It is observed that this variable varies from 0.9 to 1.

Suppose that the analyst does not want to estimate the probability distributions of independent variables A and B. Instead the analysts assumes that variables and A and B vary between the minimum of the lower bounds and the maximum of the upper bounds of the two experts' intervals. Therefore, A varies in the range [0.1, 0.6] and B in the range [0.2, 0.7]. Then lower and upper bounds of Y are 0.81 and 1.17, which is much wider than those obtained from Bayes' rule. Note that one consequence of avoiding eliciting the probability distributions of the two independent variables is the large uncertainty in the value of the dependent variable.

6.4.2.3 Application: A Bayesian approach for optimizing physical tests for reliability validation

Reliability-based design can help build safer and/or more economical products than deterministic design. Designers use analytical models to predict the performance of a design and to model uncertainties. These models always involve idealizations and approximations. Therefore, it is important to validate the reliability of the final designs. This can be done by using physical tests. First, a prior probability distribution is assumed for the modeling error. One or more prototypes of the system are tested under loading conditions that are representative of real life loads. The probability distribution of the error and the reliability of the system are updated by using Bayes' rule on the basis of the test outcome.

Tests for reliability validation are expensive. One way to reduce cost is to choose the conditions of a test so as to maximize the confidence in the estimated reliability of the tested design. For example, suppose that a designer wants to experimentally validate the reliability of a component by applying some load on it and checking if it fails. If the load is too low, the design will probably survive, but this does not prove that it is safe under real life conditions. On the other hand, if the test load is large, then the system is likely to fail, but this does to mean that the system is unsafe. This subsection presents an approach for designing physical tests that uses an approximate analytical model of the performance function of a system to select the conditions of the test so as to maximize the confidence in the estimated reliability.

In Bayesian testing, we quantify the accuracy of an analytical model by a set of parameters. These parameters represent errors in the deterministic model for predicting the performance of a system and uncertainty in the PDFs of the random variables. Designers estimate prior PDF's for these parameters, and then they update them according to the evidence from physical tests. Finally, they calculate the PDF of the probability of failure from the updated PDFs of the model parameters and the functional relation between these parameters and the probability of failure of the system. The Bayesian approach presented optimizes the test conditions to minimize the variance of the updated PDF of the probability of failure.

We will present the Bayesian approach considering a single error parameter ε for the performance function of a system, $g(\mathbf{X}, \mathrm{E})$, where $\mathbf{X}$ is the vector of random variables. This function is positive if the system survives a test and negative if it fails (Chapter 3). First a designer estimates the prior PDF of the error parameter $f_{\mathrm{E}_0}(\varepsilon)$. Then he/she tests the system and updates the prior on the basis of the test results.

The posterior PDF of the error is,

$$f_{\mathrm{E}}(\varepsilon/I = i) = \frac{P(I = i/\varepsilon) \cdot f_{\mathrm{E}_0}(\varepsilon)}{k} \tag{6.68}$$

where, I is a failure indicator function (one if system fails, zero if it survives), and $P(I = i/\varepsilon)$ is the likelihood of the outcome of the test given $\mathrm{E} = \varepsilon$,

$$P(I = i/\varepsilon) = \int \cdots \int_{\Omega} f_{\mathbf{X}}(\mathbf{x}) \, d\mathbf{x}$$

$$\text{where } \Omega = \begin{cases} g(\mathbf{X}, \varepsilon) \leq 0 & \text{if system fails} \\ g(\mathbf{X}, \varepsilon) > 0 & \text{if system survives} \end{cases} \tag{6.69}$$

The normalization constant is,

$$k = \int_{-\infty}^{\infty} P(I = i/\varepsilon) \cdot f_E(\varepsilon) \, d\varepsilon \tag{6.70}$$

The probability of failure, PF, is a function of error parameter, ε, because this parameter affects the performance function. The expected value of the probability of failure is,

$$E(PF) = \sum_{i=0,1} \left[\int_{-\infty}^{\infty} PF(\varepsilon) \cdot f_E(\varepsilon/I = i) \, d\varepsilon \right] \cdot P(I = i) \tag{6.71}$$

The posterior failure PDF of the failure probability is calculated from Equation (3.116),

$$f_{PF} (p/I = i) = \frac{f_E(\varepsilon/I = i)}{\frac{dPF(\varepsilon)}{d\varepsilon}} \quad \text{where } \varepsilon = PF^{-1}(p) \tag{6.72}$$

The posterior PDF of the probability of failure is,

$$Var_{PF} = \sum_{i=0,1} \int_{0}^{1} p^2 \cdot f_{PF}(p/I = i) \, dp \cdot P(I = i) - E(PF)^2 \tag{6.73}$$

The designer solves the following optimization problem to find the optimal test parameters,

Find the values of the controllable parameters in a test, $y_1, y_2, \ldots, y_j$

To minimize Var_{PF} \hfill (6.74)

For example, the value of an applied load is one parameter. The designer can perform one test or a sequence of tests. He/she determines the optimal test parameters for the first test by solving optimization problem (6.74). Then, based on the results from the first test, the PDF of the error can be updated using Bayes' rule. The posterior PDF of the error is used as the prior PDF for a second test. Using the same procedure, n tests can be performed until the required confidence in the estimated probability of failure is achieved. Alternatively, the designer may perform a fixed number of tests that he/she can afford.

Example 6.16:   Finding the optimum test load on a structure to estimate its reliability with maximum confidence

A structural element is subjected to a force that results in stress S. The resistance, S_U, and load, L, are random. The structural element fails if its resistance, S_U, is less than the stress S. The cross sectional area, A, is deterministic. We assume there is an error, E, in the model of the performance function, such that a positive value of the error corresponds to underestimation of the reserve strength. The PDF of the error is $f_E(\varepsilon)$. We have estimates of the PDF's of the strength, the stress and the error. The objective is to find the optimum load level in a test to estimate the probability of failure with maximum confidence.

The analytical model of the performance function is,

$$g(S_U, L, E) = S_U - \frac{L}{A} + E \tag{6.75}$$

The posterior PDF of the error is,

$$f_E\left(\varepsilon/I = i\right) = \frac{P\left(I = i/\varepsilon\right) \cdot f_E\left(\varepsilon\right)}{k} \tag{6.76}$$

where the likelihood of a failure or a success in a test are,

$$P\left(I = 1/\varepsilon\right) = P\left(S_U \le \frac{L}{A} - \varepsilon\right) = F_{S_U}\left(\frac{L}{A} - \varepsilon\right) \quad \text{and} \tag{6.77}$$

$$P\left(I = 0/\varepsilon\right) = P\left(S_U > \frac{L}{A} - \varepsilon\right) = 1 - F_{S_U}\left(\frac{L}{A} - \varepsilon\right) \tag{6.78}$$

The normalizing constant k is,

$$k = \begin{cases} \displaystyle\int_{-\infty}^{\infty} F_{S_U}\left(\frac{L}{A} - \varepsilon\right) \cdot f_E(\varepsilon)\, d\varepsilon & \text{if } I = 1 \\[2ex] \displaystyle\int_{-\infty}^{\infty} \left\{1 - F_{S_U}\left(\frac{L}{A} - \varepsilon\right)\right\} \cdot f_E(\varepsilon)\, d\varepsilon & \text{if } I = 0 \end{cases} \tag{6.79}$$

The probability of failure as a function of error is,

$$PF(\varepsilon) = \int_{-\infty}^{\infty} F_{S_U}\left(\frac{L}{A} - \varepsilon\right) \cdot f_L(l)\, dl \tag{6.80}$$

The variance of the probability of failure is given by

$$Var_{PF} = \sum_{i=0,1} \left[\int_{-\infty}^{\infty} \{PF(\varepsilon) - E(PF/I = i)\}^2 \cdot f_E(\varepsilon/I = i)\, d\varepsilon\right] \cdot P(I = i) \tag{6.81}$$

The optimum test load in a test is found solving optimization problem (6.74). We assume numerical values for the above problem to assess the effectiveness of the targeted testing approach. The PDFs of the strength, stress and the prior PDF of the error are,

$$f_{S_U}(s_u) \sim N[20, 2.5], \quad f_S(s) \sim N[15, 2.5], \quad f_E(\varepsilon) \sim N[0, 0.5]$$

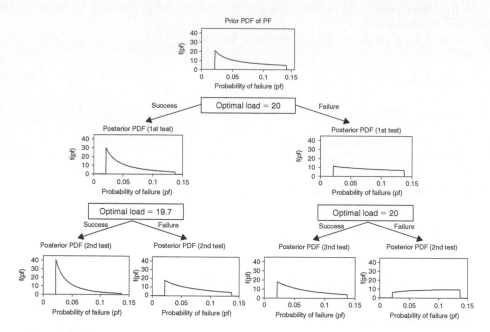

Figure 6.38 Bayesian updating of the PDFs of the probability of failure for two consecutive tests.

All units are kN. If the designer tests the structure under 25 kN load the structure will probably fail, but this does not mean that the structure is unsafe in real life. If the test load is 10 kN the structure will probably survive the test, but this does not mean that it is safe. Figure 6.38 shows the prior and posterior PDF's of the probability of failure for two consecutive tests, respectively. For the first test, the optimal load is found to be 20 kN. If the component survives the first test, this means that the model of the performance function (Equation 3.75) underestimated the reserve strength. As a result, the updated PDF shows that the probability of failure decreases compared to the prior probability of failure. On the other hand, if the component fails the first test, this means that the performance function overestimated the reserve strength. Therefore, Bayes' rule increases the probability of failure.

If the component survives the first test, the optimal stress value for the second test is 19.7 kN. But if the component fails the first test at a load of 20 kN, the optimum test load for the second test will be also 20 kN. If the component survives both tests, the Bayesian approach yields a posterior PDF indicating that it is very likely that the reserve strength was underestimated and that the probability of failure was overestimated before conducting the tests.

6.4.3   Accounting for imprecision by using probability bounds

Often, a decision maker is unsure about the probabilities of some events. For example, the decision maker is uncertain about the prior probability distribution or the likelihood function. One approach is to use probability bounds in order to account for imprecision (Aughenbaugh and Paredis, 2006 and 2007, and Aughenbaugh and Hermann, 2008). The theories of Bayesian robustness (Berger 1985, section 4.7) and imprecise probability (Walley, 1991) revolve around this concept. This approach could yield counterintuitive results when it is applied to Bayesian analysis. Specifically, when a decision maker observes one event he/she becomes more uncertain about another event.

Consider the following example. A decision maker knows that the probability of heads in one flip of a coin is 0.5. However, he/she does know anything about the joint probability of getting heads in two subsequent flips. Suppose that the first flip is a head. Find the probability of heads in the second flip given this piece of evidence.

We denote H_1 and H_2 the events "first flip is heads" and "second flip is heads" and $H_1 \cap H_2$ the intersection of these events. Since the decision maker does not know $P(H_1 \cap H_2)$ he/she will use the bounds of this probability to calculate the probability of heads in the second flip. This joint probability could vary from 0 to $\min\{P(H_1), P(H_2)\}$, which is 0.5. The conditional probability of heads in the second flip is,

$$P(H_2/H_1) = \frac{P(H_1 \cap H_2)}{P(H_1)} \qquad (6.82)$$

If the joint probability in the numerator varies from 0 to 0.5, then the conditional probability of heads in the second flip varies from 0 to 1. The same result is true for the conditional probability of tails in the second flip, $P(T_2/H_1)$. Therefore, when the decision maker finds out about the outcome of the first flip, he/she becomes clueless about the probability of getting heads in the second flip. Thus observing one event increases uncertainty about another. This paradox is called *dilation*.

6.4.3.1 Conclusion

This section presented a structured method to construct a probabilistic model of a parameter, such as the probability of failure of a system or the rate of occurrence of an event, by combining expert judgment and observed data. This approach is based on Bayes' rule. In a nutshell, this rule says that the posterior distribution of a parameter is proportional to the prior distribution times the likelihood of the observed evidence.

In order to construct probabilistic models that reflect a decision maker's beliefs and the data, the user must understand well the theory of Bayesian analysis and the problem in hand. The decision maker has to estimate a probability distribution of the underlying (or basic) random variable, such as the number of warranty claims of a product, and the number of occurrences of an earthquake. The decision maker has also to assume the type of the probability distribution of one or more parameters of this distribution, such as the rate of occurrence of an earthquake. Fortunately, there is considerable experience and guidelines for selecting a suitable probability distribution for a random quantity (Chapters 3 and 4).

This section demonstrated the use of Bayesian analysis in two cases where the evidence consists of observed data and expert opinion. In the latter case, the decision maker has to quantify the credibility of the expert(s) by estimating the probability distribution of the error in their estimates. For example, if an expert estimates a probability of failure then the decision maker has to estimate the error in this probability.

The section also explained how to use Bayesian analysis in order to validate the reliability of a design by performing physical tests. It also demonstrated how to design these tests in order to get the maximum amount of information.

People are often imprecise about the prior probability distribution and likelihood function in a decision. A decision maker must investigate the effect of imprecision on the posterior probability distribution. A natural way to do so is by using probability bounds to represent imprecision. However, the decision maker should be aware that this approach could yield counterintuitive results by amplifying the effect of imprecision on the posterior probability distribution.

6.5   Heuristics and Biases in Probability Judgments

Experts do not always act rationally, and their judgments often violate the rules of probability. One reason is that they use heuristics (convenient empirical strategies) instead of more rigorous processes to estimate probabilities. When people are under time pressure they often rely on heuristics to make judgments because they prefer effortless strategies to thorough analysis. This increases bias in their judgments.

The methodology for modeling uncertainty and making decisions presented in this book will produce poor results if the probabilistic models employed do not represent accurately the decision maker's beliefs and the available data. Often, poor estimates of the likelihood of events have detrimental effects on a company and its customers. For example, if an offshore platform designer underestimates the error in an analytical model for the stability of the platform, then he/she could design an unsafe structure. People must be aware of these biases in order to minimize their effect on decision making. A facilitator should be aware of the most common heuristics and recognize the potential of introducing bias in order to plan the elicitation process in a way that it produces unbiased judgments.

Tversky and Kahneman (1974) and Kahneman, Slovic and Tversky, 1982, studied and documented biases resulting from the use of heuristics. A summary of common heuristics and biases is presented here.

a)   Overconfidence. People tend to overestimate the frequency of them being right. For example, Alpert and Raiffa (1982) elicited 98% credible intervals of the number of foreign cars imported in US in 1968. They found that 43% percent of the assessments fell outside these intervals, instead of 2%. This means that the subjects thought that their intervals contained the right number of cars 98% of the time, but actually they did so only 57% of the time!

An overconfident decision maker tends to underestimate uncertainty. As a result, he/she may take unhealthy risks. Feedback and overfitting reduce overconfidence. When eliciting probabilities, a facilitator should give a series of calibration tests to the experts and provide them with feedback after grading each test. The facilitator should also elicit more estimates than needed to build a probabilistic model, check these estimates and bring up inconsistencies to the attention of the experts.

b)   Representativeness bias is caused by judging the probability that an object belongs to a group by comparing one or more characteristics of that object to a common characteristic of the group. People overlook the size of the group or the population in this assessment. For example, jurors may judge that a defendant committed a crime because the defendant's age, race or ethnicity match those of a small group that includes some members that have violated the law.

Experts that are influenced by representativeness bias tend to confuse a conditional probability with its inverse. For example, a physician may conclude that a patient who has a particular symptom (e.g., high fever) is very likely to have a disease that almost always causes this symptom, such as streptococcal infection. The physician confuses the conditional probability, P(streptococcal infection/high fever) with P(high fever/streptococcal infection) This is an erroneous conclusion because many diseases cause high fever, besides streptococcal infection.

c)   Anchoring bias; this is a tendency to rely too much on the initial value of some quantity and fail to adjust it when new evidence becomes available. Anchoring bias leads to decisions that can become progressively worse as the decision maker does not update his/her belief consistently. The best way a facilitator can avoid anchoring bias in the decision maker is by (1) Not giving an initial estimate of the probability and (2) giving counterexamples and/or extreme analogies. For example, consider the director of business development in a company who wants to assess the probability of success of a new product. If of all new products that the company has launched, the fraction p has been successful, the director might use this to evaluate the new product also. The facilitator can help by contrasting the new product with the ones that the company has launched in the past. The facilitator can show extreme examples like that of products that have succeeded despite poor performance or products that have failed despite the hype surrounding them. Counterexamples are a good way to convince the decision maker that his/her anchor is not necessarily correct.

d)   Availability bias; people tend to estimate the probability of an event according to how easily they remember examples of the occurrence of the event. This looks reasonable because we intuitively think that events are more likely when they are easy to remember. However, this bias results in overestimation of the probability of many events such as those that occurred recently or attracted the media attention.

In order to demonstrate this bias, consider the following questions:
Which events kill or killed more people?

1. Homicides or suicides?
2. The Atlantic Slave Trade or the First World War?
3. Shark attacks or deer caused accidents?

The correct answers are 1) suicides, 2) the Atlantic Slave Trade and 3) deer caused accidents. If you answered one or more questions incorrectly you may suffer from availability bias.

Many bad decisions are caused by availability bias. Immediately after the 9/11 terrorist attacks, many people chose to drive instead of flying because they overestimated the likelihood of dying in an airplane accident or highjack. As a result, the number of deaths from car accidents increased by about 1,000 in 2001. Availability bias caused these unnecessary deaths.

e) Motivational bias; an expert underestimates or overestimates some quantity because the consequences of one type of error are more severe than other types. In the early 1980's, NASA administrators estimated that the probability of a catastrophic failure of the space shuttle was 10^{-5} (Cooke, 1991). Later this probability was revised to about 1% in light of the two catastrophic failures in 1986 and in 2003. The NASA administrators could have suffered from motivational bias as they tried to promote the Space Shuttle program to the government and the public.

At the other extreme, decision makers often inflate the probabilities of events that could have undesirable consequences, in order to select that course of action that minimizes the impact of these events. For example, a medical doctor may tend to inflate the probability that a patient could have a serious illness and prescribe an expensive test. Weather forecasters may overestimate the probability of a storm or its severity in order to be on the safe side.

Estimation of probabilities and assessment of the consequences of uncertain events are different tasks that should be done separately. When a decision maker estimates the probability of an event, he/she should focus only on its likelihood. The decision maker should express his/her aversion to the consequences of some events (such as misdiagnosis of a serious illness) by assigning a very low utility, or high loss function, to this consequences (section 7.5).

f) Affect heuristic; often, humans make judgments driven by emotions and feelings through which they view objects or events as positive or negative. Sometimes, affective evaluations substitute rational, arduous evaluations and they result in faulty judgments. For example a driver may underestimate the risk of a fatal accident due to driving under influence of alcohol or using a mobile phone.

g) Sensitivity of the elicited probabilities to the way events are described. A psychological theory of probability assessment, support theory, stipulates that different descriptions of the same event elicit different probabilistic judgments. For example, presenting an event as the union of disjoint events makes people think it is more likely than they would if they considered the original event.

6.6 Concluding Remarks

Section 6.2 explained the concept of subjective probability. This concept is important in decision making under uncertainty because it enables a decision maker to quantify uncertainty about unique events. The subjective probability of an event is in the decision maker's mind. It is elicited by observing how the decision maker is willing to bet on this event. Specifically, subjective probability is the long-term frequency p that makes the decision maker indifferent between a ticket that pays some amount with this frequency and a second ticket that pays the same amount if and only if the event occurs. The definition of subjective probability is based on seven axioms that describe how a rational person makes decisions. It was proven that subjective probability calculations follow the same rules as those of objective probability. If someone disagrees with the theory of subjective probability, then he/she has to explain which axiom is wrong.

Subjective probability is consistent with, but more general than its objective counterpart. If one observes the results of the same experiment repeatedly then the subjective probability of the outcome converges to its long-term frequency with the number of observations.

We can only elicit probabilities of observable events. This means that after all uncertainties have been resolved one should be able to determine if the event has occurred.

The same section presented the principle of insufficient reason. Although some authors have tried to discredit it through counterexamples, this principle is necessary in decision making. We explained why some of these counterexamples do not necessarily mean that the principle of insufficient reason is not sound.

Section 6.3 described procedures to elicit probabilities of events, and probability distributions of variables. This procedure involves a facilitator and one or more subject matter experts. The section explained two procedures for testing the accuracy of elicited probabilities; calibration and falsification through observation of events that are thought not to occur. Estimating a subjective probability by directly asking an expert about it is impractical. An indirect procedure is to ask an expert for his/her maximum buying price of a lottery ticket that pays some amount if the event of interest occurs, but this procedure assumes a risk neutral expert. A second procedure that does not require this assumption is to ask the expert to compare the likelihood of the event of interest to the likelihood of an event whose long-term frequency is obvious, such as the probability that the arrow of a wheel of fortune will stop in a sector. The section also presented two methods for estimating probabilities of rare events that use decomposition and statistics of extremes.

We construct probability distributions in three steps. First, we select a parametric family of distributions. Then we elicit statistical summaries and finally determine the values of the parameters by fitting the distribution to the elicited data. Eliciting more summaries than needed to determine the parameters and providing experts with feedback improve the accuracy of the estimated probabilities.

There are two sources uncertainty in elicited probabilities or distributions:

a)   Imprecision in the values of the elicited summaries.
b)   Uncertainty about how well the fitted probability distribution represents the expert's beliefs. For example, many distributions fit the data about the projected demand for a new car model well, but only one could represent a marketing expert's beliefs.

Sensitivity analysis is one approach to quantify uncertainty in an elicited distribution. For this purpose, an expert determines a class of admissible probability distributions that are consistent with the decision maker's beliefs. The decision maker studies the effect of the choice of the probability distribution from this class on the optimum course of action and its consequences.

A decision maker can use a second-order model to represent uncertainty in a distribution. This model is a probability distribution whose parameters are random variables with their own distributions.

Section 6.4 explained Bayesian analysis, which is a formal method to estimate probability distributions by combining information from experts and data. This is a structured method that combines information from multiple sources correctly by using laws of probability. In this method, the decision maker constructs a second order probabilistic model for the random variable of interest. She has to choose a parametric family of probability distributions of this variable. The decision maker also assumes the type of the probability distribution of one or more parameters of this distribution, such as the rate of occurrence of an earthquake.

Section 6.5 explained that people are generally poor estimators of probabilities. One reason is that they use heuristics (convenient empirical strategies) instead of more rigorous processes to estimate probabilities. Another reason is that people are biased by their own interests. It is important to be aware of sources of bias in order to minimize their effect.

Questions and Exercises

Types of uncertainty

6.1   Identify if the following are examples of epistemic or random uncertainty or both

   a) What is the arrival time of flight 2212 at the Las Vegas airport?
   b) How many people will take a cab to go to a casino from the airport?
   c) Will a given person taking the cab have a flat tire?
   d) Will a given person playing the roulette win the first round?
   e) Is there going to be a solar eclipse in the next 20 days in Ann Arbor, Michigan?
   f) What will be the time in a particular clock when its battery dies?
   g) How many times will wine be spilled in a big party?
   h) How many lemon trees were in the garden of the first author's family house?
   i) Each lemon on one of the trees in the author's house had different size. What type of uncertainty does this variation cause?

Axiomatic definition of subjective probability

6.2   A teacher wants to assess a student's belief that he would pass an exam. He offers to sell him a ticket which pays $10 if he passes the test and zero otherwise. The student says that he will pay up to $5 for the ticket. What probability does the student think he has of passing the test?

6.3   After the student responds to the professor's question, the professor offers to sell him a ticket that pays $10 if he does not pass the exam. The student says that he is willing to pay up to $8 for the ticket. The professor says that the student's responses are irrational, when considered together. Justify this statement.

6.4   The student in the above problem is ready to pay $25 if the ticket paid $100 if he passed the test and zero if the student failed. Is his belief about the probability of his passing the test same as before? Why is there a difference, if at all? What would you suggest the teacher do to correctly assess the student's belief about the probability?

6.5   Two balls have radii between 0.8 inch and 1.2 inch and between 0.9 inch and 1.4 inch, respectively.

   a) What are the PDFs of their volumes? Do we have enough information to assess the PDF's? What approach should one use?
   b) If the balls were to be melted and re-formed into a bigger ball what would be the PDF of the radius of the bigger ball?

6.6   Now you know that the most likely values (mode) of the radii for the two balls in problem 6.5 are 0.9 and 1.1. How does your answer change in light of this information? Find the probability that the ball with smaller mean radius would pass through a round hole of 1 inch radius and the bigger ball will not.

6.7   You ask a friend to appraise two tickets that will be worth $100 if the Dow Jones Industrial Index closes above 14,000 points, and below or at 14,000, one year from now, respectively. Your friend cannot decide which ticket is worth more, and he believes that each ticket is worth less than $10 because the value of the index is highly unpredictable. Do you think that your friend's responses are rational?

6.8   A facilitator asks a urologist the following question in order to elicit the probability that a randomly chosen 50-year old male who takes a PSA test could test positive.

Consider a lottery ticket that pays $1,000 with probability p and zero with probability $1 - p$, and a second that pays the same amounts if a 50-year old male tests positive and negative, respectively. For what probability p are you indifferent between the two tickets? The urologist says that he is indifferent for a probability of 0.2.

| This ticket is worth $1,000 with probability p and zero with $1 - p$. | This ticket is worth $1,000 if a 50-year old tests PSA positive and zero otherwise. |

a) What is the urologist's probability that the 50-year male would test positive?
b) What probability makes the urologist indifferent between the first ticket and a second that pays $1,000 if a 50-old male tests negative, based on the urologist's response to the first question?
c) Now the urologist compares the two tickets in question a) for a 40-year old male. He says that he is indifferent between the two tickets for a probability $p = 0.4$. He also says that the younger a patient the less likely he is to exhibit prostatic abnormalities. Are the urologist's answers to questions a and c consistent?

6.9   The facilitator in the previous problem asks the urologist the following question:
Suppose that the 50-year old patient takes the above PSA test. No action is taken in the result is negative. If the test is positive he will have a biopsy. Consider the following ticket:

- If the patient tests PSA negative the owner gets a refund of the ticket price.
- If the patients tests positive in both the PSA and biopsy tests the ticket is worth $1,000.
- If the patient tests PSA positive and gets a negative biopsy test, this ticket is worthless.

For what probability p are you indifferent between the above ticket and a second that pays $1000 with probability p and zero otherwise?

The urologist estimates that this probability is 0.1. What event's probability can the facilitator infer from this elicitation?

6.10   Estimate the probabilities of the following events based on the results in the above two elicitation problems:

a) A 50-year old male could test positive in both the PSA <u>and</u> the biopsy test.

b) A 50-year old male could test positive in the PSA <u>and</u> negative in the biopsy test.

6.11   What ticket should the urologist consider in order to estimate probabilities in the previous problem?

6.12   Two urns contain six black and two white balls (urn 1) and two white and two black balls (urn 2). You select one urn blindly, and then pick up a ball from it. In order to estimate the probability that the ball is from the first urn, you compare the tickets below:

> This ticket is worth
> $1,000 with probability p
> and zero otherwise.

> This ticket is worth
> $1,000 if you select a ball
> from the first urn and zero
> otherwise.

A probability $p = 0.667$ makes you indifferent between the two tickets.

Find the probability that the ball selected is black. Assume risk-neutral behavior.

Eliciting expert's judgments in order to construct models of uncertainty

Elicitation process

6.13   Take the two calibration tests in section 6.3.2. Do the results suggest that you are overconfident?

6.14   What questions should you ask yourself when answering each question in the above tests in order to avoid overconfidence?

Eliciting probabilities

6.15   You want to elicit your friend's subjective probability that the Detroit Pistons will win the next basketball game against Boston Celtics. You ask her for the maximum price she is willing to pay for a ticket that pays $100 if the Detroit Pistons win. Your friend is willing to pay up to $60 for this ticket. You also ask your friend how much she is willing to bet for and against Detroit Pistons (see Figure 6.12). Your friend is willing to take the risk of losing $4 if Pistons win, if she would win $6 if they lose. Assume that she is risk neutral.

a) Estimate your friend's probability that Pistons will win on the basis of the two bets. Are the two bets consistent?

b) If your friend is willing to take the risk of losing $10 if the Pistons lose, how much should she expect to win if the Pistons win?

6.16   A doctor in an emergency room estimates the cumulative probability of the life a critically injured patient considering bets that the patient could survive in the next few days with bets of a probability wheel (see Figure 6.13). The doctor's judgments are shown below,

Event	Angle that sector A' subtends
Patient will survive for more than a week	180°
Patient will die during the second week	144°
Patient will survive for more than two weeks	36°

a) Are the probabilities corresponding to the above judgments consistent?

b) Calculate the probability that the patient could die before the second week.

Eliciting probabilities of rare events

6.17   A healthy young adult wants to estimate the probability of being hospitalized during the next year. Explain how to estimate this probability by decomposing the event "hospitalized" into "hospitalized due to a car accident" and "hospitalized due to other causes." Construct a failure tree and explain how the young adult could estimate the probability of each event in order to estimate the probability of hospitalization.

Hint: The young adult could estimate the probability of hospitalization by estimating the probabilities of hospitalization conditioned upon the events "car accident" and "no car accident."

Eliciting probability distributions

6.18   In problem 6.16, the prognosis if the patient survives the first two weeks is excellent; the doctor estimates that if the patient survives for more than two weeks, then the conditional probability of surviving for more than six months is 0.99 and for more than a year is 0.98. Estimate and plot the cumulative probability distribution function of the patient's remaining life.

6.19   Joint probability distributions are n-increasing functions. For a two-variable case $(n=2)$ this means that $F_{X,Y}(x_2, y_2) - F_{X,Y}(x_2, y_1) - F_{X,Y}(x_1, y_2) + F_{X,Y}(x_1, y_1) \geq 0$ for all $\{x_2, y_2\} \geq \{x_1, y_1\}$. What would happen if this condition were violated?

6.20   A test engineer determines that there is a crack in a weld. The engineer asks an expert to estimate the length. The expert says that she believes that the length of the crack could be no greater than 0.11 mm with probability 0.15 and no greater than 0.175 mm with probability 0.9.

The engineer believes that a lognormal PDF represents well the crack length,

$$f_X(x) = \frac{1}{x \cdot \sigma \sqrt{2\pi}} e^{-\frac{1}{2} \left[\frac{\ln(x) - \mu}{\sigma} \right]^2} \quad \text{for } x > 0$$

where μ and σ are the mean value and standard deviation of the logarithm of the length. However, the engineer also wants to represent the length with an alternative distribution that has long tails. He decides to use the Cauchy PDF,

$$f_X(x) = \frac{1}{\pi \cdot s \left\{ 1 + \left(\frac{x-l}{s} \right)^2 \right\}}$$

where l and s are the location and scale parameters.

a) Estimate the parameters of the two candidate distributions based on the expert's estimates of the 0.15 and 0.9 quantiles of the length.

b) Represent the crank length by using the epsilon-contamination model in Equation (6.39), where $f_{\Theta_0}(\theta)$ and $q_\Theta(\theta)$ are the lognormal and Cauchy PDFs. Plot PDF $f_{\Theta_0}(\theta)$ and $q_\Theta(\theta)$ for few values of parameter ε from 0 to 0.1.

c) The engineer wants to know the probability that the crank length could exceed 0.25 mm. Calculate this probability when parameter ε varies from 0 to 0.1.

6.21 Prove the following propositions are equivalent,

$$P(X > X_{0.5}/Y > Y_{0.75}) > P(X > X_{0.5}/Y > Y_{0.5}), \text{ and}$$
$$P(X > X_{0.5} \cap Y > Y_{0.75}) > P(X > X_{0.5} \cap Y_{0.75} \geq Y > Y_{0.5}).$$

6.22 In example of the estimation of the joint CDF of the demand of the Hybrid/Diesel car in US and Canada, estimate the probability that both the demands in US and Canada will be less than or equal to the 0.25 quantiles by using the copula model developed. Compare this probability to the probability of the same event if demand in the two countries were statistically independent.

6.23 The table below shows five fractiles of the annual wave height and wind speed at a location in Lake Erie. An expert estimates that the joint probability of both the wave height and the wind speed being no greater than their 0.5-fractiles is 0.4. In addition, the expert estimates that the joint probability of the wave height being no greater than its 0.5-fractile and the wind speed being no greater than its 0.75-fractile is 0.65.

Fractile	Annual wave height (m)	Annual wind speed (m/sec)
0.05	4	15
0.25	4.2	17
0.5	4.8	19
0.75	5.2	20
0.95	5.4	20.5

a) Check if the expert's estimates are consistent with the laws of probability.

b) Assume that the dependence of the wave height and wind speed is represented by Frank's copula. Estimate the value of parameter θ and Kendall's correlation coefficient using the approach in example in section 6.3.4. Are the two quantities positively dependent?

6.24 The probability, p, of a tip-up flip of a thumbtack follows a beta distribution (section 3.2.1.2) with parameters $\beta_1 = 3$ and $\beta_2 = 7$. We flip the thumbtack 10 times.

    a) Find the probability mass function of the number of heads in three cases where p is equal to 0.2, 0.4 and 0.6.

    b) Calculate the conditional probability that the number of tip-up flips is less than or equal to 5 in each of the three cases in question a.

    c) Write an equation for the unconditional probability that the number of tip-up flips is less than or equal to 5. Calculate this probability by using numerical integration.

Second order models

6.25 An accident in a power plant can be caused by human error or electrical failure. The probability of an accident conditioned upon human error is 0.005 and the probability of an accident conditioned on an electrical failure is 0.01. The probability of an electrical failure in a year is 10^{-6}. An expert estimates that the probability of a human error could be as low as $0.5 \cdot 10^{-4}$ and as high as 10^{-4}.

    a) Write an equation for the probability of failure of the power plant as a function of the probability of human error.

    b) Calculate the probability of failure of the power plant for the bounds of the probability of human error that the expert provided using the equation in a.

6.26 In the above problem, the expert believes that the probability of human error follows a beta distribution and her estimates are the 0.05- and 0.95-quantiles of this probability.

    a) Calculate and plot the PDF of the probability of human error.

    b) Calculate and plot the PDF of the probability of failure of the plant.

    Hint: the probability of failure of the plant is a function of the probability of human error. Use the method for calculation of the PDF of a function of a random variable in section 3.3.1.

Bayesian Analysis

6.27 Assess the probability distribution of the probability that a randomly selected car is blue in color using a two parameter beta distribution (Section 3.2.1.2). Then go and look at a few cars and update this prior using the Bayesian method.

6.28 A doctor says that the probability of surviving a certain kind of cancer is normally distributed with mean 0.2 and standard deviation 0.04. Another doctor puts these values at 0.4 and 0.05. Assume you started with a uniform distribution for this probability with the interval of 0 and 0.5. What is the posterior CDF of the probability of surviving this cancer?

6.29 The manager of an electronic parts supplier believes that the probability of failure of a hard disk before 500 hours of operation can be as low as 0.005 and as high as 0.2. The supplier tested 100 parts for 500 hours and observed no failures. Calculate the updated probability of failure assuming a four parameter beta prior with location parameters equal to the lower and upper bounds of the probability of failure (Section 3.2.1.2). Assume that the other two parameters are $\beta_1 = \beta_2 = 2$.

6.30 Solve the above problem assuming a uniform prior probability distribution from 0.005 and 0.2.

6.31 A scientist wants to estimate the diameter of a carbon tube nanofiber. The scientist's best guess is that the diameter is 20 μm but she feels that it could be as low as 10 and as high as 40 μm. The scientist measures the diameter and finds that it is equal to 25 μm. However, she recognizes that the measurement method is imprecise and that the measured value could be in error by up to 10 μm. Assume that the prior PDF and the measurement error are both triangular. Find the posterior distribution of the diameter.

6.32 The PDF of a crack being present in weld with length l is lognormal with parameters $\mu = -2$ mm, $\sigma = 0.2$ mm,

$$f_L(l) = \frac{1}{l \cdot \sigma\sqrt{2\pi}} e^{-\frac{1}{2}\left[\frac{\ln(l)-\mu}{\sigma}\right]^2} \quad \text{for } l > 0$$

where the crack length is measured in mm.

The probability of detection of a crack by using a nondestructive test is given by the following equation,

$$P_{Detection}(l) = 1 - e^{-\left(\frac{l}{\bar{l}}\right)^2}$$

where $\bar{l} = 0.1$ mm.

A test engineer inspected the weld using the above nondestructive test and detected not cracks.

a) Derive the posterior PDF of a crack length given that the test did not detect a crack.

b) Calculate the probability that a crack with length greater than 0.2 mm is present in the weld before and after the test.

6.33 A structural engineer wants to estimate the probability of failure of a structure using Monte Carlo simulation. Before performing simulation, she estimates that this probability is most likely to be equal to 0.05 but it could be as low as zero and as high as 0.2. Because simulation is expensive, she wants to estimate the number of simulations that she needs in order to estimate the failure probability in a way that the half width of the 95% credible interval of the estimated probability is 0.01, before performing simulation.

For this purpose, the analyst assumes a uniform prior from 0 to 0.20 for the failure probability and updates it based on the simulation results using Bayes' rule. In order to compute the likelihood of the failure probability she assumes that if she performs n simulations, she will get $0.05n$ failures.

Write an equation for the posterior PDF of the failure probability. Calculate the bounds of the 95% central credible interval for 100, 400 and 1,600 simulations.

6.34 Calculate the 95% Wald's confidence intervals (Chapter 4, section 2) of the probability of failure in the above problem. Compare these intervals with those from Bayes' rule. How does the difference between the results of the two methods vary with the number of simulations? Explain the observations from this comparison.

6.35 An expert estimates that the probability of a failure is $p = 0.05$. The expert estimates that the equivalent prior sample size that reflects his confidence in the estimate is $\hat{n} = 200$ (chapter 3, section 3.2.1.2). Repeat the Bayesian approach for calculation of credible intervals in problem 6.33 using a beta prior with parameters $\beta_1 = p\hat{n}$, $\beta_2 = (1 - p)\hat{n}$.

6.36 Solve the above problem for $\hat{n} = 100$ and $\hat{n} = 1000$.

Heuristics and biases

6.37 Consider the following conversation between Steve and John.

Steve: "I would rather change oil in my car myself; I don't trust mechanics after what Jenna's regular mechanic did"

John: "What did he do?"

Steve: "Well, she took the car to the shop because she heard a rubbing noise coming from her tires. The mechanic was adamant that it was the bearings. Even after the inspection hinted that it was probably the brake rotors, the mechanic insisted that she change the bearings. She still has the same problems after spending $500."

John: "That's too bad. What is she going to do now?"

Steve: "She said she is going to fix it herself."

What cognitive biases *might* Steve, the mechanic, John and Jenna suffer from?

6.38 When most people take the probability calibration tests in section 6.3.2 they underestimate the width of confidence intervals. What bias or biases they suffer from?

6.39 A physician just started working in a US hospital after being a staff physician in Southern Asia for 5 years. The physician examines a patient who has high fever, chills and sweats for a two days, and concludes that the patient has malaria. Another physician examines the patient and disagrees with the original diagnosis. After getting the results of some tests, both physicians agree that the patient has flu. What cognitive bias did the first physician suffered from?

6.40 How would you coach people who take calibration tests in order to help them reduce their bias?

Chapter 7

Decision Analysis

7.1 Introduction

We all make choices that greatly affect our personal and professional lives. Similarly, the welfare of corporations and nations hinges on the ability of business and political leaders to make good decisions. These decisions are complex, involve significant uncertainty, and they depend on the decision maker's preferences, which are difficult to quantify.

Chapter 1 explained that design of a product is a sequence of decisions under uncertainty. Designers want to choose the values of engineering and business attributes of a product to maximize an objective function, such as profit. For example, a car manufacturer wants to choose the wheelbase, the overall length, the engine type and power, and the car price in order to maximize the company's profit. There is significant uncertainty in the cost, performance and demand of the final design in the early stages of its development. The profit from the product depends on the designers' decisions and on the resolution of uncertain events such as the market conditions, the price of gas and the state of the economy during the life of the car model.

People generally take poor decisions because they rely too much on intuition or empirical rules. This chapter presents decision analysis, which is a structured approach for making better decisions using common sense rules (axioms) and mathematics. This approach enables the decision maker to quantify his/her preferences and beliefs about the likelihood of the outcomes of uncertain events, and to select the best course of action given these preferences and beliefs. Decision analysis is not a descriptive but a normative approach; it prescribes common sense rules for making choices – it does not describe nor explain how people make choices in real life.

Decision analysis does not impose someone else's preferences on the decision maker. Although decision analysis does not guarantee a good consequence for every decision, it has better long term average performance than ad-hoc approaches or approaches based on empirical rules. A main reason is that it encourages a decision maker to think hard about the decision and make decisions that are consistent with his/her preferences and beliefs.

7.1.1 Examples of Decision Problems

Example 7.1:   Drill for oil

You own a site in an oil rich region. You believe that there is a 50% chance that there is oil in the site. Drilling is expensive, and you do not know in advance how much oil is in the site. You can go ahead and drill (option 1) or you can hire a geologist to perform a seismic test (option 2). The results of the test will give you a better idea about how much oil is in the site, but the test is costly. A third option is to sell the land.

The cost of drilling is $70,000 and the cost of the test is $10,000. If you find oil, the revenue depends on the amount; you will make $120,000 if the well is "wet" and $270,000 if it is "soaking." A real estate agent estimated that you can sell the land for $20,000 of which you will pay $1,500 to the agent. The agent said that the land could still be sold for the same price after performing a seismic test. However, nobody would buy the land after drilling and not finding oil.

Should you sell, drill or consult a geologist? If you opt to consult a geologist what should you do after getting the geologist's assessment?

Example 7.2:   City Evacuation

A hurricane is approaching New Orleans. The mayor has three options: order evacuation, wait for a weather forecast or do nothing. If the hurricane hits the city and the residents stay, there will be a significant loss of life and property. However, a massive evacuation will also be costly; it will cause loss of life from traffic accidents, distress of the elderly and loss of business. Finally, the political cost of an unnecessary evacuation could be significant.

In order to make a decision, the mayor needs to quantify the probability of the hurricane hitting the city and the accuracy of the forecast. The weather forecast will be issued shortly, so there is minimal risk that the hurricane will hit the city while the mayor is waiting for the forecast. The probability of the hurricane hitting the city is believed to be 70% based on historical data. In the past, the forecaster has correctly predicted that the hurricane will land in the city 9 out 10 times when a hurricane actually hit the city. In addition, the forecaster issued a false alarm 3 out of 10 times (i.e. predicted that the hurricane would hit the city in cases where the hurricane missed it.).

Should the mayor order immediate evaluation or wait for the weather report? If the mayor decides to wait, what should he do when he/she gets the weather report?

Example 7.3:   Grape harvesting decision

A wine producer is growing grapes. Just two days before harvesting, a weather forecast indicates the possibility of a storm. The producer can harvest early and produce wine of ordinary quality that yields a profit of $5 per bottle. If he decides to wait, there are three possibilities. 1) The storm may strike the farm and completely destroy the grapes. 2) The storm could degenerate to a warm, light rain, which is benign to the grapes. The texture of the grapes will improve and as a result, the farmer will produce premium wine that is highly sought by connoisseurs and yields $100 profit per bottle. Besides the high revenue, selling high quality wine could benefit the producer's reputation, thereby allowing him to increase his prices in the future. 3) If the storm misses the farm, then the farmer will produce good quality wine yielding $15 profit per bottle.

Should the farmer harvest immediately or wait?

Example 7.4:   Seeking funding for a risky venture

This decision problem involves two decision makers, an inventor and a financier. Here, we present the problem as viewed by the inventor. The inventor has an idea about a new mechanism for an automated factory and seeks funding from a financier to develop

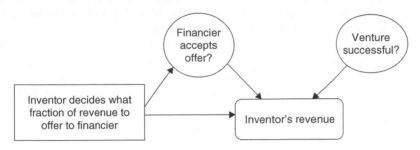

Figure 7.1 Influence diagram of inventor's decision.

this mechanism. For this purpose, the inventor offers the financier a fraction of the project revenue. The inventor does not know if the financier will accept the offer. Of course there is uncertainty about the success or failure of the project for developing the mechanism, which affects the revenue from the venture. The inventor estimates a probability of success of the project, and so does the financier. These probabilities are different and the decision makers do not know each other's assessed probability of success of the venture.

Both decision makers want to maximize their profit from the risky venture. The inventor wants to decide what fraction of the revenue from the project to offer to the financier, while the financier wants to decide whether to accept the offer. Figure 7.1 summarizes the financier's decision. In this diagram, a box represents a decision, a circle an uncertain event and a rounded box represents the consequence of the decision. Arrows indicate dependence, not sequence. According to this figure, the inventor's revenue depends on his decision, the financier's decision and the success of the venture. The financier's decision in turn is affected by the fraction of the revenue that the inventor offers to the financier. Finally, the inventor is uncertain about the financier's response to the inventor's offer and about the success of the venture.

Should the inventor make an offer to the financier? If yes, then what fraction of the revenue should he offer?

7.1.2 *Elements of Decision Problems and Terminology*

All decisions include six elements.

1. Values and objectives

Decision makers have a single or multiple objectives. In the search for oil and the inventor-financier examples, the decision makers have one objective: maximize profit. There are multiple objectives in the city evacuation and the grape harvesting decisions. In the former decision, the mayor wants to minimize the deaths and injuries and the political cost from a decision that can have bad consequences. In the latter decision, the wine producer wants to maximize profit but may also want to improve his company's image.

Objectives can be classified into *fundamental* and *means* objectives. The former type explains what the decision maker wants to accomplish, while the latter explains how.

As an example, the farmer's main objective in the grape harvesting problem could be increasing his long term profit while he might think that improving his company's image will help him achieve that (thus is a means objective.)

In ethics, the term *value* refers to any object or quality that is desirable to a decision maker as a means or as an end. A person's values determine his/her objectives in a particular decision. Values are subjective; when two decision makers face the same decision they might have different objectives. In the grape harvesting decision, one producer may only care about his profit, while another may care for his/her company's reputation besides profit.

2. Alternative courses of action (options)

The decision maker can choose among a set of alternatives. In the oil search decision, alternatives include selling the land, drilling for oil and conducting a seismic test. The harder we think about a decision the richer we make our suite of options. More importantly, we conceive creative options that reduce risk and increase payoff. In the drill for oil decision, the owner may realize that he/she could reduce risk by seeking funding for drilling from a venture capitalist in return for a portion of the potential revenue from oil. Alternatively, the owner could simply wait for one year and see if the price of land or oil changes.

3. Uncertain events that affect the consequences of the actions

All of the above decisions are risky, that is, the consequences of each course of action are uncertain. The reason is that the consequences of each course of action depend on the outcomes of uncertain events and these outcomes are unpredictable. In the search for oil decision, the owner's payoff depends on the amount of oil in the site and the test result. The owner cannot predict which outcome will materialize before drilling and testing.

If the owner chooses to drill without performing a seismic test, he/she will make a profit that could be as high as $200,000 if there is oil. However, he/she could lose up to $70,000 if there is no oil. If the owner first performs a test and then decides whether to drill, his/her profit could be as high as $190,000 and his/her loss as high as $80,000. In order to make an informed choice, the decision maker should determine the risk profile of each option, which shows all possible payoffs and their probabilities.

A good decision can have bad consequences, and a poor decision can have good consequences. The reason is that uncertain events affect the consequences of the decision. Therefore, it is wrong to judge a decision only on the basis of its consequences.

4. Information about the consequences of each choice

A decision maker should make a choice by comparing the consequences of each alternative course of action. Since the consequences are unpredictable, the best that the decision maker can do is to estimate their relative likelihood. The decision maker needs to determine the following information in order to make a good decision in the presence of uncertainty.

- Identify all uncertain events that affect the consequences of a decision and all possible outcomes of these events.

Table 7.1 Payoff of each course of action in grape harvesting decision.

Action\Outcome	Storm hits farm (0.3)	Storm misses farm (0.4)	Strom degenerates to warm rain (0.3)
Harvest early	$5	$5	$5
Wait	−$3	$15	$100

- Determine the consequences of each alternative course of action for each possible outcome of the uncertain events.
- Estimate the probabilities of the outcomes of the uncertain events.

In this subsection, we will explain how to accomplish the first two tasks. The next subsection will explain how to estimate probabilities.

In the drill for oil example, the owner determines the possible outcomes of a seismic test and amount of oil in the site based on consultation with experts. A geologist tells the owner that the seismic test can show no structure, an open or a closed structure. These test outcomes depend strongly on the amount of oil in the site. In addition, the owner concludes that there are three possibilities for the amount of oil; the land could be dry (has no oil) or wet (has some oil) or soaking (has a lot of oil). An oil trader helps the owner determine the revenue from the site when the site is wet or soaking. Finally, a realtor estimates the time that the land will stay in the market and the selling price.

The farmer in the grape harvesting decision obtains information from the weather forecast. In addition, the farmer relies on his experience. Table 7.1 shows an example of information that helps the farmer decide whether to harvest immediately or wait. The table shows the possible consequences of each course of action, and the probabilities of these consequences in parentheses. According to this table, if the farmer harvests the grapes early to avoid the consequences of the storm then he/she will make a profit of $5 per bottle of wine regardless of the outcome of the storm. If the farmer waits, then he/she will make a profit of $15 if the storm misses the farm and $100 if the storm degenerates to warm rain. However, he/she will lose $3 if the storm hits the farm.

According to the above table, the farmer is very uncertain about the storm path because all three outcomes of the storm are almost equally likely. Imagine that the farmer obtains more accurate information. For example, if a very experienced and reliable meteorologist tells the farmer that the probability of the storm degenerating to warm rain is practically one, then the farmer will harvest after the storm and make a fortune from selling premium wine. In sections 7.2 and 7.3 we will learn how to quantify the value of information in a particular decision on the basis of the increase in the profit resulting from the information.

In order to develop predictive models in the city evacuation and the inventor-financier decisions, we need additional information. In the city evacuation example, we need to develop models for predicting the capacities of levies, the effect of the storm on the population, the traffic patterns and the number of accidents in case of an evacuation. Sophisticated mathematical models for predicting hurricane damage can be developed using past data and experience.

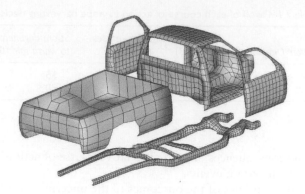

Figure 7.2 Predictive model for sound pressure level in a truck.

Some predictive models are very complex and expensive. For example, automotive manufacturers develop very detailed finite element models of a car in order to predict the effect of design decisions on the noise and vibration of the car and its performance in a crash. Figure 7.2 shows a car model for predicting the sound pressure level at the driver's ear for a given road excitation. Such a model could have hundreds of thousands or even millions of degrees of freedom.

5. Probabilities of outcomes of uncertain events

In order to make an informed decision, a decision maker needs to estimate the likelihood of the outcomes of the uncertain events that affect the consequences of the decision. For example, in the inventor-financier decision, the inventor must estimate the likelihood of success of the project and the financier's estimate of the same likelihood. If the inventor is confident that the project will succeed and has convinced the financier about that, then the inventor will offer a low fraction of the revenue. On the other hand, if the financier seems unconvinced, and the inventor badly needs funding, the inventor would offer a higher fraction to secure funding.

Chapters 2 and 6 explained that objective probability is not applicable to most decisions under uncertainty. The main reason is that most decisions involve one-time events, and objective probability is not applicable to such events. For example, in the drill for oil decision, the revenue from drilling the land depends on the amount of oil in the site. Although, usually there is data from other sites in the same region, each site has unique characteristics. In addition, the data could be obsolete because the amount of oil depends on how much oil has been already extracted from neighboring sites.

Therefore, besides observations, decision makers should rely on their beliefs to assess the likelihood of the outcomes of uncertain events. We elicit a person's subjective probability of an outcome by observing how that person bets on or against that outcome. A person's subjective probability of an outcome is the fair price of lottery ticket that pays $1 if the outcome occurs and zero otherwise. In example 7.3, if the farmer thinks that a lottery ticket that pays $1 if the storm hits his farm has a fair price of $0.3 this means that his subjective probability of this outcome is 0.3. It is important though, to keep the amounts in a bet low to ensure risk neutral attitude.

Table 7.2 Profit of each course of action in harvesting decision. Probabilities of each outcome about the path and effect of the storm are in parentheses.

Action\Outcome	Storm hits farm (0.3)	Storm misses farm (0.4)	Strom degenerates to warm rain (0.3)
Harvest early	$50,000	$50,000	$50,000
Wait	−$30,000	$150,000	$1,000,000

6. A decision criterion

Alternative courses of action must be evaluated and compared according to their consequences. Once the decision maker estimates the consequences of all alternative courses of action and their probabilities, the decision maker needs a decision rule in order to select the best alternative. This rule should account for both the consequences of each alternative and the decision maker's attitude toward risk.

Decision makers face two issues in this task:

a) Determine the best course of action given the risk profile of each action, that is, a list of potential payoffs and their probabilities.

b) Determine the best course of action when the decision maker has multiple objectives. In this case, the outcomes of each alternative are characterized by multiple attributes corresponding to each objective.

First, we will discuss how to address the first issue. Consider payoff Table 7.2 in the grape harvesting decision. This table assumes that the farmer will sell 10,000 bottles of wine. If the farmer decides to wait for a few days before harvesting despite the weather report, he will make a high profit if the storm misses the farm or even become rich if the storm degenerates to warm rain. This option is risky though; it may result in a sizeable loss of money if the storm hits the farm. Most people are risk averse; they prefer a certain payoff to an option that provides the opportunity of a windfall but also the risk of a sizable loss.

Decision makers need criteria to make risky decisions consistently. These criteria must be based on common sense rules. We will develop such criteria in section 7.4.

7.1.3 Single-Step Decisions and Sequential Decisions, Planning Horizon

Many decisions are sequential. In the inventor-financier decision, we assumed that the financier only accepts or rejects an offer; he/she does not counteroffer. In reality, the inventor and financier make a sequence of decisions rather than a single one. Specifically, the inventor makes an initial offer that the financier accepts, rejects, or counteroffers a fraction of the revenue. The inventor in turn responds by accepting the offer, rejecting it or making a second offer. Often, this sequence is repeated until the two sides settle or reach an impasse. Figure 7.3 shows this sequence of decisions. Similarly, in the grape harvesting decision, the farmer may decide to harvest the grapes immediately or wait and update his decision every day based on the weather report.

In a sequential decision, the decision maker must consider the entire sequence of decisions simultaneously and choose an optimum strategy that specifies the best choice

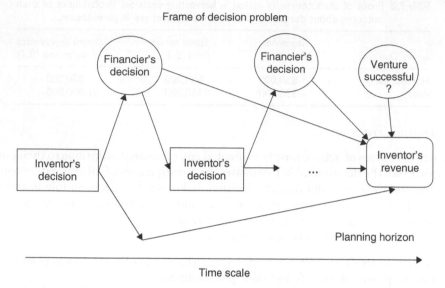

Figure 7.3 Sequential decisions in inventor-financier negotiation.

in every step of the sequence. In the drill for oil decision, the site owner should decide what to do if the seismic test is favorable or unfavorable before deciding whether to test, drill anyway or sell. In the inventor financier negotiation, the inventor should think how to respond to a financier's counteroffer before deciding how much to offer initially to the financier.

However, there is generally a limit, called planning horizon, to the number of decisions that the decision maker can make simultaneously. Beyond this limit uncertainty becomes too large or the number of options overwhelms the decision maker. Decisions outside the planning horizon are postponed until uncertainty is resolved further.

In the early stages of a decision, we frame the problem. Frame is a mental boundary that separates those aspects of a decision problem that one considers from those that one neglects. Figure 7.3 shows the frame of the inventor financier decision.

7.1.4   Steps of the Decision Process

Figure 7.4 shows the steps of a decision process. First the decision maker must **define the decision problem** and decide what decisions to make now and what decisions to make later. This important first step is called metadecision (the process where one decides what to decide). The decision maker also determines what aspects of the problem to consider and what to neglect. The boundary separating the two groups is the **frame** of the decision problem.

In the second step, the decision maker **identifies his/her options** or alternative courses of action. These first two steps are critical because it is important to solve the right problem, consider all of its important aspects and also consider a sufficient number of creative alternatives.

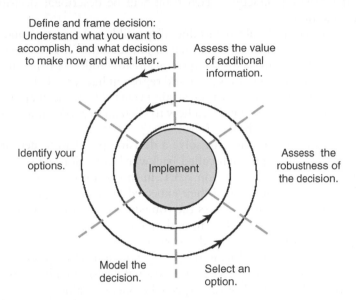

Figure 7.4 Decision analysis process.

The next step is to **structure the decision**, that is, to develop an analytical model of the decision. This involves understanding and representing preferences and uncertainties, and developing a predictive model for the consequences of a decision. Decision trees, and influence diagrams are tools that help one understand and structure the decision problem. The last two steps are to a) **solve the problem** in order to determine the best course of action and b) assess the robustness of this conclusion by performing **sensitivity analysis.**

Additional information can improve a decision. In the drill for oil problem (example 7.1), an unfavorable expert report could protect the owner from loss and a favorable one help the owner make a windfall. We will learn how to **assess the value of information** from a test for a particular decision in sections 7.2 and 7.3.

Decision making is iterative. The decision maker should review the solution of the decision process and evaluate the conclusions. The decision maker should redefine the problem and solve it again if needed.

Subjectivity is important in decision making. All of the above tasks, excepting the choice of the best course of action given the decision maker's beliefs and preferences, require eliciting judgments.

7.1.5 Outline of this chapter

This chapter explains principles and methods to address challenges in defining, modeling and solving a decision problem;

- Section 7.2 presents guidelines for framing and structuring a decision. The same section explains how to assemble into a model the six elements of a decision presented in the introduction. Two tools will be described; decision trees and influence diagrams.

A decision maker should determine all possible outcomes of uncertain events that affect the consequences of an action and quantify his/her beliefs about the likelihood of these outcomes. Chapter 6 already addressed the latter issue. We explained that the decision maker should represent his/her beliefs by the subjective probabilities of the outcomes of uncertain events. The same chapter defined subjective probability. It also explained how to elicit an expert's subjective probability and how to fuse it with observations.

- Section 7.3 demonstrates how to solve a decision problem to determine the best course of action. An approach called backward induction for solving a decision problem is presented in subsection 7.3.1. In this section, we compare alternative courses of action on the basis of their expected monetary values.

- Section 7.4 explains how to assess the robustness of a decision to imprecision in parameters such as the payoff of an option and or the probability of an outcome.

- It is important to represent the decision maker's risk attitude. Section 7.5 will address this issue by introducing the utility function. Using common sense rules (axioms), this section will prove that a rational decision maker should maximize the expected utility instead of the expected monetary value.

We used software Precision Tree for Excel to structure and solve decision problems. This program was developed by Palisade Corporation.

Refer to Clemen (1997) for a comprehensive treatment of decision analysis. French (1986) is a readable book on decision theory with emphasis on its axiomatic foundation. Ang and Tang (1984) explain concepts and tools for making decisions under risk in civil and other engineering fields.

7.2   Framing and Structuring Decisions

This section presents guidelines for defining a decision problem by identifying the crux of the decision. It also explains how to adopt a perspective that includes all important aspects of the decision. These are the most important and challenging steps in decision making. Then, the section explains how to structure the decision and presents the requisite tools for this purpose.

7.2.1   Define and frame a decision

It is always critical to consider and solve the right decision problem. The first step in a decision is to define the problem. In this step the decision maker addresses the following issues:

1.   What is the pivotal point of the decision?
2.   What are my objectives?
3.   Which step of the decision is the most critical?
4.   Which step requires more resources (including time)?
5.   Do I have the authority to make this decision?
6.   Do I have the resources to make the decision?

You may dramatically improve the consequences of a decision if you define the decision correctly. The following examples demonstrate this point.

Example 7.5: Testing a car bumper

A supplier developed a car bumper for an automotive manufacturer. The supplier had commissioned a small company to test the structural integrity of the design through a standard test. The supplier had reached a tentative agreement with the testing company for this purpose.

The company's design and release engineers constructed six prototypes of this design. When these engineers contacted the testing company, they found out that the manager who negotiated the tentative agreement had left. The new manager asked for a considerably higher price for the test, which the supplier would not be able to afford. The manager would not back-off.

The supplier's engineers met with their managers to decide how to address this issue. They considered the following options,

- Go to court.
- Request proposals from other companies for the test.
- Explain to the customer the situation and seek additional funding to perform the test.

After some discussion, most participants agreed that the above options were impractical because they were both expensive and time consuming.

Then one engineer stated that she was confident that the new design would pass a structural test on the basis of her CAE (Computer Aided Engineering) analysis. Specifically, stresses calculated from a finite element model under the loading conditions in the standard test were well below the endurance limit of the bumper material. She believed that the new design would do much better under a tough real-life test than a previous bumper design that the supplier had provided to the same automotive manufacturer. This previous design had worked well for several years in the field. Another engineer agreed that the new design was stronger than the old one, and added that he kept six prototypes of the old design in his office.

Based on the two engineers' statements, the group changed the question that they tried to address from,

> "How should we perform the standard test within the available budget and time constraints?"

to a broader one:

> "How should we verify that the new bumper is less likely to fail than the old one?"

The group designed and successfully performed the following inexpensive test. They drove their cars to a deserted, dilapidated parking lot, tied the prototypes of the new and old bumpers to the back of their cars and drove around the parking lot at different speeds. Then they examined carefully, after a few collisions, the new bumpers and compared them with the old ones. The new bumpers were in much better condition than the old ones. This evidence confirmed that the new design was stronger than the old.

These test results convinced both the supplier and the automotive manufacturer that the new bumper was strong enough. The automotive manufacturer accepted the new design. This design performed well in the field.

Some people make poor decisions because they have a narrow view of a decision problem. This example shows that considering the right question in a decision increases the likelihood that a decision maker will enjoy a good outcome.

Example 7.6: Defining and framing the drill for oil decision

Consider the drill for oil example 7.1 where a land owner in an oil rich region wanted to decide whether to drill for oil or sell the land. After some introspection, the owner concluded that both options were unsatisfactory. It would be too risky to drill the land. On the other hand, he could lose the opportunity to make a fortune if he/she sold it. Consulting an expert would also be expensive, and he/she would still face a risky decision after getting the test results.

After thinking about the problem, the owner realized that **the important question was not whether to sell the land or drill, but how to make the most money from the land while minimizing risk.** The owner realized that there were better options; the owner could ask an investor to finance the seismic test and drilling in return for a portion of the potential revenue from the oil. Or he could postpone the decision and reconsider it after a year. In the meantime, a neighbor could drill and find oil, which would make it almost certain that there would be oil in the decision maker's site. In this case, the owner could sell the land for a much higher price.

Example 7.7: How Pepsi Cola increased sales by addressing the right question (Russo and Schoemaker, 2002)

In the 1970s, Pepsi Cola unsuccessfully tried to increase its market share. A main reason was that Pepsi Cola concentrated on improving the design of its bottle. Eventually, Pepsi Cola's management concluded that it would be impractical to design a bottle that customers would like better than Coca-Cola's bottle. The reason was that Coca-Cola's bottle had been firmly established in the mind of consumers worldwide and had become an American icon. Finally, the management realized that the question "how can we design a bottle that is more attractive than Coca Cola's?" was the wrong one. Instead, they changed the question to "how can we minimize Coca Cola's advantage derived from the shape of its bottle?"

After conducting a lengthy consumer survey, Pepsi Cola's management concluded that customers were consuming increasing amounts of soda. In fact, they would drink as much as the company would convince them to buy[1]. The management predicted that, in the near future, the capacity of a bottle would become more important than its shape, to most customers. Pepsi Cola was able to influence customers and increase its market share significantly at Coca Cola's expense by designing containers with large capacity.

---

[1] Decision Analysis is considered amoral. It does not question the preferences or motives of the decision maker. In the given example, decision analysis will help make a good decision without questioning whether too much soda consumption is good or bad for people's health. Ethics play an important part in engineering and social decision making and thus should be considered in the preferences of the decision maker. Once that happens, decision analysis will help incorporate it while finding optimal decisions.

Many decisions produce disappointing consequences because decision makers try to solve the wrong problem, like Pepsi Cola's management did initially in the early 1970's. You should take your time to identify the basic point of a decision. Carefully examine the question that you are trying to answer, consider alternative questions and make sure you are considering the right one.

A *thinking frame* of a situation is a mental boundary that encompasses some aspects of the situation while ignoring others. When we frame a decision problem, we adopt a particular perspective of it. Different people have different frames for the same problem. For example, one student in a decision analysis class views the class as an opportunity to learn how to think analytically and make better decisions based on common sense rules. Another student views the class only as a requirement for graduation. One student, views his/her classmates as friends and colleagues, while another views them as rivals with whom he/she will compete after graduation for a good job.

Frames simplify a decision and make it practical to analyze it because they allow us to focus on the important aspects of the decision. However, adoption of a narrow frame creates blind spots (important aspects of a situation that we ignore). This has detrimental effects on the consequences of the decision.

When facing a decision problem it is critical to frame it properly. Ask yourself the following questions in order to construct a proper frame,

- What decisions should I make now and what decisions should I make later?
- What are the most important objectives to aim for?
- How will I measure success toward these objectives?
- To what aspects of the decision should I pay attention?
- What aspects should I ignore in order to make the problem tractable?
- What are my options?
- Who will be affected by the decision? Who will support it and who will oppose it?

Example 7.8: Design of a car joint

Joints are parts of a car body made of steel plates fastened by spot welds. These components significantly affect the noise, vibration and harshness (NVH) performance and crashworthiness. Figure 7.5 shows the joint connecting the B-pillar to the rocker of a family sedan.

A new engineer was asked to design the joint of a new car model. The engineer's supervisor told him that the overall merit of a joint design depends on its static stiffness and crashworthiness. In addition, it should be feasible to stamp the joint. The engineer adopted a narrow perspective of the design problem focusing primarily on NVH performance. Two reasons for this were that his/her expertise was on structural dynamics, and that he was a member of an NVH group. In addition, communication between the NVH, crashworthiness and stamping groups was limited. The engineer developed and validated a model for predicting the NVH characteristics of the joint. The engineer also talked to crash analysis and stamping engineers and established some simple constraints on the dimensions. Then he optimized the design to minimize the mass, while meeting requirements for NVH, crash and stamping feasibility.

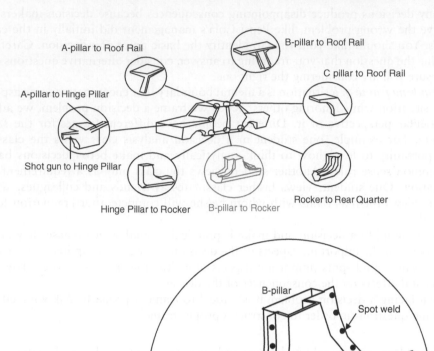

A-pillar to Roof Rail

A-pillar to Hinge Pillar

Shot gun to Hinge Pillar

Hinge Pillar to Rocker   B-pillar to Rocker

B-pillar to Roof Rail

C pillar to Roof Rail

Rocker to Rear Quarter

B-pillar

Spot weld

Rocker

Figure 7.5 Automotive joint.

Unfortunately, when the engineer presented his design to the management, his design was criticized for poor crashworthiness. Experts claimed that the beams of the joint would not absorb enough energy in a crash because they would snap easily. This happened because the engineer had reduced the plate gages and increased the cross sectional areas in order to meet stiffness requirements and keep mass low, without considering the attendant deterioration in crash performance.

The above is an example where a decision maker made a poor decision because he/she adopted a narrow frame. The engineer should have invested much more time to understand the importance of crashworthiness and stamping feasibility and develop and use models for predicting this performance accurately when making his/her design decisions.

Fundamental vs. Means Objectives

A decision maker should know what he/she wants to accomplish. Objectives can be classified into fundamental and means objectives (Clemen, 1997). Fundamental objectives express what the decision maker wants to accomplish, while means objectives explain how the decision maker will accomplish the fundamental objectives. In the city evacuation example, in subsection 7.1, fundamental objectives include:

- Prevent loss of life.
- Minimize injuries.
- Prevent property damage.

Means objectives explain how the mayor will accomplish the above objectives. These include:

- Provide residents with reliable and readily accessible transportation.
- Provide accurate and timely information to all residents.
- Provide clear guidance and effectively coordinate different agencies such as police, transportation authority, and coast guard.
- Facilitate and ensure effective interagency communication.

It is important to measure the degree to which the consequences of a decision satisfy the objectives. In automotive design, objectives are measured in terms of attributes, such as money, market share, and sound pressure level in a car cabin. The above three attributes have a natural scale. However, there is no natural scale for measuring the worth of the consequences of some decisions, when the consequences involve subjective attributes. For example, it is challenging to measure, with a single number, attributes such as a company's reputation, the style, the quality of the exhaust sound or the steering feel of a car.

The worth of subjective attributes could be quantified in customer clinics. These clinics must be organized carefully; participants must be trained before being asked to evaluate products. First, organizers must define in detail what terms such as best, average and poor performance mean. Here is an example of defining a scale for the quality of noise, vibration and harshness performance of a car.

> **Best:** Ride is cushy and extremely comfortable. Even sharp road irregularities are absorbed nicely and are muted. There are no side motions or jiggles. Highway ride is serene even at 80 mph. It is easy to hear conversations in cabin even under strong acceleration.

> **Average:** The car rides smoothly on good highways, but bumpy secondary roads deliver firm kicks, rubbery jiggles and sideway motions. The cabin is quiet in good highways at speeds up to 60 mph but becomes noisy above that speed.

> **Worst:** Ride is harsh and jerky. Even small road irregularities elicit sharp motions. It is hard to have a conversation in the car above 45 mph.

The participants should also be trained by letting them use products with excellent, good and unacceptable attributes in these clinics.

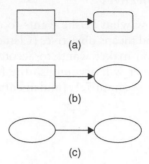

Figure 7.6   Dependence of elements of a decision
   (a) Decisions (box) have consequences (rounded box).
   (b) Decisions (box) affect the probabilities of the outcomes of an uncertain event
       (ellipse).
   (c) The outcome of an event (left ellipse) affects the probabilities of the outcomes of
       another uncertain event (right ellipse).

Many decisions involve multiple objectives. For example, many companies want to maximize profit and market share, and improve their reputation simultaneously.

7.2.2   Structure a Decision Problem

Once a decision maker has defined the elements of a decision, the decision maker should structure these elements into a logical framework. Structuring a decision problem is the process of specifying an analytical model representing a vaguely defined decision problem. Influence diagrams and decision trees are useful tools for assembling the elements of a decision into a logical framework. These tools contain identical information. One can convert an influence diagram to a decision tree and vice versa Clemen (1997). We describe each tool in the subsections below.

Influence diagrams

Influence diagrams show how the elements of a decision affect each other. In these diagrams, elements are represented by three types of nodes:

1.   *Calculation* and *consequence nodes*: Rounded boxes represent both calculations and consequences.

   Arcs toward a particular node show which elements affect the element represented by the node (Figure 7.6). An arc from a decision node to a consequence node shows that the choice represented by the decision node affects the consequences described by the consequence node (Figure 7.6a). An arc from a decision node to a chance node shows that choices of the decision node affect the probabilities of the uncertain event represented by the chance node (Figure 7.6b). An arc from an ellipse to another ellipse shows that the outcome of the uncertain event represented by the first ellipse affects the probabilities of the uncertain event represented by the second (Figure 7.6c). Tables

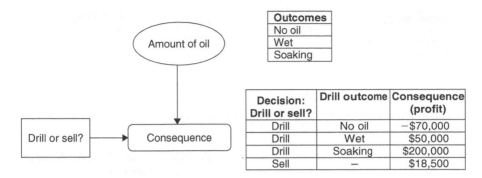

Outcomes
No oil
Wet
Soaking

Decision: Drill or sell?	Drill outcome	Consequence (profit)
Drill	No oil	−$70,000
Drill	Wet	$50,000
Drill	Soaking	$200,000
Sell	–	$18,500

Figure 7.7 Drill for oil decision.

next to the nodes show the available alternative courses of action, all possible outcomes of uncertain events and the resulting consequences of each action.

The influence diagram in Figure 7.7 illustrates how the owner's decision and the outcomes of an uncertain event influence the owner's consequences in the drill for oil decision (example 7.6). This figure considers only the options of drilling and selling. According to this diagram, the owner's payoff depends both on a) his/her choice to drill or sell, and b) the amount of oil in the site.

From the influence diagram, the owner can determine the consequences of each choice for each possible outcome of the uncertain event. This determination does not require any additional information, interpretation or judgment. Every analytical model of a decision must pass this test.

Now we consider the same drill for oil decision but with the additional option to perform a seismic test. The test has three possible outcomes: no structure, open structure and closed structure. The amount of oil in the site influences the likelihood of these outcomes; if the site is soaking (there is a lot of oil) then it is more likely that the outcome of the test is "closed structure" than "open structure" or "no structure". This information from the test is valuable because it can help the owner of the site decide whether to drill or sell. This information does not remove the uncertainty entirely though because the test could indicate a closed structure, even if the site is dry.

Figure 7.8 explains the decision with the option to test. An uncertain event represents the test. The arc from uncertain event "amount of oil" to the event "test" indicates that the probabilities of the outcomes of the test depend on the amount of oil in the site.

One should be cautious when interpreting the numbers in the consequence table in Figure 7.8. According to this table, one may think that drilling without testing always yields a larger payoff than testing before drilling. However, the test results enable the owner to predict the amount of oil in the site with high confidence and select that option with the highest payoff. For example, if the test indicates a closed structure, and the owner knows that the probability that there is oil in the site is 0.999 in view of this test result, then the owner will seize the opportunity to make a fortune by drilling.

Influence diagrams represent a decision more compactly than decision trees. Often, it is easier to explain to someone a decision problem by using an influence diagram than

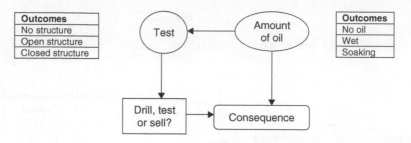

Decision: Test?	Decision: Drill or sell?	Drill outcome	Consequence
Yes	Drill	No oil	−$80,000
Yes	Drill	Wet	$40,000
Yes	Drill	Soaking	$190,000
Yes	Sell	–	$8,500
No	Drill	No oil	−$70,000
No	Drill	Wet	$50,000
No	Drill	Soaking	$200,000
No	Sell	–	$18,500

Figure 7.8 Drill for oil decision with testing option.

a decision tree. On the other hand, influence diagrams hide important information, such as the sequence in which decisions are made. Also, in complex decisions, it is hard to construct influence diagrams.

Decision trees

A decision tree graphically displays the sequence of choices and uncertain events in a decision. A decision tree consists of decision nodes (squares) and chance nodes (circles). These elements are connected by arcs, like in influence diagrams. Decision trees are evaluated from left to right.

Arcs in decision trees show the sequence in which choices are made and uncertain events are resolved – not dependence. For example, an arc emanating from a decision node and pointing toward a chance node indicates that the decision is made first and then the uncertain event is resolved. Each path of a decision tree corresponds to a possible scenario for a particular sequence of decisions and uncertain events.

We present below three examples of decision trees representing the drill for oil problem and a sequence of decisions of a product manufacturer.

Example 7.9:   Constructing a decision tree for drill for oil problem

First consider the drill for oil decision without the option of performing a seismic test. First, the owner decides whether to drill the site or sell it. If the owner decides to drill then he/she will pay $70,000. The consequences of this decision depend on the amount of oil in the site. If there is no oil then the owner will lose this amount. In addition, the owner will be unable to sell the site. If the site is wet then the owner will collect

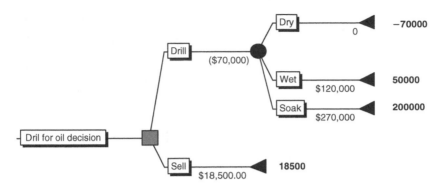

Figure 7.9 Decision Tree for drill for oil decision. Numbers under branches show the cost or payoff associated with the option or the outcome represented by the branch.

$120,000 and his/her profit will be $50,000. If the site is soaking then he/she will make a profit of $200,000. If the owner sells the site instead of drilling, he/she will make a profit of $18,500 (sale price minus realtor's commission).

The decision tree in Figure 7.9 illustrates the owner's decision. This figure was created by using software *Precision Tree*, version 5.5, developed by Palisade Corporation. Squares and circles represent decisions and uncertain events. Triangles are terminal nodes. Labeled rectangles explain options or outcomes of uncertain events. A number under a branch shows the payoff associated with the corresponding option or outcome. For example, number "($70,000)" under branch labeled "Drill" shows that the owner has to pay $70,000 to drill the land. The owner's net profit or loss for each scenario is shown by the number next to the terminal node (triangle). For example, number "−70000" next the end of branch "Dry" shows that if the owner drills the site and finds no oil then the owner will incur a $70,000 net loss.

Note that the influence diagram in Figure 7.7 and the decision tree in Figure 7.9 contain identical information.

Consider the same decision but with the option of performing a seismic test before deciding whether to drill or sell. Figure 7.10 shows the decision tree for this case. First, the owner decides whether to consult a geologist who will perform a test, drill anyway or sell. If the owner decides to skip the test, then he/she will have to decide whether to drill or sell. The lower two branches of the decision tree in Figure 7.10 show the consequences of these two options. The part of the decision tree corresponding to these two options is identical to the one in Figure 7.9.

If the owner performs a test first then he/she will pay $10,000 for it. After the owner studies the test results, he/she will decide whether to sell or drill. The upper branch (labeled "Test") of the decision tree in Figure 7.10 shows the possible consequences of this decision. If the owner decides to sell the land, then he/she will make a profit of $8,500 (profit from sale minus cost of test). If the owner decides to drill, then his/her profit will be $40,000 if the land is wet or $190,000 if it is soaking. The owner will lose $80,000 (test plus drill costs) if there is no oil. Note that this profit profile is identical to the one in Figure 7.8.

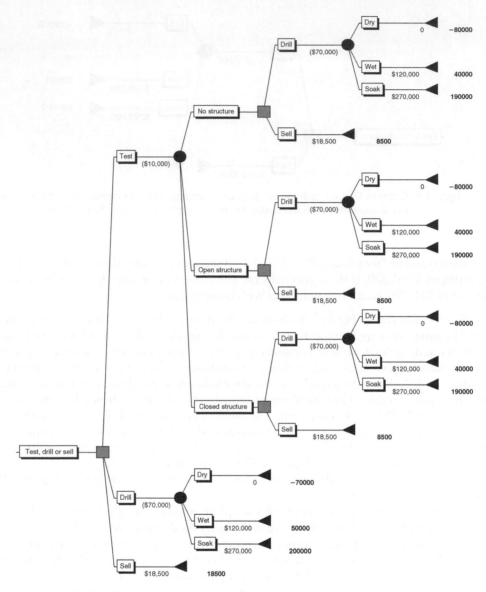

Figure 7.10 Decision tree including the option of performing a seismic test.

The decision trees in Figures 7.9 and 7.10 do not provide sufficient information in order to enable the land owner to select the best course of action. In addition to this information, the owner needs the probabilities of all possible outcomes of the uncertain events. Specifically, the owner needs the following information,

- the three probabilities of the outcomes of the test (no structure, open and closed structure),

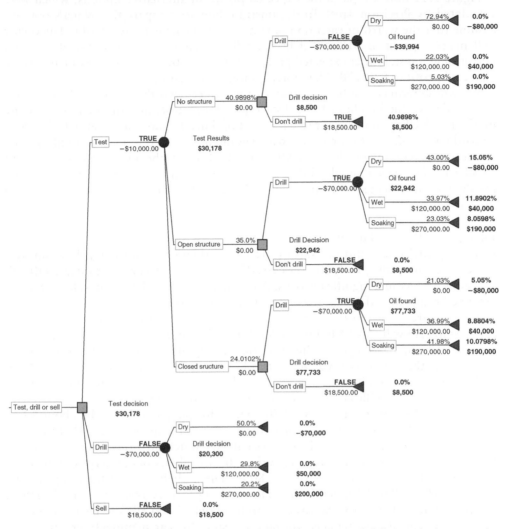

Figure 7.11 Drill for oil decision with probabilities of outcomes of drilling and of test results. Braches of the decision nodes representing the best strategy are marked "TRUE." Branches representing suboptimal decisions are marked "FALSE."

- the three unconditional probabilities of the site being dry, wet or soaking, and
- the nine conditional probabilities of the site being dry, wet or soaking conditioned upon the test results (no structure, open and closed structure)

Figure 7.11 shows the same decision tree as in Figure 7.10 but with estimates of the probabilities of each uncertain event. These probabilities are above the branches representing the uncertain events in the decision. For example, number "72.94%" above the branch on the upper right corner of Figure 7.11 shows that the probability that the site is dry, given that a test found no structure, is 72.94%.

Figure 7.11 also compares the expected profits of alternative options, which were calculated by Precision Tree. To demonstrate how to compare the owner's options, assume that the worth of each option is measured by its expected profit. The owner will make a profit of $18,500 if he/she sells the land. If the owner drills without performing a test, his/her expected profit will be $20,300. A seismic test increases the expected profit to $30,178. Therefore, testing is the best option.

We will determine the best strategy of the owner from the results in Figure 7.11. In this figure, the braches of the decision nodes representing the best strategy are marked "TRUE." Branches representing suboptimal decisions are marked "FALSE." According to Figure 7.11, first, the owner should perform a test. If the test indicates no structure, then the owner should sell the land. In this case, the owner will make a profit of $8,500. The owner should drill if the test indicates an open or a closed structure. He/she will make an expected profit of $22,942 in the former case and of $77,733 in the latter.

Example 7.10:   Product planning

Increasing fuel prices and pressures to reduce carbon emissions compel automakers to develop innovative, fuel efficient cars. This requires automakers to make difficult decisions and assume significant risks. A car manufacturer plans to develop either an extended range hybrid car or a very small car (econobox).

The venture involving the extended range hybrid is riskier than that of the small car. The hybrid will cost $5 billion to develop while the small car will costs $2 billion. Success of both cars hinges on the gas price during their lifetimes. Product analysts expect that the hybrid car will generate $12 billion if the gas price is about $4 per gallon, but only $1 billion if the price is about $3 per gallon. The small car will generate $5 and $3 billion, respectively, for the above gas prices.

Experts believe that there is a 50/50 chance for the gas price to be about $3 or $4 during the lifetime of the two car models. However, experts believe that they will become more confident about the gas price if they watch how this price evolves over the next year. If gas costs $3 after a year, then the probability of the gas price remaining $3 over the lifetime of the cars will be 80%. The probability that the gas price will increase to $4 will be 20% in this case. However, if gas costs $4 after a year, then there is a probability of 20% for the gas price to be about $3 and 80% to be about $4.

Development of both models will take 2 years. The car manufacturer can decide which car to develop now. However, because development of the electric car is very risky, the automaker considers the option to start developing both cars now and discontinue one car program when the manufacturer obtains updated gas price forecasts. This will not increase the development cost of either car.

The decision tree in Figure 7.12 shows the sequence of the two decisions. First, the manufacturer decides whether to select one car for development now or wait and watch fuel prices for a year. If the manufacturer postpones the decision for a year, the manufacturer will be able to predict the gas price accurately and make an informed decision. However, by postponing the decision about which car to develop, the manufacturer will lose the $1 billion invested in the development of the discontinued model.

The optimum product development strategy was determined assuming that the automaker seeks to maximize the expected profit. It is observed from Figure 7.12

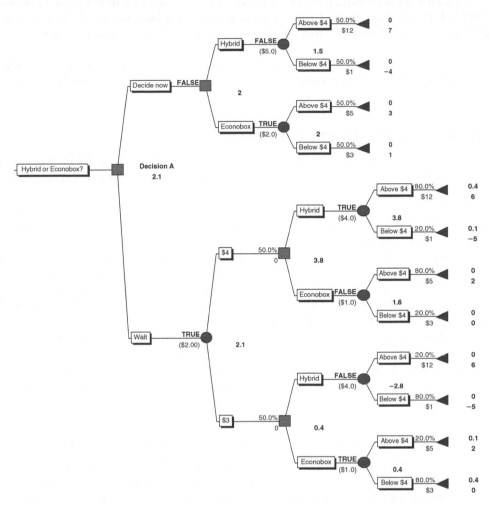

Figure 7.12 Decision to develop small car or long range hybrid car. All amounts are in billions of dollars. Braches of the decision nodes representing the best strategy are marked "TRUE." Branches representing suboptimal decisions are marked "FALSE."

that the best strategy is to postpone the decision for a year and start developing both cars right now. Then develop the hybrid car if gas costs $4 a gallon after a year or the small car if it costs $3.

However, the decision is very sensitive to the assumptions of the problem. The expected return of deciding immediately to develop the small car results in an expected monetary return of $2 billion that is only slightly lower than postponing the decision. The automaker should reconsider the decision after collecting additional information about costs and revenues. Furthermore, given the high stakes of the decision, the automaker should account for the stakeholders' risk tolerance in this decision.

A decision model should pass the following *clarity test* in order to be useful (Howard, 1988). Imagine that, in the future, perfect information becomes available regarding all aspects of the decision. Then the consequence of each alternative course of action should become known.

Decision makers should construct a *requisite model* of the decision problem. This is a model that contains everything essential for solving the decision problem but no superfluous information.

7.3   Solving Decision Problems

The previous section explained how to construct an analytical model of a decision and ensure that the model contain all the information needed in order to make the best decision. This effort produces a decision tree or an influence diagram that represents the decision problem. This section presents an approach to determine the best choice according to the decision make's beliefs about the likelihoods of the outcomes of uncertain events and his/her preferences about the consequences.

In a decision, the choice of the best alternative course of action depends on the decision maker's attitude toward risk. Decision making requires a decision criterion that will enable the decision maker to select the best course of action according to his/her attitude toward risk. In this section, we consider decisions in which the consequences are evaluated by their monetary profit. We assume that the best option is the one with the maximum expected monetary profit. This rule is realistic if the same risky decision is repeated many times. Then the average payoff converges to the expected profit of the decision with the number of repeated decisions. In addition, many decision makers evaluate the worth of gambles involving small amounts according to their average profit.

7.3.1   *Backward induction (or folding back the decision tree)*

The method of backward induction is rigorous and simple. A decision tree consists of decision nodes and chance nodes representing gambles. The key idea is to replace the chance nodes with the *certainty equivalent* of the gambles that these nodes represent. Certainty equivalent of a gamble is the sure amount that is equivalent to the uncertain payoff of the gamble. Then the decision maker finds the best option by comparing the certainty equivalents of the alternative options.

For simplicity, this section assumes that the certainty equivalent is equal to the expected monetary value of the gamble. In section 7.4 we will show that the expected monetary value criterion does not always represent a decision maker's attitude toward risk and examine alternative decision rules.

First, we replace all chance nodes on the right end of the tree with their certainty equivalents. This step enables an easy choice between options with sure payoffs (which are equal to the certainty equivalents of the options). After these choices are made we replace the decision nodes with their highest payoffs. In one-step decisions, the best option is revealed after completing these two steps and this completes the solution.

In sequential decisions, we repeat the two steps of replacing chance nodes with their certainty equivalents and decision nodes with their highest payoffs until we can determine the best sequence of choices.

Example 7.11:   Choice between two risky options
A decision maker considers two gambles: 1) A fair coin will be flipped once. Buy a
ticket for $5 that is worth $10 if the coin lands heads up in the flip and zero otherwise. 2) A thumbtack will be flipped once. Buy a ticket for $5 that is worth $9 if a
thumbtack lands with its pin up and $1 otherwise. After examining the thumbtack,
the decision maker estimated that the probability of the thumbtack landing with its
pin up is 0.6. Construct a decision tree representing the choice between the gambles
and the consequences and find the optimum choice.

Solution:
Figure 7.13 illustrates this decision. Betting on the coin landing heads up could yield
$5 profit or loss, each with 50% probability. Betting on the thumbtack landing with
its pin up could yield $4 profit with probability 60% and $4 loss with probability
40%. The two numbers to the right of the chance nodes representing the gambles with
the coin and the thumbtack are the probabilities of the branches and the expected
profits of the two gambles. For example, the numbers next to the branch representing
the outcome that the thumbtack will land with its pin up show that there is a 60%
probability that this outcome will occur, and that if the decision maker had bet on this
event, he/she would make $4 profit.

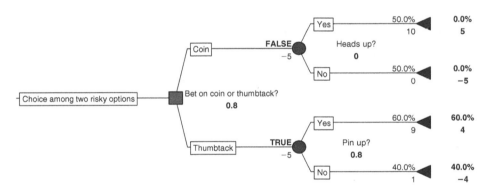

Figure 7.13 Decision to bet on coin or thumbtack.

The lower branch of the decision tree shows the optimum solution. The word
"TRUE" above the branch corresponding to the option to bet on the thumbtack
landing pin up indicates that this is the best option. The word "FALSE" above the
branch corresponding to the option to bet on the coin landing heads up shows that
this is not the best option. In addition, the zero values of the probabilities next to
the branches representing the outcomes of this option indicate that the option is not
optimal.

Figure 7.14 shows the decision tree after replacing the two gambles with their
expected profits. Betting on the coin landing heads up has an expected profit of
$0 while betting on the thumbtack landing with its tip up has an expected profit
of $(\$9 - \$5) \cdot 0.6 + (\$1 - \$5) \cdot 0.4 = \$0.8$. Therefore, the thumbtack is the best bet.

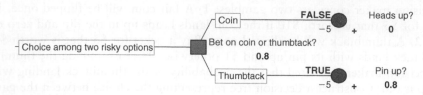

Figure 7.14 Folded decision tree.

Example 7.12: Sequential decisions in software development

The owner of a small company considers developing new software. Development will cost $2 million, and it is likely to be successful. The owner estimates that the probability of success is 70%, after consultation with the company's system analysts. If the software development is successful, then the small company can sell the software to a bigger company for $5 million or directly to consumers. The latter option requires an additional investment of $1 million for marketing. The revenue from selling the software to consumers depends on the demand; it will be $20 million if demand is high, $5 million if it is medium and only $0.2 million if it is low. The probabilities of high, medium and low demand are 20%, 55% and 25%, respectively. Figure 7.15 illustrates the owner's decision.

Solve this problem to help the owner decide whether to fund the software development. In case that the owner decides to do so, determine whether he/she should sell the program to a bigger company or directly to consumers.

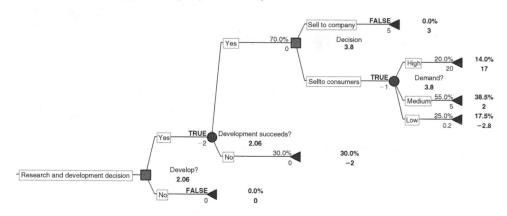

Figure 7.15 Software development decision.

Solution:

In backward induction, the owner makes the second decision first, that is, decides to whom he/she will sell the software once it has been developed. Then, the owner estimates the expected profit from the venture and this information makes the solution to the first decision (develop the program or not) obvious.

The expected profit from selling the software to consumers is $(\$20 - \$3) \cdot 0.20 + (\$5 - \$3) \cdot 0.55 + (\$0.2 - \$3) \cdot 0.25 = \$3.8$ million. Then it becomes obvious that the owner should sell the software to customers directly at a $3.8 million profit instead of selling it to a bigger company at a profit of $3 million. Figure 7.16 shows the decision tree after the owner has decided to sell the software directly to customers.

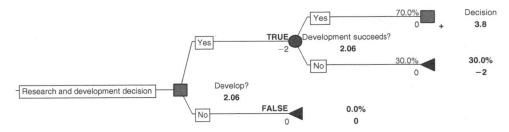

Figure 7.16 Decision tree after deciding to sell the software to customers.

Now the owner calculates the expected profit of the venture and decides whether to develop the software. The expected profit is $3.8 \cdot 0.70 - \$2 \cdot 0.3 = \$2.06$ million. The option of developing the software is better because its expected profit is positive.

The owner concludes that the optimum strategy is to develop the software and sell it directly to customers if the development is successful.

Risk profile of a decision

Risk profile of a risky decision is the probability mass function of the profit. That is a list of the possible profits or losses resulting from the decision and their probabilities. Risk profiles provide decision makers with insights into the risks that each alternative course of action entails. Expected profits hide this information.

Figure 7.17 shows the risk profile of the software development venture. Software development could yield $17 million with 14% probability and $2 million with 38.5%

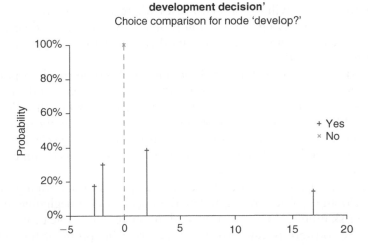

Figure 7.17 Risk profile of decision to develop software (solid bar) and do nothing (dashed bar).

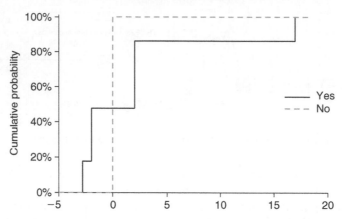

Cumulative probabilities for decision tree 'research and development decision'

Choice comparison for node 'develop?'

Figure 7.18 Cumulative risk profiles of decision to develop software (solid bar) and do nothing (dashed bar).

probability, but it could also result in possible losses of $2 million and $2.8 million with probabilities 30% and 17.5%, respectively. The owner can also compare the cumulative risk profiles of the decisions in Figure 7.18 to understand the risk that he/she assumes. The two risk profiles indicate that if the company decides to develop the software it has the opportunity to make a lot of money but also assumes the risk of losing a sizeable amount.

This example demonstrates the limitations of the expected monetary value as a measure of the worth of a risky venture. Most small companies would not assume the risk of losing $2 million to $2.8 million with almost 50% probability.

In the following, we will explain two cases where a decision maker can eliminate some options, before solving the decision problem, by comparing their risk profiles. This occurs when one option dominates another. There are two types of dominance, *deterministic* and *probabilistic*.

Deterministic dominance: Option A dominates deterministically option B if the worst consequence of A is at least as good as the best consequence of B. For example, if the costs of software development and marketing were negligible, then developing the software would dominate the option of not developing it. When one option deterministically dominates another option, then the decision maker must select the dominating option regardless of the probabilities of the outcomes of the uncertain events and the decision maker's attitude toward risk.

Figure 7.18a compares the risk profiles of the options of developing the software, and not developing it, if developing and marketing costs were negligible. It is observed that the two risk profiles do not overlap, and the smallest possible profit of the option to develop the software is equal to the highest possible profit of the option not to develop it.

Probabilities for decision tree 'research and development decision'
Choice comparison for node 'develop?'

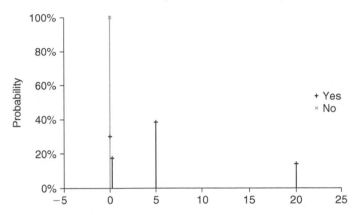

Figure 7.18a Option of developing software (three right bars) dominates deterministically the option of not developing it (leftmost bar).

Probabilistic dominance: Option A probabilistically dominates option B if the profit of option A is more likely to exceed any target value than the profit of option B. For example, suppose that you get $10 if a fair coin lands heads up or a thumbtack lands with its tip up in one flip. The probability of the latter event is 0.3. The option to bet $5 on the fair coin probabilistically dominates the option to bet the same amount on the thumbtack.

When one option probabilistically dominates another, then the decision maker must select the dominating option regardless of the decision maker's attitude toward risk. However, this type of dominance is sensitive to the probabilities of the outcomes.

If an option probabilistically dominates another option then the cumulative risk profile of the dominated option will cover completely the cumulative risk profile of the dominant option. This is observed in Figure 7.19, which shows the cumulative risk profiles of the options to bet on the coin and to bet on the thumbtack.

Example 7.13: Using information from tests to reduce uncertainty
A major household product manufacturer plans to develop an innovative squeegee design. The company commissioned a consultant to check if the design could break easily under heavy use. The consultant reported that the design has adequate strength. However, one of the company's engineers, who reviewed the consultant's calculations, is concerned that the consultant might have made an erroneous assumption. The engineer estimated that there is a 25% probability that the consultant's assumption is wrong.

If the consultant's assumptions are valid, the company will make $100 million revenue from marketing the squeegee. The profit is $50 million because the total cost of developing, manufacturing and advertising the product is $50 million. However, if the consultant's assumption is invalid then the squeegee will generate only $10 million revenue, and the company will lose $40 million.

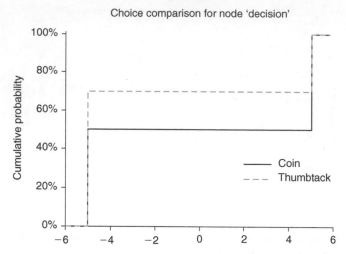

Cumulative probabilities for decision tree 'coin or Thumbtack?'

Choice comparison for node 'decision'

Figure 7.19 Probabilistic dominance.

The company can construct and test 10 prototypes. The design will pass the test if no prototype breaks. The probability that the design will pass the test, given that the consultant's assumptions are right, is 0.98. If the assumptions are wrong, the probability of the same outcome is only 0.1. The test will cost $100,000.

The manager of the product development considers the following options:

1. Develop the product without testing.
2. Cancel the product development.
3. Test the product and then decide whether to develop it.

In the latter case, the manager should decide upfront what to do in both cases where the squeegee passes and fails the test.

In order to define the decision problem, we need a) the probability that the squeegee will pass the test, and b) the conditional probabilities that the consultant's assumptions are right given that the squeegee passed and failed the test. We define the following symbols representing the outcomes of the test and the validity of the assumptions:

V: the assumptions are valid
V^C: the assumptions are not valid
P: the product will pass the test
P^C: the product will fail the test

The probability of passing the test is equal to the conditional probability of the product passing the test given that the assumptions are valid times the probability that the assumptions are valid plus the conditional probability of the product passing the test given that the assumptions are invalid times the probability that the assumptions are invalid,

$$P(P) = P(P/V)P(V) + P(P/V^C)P(V^C) = 0.98 \cdot 0.75 + 0.1 \cdot 0.25 = 0.76$$

The conditional probability that the assumptions are valid given that the test was successful is calculated by using Bayes' rule,

$$P(V/P) = \frac{P(P/V)P(V)}{P(P)} = \frac{0.98 \cdot 0.75}{0.76} = 0.9671$$

Similarly, the conditional probability that the assumptions are valid given that the test was unsuccessful is,

$$P(V/P^C) = \frac{P(P^C/V)P(V)}{P(P^C)} = \frac{0.02 \cdot 0.75}{0.24} = 0.0625$$

The probabilities that the assumptions are invalid given that the product passed the test, and the assumptions are invalid given that the product failed the test are,

$$P(V^C/P) = 1 - P(V/P) = 0.0329$$
$$P(V^C/P^C) = 1 - P(V/P^C) = 0.9375$$

The test is very powerful according the above probabilities because the test results enable the manager to predict with high confidence if the assumptions are valid. Suppose that the product passed the test. This increases the probability that the assumptions are valid from 0.75 to 0.9671. On the other hand, if the product fails the test, the manager becomes almost certain that the assumptions are invalid.

Figure 7.20 shows the tree representing the manager's decision. The first step for solving this problem is to replace the three gambles of developing the product in the three cases where the product passed and failed the test and in the case where the product was not tested with their certainty equivalents. For this purpose, we replace the chance nodes on the right end of the tree with their expected profits. Consider the case where the company decided to test the product. If the consultant's assumptions are valid then the profit will be equal to the revenue minus the development and testing costs,

$$\text{Profit} = \$100 - \$50 - \$0.1 = \$49.9$$

If the assumptions are invalid then the profit will be,

$$\text{Profit} = \$10 - \$50 - \$0.1 = -\$40.1$$

Then the expected profit of the chance node representing the development of the product given that it passed the test is, $(\$100 - \$50 - \$0.1) \cdot 0.9671 + (\$10 - \$50 - \$0.1) \cdot 0.0329 = \$46.94$. The expected profit of the gamble of developing the product given that it failed the test is: $(\$100 - \$50 - \$0.1) \cdot 0.0625 + (\$10 - \$50 - \$0.1) \cdot 0.9375 = -\$34.475$. Finally, the expected profit from developing the product without

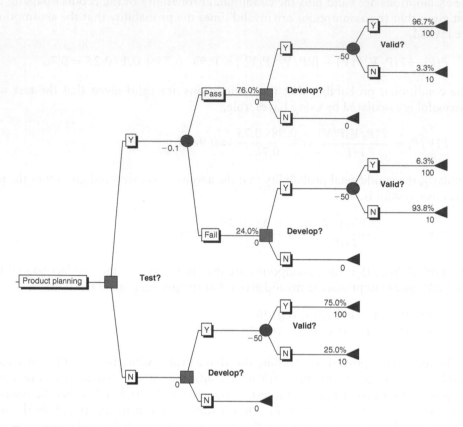

Figure 7.20 Model of squeegee development decision.

testing is: ($100 − $50) · 0.75 +($10 − $50) · 0.25 = $27.5. Figure 7.21 shows the decision tree after replacing the three chance nodes on the right with the expected values of the profit from the gambles that these nodes represent.

The next step is to decide whether to develop the product in the two cases where the product passed and failed the test and in the case where no test was performed. If the product passes the test then the manufacturer should develop the product because this option has higher expected profit than the alternative option of canceling its development. On the other hand, if the product fails the test, then it is better to cancel the development of the product because it is almost certain that the company will lose money from it. Even if the company does not test the product, it should still develop it because it is expected to make a profit from this option. Figure 7.22 shows the decision tree after making these three decisions.

The last step is to replace the chance node representing the test of the product with its expected profit. Figure 7.23 shows the resulting decision tree. The expected profit from the test is: 0.76 · $46.939 − 0.24 · $0.1 = $36.65. Now the decision is between two risky options: test or not test. It is better to test the product because this option has higher expected profit.

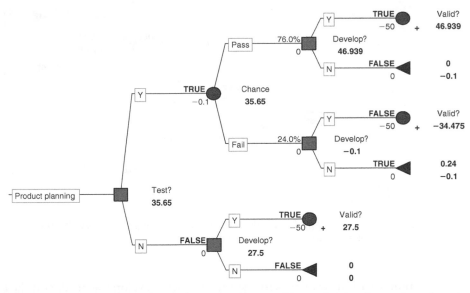

Figure 7.21 Decision tree after replacing the three gambles to develop the product given that it passed the test, failed it and was not tested with their expected profit.

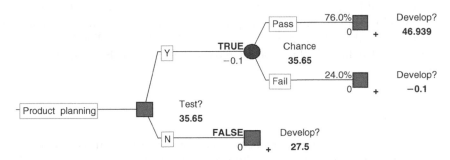

Figure 7.22 Decision tree in Figure 7.21 after replacing the decision nodes representing the product development with their maximum expected profit.

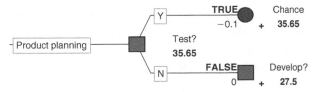

Figure 7.23 Decision tree in Figure 7.22 after replacing the chance node of the test with its expected profit (It is better to test the product according to this figure).

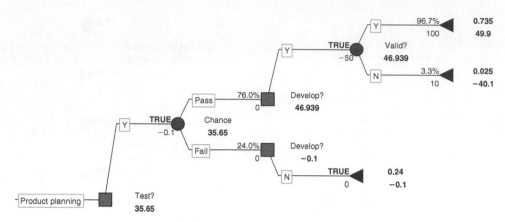

Figure 7.24 Optimum strategy for product testing and development. Test the product first. Develop it if it passes but do not develop it if it fails.

The optimum decision strategy is to test the product. Develop the product if the test is successful, and discontinue it if it is unsuccessful. Figure 7.24 shows the optimal strategy.

However, the optimum strategy is very risky according to Figure 7.24 because it can result in a big loss or profit. The worst case scenario will occur if the product passes the test although the consultant's assumptions are invalid. In this case, the company will lose $40.1 million. If the product fails the test and the company cancels the development, then the company will lose only $100,000. Finally, the company will make $49.9 million if the product passes the test and the company decides to develop it.

The value of information

Information enables people to make better decisions. Information economics is a valuable tool for supporting decision making by helping designers manage testing and computing costs (Schlosser, and Paredis, 2007). Information economics is the systematic study of measuring the value of information in support of a decision. A basic concept in this study is that the value of information from running a test or a predictive model is equal to the gain from using the information to select the best course of action.

In the oil exploration example, if an expert could tell the owner that his/her property was dry then the owner would sell the land and save money in testing and drilling. Here we present a method for measuring the value of information in a particular decision. This method is based on two principles:

1)   The value of information depends on the particular decision.
2)   The value of information is equal to the increase in the expected profit resulting from using the information to improve the decision.

Specifically, the value of information from a test is,

$$EVI = EMP_{\text{with test}} - EMP_{\text{no test}} \tag{7.1}$$

where EVI is the expected value of the information from the test, $EMP_{\text{with test}}$ is the expected profit of the decision using the information from the test, and $EMP_{\text{no test}}$ the expected profit without using it. The cost of the test is not considered in this calculation.

The net expected value of the test is the value of the test minus the cost of the test,

$$NEVI = EVI - TC \tag{7.2}$$

where $NEVI$ is the net expected value of the information from the test and TC its cost.

Consider the decision for developing the squeegee in Example 7.13. First, we will calculate the value of an ideal test that always predicts correctly if the consultant's assumptions are valid. The test cost will not be considered in this step. If the company knew that the assumptions were valid, then it would develop the product and make a profit of $50 million. This would happen 75% of the time, because the probability that the assumptions are right is 0.75. In the remaining 25% of the time the company would not develop the product and avoid any loss. Therefore, the expected profit with perfect information from the test is,

$$EMP_{\text{with perfect information}} = \$50 \cdot 0.75 = \$37.5$$

If there was no information then the company would always develop the product because this decision has a positive expected profit (see Figure 7.21). Then the company would make $50 million profit 75% of the time, and lose $40 million the rest 25% of the time. Therefore,

$$EMP_{\text{no test}} = \$50 \cdot 0.75 - \$40 \cdot 0.25 = \$27.5$$

Therefore, the value of perfect information is $10 million. This amount is equal to the expected loss that the company would avoid if it could predict with certainty when the consultant's assumptions were invalid. The company should not pay more than the value of perfect information for a test.

We will calculate the value of the information from the actual test, which is not perfectly accurate. This value will be calculated from Equation (7.1). From Figure 7.21, the expected monetary profit if we perform the test is,

$$EMP_{\text{with information from test}} = \$35.75$$

This amount does not include the test cost.

The expected profit without performing the test is,

$$EMP_{\text{no test}} = \$27.5$$

The value of the imperfect test is, $\$35.75 - \$27.5 = \$8.25$ million. This value is lower than the value of perfect information, which is $10 million. The net value of the test is obtained by subtracting the test cost: $NEVI = \$8.15$ million. Clearly, the test is worth its cost.

Note that the value of the information from a test depends on both the accuracy and the impact of the information on the decision. For example, if the squeegee were too expensive to develop anyway, then the test to find if the consultant made the right assumptions would be useless.

This example brings about the important concept of a _material distinction._ A material distinction or choice exists when the decision maker would choose one option over the other under prescribed conditions but will switch if the conditions change. In this example, the term "conditions" applies to the probabilities of all possible values of the profit. When distinctions are not material, deterministic dominance exists as presented earlier in this section and the decision maker will always prefer one option regardless of the probabilities of the outcomes of a decision. Information is useless by default when deterministic dominance exists and no calculations need to be carried out.

Challenges and advances in solving practical decision problems in engineering design

In most practical engineering design decisions, there are many alternative designs. Often, the backward induction method cannot solve these problems because the computational cost is inordinate. Three reasons are:

a) Many alternative designs must be evaluated. The number of alternative designs becomes too large when there are many decision (or design) variables (e.g., 50). This problem is exacerbated by continuous variables, which can allow for infinite number of solutions. Either complicated integrals are involved or discretization needs to be performed for tractability.
b) Many uncertain events influence the performance of each design.
c) The performance of a design is predicted using expensive complex models, such as the one in Figure 7.2. Such models may take between few minutes to several days of CPU time to predict the consequences of a design decision on the performance of one design.

In order to demonstrate the difficulty of solving such decision problems, consider that we want to minimize the expected value of an objective function for the truck in Figure 7.2. This function depends on the weight of the truck and the sound pressure level in the cabin at 60 miles per hour. This truck has 40 decision variables, which are the mean values of its plate gages. The true values of the plate gages are random because of manufacturing variability.

Consider that the mean value of each plate gage can only assume five equally spaced values. In order to account for variability, we assume that the true value of each plate gage can be equal to its mean value or deviate from it in both directions by a fixed increment. Therefore, each plate gage can assume three values with probability 1/3 each.

Suppose that we try to solve this design problem by backward induction. A decision maker must consider $5^{40} = 9.1 \times 10^{27}$ options (alternative designs) in this decision problem. Each design corresponds to one combination of the mean values of the plate gages. To compute the mean value of the objective function for one design alternative, we should calculate the sound pressure level and the weight for $3^{40} = 1.2 \times 10^{19}$ combinations of plate gages. Assume that it takes 210 seconds to analyze each design using a commercial finite element analysis program such as MD-NASTRAN. Then it will take 2.3×10^{49} seconds or 2.7×10^{44} days CPU time to compute the objective function for all design alternatives. In comparison, this is orders of magnitude more than the current estimated age of the universe!

Therefore, it is too expensive to solve such design decision problems by direct application of the methods described in section 7.3. Efforts to reduce the cost of solving such decision problems include,

a)   Development of deterministic efficient methods for calculation of the performance attributes of a particular design given the values of its design variables. Such methods use response surfaces, local approximation techniques, and reanalysis methods (Zhang et al. 2009). The latter include the combined approximation, the parametric reduced order and the modified combined approximation methods.

b)   Development of efficient probabilistic methods for calculation of the mean values the performance attributes of a given design, such as variance reduction techniques in Monte Carlo simulation (Law, 2007) and probabilistic reanalysis (Farizal and Nikolaidis, 2007, Fonseca, et al. 2007).

c)   Instead of analyzing and comparing all alternative designs, we find the optimum by analyzing a small subset of these designs. This is done by using efficient methods for exploration of the design space, such as genetic algorithms (Gen, 2000), simulated annealing, branch and bound methods and gradient-based methods (Vanderplaats, G., 1999).

Advances in the above three fronts have made it feasible to solve practical design decision problems.

7.4   Performing Sensitivity Analysis

7.4.1   Introduction

The objectives of sensitivity analysis are to identify those variables that significantly affect the consequences of a decision, and to appraise the quality of the optimum decision strategy by assessing its robustness to changes in the variables. In this step, the decision maker analyzes the effect of the problem variables on the consequences of a decision and on the optimum strategy. The problem variables include revenues, costs and probabilities.

We perform sensitivity analysis before solving the decision problem in order to construct a requisite decision model. This is a model that contains everything essential for solving the decision problem but no superfluous information. We also perform sensitivity analysis after solving the problem in order to decide whether to accept the optimum decision strategy or to refine the decision model and solve it again.

There are various ways to perform sensitivity analysis. This section presents and demonstrates procedures for sensitivity analysis. First, we study how to assess the sensitivity of a decision to the problem definition and structure. Then, we review procedures to assess the effect of the problem variables on the expected profit. Finally,

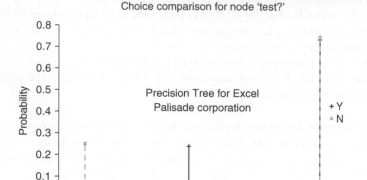

Figure 7.25 Risk profiles of options to develop product without testing prototypes (dashed bars) and testing prototypes before development (continuous bars).

we explain how to analyze the effect of imprecision in probabilistic models on the selected optimum strategy.

7.4.2 Sensitivity to the definition, framing and structure of the problem

In section 7.2, we explained that the definition of a problem is critical to the quality of the decision. In the beginning of the decision analysis, the decision maker must understand the situation and correctly identify the decision problem. It is important for the decision maker to consider the following questions during this process;

- What is the catalytic question in this decision?
- What do I want to accomplish?
- Am I working on the right problem?
- Are there better options than the ones already considered?

After solving the decision problem, the decision maker should consider the above questions and redefine and solve the problem again if he/she feels that this will improve the decision. We will demonstrate how to appraise the quality of the decision model in the problem with the household product manufacturer in Section 7.3, example 7.13. In this problem, the manufacturer wants to decide whether to a) test prototypes of a new design of a squeegee before deciding whether to develop the product, or b) whether to skip the test.

Figure 7.25 shows the risk profiles of the strategies of developing the product without testing and of testing prototypes first. It is better to test the product because testing is affordable and dramatically reduces the probability of losing $40 million from 25% to 3%. Practically, the "test" option deterministically dominates the option

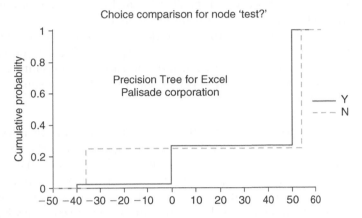

Cumulative probabilities for decision tree 'Product Planning'

Choice comparison for node 'test?'

Precision Tree for Excel
Palisade corporation

——— Y
– – – N

Figure 7.26 Cumulative risk profiles of the options in Figure 7.25.

of skipping the test according to the cumulative risk profiles of these options in Figure 7.26.

Figure 7.25 shows that both strategies are very risky because the company will either make a large profit or incur a big loss by following both strategies. Although a manager could significantly reduce the risk of loss by performing a relatively inexpensive test, the manager could still hesitate to assume the risk of losing $40 million.

The manager ought to redefine the decision problem to reduce the risk and the size of the potential loss. The manager could revise the original question "test the product or develop the product without testing?" to "how can my company make a big profit at a low risk?" The manager could also consider more options such as those below:

- Develop a conventional, affordable design instead of the original innovative design.
- Reduce the cost of the original design by outsourcing its development.
- Share the cost of development of the original design with another company.
- Develop a different product such as a glass cleaner.

In the early stages of product development, there is significant imprecision in the estimates of the problem variables. For example, decision makers are uncertain about the cost of a product and its revenue. In addition, a decision maker is imprecise about the probabilities of the outcomes of uncertain events. It is important to quantify this imprecision and assess its effect on the expected profit. The simplest way to model imprecision is by using intervals. For example, the decision maker could believe that the development cost could be as low as $37.5 million and as high as $62.5 million. This corresponds to a ±25% variation of the cost from its nominal estimate, which is $50 million.

The decision maker must also quantify the imprecision in the probabilities of the outcomes of uncertain events. Here we explain that in practice, people quantify imprecision in probabilities of outcomes by using probability bounds. As Chapters 2 and 6 explained, a risk neutral decision maker's subjective probability of an outcome is the fair price of a ticket that is worth \$1 if the outcome occurs and zero otherwise. Fair price of a ticket is the price for which the decision maker would happily agree to both buy and sell the ticket. When the decision maker is imprecise about the probability of an outcome, he/she would buy the above ticket for a lower price than the selling price. The difference between the selling and buying prices reflects the decision maker's aversion to ambiguity in probabilities. Then, the decision maker quantifies uncertainty in an outcome using a lower and an upper probability, which are equal to the buying and selling prices of the ticket.

Example 7.14:   How to quantify the effect of imprecision in product development
Here we demonstrate how to assess the effect of imprecision in the variables of the squeegee decision problem on its expected profit. Table 7.3 shows the range and the base values of the variables in this problem. We will estimate the minimum and maximum values of the expected profit when the variables vary in their ranges in Table 7.3. Calculation of these values requires the solution of two optimization problems to minimize and maximize the expected profit of each option. Fortunately, in the squeegee development decision problem, we can identify the worst case and best case scenarios and calculate the expected profit for these scenarios, thereby avoiding the cost of numerical optimization. The worst case scenario materializes when the revenues and costs are equal to their lower and upper bounds, respectively. In the worst case scenario, the probability that the consultant's assumptions are valid is equal to its lower bound and so are the probabilities of the test yielding the correct conclusion. These three probabilities are equal to their upper bounds in the best case scenario.

Table 7.3 Ranges of values of variables in the squeegee development decision. Amounts are in millions of dollars.

Variable	Base value	Low	High
Revenue, valid assumptions	100	75	125
Development cost	50	37.5	62.5
Revenue, invalid assumptions	10	7.5	12.5
Test cost	0.1	0.075	0.125
$P(V)$	0.75	0.6	0.9
$P(P/V)$	0.98	0.95	0.99
$P(P/V^C)$	0.1	0.05	0.2

A decision maker calculates the expected profit of the two options (test and do not test) for the three scenarios in Table 7.3 by inserting the values of the variables and probabilities in the decision tree in Figure 7.20 or by using probability calculus. The expected profit in the best case scenario is calculated as follows. The probability of the product passing the test is,

$$P(P) = P(P/V)P(V) + P(P/V^C)P(V^C) = 0.99 \cdot 0.9 + 0.05 \cdot 0.1 = 0.896$$

The conditional probability that the assumptions are valid given that the product passed the test is calculated by using Bayes' rule,

$$P(V/P) = \frac{P(P/V)P(V)}{P(P)} = \frac{0.99 \cdot 0.9}{0.896} = 0.9944$$

Similarly the conditional probability that the assumptions are valid given that the test was unsuccessful is,

$$P(V/P^C) = \frac{P(P^C/V)P(V)}{P(P^C)} = \frac{0.01 \cdot 0.9}{1 - 0.896} = 0.0865$$

The expected profit of the test option is,

$$E(\text{Profit}) = E(\text{Profit}/P) \cdot P(P) + E(\text{Profit}/P^C) \cdot P(P^C)$$

The expected profit given that the product passed the test is,

$$E(\text{Profit}/P) = E(\text{Profit}/V \cap P) \cdot P(V/P) + E(\text{Profit}/V^C \cap P) \cdot P(V^C/P)$$

where $E(\text{Profit}/V \cap P)$ is the conditional expectation of the profit given both that the product passed the test and that the assumptions are valid. Conditional expectation, $E(\text{Profit}/V^C \cap P)$, is defined similarly.
The expected profit given that the product failed the test is,

$$E(\text{Profit}/P^C) = E(\text{Profit}/V \cap P^C) \cdot P(V/P^C) + E(\text{Profit}/V^C \cap P^C) \cdot P(V^C/P^C)$$

Therefore, the decision maker can calculate the expected profit from the equation below,

$$\begin{aligned} E(\text{Profit}) = &\{E(\text{Profit}/V \cap P) \cdot P(V/P) + E(\text{Profit}/V^C \cap P) \cdot P(V^C/P)\} \cdot P(P) \\ &+ \{E(\text{Profit}/V \cap P^C) \cdot P(V/P^C) + E(\text{Profit}/V^C \cap P^C) \cdot P(V^C/P^C)\} \\ &\cdot P(P^C) \end{aligned}$$

Consider that the product passed the test. The expected profits in the two cases where the assumptions are valid and invalid are equal to the revenue minus the development and test costs:

$$E(\text{Profit}/V \cap P) = \$125 - \$37.5 - \$0.075 = \$87.425$$
$$E(\text{Profit}/V^C \cap P) = \$12.5 - \$37.5 - \$0.075 = -\$25.075$$

The profit in the other two cases where the product failed the test is equal to minus the test cost.

$$E(\text{Profit}/V \cap P^C) = E(\text{Profit}/V^C \cap P^C) = -\$0.075$$

Table 7.4 Expected profit for base, best and worst case scenarios. Amounts are in millions of dollars.

Variable	Base Scenario	Best-case scenario	Worst-case scenario
Revenue, valid assumptions	100	125	75
Development cost	50	37.5	62.5
Revenue, invalid assumptions	10	12.5	7.5
Test cost	0.1	0.075	0.125
$P(V)$	0.75	0.9	0.6
$P(P/V)$	0.98	0.99	0.95
$P(P/V^C)$	0.1	0.05	0.2
Strategy A: expected profit	35.65	77.76	2.6
Strategy B: expected profit	27.5	76.25	−14.5

We plug in the values of the probabilities and expected profit into the above equation for the expected profit in order to find the expected profit for strategy "test and then decide whether to develop the product",

$$
\begin{aligned}
E(\text{Profit}) &= [\$87.425 \cdot 0.9944 - \$25.075 \cdot (1 - 0.9944)] \cdot 0.896 \\
&\quad +[-\$0.075 \cdot 0.0865 - \$0.075 \cdot (1 - 0.0865)] \cdot (1 - 0.896) \\
&= \$77.76
\end{aligned}
$$

The expected profit of the option to skip the test is,

$$
\begin{aligned}
E(\text{Profit}) &= E(\text{Profit}/V)P(V) + E(\text{Profit}/V^C)P(V^C) \\
&= (\$125 - \$37.5) \cdot 0.9 + (\$12.5 - \$37.5) \cdot 0.1 \\
&= \$76.25
\end{aligned}
$$

Table 7.4 shows the expected profit for each scenario. Two conclusions can be drawn from this table. First, imprecision in the variables induces a large imprecision in the expected profit. For example, the expected profit of the option to test the product to find out if the assumptions are valid can be as low as $2.6 million and as high as $77.76 million. Second, the option to test is better than the option not to test the product regardless of what scenario materializes.

However, the last observation does not prove that testing the product is better than not testing it for all possible combinations of values of the variables and probabilities in their ranges in Table 7.4. In order to check if the option to test the product is always better, we should minimize the difference of the expected profits of the two options while the values of variables and probilities are constrained to vary in their ranges.

In general, we must identify those sources of imprecision that significantly affect the expected profit and collect additional information about these variables in order to reduce imprecision. The next subsections present procedures to assess the effect of imprecision in the variables and the probabilities of the profits and identify important variables.

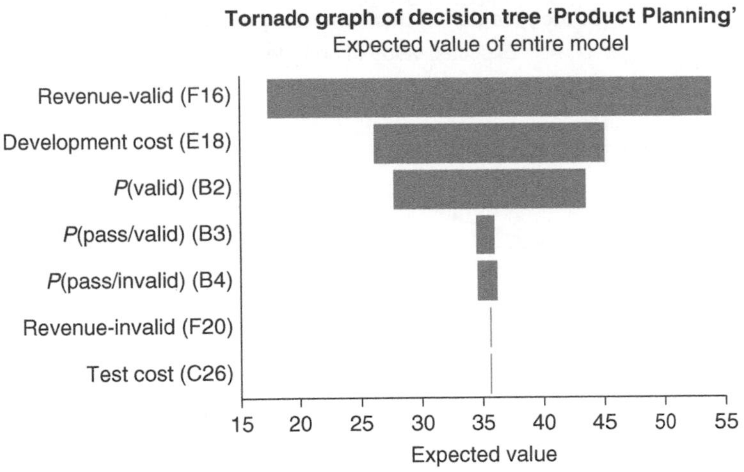

Tornado graph of decision tree 'Product Planning'
Expected value of entire model

Figure 7.27 Tornado diagram of optimum strategy (test first, and then develop the product only if it passes the test).

7.4.3 One-way sensitivity analysis

In a one-way sensitivity analysis, a decision maker investigates the effect of each variable on the expected profit and the selection of the optimum strategy by varying one variable at a time. This type of analysis accounts for the effect of the variation of one variable and it ignores the effect of simultaneous variation of many variables. The results of this analysis are presented in *tornado diagrams* and *strategy regions*. Tornado diagrams compare the effect of each variable on the expected profit of a course of action. Strategy regions display the variation of the expected profit of alternative strategies with each variable.

The rest of this subsection demonstrates the application of tornado diagrams and strategy regions in the squeegee development decision. We created these diagrams by varying the problem variables one at a time in their ranges in Table 7.3.

Tornado diagrams help decision makers identify the most influential variables for a particular option. Figure 7.27 is a tornado diagram of the optimum strategy in the squeegee development decision, which is to test the product first and then develop it only if it passes the test. The figure shows that the revenue from the product development is the most important variable, followed by the development cost, and the (unconditional) probability that the consultant's assumptions are valid. The decision maker should invest resources to accurately estimate these quantities. The conditional probabilities that the product will pass the test given that the consultant's assumptions are valid and invalid, respectively, also affect the expected profit but to a lesser degree than the previous variables. The test cost and the revenue, given that the assumptions are invalid are unimportant.

A decision maker can study the effect of the problem variables on the selection of the best strategy. Figure 7.28 shows the variation of the expected profit of the optimum strategy (test and then decide whether to develop the product) and the strategy to

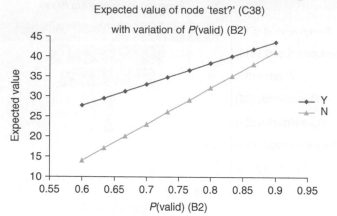

Figure 7.28 Variation of expected profit of strategies "test and then decide whether to develop" and "decide without testing" with the probability that the consultant's assumptions are valid.

skip the test with the probability that the consultant's assumptions are valid. It is observed that the expected profits of both strategies increase with the probability that the assumptions are valid, as expected. The expected profit of the decision to proceed without performing a test increases faster than the profit of the decision to test. The reason is that the value of the information from the test is small when the decision maker is very confident in the consultant's assumptions, before the test.

7.4.4 Two-way sensitivity analysis

In a two-way sensitivity analysis, a decision maker varies two variables simultaneously and observes the effect on the expected profit of one or more strategies. The results are displayed in three-dimensional plots, or in regions in which alternative strategies are optimum plotted in the two-dimensional space of the variables.

Figure 7.29 shows the expected profit of the optimum strategy in the squeegee decision as a function of the two most important variables. The expected profit increases with both quantities. It is observed that the rate of increase of the expected profit with respect to each quantity is not affected significantly by the value of the other.

Figure 7.30 shows the regions where strategies "test" and "do not test" are optimum in the space of the two variables representing the revenue when the consultant's assumptions are valid and the *probability* that these assumptions are valid. This plot was constructed by calculating and comparing the expected profit of both strategies "test" and "do not test" for pairs of values of the above two variables on a grid covering the range of these variables. Strategy "test" is optimal over the entire region. One reason is that the test is both accurate and affordable.

Figure 7.31 shows the same results as Figure 7.30 when the test costs $10 million. This figure shows that the selection of the optimum strategy depends on the unconditional probability that the consultant's assumptions are valid. For probabilities less

Sensitivity of decision tree 'product planning'
Expected value of node 'test?' (C38)

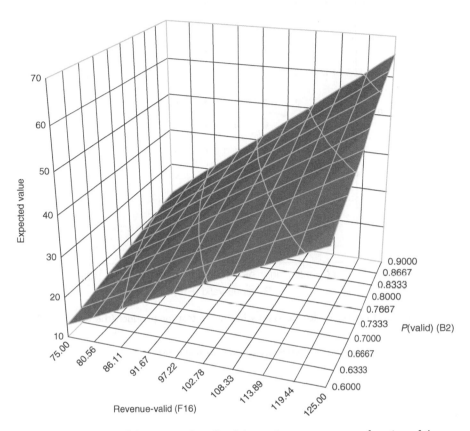

Figure 7.29 Variation of the expected profit of the optimum strategy as a function of the revenue given that the consultant's assumptions are valid and the probability of this outcome.

than 73%, testing is still better than making a decision without testing. However, if this probability exceeds 73%, the decision maker does not get his/her money's worth from the test and is better off skipping it.

7.4.5 *Sensitivity of the selection of the optimum option to imprecision in probabilities*

It is difficult for a decision maker to estimate the probability of an event, when the decision maker has limited information or the event is rare. When decision makers are imprecise about the probabilities of the outcomes of uncertain events, they often prefer to express their beliefs in terms of probability bounds instead of a single probability. For example, suppose that a facilitator asks an expert for the probability that a serious accident will occur in a nuclear power plant in the US during the next 10 years. A range

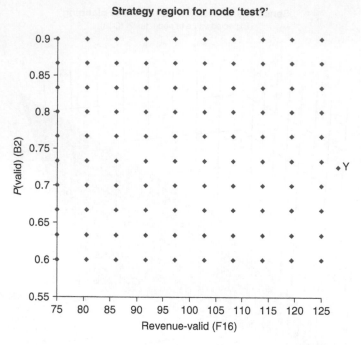

Figure 7.30   Strategy region for decision whether to test the product (the test costs $100,000).

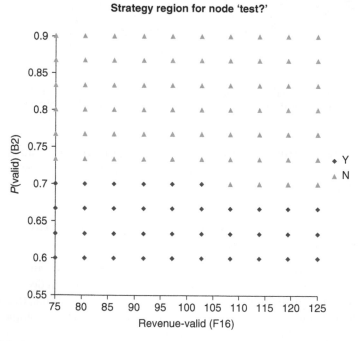

Figure 7.31   Strategy region for the decision whether to test the product (test costs $10 million).

of probabilities, such as $[10^{-12}, 10^{-6}]$ could represent better the decision maker's beliefs than a precise probability.

When we specify probabilities up to their bounds, we can only determine the expected profits of alternative courses of action up to their bounds. In this case, we need a decision rule in order to rank alternative courses of action and make decisions.

We could compare two options A and B according to the extreme values of their expected profits if the probabilities of the uncertain events vary between their bounds. Specifically, if the minimum expected profit of option A is still higher than the maximum expected profit of option B, then we will conclude that option A is better, and vice versa. We cannot decide which option is better when the ranges of the expected profits of the two options overlap.

However, this approach exaggerates the ranges of the expected profits, thereby leading frequently to indecision. The reason is that the approach varies independently the same probability, when finding the minimum and maximum values of the expect profit. For example, consider that we solve the following pair of optimization problems,

Find $p_1, \ldots, p_n$
to minimize $E(\mathrm{Profit_A})$ $\qquad$ (7.3)
so that $p_{i\min} \le p_i \le p_{i\max}$ $i = 1, \ldots, n$

and,

Find $p_1, \ldots, p_n$
to maximize $E(\mathrm{Profit_B})$ $\qquad$ (7.4)
so that $p_{i\min} \le p_i \le p_{i\max}$ $i = 1, \ldots, n$

Variables $p_1, \ldots, p_n$ are the imprecise probabilities, $p_{i\min}$ and $p_{i\max}$ their bounds, and $E(\mathrm{Profit_A})$ and $E(\mathrm{Profit_B})$ the expected profits of the two options. An optimizer changes probability p_i independently when solving each problem (7.3) and (7.4), while in reality this probability has the same value in both scenarios.

An unbiased comparison of two options should rely on a measure of the <u>difference</u> of their profits. Walley (1991) suggested the following rule for comparing two options A and B:

- Option A is better than B if and only if the minimum value of the expectation of the difference of the utilities of options A and B is positive.
- Option B is better than A if and only if the maximum value of the expectation of the difference of the utilities of options A and B is negative.
- Otherwise, there is not enough information in order to decide which option is better.

We adopt this decision rule in order to compare alternative options, when each option is valued according to the imprecise expectation of its profit. Figure 7.32 explains the proposed rule. When the minimum expected value of the difference of the profits of alternative options A and B is positive then option A is preferable (Case 1). When the maximum of the same difference is negative, then option B is preferable (Case 2). Otherwise, we cannot conclude which option is better (Case 3).

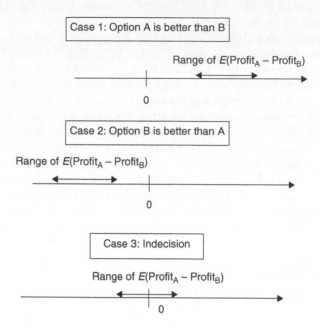

Figure 7.32 Decision rule for imprecise probabilities.

In order to compare two options in a decision where probabilities are imprecise, we find the extreme values of the expected value of the difference of the profits of the two options. For this purpose, we formulate and solve the following pair of optimization problems,

Find $p_1, \ldots, p_n$
to minimize(maximize) $E(\text{Profit}_A - \text{Profit}_B)$ (7.5)
so that $p_{i\min} \leq p_i \leq p_{i\max}$ $i = 1, \ldots, n$

where $p_1, \ldots, p_n$ are the imprecise probabilities in this decision problem, and $p_{i\min}$ and $p_{i\max}$ their lower and upper bounds. This optimization problem should be solved using a global optimizer.

Example 7.15: How to make a decision when probabilities are imprecise
The decision maker in examples 7.13 and 7.14 knows only the bounds of the probabilities of the following events:

a)   The consultant's assumptions are valid
b)   The product will pass the test given that the consultant's assumptions are valid
c)   The product will pass the test given that the assumptions are invalid.

Table 7.3 specifies the bounds of these probabilities. Figure 7.33 shows the expected profits of options "test" and "not test" for the values of the probabilities $P(V)$, $P(P/V)$

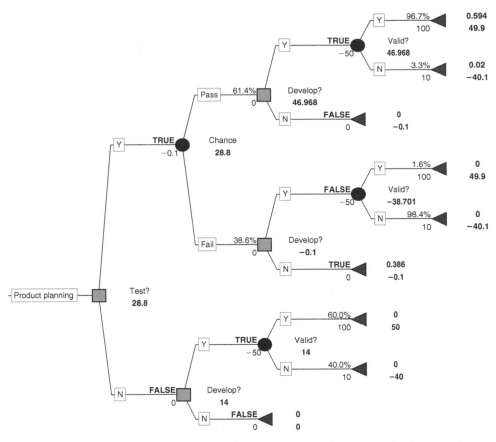

Figure 7.33 Scenario in which the value of the test is maximum in squeegee development decision (Probabilities are specified up to their bounds in Table 7.3.).

and $P(P/V^C)$ that maximize the expectation of the difference of the profits of these options. This case occurs when the decision maker has low confidence in the consultant's assumptions and high confidence in the test. The expectation of the difference of the profits of the "test" and "no test" strategies is 14.8 million. The minimum value of this expectation is 0.85 million according to Figure 7.34. Therefore, option "test and then decide" is the best even when probabilities are imprecise.

Note that the extreme values of the expected profit of option "test" are $25.2 million and $44.25 million, while the corresponding values of option "no test" are $14 million and $41 million. The ranges of the expected profits of the two options overlap. If we had tried to compare the above options only on the basis of these ranges, we would have been unable to find the best option. This comparison exaggerates the effect of imprecision on the choice of the optimum option because it varies the probabilities of the outcomes independently, which is unrealistic. The correct way to compare alternative options is by using the decision rule in Figure 7.32.

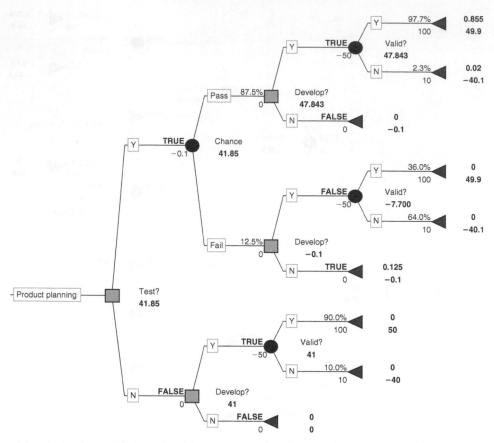

Figure 7.34 Scenario in which the value of the test is minimum in squeegee development decision
(Probabilities are specified up to their bounds in Table 7.3.).

Perspectives on modeling imprecision in probabilistic models

Some authors criticize the use of imprecise probability in decision making because this
approach can lead to indecision. They argue that a rational decision maker must be
able to estimate the probability of an uncertain event if he/she is motivated to consider
the decision problem seriously (French 1986, O'Hagan and Oakley, 2004). The logical
framework for estimating probabilities explained in Chapter 6 of this book, can help
a decision maker in this task. When facing a decision problem, one must select an
option; even when he/she does nothing, he/she selects inadvertently the default option.
Instead, the decision maker should evaluate his/her options and choose the best one
based on his/her beliefs and the available information.

Walley (1991), de Cooman and Troffaes (2004), argued that imprecise probability
reflects the fact that, sometimes, people do not have enough information in order to
estimate probabilities precisely. A decision approach based on imprecise probability
does not force a decision maker to make assumptions that are not supported by the

available evidence, and may alert a decision maker that he/she needs to collect more information when it is required.

Aughenbaugh, and Paredis (2006, 2007) proposed a method for making design decisions when probabilities are imprecise. They showed that an imprecise probability method could protect a decision maker from large losses resulting from poor estimates of the probabilities. They argued that the classical probabilistic method is more vulnerable to errors in probability estimates than a method using imprecise probabilities.

Haftka et al. (2006) introduced an approach for testing methods for decision-making under uncertainty. The testing approach uses data for which the generating mechanism and probability distribution is unknown. The approach simulates a very large number of decisions and outcomes of uncertain events to test methods. Pandey and Nikolaidis (2008) presented a procedure that follows the above approach to experimentally test methods for decision under uncertainty. They considered a decision problem involving two decision makers, an inventor and a financier who want to develop and market a new device. Real-life data for simulating the outcomes of the project were collected using 133 slider-crank mechanisms that undergraduate students constructed. The mechanisms were constructed and measured to simulate the entire risky venture of developing and marketing a new device or product. The data obtained was used to simulate thousands of decisions of the inventor and the financier on a computer, using different methods for decision under uncertainty; standard probability, imprecise probability and Bayesian probability. These methods were then judged on the basis of the expected utilities that they produced when used by the two decision makers and also of their sensitivities to changes in the amount of available information or the risk attitudes of the decision makers. In general, the imprecise probability method was found insensitive to the changes in the problem parameters. A Bayesian probability method, although sensitive to prior models of uncertainty, produced the highest utilities in general, and it was also found insensitive to the information level of a decision maker, when a good prior was chosen.

7.5 Modeling Preferences

7.5.1 *Motivation*

Up to this point, this chapter has presented a methodology for decision making that measures the consequences of a course of an action by its expected profit. However, this measure does not really represent the worth of a course of action to a decision maker because different people have different attitudes toward risk.

Consider the drill for oil decision in section 7.2.2. The owner of the land has to decide whether to sell the land or drill for oil. This is a risky decision; if the owner sells the land he/she will make a sure profit of $18,500, and if he/she performs a seismic test before deciding whether to drill he/she could earn up to $190,000 or lose $80,000. Although, the expected profit of drilling and testing is higher than that of selling the land, many people would prefer selling the land than risking $80,000.

Bernoulli demonstrated that the expected profit does not represent people's preferences. He invented a game that although has infinite expected payoff, very few

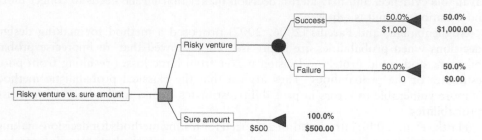

Figure 7.35 Determining a decision maker's attitude toward risk by comparing a risky venture to a sure amount equal to expected payoff of the risky venture.
- A Risk neutral decision maker is indifferent between risky venture and sure amount.
- A Risk averse decision maker prefers sure amount.
- A Risk seeking decision maker prefers risky venture.

people would pay more than $50 to play once. This example is called Saint Petersburg's paradox. Consider the following game. A rich friend of yours, who is honest and always pays his/her debts, flips a fair coin. If the coin lands heads up you get $2 and the game stops. Otherwise your friend flips again and if the coin lands heads up this time, you get $4 and the game stops. Otherwise your friend flips again until he/she gets heads up. In the nth flip, your friend pays you 2^n if the coin lands heads up and the game stops; otherwise he/she flips the coin again. The expected payoff of this game is infinite,

$$\text{Expected payoff} = \sum_{i=1}^{\infty} \left(\frac{1}{2}\right)^i 2^i = \sum_{i=1}^{\infty} 1 = \infty \tag{7.6}$$

and so is the expected profit of this game for any finite ticket price.

If people were making decisions on the basis of the expected profit then they would pay all the money they could gather to play this game once. However, most people would not pay more that $50. The reason is that a player's payoff from this game will be less than or equal to $8 with probability 0.875, less than or equal to $16 with probability 0.94 and less than or equal to $32 with probability 0.97.

The worth of a risky venture to a person depends on the risk profile of the venture and the person's risk attitude. The behavior of a person who faces a risky decision is classified into risk averse, risk neutral and risk seeking (Figure 7.35). Most people are risk averse; they prefer a certain amount to a lottery with the same expected monetary value. For example, most people prefer a certain amount of $500 to a lottery ticket that pays $1,000 or $0, both with probability 0.5. We exhibit risk aversion when we buy insurance against a potential loss such as wrecking of our car. We pay a premium to avoid the risk of losing our car, although the premium is higher than the expected value of the loss.

A risk neutral person is indifferent between a risky venture and a sure amount equal to the expected monetary value of the venture. On the other hand, some people are risk-seeking in some situations. They buy a state lottery ticket for $1 although the expected monetary value of this ticket is often less than 50 cents.

In order to make choices among alternative courses of action we need a function that transforms monetary values to a measure that incorporates our attitude toward risk. This chapter presents a logical framework for developing such a function. First, section 7.5.2 presents three criteria for selecting a course of action: the max min, max max and maximum likelihood criteria. Often people make decisions using these criteria. However these criteria could lead to irrational decisions. Section 7.5.3 presents the utility function for measuring the worth of a risky venture and demonstrates how to use it in decision making. See Clemen (1997), French (1986) and Thurston (2006A) for a presentation of the utility function.

7.5.2 Simple criteria for decision making

Max-min criterion

When a decision maker is imprecise about the probabilities of the outcomes that affect the consequences of a decision, the decision maker may prefer to compare the worth of each course of action on the basis of its minimum possible profit. The best course of action is the one with the maximum payoff, in the worst case scenario. This criterion for selecting the best course of action is called max-min criterion.

The motivation for using the max-min criterion is that it guarantees a minimum profit. Moreover, this criterion does not require assessment of probabilities. On the other hand, a decision maker who uses this overly cautious criterion will miss many opportunities to make a large profit.

We demonstrate the max-min criterion in the drill for oil example. A land owner compares three options in this example: sell the land, drill and test (Figure 7.10). The minimum profit of options "sell" and "drill" are $18,500 and −$70,000. If the owner considers a seismic test, then he/she will always sell the land regardless of the test result. The reason is that the minimum profit of the test option is $8,500. Therefore, selling the land is the best option.

Max-max criterion

The max-max criterion is based on the principle that a course of action is as good as its profit under the best case scenario. An opportunistic decision maker might use this criterion to evaluate his/her options. In the drill for oil example, the maximum profits of options "sell", "drill" and "test" are $18,500, $200,000 and $190,000, respectively (Figure 7.10). Therefore, the owner should drill without performing a seismic test according to this criterion.

This criterion guarantees that the decision maker will never lose an opportunity to make a big profit. On the other hand, it does not protect the decision maker from a disaster because it neglects scenarios that can lead to a large loss, even if these scenarios are very likely.

Maximum likelihood criterion

The principle on which this criterion is based is that a course of action is as good as its profit under the most likely scenario. Many everyday life decisions, such as deciding whether to take an umbrella or wear a coat in the morning, are made according to this

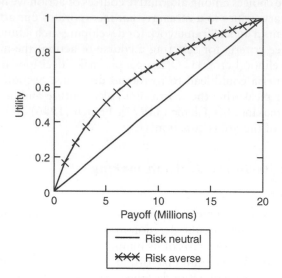

Figure 7.36 Utility functions of a payoff of a risky venture for risk neutral and risk averse decision
makers.

criterion. We guess what is more likely to happen (rain or sunshine) and then choose
the option that yields the best consequence.

In the drill for oil example (Figure 7.11), if the owner decides to drill the most likely
scenario is to find no oil. This will cause a loss of $70,000. If the owner decides to
perform a test, then it is most likely that he/she will get an unfavorable result (No structure). In this case, the owner will sell the land because the most likely scenario is that
there is no oil. The profit in this case is $8,500 (profit from sale of the land minus cost of
the test). Therefore the profits of options "sell", "drill" and "test" under the most likely
scenarios are $18,500, −$70,000 and $8,500, respectively. Consequently, selling the
land is the best option. A drawback of this criterion is that it ignores many scenarios,
even if the consequences of these scenarios are highly desirable or catastrophic.

An important limitation of the above three criteria is that they ignore the likelihoods
of all or some of the outcomes.

7.5.3 *Utility*

Utility is a monotonically increasing function measuring the value of an amount of
money to a decision maker. This function translates money into units of utility (utils).
Although utility is defined up to a linear transformation, we scale utility over the
range of amounts of money in a decision so that the minimum and maximum amounts
have utilities zero and one, respectively. Figure 7.36 shows the shape of the utility
function of risk neutral and risk averse decision makers. The utility function is a
straight line for a risk neutral decision maker. This indicates that the marginal utility
(increase in utility per unit increase in the amount of money) is constant for a risk
neutral decision maker. Most people are risk averse; their utility function is concave
indicating that their marginal utility decreases with the amount of money. Indeed,

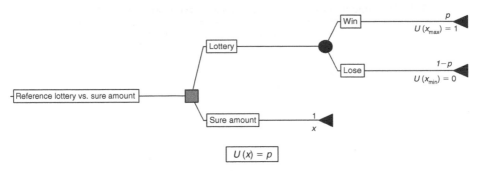

Figure 7.37 Finding the utility of certain amount *x* by comparing it to a reference lottery that pays two amounts that have utilities one and zero.

earning $100,000 improves an average person's life more than earning an additional $100,000.

How do we use utility to choose among alternatives with uncertain payoffs? The next subsection will introduce common sense rules (axioms) that dictate a rational person's choices. From these rules, it will follow that an alternative is as good as its expected utility – not its expected monetary value. Therefore, the following propositions are true;

---

1. Alternative course of action A is at least as good as B if and only if the expected utility of A is greater than or equal to that of B.
2. A decision maker is indifferent between alternatives A and B if and only if these alternatives have equal expected utilities.

Therefore, a decision maker should choose the option with maximum expected utility among the available options.

Utility can be thought of as a probability. Consider that the utility function is scaled so that the utilities of the highest and lowest payoffs in a risky venture are 1 and 0, respectively. Then, the following lemma is true.

Lemma 1: The utility of a sure amount to a decision maker is equal to the probability p that makes the decision maker indifferent between this amount and a reference lottery (Figure 7.37) that pays the highest possible payoff with probability p and the lowest one with probability $1 - p$.

Proof: Consider the lottery that pays the highest possible payoff x_{max} with probability p and the lowest one x_{min} with probability $1 - p$. The decision maker is indifferent between a sure amount x and the lottery. Then, the expected utility of the lottery is equal to the utility of amount x,

$$U(x) = EU(lottery)$$
$$U(x) = p \cdot U(x_{max}) + (1 - p) \cdot U(x_{min}) \tag{7.7}$$
$$U(x) = p$$

where $U(x)$ is the utility of x.

Table 7.5 Amount of money after 12 years after accounting for inflation.

Investment	Market conditions	Final cash position ($1,000)	Utility
Fixed return account	–	$478	0.5
Stock fund	Bull market	$1,558	1
Stock fund	Average	$678	0.7
Stock fund	Bear market	$216	0

Therefore, the utility of amount x is equal to the expected utility of the reference lottery. Q.E.D.

A decision maker's utility is specified by using a lookup table, a mathematical function or a model fit to data. In order to make a risky decision, we only need the utility function over the range of the possible payoffs of the venture. A lookup table presents the utilities of all possible payoffs. For example, Table 7.5 presents all values of an investor's final cash position from two alternative investments and their utilities.

An alternative representation of utility is by assuming a mathematical form such as the exponential and the logarithmic functions. The exponential function is,

$$U(x) = 1 - e^{-\frac{x}{R}} \tag{7.8}$$

where x is the payoff and R is the decision maker's *risk tolerance*. The more conservative the decision maker is the lower is the value of R. We can estimate the risk tolerance from elicited utilities of the decision maker for different payoffs. The utility of an amount becomes one only when this amount is equal to infinity. We can normalize the expression in this equation by the maximum so that this value becomes 1.

Finally, we can fit a model such as a polynomial or an expression for the cumulative function of a common probability distribution to elicited values of the decision maker's utilities (Abbas and Howard, 2005).

Certainty equivalent and risk premium

For each gamble, there is a certain amount that is equivalent to that gamble. This amount is called *certainty equivalent* (CE) of the gamble. Consider a gamble with possible payoffs x_i, $i = 1, \ldots, n$ with corresponding probabilities p_i, $i = 1, \ldots, n$. The certainty equivalent of the gamble is,

$$CE = U^{-1} \left\{ \sum_{i=1}^{n} p_i U(x_i) \right\} \tag{7.9}$$

where $U(x)$ denotes the utility of x and $U^{-1}(c)$ is the inverse utility function, i.e. the amount of money whose utility is equal to c.

Figure 7.38 illustrates the calculation of the certainty equivalent of a gamble that pays $20 million and 0, both with probability 0.5. The utility of $20 million is equal to

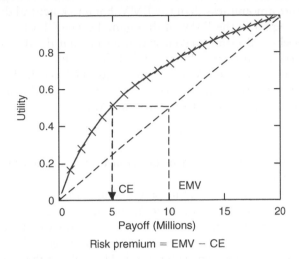

Risk premium = EMV − CE

Figure 7.38 Calculation of the CE of a gamble that pays $20 million and $0, each with probability 0.5.
- CE is the certain amount that is equivalent to the gamble.
- Risk premium is the maximum amount that the decision maker would pay in order to receive the EMV of the gamble for certain, thereby avoiding gambling.

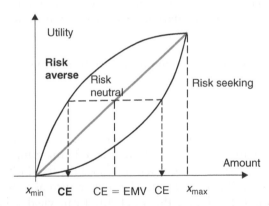

Figure 7.39 Relation of CE and EMV of a gamble that pays x_{min} and x_{max} with probability 0.5.
- Risk averse decision maker: **CE** < EMV
- Risk neutral decision maker: CE = EMV
- Risk seeking decision maker: CE > EMV.

one, while the utility of $0 is zero. The expected utility of the gamble is 0.5. Therefore, the CE of the gamble is $5 million, because this amount has utility 0.5. Note that the CE of the gamble is smaller than its EMV. The reason is that the decision maker is risk averse; a sure amount equal to the EMV of a gamble is better than the gamble itself to the decision maker.

Figure 7.39 shows the relation of the CE and the EMV for three decision makers with different risk attitudes. The CE of a gamble is lower than its EMV for the risk averse decision maker. This means that this decision maker would exchange the opportunity to gamble for a sure amount less than the EMV. For a risk neutral decision maker, the CE of a gamble is equal to the EMV of that gamble. The CE of a gamble exceeds its EMV for a risk seeking decision maker, which means that the decision maker prefers gambling than receiving for certain the EMV of the gamble.

The difference between the EMV of a gamble and its CE is called *risk premium*. This is the maximum amount that a risk averse decision maker would pay is order to avoid gambling and get a certain amount equal to the EMV.

Examples

Example 7.16:   Making an investment decision using expected monetary value and expected utility criteria
(All amounts in this example are in thousands of dollars.)

An investor has saved $400 and wants to invest them for her retirement. She plans to retire after 12 years at which time she will transfer the money to a fixed return account and start spending it. The investor wants to decide whether to invest the money in a fixed return account or a domestic stock fund. The yield of the fixed return account is 4.5%. The yield of the stock market fund over the 12 year period depends on the market conditions, and it can be 15% in a bull market, −2.5% in a bear market and 7.5% if the market performance is average. The annual inflation rate is expected to be 3%.

The investor estimates that the probabilities of a bull and a bear market are both 0.33 and the probability that the market performance will be average is 0.34. Which investment is better?

Equation (7.10) shows how to compute the inflation adjusted cash position x after 12 years,

$$x = (1 + i - i')^n x_0 \tag{7.10}$$

where i is the yield rate, i' the annual rate of inflation, n the investment period (years) and x_0 the initial investment. Table 7.5 shows the inflation adjusted amount of money of the investor after 12 years for the stock fund and the fixed return account.

The investor faces a difficult decision. The fixed return account is safe but it yields a low return. On the other hand, the stock fund can yield a high return but also a significant loss. Indeed, it is difficult to retire with $216. We will select the best investment using both the expected monetary value and the expected utility criteria.

Figure 7.40 shows the decision tree using the EMV criterion. The EMV of the stock fund and the fixed return account are $816 and $478, respectively. Therefore, the stock fund is the best choice according to this criterion.

In order to select an investment using the expected utility criterion we need to elicit the investor's utilities of all possible cash positions from the two investments. Suppose that we want to estimate the utility of $478. First we assign a utility of 0 to the minimum cash position and 1 to the maximum cash position. Then we ask the investor to compare a lottery that yields $1,558 with probability p and $216 with probability $1 - p$ with a sure amount of $478 (Figure 7.37). If p is small, the investor will prefer the sure amount. We increase p incrementally until the investor becomes indifferent between

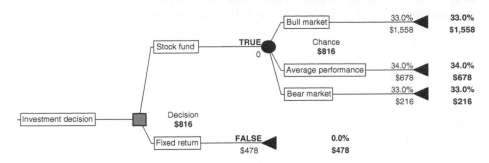

Figure 7.40 Choosing among two investment opportunities using the EMV criterion.

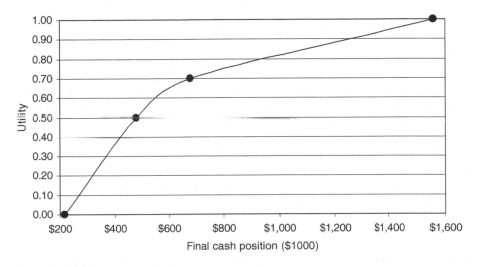

Figure 7.41 Utility curve in the investment example.

the lottery and the sure amount. Then, the utility of $478 is equal to probability p (Equation 7.7). Figure 7.41 shows the utility function constructed by using this method. The utilities of $478 and $678 are 0.5 and 0.7 according to this figure.

The expected utility of the stock fund is,

$$EU(\text{stock fund}) = 0.33 \cdot U(\$1,558) + 0.34 \cdot U(\$678) + 0.33 \cdot U(\$216) = 0.568$$

The expected utility if the fixed return investment is,

$$EU(\text{fixed return}) = U(\$478) = 0.5$$

Therefore, the investor should invest in the stock fund based on her estimated probabilities and utilities.

Example 7.17: Finding certainty equivalent and risk premium
A decision maker's utility is given by the following equation,

$$U(x) = 1 - e^{-\frac{x}{2,000}} \tag{7.11}$$

Find the CE and the risk premium of a gamble that pays $8,000 with probability 0.8 and zero otherwise.

Solution:
The utility of the certainty equivalent is equal to the expected utility of the gamble,

$$U(CE) = 0.8 \cdot U(\$8,000) + 0.2 \cdot U(0) = 0.7853$$

Therefore, $CE = -\$2,000 \cdot \ln(1 - 0.7853) = \$3,077$. This means that the decision maker is indifferent between a sure amount of $3,077 and a gamble that pays 8,000 with probability 0.8 and zero otherwise.
The risk premium is the difference of the EMV of the gamble and the CE,

$$\text{Risk premium} = \text{EMV} - \text{CE} = 0.8 \cdot \$8,000 - \$3,077 = \$3,323$$

This means that if the decision maker has a certain amount of $6,400, we have to pay him/her $3,323 in order to convince him/her to exchange the certain amount with the opportunity to win 8,000 with probability 0.8.

Example 7.18: Considering the decision maker's risk attitude in software development
Here we solve the software development decision in example 7.12 in section 7.3.1 using the following utility function,

$$U(x) = 1 - e^{-\frac{x}{3}} \tag{7.12}$$

where x is the profit in millions (Figure 7.42). The utility function in Figure 7.42 suggests that the company is very risk averse, especially when dealing with losses. Indeed, utility drops steeply for negative amounts.

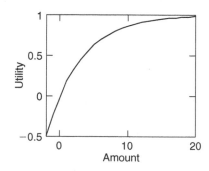

Figure 7.42 Utility of software company profit.

Solution:
Figure 7.43 shows the decision tree where the numbers next to the terminal nodes are the probabilities and the utilities of the net profit in the corresponding scenario. For example, if the company successfully develops the software and sells it to another company, then it will make a profit of $3 million. This profit has utility 0.45. The probability that this scenario will materialize is 0.7 (or 70%). If the company decides to sell the software to the customers, and demand is high, then it will make a profit of $17 million ($20 million sales revenue −$3 million marketing and development cost). This profit has utility 0.97, as indicated by the lower number next to the terminal node. The number next to the terminal node shows that the probability that the company will make a profit of $17 million, given that the product development process is successful, is 0.2.

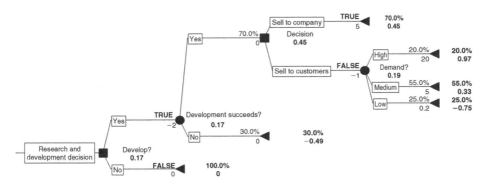

Figure 7.43 Software development decision.

Table 7.6 shows the possible values of the profit that the company can make in this venture and their utility values. These numbers show that the regret from losing $2.3 million is stronger than the delight from earning the same amount.

Table 7.6 Utilities of amounts in software development decision.

Amount ($ million)	Utility
−2.3	−0.751
−2	−0.492
2	0.33
3	0.451
17	0.967

First we compare the options of selling the software to another company and selling the software directly to customers. The expected utility of the profit from selling the software is, $EU(\text{Sell Customers}) = 0.2 \cdot 0.97 + 0.55 \cdot 0.33 + 0.25 \cdot (-0.75) = 0.19$. If the software manufacturer sells the program to another company, it will make a profit of $3 million, which has utility 0.451. The company prefers selling the software to another company, because this option has higher utility than selling it directly to customers.

The expected utility of the option of developing the software is $EU(\text{Develop software}) = 0.7 \cdot 0.451 - 0.3 \cdot 0.49 = 0.17$. Since this expected utility is positive, the company should develop the software. The optimum strategy is: Develop the program; sell it to a company if the development is successful.

The above optimum strategy is different than the one that maximizes the expected monetary value of the profit in Example 7.12, section 7.3.1. When making decisions on the basis of the expected utilities of alternatives, the company prefers making a sure profit of $3 million to selling the program directly to customers, which is risky. The reason is that the company is risk averse as evidenced by its utility function in Figure 7.42. The EMV criterion cannot account for the company's risk aversion.

Example 7.19:   Considering the decision maker's risk attitude when making investments

Here we revisit the inventor-financier decision problem (example 7.4) described in section 7.1.1. The inventor has an idea on how to build a slider-crank mechanism and wants the financier to fund the project. The inventor in return offers a fraction, r, of the expected revenue to the financier. The financier has to choose between funding the mechanism project or invest in a bank with a fixed rate of return. Both the financier and the inventor are making decisions under uncertainty as they cannot guarantee the success of the mechanism. A slider-crank mechanism is a single degree of freedom, four bar mechanism as shown in Figure 7.44. The performance characteristics are the stroke and the imbalance angle which determines the time-ratio of the mechanism. Table 7.7 lists the parameters of the decision problem. The inventor wants to find what fraction r to offer the financier.

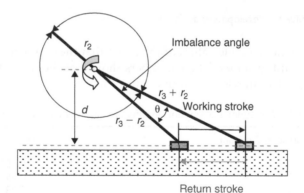

Figure 7.44 A slider-crank mechanism.

Solution:
From Table 7.7 we conclude that if the inventor offers a fraction r to the financier, his profit if the project succeeds is $\$1000(1 - r)$ and if the project fails is $\$10(1 - r)$. Similarly if the project succeeds the financier's profit is $\$1000r - \$50$, and if it fails is $\$10r - \$50$. The financier needs to make sure that the expected utility from the venture exceed the expected utility of a fixed investment, i.e. if he had invested the $50 in the bank at 10% interest rate. Both the inventor and the financier estimate the probability of success of the mechanism, p, from a small number of test mechanisms that the inventor built. The inventor should offer the financier the minimum fraction that is acceptable to the financier. Therefore, the expected utility of the financier's profit from

Table 7.7 Parameters used in the inventor-financier decision problem.

Parameter	Value
Risk tolerance of the investor	$200
Cost of manufacture	$50
Revenue if the mechanism succeeds	$1000
Salvage value if the mechanism fails	$10
Interest rate during the period of development	10%

the venture should be equal to the utility of the investment in the bank for this fraction. The equation that the inventor has to solve to find the optimal r is:

$$pU(1000r - 50) + (1 - p)\, U(10r - 50) = U(50 + 0.1 \cdot 50) \tag{7.13}$$

where the utility function of the financier (known to the inventor) is given by:

$$U(x) = 1 - e^{-\frac{x}{R}}$$

Equation 7.13 is solved numerically to find the optimal fraction that the inventor should offer the financier. Figure 7.45 shows the result of solving this problem for different values of the probability of success. As expected the fraction to be offered to the financier drops as the probability of success increases because even a small fraction is enough to convince the financier to fund the project. It is also seen that the project is infeasible for probability of success approximately below 0.4. This means that the financier would make more profit if he invested the money in the bank even if he was offered 100% of the revenue. Of course, a financier with higher risk tolerance will accept lower fractions of the revenue than the ones in Figure 7.45.

In terms of design insights, the inventor should understand that there is a cutoff probability of success below which the project will not be funded, regardless of the revenue that the inventor offers the financier. He should try to improve his design or use precise manufacturing methods. He might also try to find a different financier with higher risk tolerance.

Example 7.20:   Same as example 7.19 but the inventor has enough information to calculate the probability of success
The inventor in the previous example has now narrowed down the design. He believes that the best design that would fulfill the purpose of the mechanism has a stroke of 60 mm and an imbalance angle of 20 degrees. A deviation of more than 3 mm in stroke or 2.5 degrees in the imbalance angle will render the mechanism useless. He also decides to make the connecting rod three times the length of the crank, i.e. $r_3 = 3r_2$, as shown in Figure 7.44. He proposes that the construction method would be such that the lengths of the connecting rod and the crank and the offset d are independent design variables. He realizes that manufacturing variability leads to uncertainty in the design variables such that they are independently normally distributed with mean at the nominal value and the standard deviation equal to 4% of the mean. Find the optimum fraction that he should offer with this new information.

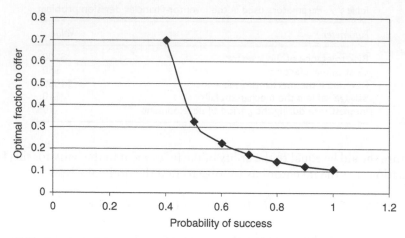

Figure 7.45 Fraction of the revenue offered to the financier as a function of the probability of success.

Solution:

First we need the expressions of the stroke and the imbalance angle of the mechanism:

$$\text{Stroke} = \sqrt{(4r_2)^2 - d^2} - \sqrt{(2r_2)^2 - d^2} = 60 \text{ mm} \tag{7.14}$$

$$\text{Imbalance Angle} = \sin^{-1}\left(\frac{d}{2r_2}\right) - \sin^{-1}\left(\frac{d}{4r_2}\right) = 20 \text{ deg} \tag{7.15}$$

Solving equations 7.14 and 7.15 simultaneously we find that $r_2 = 26.92$ mm, $r_3 = 80.78$ mm and $d = 33.06$ mm. These are the nominal values. Next we generate normal random numbers with means equal to these values of the variables and standard deviation equal to 4% of the value. Monte-Carlo simulation is performed using these parameters. For each set of realizations of the random numbers, the values of stroke and imbalance angle are calculated. These values are tested against that maximum deviation allowed for the stroke and imbalance angle, which are 3 mm and 2.5 degrees respectively. It is seen that about 56% of the mechanisms succeed in about 2000 realizations. Using this information and from Figure 7.45 we conclude that the inventor should offer 25% of the revenue to the financier.

7.5.4   Axioms of utility

Von Neumann and Mongenstern (1947) presented a logical framework based on an axiomatic foundation for making risky decisions. This subsection presents six axioms (common sense rules) that prescribe how a rational person should make risky decisions. Then the subsection proves that this person should maximize the expected utility when making decisions with uncertain payoff. Finally, the subsection presents procedures for elicitation of a person's utility, which are based on the six axioms.

It is easy to defend a theory or an approach that has a strong axiomatic foundation. Suppose that an opponent of utility theory claims that people should not use utility.

You can show the opponent the axioms and ask him/her to explain to you which one is wrong and why.

Some authors have criticized the axioms of utility by arguing that people often violate the precepts of utility theory when making risky choices. Moreover, in some example decision problems, a decision maker who seeks to maximize expect utility makes decisions that look counterintuitive. These examples are called *paradoxes* and are presented and discussed in the end of this subsection.

The axioms of utility prescribe an ideal behavior. In reality, people's behavior deviates from this ideal. For example, often decision makers can only roughly approximate the certainty equivalent of a risky venture or the indifference probability of a reference lottery and a sure amount (see continuity axiom below). This does not reduce the value of decision theory though. If a decision maker tries to act consistently with the ideal behavior prescribed by utility theory then the decision maker will make informed choices that are consistent with his/her preferences and beliefs and will enjoy better outcomes on average than a decision maker who uses other approaches.

Axiom 1: Complete and Transitive Ordering

The decision maker can order all possible consequences of a decision from best to worst, with ties allowed. The ordering is transitive.

According to this axiom, the decision maker can compare two certain amounts and decide that he/she prefers one to the other or that he/she is indifferent between them. This axiom is obviously true when money is the attribute because more money is always better than less money.

Axiom 2: Continuity

The decision maker is indifferent between a *reference lottery* that pays two possible amounts and a sure amount for some probability of the highest amount in the reference lottery. Suppose that the decision maker compares a sure amount $B to a lottery that pays amounts $A and $C with probabilities p and $1 - p$, respectively, where $\$A \geq \$B \geq \$C$. Then, there is always a probability p of earning the largest amount $A for which the decision maker is indifferent between the sure amount and the lottery. This probability is called *indifference probability*.

Some people may contest this axiom. A decision maker could claim that there is no indifference probability for a lottery because one of the consequences of playing the lottery is totally unacceptable. For example, consider that an investment advisor asks an investor to compare an investment that yields a profit of $1 million with probability p and a loss of $500,000 with probability $1 - p$ to a certain amount of $200,000. The investor could say that he/she cannot consider losing $500,000 even with infinitesimal probability. The advisor may explain that it is impractical to avoid risks in real life. Everyday, people assume risks that could result in worst outcomes than losing half a million dollars. For example people drive their cars or cross streets although there is a small chance to die in an accident. If the investor assumes these risks in every day life then he/she must also assume the risk of losing half a million with some low probability.

Another argument against this axiom is that a decision maker could be unable to determine the exact value of the indifference probability. However, if the decision maker considers the consequences of the decision and his/her preferences carefully, he/she should be able to estimate the indifference probability with reasonable accuracy.

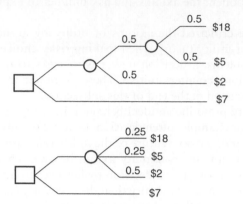

Figure 7.46a   Reduction of a compound lottery. The decision maker reduces the compound lottery in the upper panel to a simple one in the lower panel using standard probability calculations.

Axiom 3: Substitution
There are two substitution axioms:

- First Axiom: Reduction of a compound lottery
  A decision maker is indifferent to substitution of a compound lottery (a lottery whose outcomes are lotteries too) with a simple lottery obtained from the compound lottery by probability calculations.
- Second Axiom: Replacement of a sure amount with an equivalent uncertain event
  A decision maker is indifferent to substitution of a lottery for its certainty equivalent.

Figure 7.46a demonstrates the substitution of a compound lottery with a simple one. This substitution does not affect the decision maker's preferences because both the compound and the simple lotteries yield the same payoffs with the same probabilities. For example, the probability of winning $18 in both lotteries is 0.25.

Critique: The first substitution axiom is debatable. A decision maker may prefer the simple lottery in the lower panel (Figure 7.46a) to the compound lottery in the upper panel because of imprecision in the probabilities of the payoffs. Suppose that he/she believes that all probabilities could vary from their nominal value of 0.5 by ±10% as shown in Figure 7.46b. The bounds of the probabilities of the payoffs of the reduced lottery are calculated by using interval arithmetic,

$$P(\$18) \in [0.45^2, 0.55^2] \Leftrightarrow P(\$18) \in [0.2035, 0.3025]$$

$$P(\$5) \in [0.45^2, 0.55^2] \Leftrightarrow P(\$5) \in [0.2035, 0.3025]$$

$$P(\$2) \in [0.45, 0.55]$$

Figure 7.46b shows the reduced lottery with probabilities obtained by using probability bounds analysis. There is greater imprecision in the consequences of this lottery than in a reduced lottery whose probabilities vary by ±10%. Therefore, a decision

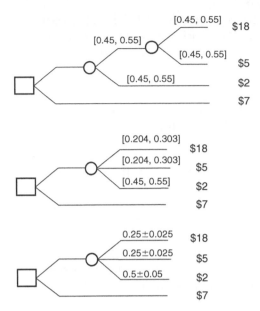

Figure 7.46b Critique of first substitution axiom. Reduction of a compound lottery to a simple one changes the decision maker's preferences.

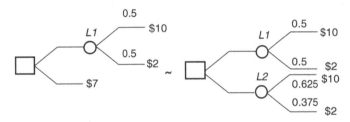

Figure 7.46c The decision tree in the left panel is equivalent to the lottery in the right panel that is obtained by replacing the sure payoff of $7 with lottery L2.

maker would prefer the simple lottery to the compound one. This means that reducing a compound lottery to a simple one by using standard probability calculations may change the decision maker's preferences. The reason is that information about the imprecision in the probabilities of the consequences is lost when the decision maker uses standard probability calculations to reduce the compound lottery to the simple one.

Figure 7.46c illustrates how to use of the second substitution axiom to replace a certain payoff of $7 with equivalent lottery L2. This operation increases complexity but it allows for an 'apples-to-apples' comparison of risky ventures. Note that we selected lottery L2 so that it yields the same payoffs as lottery L1, but with different probabilities. This substitution enables us to conclude that lottery L2 stochastically dominates L1. We could not reach this conclusion without using the second substitution axiom.

Axiom 4: Monotonicity

Suppose that payoff $A is greater than $B. Then, between two reference lotteries that yield these two payoffs, the decision maker prefers the one that yields $A with higher probability. This axiom says that a decision maker prefers a lottery that stochastically dominates another lottery.

Axiom 5: Invariance

All that is needed to determine a decision maker's preferences among lotteries are the payoffs and the associated probabilities.

Axiom 6: No Hell, No Heaven

No consequence is infinitely bad or good. If one does not accept this axiom then one can be manipulated easily. Suppose that a stranger knocks your door on a Sunday morning and asks you to join his religion. The stranger says that if you refuse then you will go to hell after death. You believe that the probability that the stranger is right is infinitesimal, but not zero. Since hell has utility $-\infty$ (by definition) you must join the stranger's religion because refusing to do so has expected utility $-\infty$. Similarly, if one considers dying to have utility $-\infty$, one should never step outside or do anything that will have even an infinitesimal probability of resulting in death.

Expected utility criterion

If a decision maker accepts the above axioms, then the decision maker can construct a function $U(x)$ representing his/her preferences for different amounts of money x. The decision maker should use the expected utility criterion below to select the best course of action among alternatives. We will prove the following theorem to support this proposition.

Expected utility theorem: Among n lotteries, a rational decision maker must choose the one with maximum expected utility.

We will follow the following strategy to prove this theorem. We will consider the lottery in Figure 7.47 (upper panel) that pays $x_1, \ldots, x_r with probabilities $p_1, \ldots, p_r$, respectivelly, where $x_1 \geq $x_2 \geq \cdots \geq x_r. Then, we will show that a rational decision maker is indifferent between this lottery and a reference lottery that pays the maximum possible amount with probability equal to the expected utility of the original lottery, $EU = \sum_{i=1}^{r} p_i U(x_i)$, and the lowest amount with the remaing probability. Then among two lotteries, the one with largest expected utility is the best because it is more likely to yield the best prize than the other lottery.

According to the continuity axiom, for any amount $x between the minimum and maximum payoffs, there is a probability that that makes a decision maker indifferent between the sure amount $x and the reference lottery. According to Figure 7.37, we <u>define</u> the utility of amount $x to be equal to the indifference probability,

$$U(x) = \text{indifference probability} \tag{7.16}$$

From the second substitution axiom, we can replace each amount of the simple lottery in Figure 7.47 (upper panel) with a lottery that pays the highest and lowest

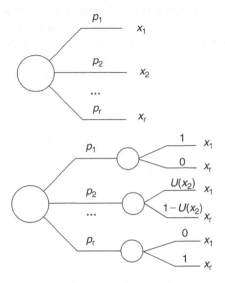

Figure 7.47 Substitution of a simple lottery with an equivalent composite lottery involing only the maximum and minimum payoffs of the simple lottery.
By definition, $U(x_i) =$ indifference probability for sure x_i and a reference lottery that pays the highest and lowest prizes.

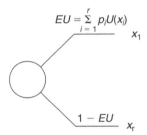

Figure 7.48 Reduction of the composite lottery to a reference lottery that yields the highest prize with probability equal to the expected utility and the lowest with the remaining probability.

possible amounts without changing the decision maker's preferences. For example, Figure 7.47 shows that we can replace amount $\$x_2$ with a reference lottery that pays of $\$x_1$ with probability equal to the utility of $\$x_2$ and the lowest amount with the rest of the probability. This substitution of sure amounts with reference lotteries results in the composite lottery in Figure 7.47 (lower panel).

Now we reduce the composite lottery in Figure 7.47 to the reference lottery in Figure 7.48 using the first subsitution axiom. This lottery pays the maximum amount $\$x_1$ with probability equal to the expected utility

$$EU = \sum_{i=1}^{r} p_i U(x_i) \tag{7.17}$$

In conclusion, every simple lottery that pays different amounts of money is equivalent to a reference lottery that pays the highest amount with probability equal to the expected utility of the simple lottery and the lowest amount with the remaining probability. A consequence of this conclusion is that a decision maker can compare lotteries by reducing them to their equivalent reference lotteries and compare their expected utilities. The decision maker should select the lottery with highest expected utility because this lottery yields the best prize with higher probability than the others (axiom 4). Q.E.D.

Invariance of utility to a linear transformation with a positive scaling constant
We can multiply a utility function by a positive factor and add a constant to it without changing the decision maker's preferences. Specifically, we will show that if L_1 and L_2 are two lotteries, and $U'(x) = aU(x) + b$, $a > 0$, then,

$$EU(L_1) > EU(L_2) \Leftrightarrow EU'(L_1) > EU'(L_2) \tag{7.18}$$

Proof

$$EU(L_1) = \sum_{i=1}^{r_1} p_i^1 U(x_i^1) \quad \text{and} \quad EU(L_2) = \sum_{i=1}^{r_2} p_i^2 U(x_i^2)$$

where x_i^1 are the r_1 possible payoffs of lottery L_1 and p_i^1 their probabilities. According to the expected utility thorem,

$$EU(L_1) > EU(L_2) \Leftrightarrow$$
$$\sum_{i=1}^{r_1} p_i^1 U(x_i^1) > \sum_{i=1}^{r_2} p_i^2 U(x_i^2)$$

We scale both sides of the above inequality by positive factor a and then add b to both sides. Then,

$$EU(L_1) > EU(L_2) \Leftrightarrow$$
$$a \cdot \left[\sum_{i=1}^{r_1} p_i^1 U(x_i^1) \right] + b > a \cdot \left[\sum_{i=1}^{r_2} p_i^2 U(x_i^2) \right] + b \Leftrightarrow$$
$$\left[\sum_{i=1}^{r_1} p_i^1 a \cdot U(x_i^1) \right] + b > \left[\sum_{i=1}^{r_2} p_i^2 a \cdot U(x_i^2) \right] + b \Leftrightarrow$$
$$\sum_{i=1}^{r_1} p_i^1 a \cdot U(x_i^1) + \sum_{i=1}^{r_1} p_i^1 \cdot b > \sum_{i=1}^{r_2} p_i^2 a \cdot U(x_i^2) + \sum_{i=1}^{r_2} p_i^2 \cdot b \Leftrightarrow$$
$$\sum_{i=1}^{r_1} p_i^1 \cdot [a \cdot U(x_i^1) + b] > \sum_{i=1}^{r_2} p_i^2 \cdot [a \cdot U(x_i^2) + b] \Leftrightarrow$$
$$EU'(L_1) > EU'(L_2)$$

Q.E.D.

Estimation of utility

The objective of this subsection is to present two methods for estimation of a person's utility and demonstrate these methods on an example. The two methods compare a certain amount of money to a reference lottery. The first method estimates the indifference probability of the lottery. That is, the probability of the highest prize that makes the person indifferent between the lottery and the certain amount. The second method, finds the certainty equivalent of the reference lottery. That is, the certain amount that the person believes that is equivalent to the lottery.

When presenting each method, we will consider a *facilitator* who helps a decision maker estimate the utilities of different amounts of money. The facilitator is an expert in decision analysis, while the decision maker could be more familiar with the particular decision *problem* than the facilitator. He/she asks the decision maker questions, and estimates utilities of different amounts of money based on the decision maker's responses. The facilitator should elicit the utility of the same amount by using different sets of questions to ensure consistency. After checking the utility estimates the facilitator should provide feedback to the decision maker.

Example 7.21:   Using indifference probability method to estimate utility function
Suppose that the minimum and maximum payoffs in a decision are $0 and $100. The facilitator assigns utilities values equal to 0 and 1 to these amounts. This is an arbitrary assumption that is made for convenience. In order to estimate the utility of $50 the facilitator asks the decision maker to consider the following reference lottery,

$$\text{Win \$100 with probability } p$$
$$\text{Get nothing with probability } 1 - p \qquad (7.19)$$

Then the facilitator asks the decision maker to adjust the probability p until the decision maker becomes indifferent between the lottery and a certain payoff of $50. Then, according to the expected utility theorem,

$$U(\$50) = \text{Expected utility of reference lottery}$$
$$= pU(\$100) + (1 - p)U(0)$$
$$= p$$

For example, if the decision maker says that his/her indifference probability is $p = 0.65$, then $U(\$50) = 0.65$.

Using this method, the facilitator elicits the utilities of given amounts of money by repeating this process.

Example 7.22:   Certainty equivalent method
Like in the indifference probability method, the decision maker compares a reference lottery to a certain amount, but the decision maker adjusts the certain amount instead of the probability of the best prize. This method, finds the amounts that have given utility values. For example, suppose that the facilitator wants to elicit the certain

amount that has utility 0.75. Then, the facilitator asks the decision maker to find the certain equivalent of the following lottery,

Win $100 with probability 0.75

Get nothing with probability 0.25 $\qquad$ (7.20)

Suppose that the decision maker says that he/she is indifferent between $60 and the reference lottery. Then, $U(\$60) = 0.75$.

Feedback-and-correction

Feedback-and-correction is a useful practice to ensure that the estimated utility function truly represent the decision maker's preference toward risk. The facilitator should elicit more estimates from the decision maker than are needed to estimate the utility of an amount. The facilitator should check the estimates for consistency. If the estimates are inconsistent, then the facilitator must bring this to the decision maker's attention.

Example 7.23:   Correcting a decision maker's utility function

According to the decision maker's responses in examples 7.21 and 7.22 in this subsection the utilities of $50 and $60 are 0.65 and 0.75 respectively. The facilitator asks the decision maker for the probability that makes him/her indifferent between a sure $50 and the reference lottery

Win $60 with probability p

Get nothing with probability $1 - p$ $\qquad$ (7.21)

This probability should be close to one because the high prize of the lottery is approximately equal to the certain amount. Suppose that the decision maker is indifferent for a probability $p = 0.9$. Then the facilitator can check the three judgments of the decision maker as follows. The expected utility of reference lottery (7.21) is equal to the utility of a sure $60,

$$pU(\$60) = U(\$50) \qquad (7.22)$$

Therefore, $U(\$50) = 0.9 \cdot 0.75 = 0.675$. The facilitator estimated that the utility of $50 is equal to 0.65 from the comparison of reference lottery (7.19) to $50. These two estimates of the utility are almost equal. But because a single estimate of utility is needed, the facilitator should ask the decision maker to either adjust the indifference probabilities of lotteries (7.20) or (7.21) or the CE of $60 so that the three estimates become consistent.

Utility estimation: A case study

This example demonstrates how a financial advisor helps an investor estimate her utility function. The investor considers investing her retirement savings in a mutual

Table 7.8 Impact of the investor's final cash position when she will retire on her lifestyle.

Amount (thousands)	Quality of life	Comparison of lifestyle to the one in the row below
$1,800	Great. Unless something really bad happens, I will have more than enough money to maintain my present lifestyle. I will keep my house; buy fancy furniture and a luxury car. I will travel both in the US and abroad.	Having an additional $700,000 will not make a significant difference in my life.
$1,025	Very good. I will afford to keep my house, buy a nice car and travel in the US and abroad.	My lifestyle will be better than in the row below but the difference will not be large.
$640	Good. I will probably sell my house and move to a condo in a good neighborhood. I will afford a decent car and some travel.	My lifestyle will be considerably better than in the row below.
$450	OK. I will have to sell my house and move to a small apartment in a reasonable neighborhood. I will probably have to get a part time job for a few years. I may be able to buy a small used car.	Having an additional $200,000 will make a big impact on my life.
$250	This will be awful. In less than 10 years my savings will be gone. I will have to sell my house, my car and most of my belongings. I will have to find a job. I am afraid that I will be unable to support myself and become a burden to my family and friends. I may even end up in a homeless shelter.	

fund. She plans to retire after 12 years at which time she will transfer her money to a fixed return account and start spending it. Based on historical data, the financial advisor believes that the investor's final cash position after 12 years, adjusted for inflation, can be as low as $250,000 and as high as $1,800,000.

In order to coach the investor, the advisor asks her to describe her lifestyle during retirement for the five possible cash positions in Table 7.8. This table shows the investor's responses to the financial advisor's questions.

Next the advisor asks the investor to compare a series of reference lotteries to the sure amounts in Table 7.8. All amounts will be paid after 12 years, when the investor plans to retire. The smallest and largest amounts are assigned utilities of 0 and 1, respectively. First, the financial advisor asks the investor to consider the following reference lottery,

$1,800,000 with probability p

$250,000 with probability $1 - p$ (7.23)

The financial advisor asks the investor for what probability she is indifferent between the lottery (7.23) and a sure amount of $1,025,000. The investor says that her indifference probability is 0.9. According to this answer, $U(\$1,025,000) = 0.9$.

Then the investor estimates the indifference probability between a sure $640,000 and the reference lottery (7.23). Her estimated indifference probability is 0.7, which suggests that $U(\$640,000) = 0.7$.

The financial advisor also asks the investor to estimate the indifference probability for $640,000 and the lottery,

$1,025,000 with probability p

$250,000 with probability $1 - p$ (7.24)

Although this question is redundant, it helps the advisor check the consistency of the investor's answers. The indifference probability of the investor is 0.8.

Now the advisor can calculate the relation between $U(\$1,025,000)$ and $U(\$640,000)$ from the answer to the last question. The utility of $640,000 is equal to the expected utility lottery (7.24),

$U(\$640,000) = 0.8U(\$1,025,000) \Rightarrow$

$U(\$640,000) = 0.8 \cdot 0.9 \Rightarrow$

$U(\$640,000) = 0.72$

This estimate of the utility of $640,000 is very close to that from the comparison of lottery (7.23) to this sure amount of money. However, the advisor asks the investor to adjust at least one of the elicited utilities of the indifference probability of lottery (7.24) so that the value $U(\$640,000)$ becomes unique. After reviewing her responses, the investor changes the indifference probability between lottery (7.24) and the sure amount from 0.8 to 0.78. This means that $U(\$640,000) = 0.7$.

In order to estimate the utility of $450,000 the investor estimates the indifference probability between lottery (7.23) and $450,000. The investor estimates that this probability is 0.4, which suggests that $U(\$450,000) = 0.4$. In order to check this estimate, the investor also estimates the indifference probability between $450,000 and lottery (7.25) below,

$640,000 with probability p

$250,000 with probability $1 - p$ (7.25)

The investor's indifference probability is 0.6. This estimate suggests that,

$U(\$450,000) = 0.6U(\$640,000) \Rightarrow$

$U(\$450,000) = 0.42$

After some introspection, the investor decides that the utility of $450,000 is equal to 0.4.

Table 7.9 summarizes the results of this exercise for estimation of the utility.

Table 7.9 Investor's utility.

Amount (thousands)	Utility
$250	0
$450	0.4
$640	0.7
$1,025	0.9
$1,800	1.0

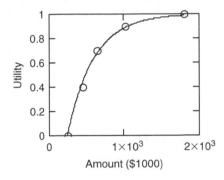

Figure 7.49 The investor's utility.

The advisor fitted an exponential expression (Equation 7.8) to the utility estimates in Table 7.9. The investor's risk tolerance R was found to be $348,000 by minimizing the sum of the squares of the differences between the utility values in Table 7.8 and the values from Equation (7.8). Figure 7.49 shows the fitted utility function and the data in Table 7.9. The exponential expression fits the values of the utility in Table 7.9 very well.

Paradoxes

A paradox is an assertion that, although based on a valid deduction from acceptable premises, sometimes contradicts commonly accepted opinion. Although the axioms of utility are compelling, they fail to describe the way people make decisions in certain situations. Moreover, although these decisions violate the precepts of utility theory, they seem reasonable. We describe three examples in this subsection and try to explain why people's behavior contradicts utility theory. It is important for decision makers to be aware of these situations in order to avoid making irrational choices.

Framing effect

It has been observed repeatedly that the way a decision problem is posed to a decision maker affects the decision maker's choice (Tversky and Kahneman, 1981). The following example demonstrates this point (Clemen, 1997).

There is an influenza outbreak in a community that is expected to kill 1000 people. The authorities consider two alternative programs that are presented using two different frames:

Frame 1

- Program A will save 800 people for sure.
- Program B could save all 1,000 people with probability 0.8 and no one with probability 0.2.

Note that frame 1 presents the problem as choosing between two opportunities to save lives. One of the authors (Nikolaidis) asked his students in a senior Operations Research class, to compare these options. The students agreed that program A is better. The students' responses are consistent with reported responses of other people who compared the same options. It seems reasonable to choose program A over B because the certainty of saving 800 people is preferable to the risk of letting all 1,000 die, with 20% chance.

The same decision is presented as a choice among two potential losses as follows:

Frame 2

- Program C: 200 people will die for certain.
- Program D: None will die with probability 0.8 but 1,000 could die with probability 0.2.

When the students compared these options, they preferred program D. They thought that program C would be unethical because it could be viewed as sentencing 200 people to death. Again this response is consistent with other people's responses observed in controlled experiments. However, when considered together, the observed choices are irrational because the two programs are identical in frames 1 and 2, they are only presented differently.

This irrational behavior can be explained by the tendency of most people to be risk averse when comparing opportunities with positive outcomes and risk seeking when comparing potential losses. That is, most people prefer a sure profit to an opportunity to make a bigger profit. However, in order to avoid a loss the same people take a risk of incurring bigger losses. Loss aversion explains why many managers continue to invest in ventures that are almost certain to fail once these managers have already invested some money in the ventures.

Certainty effect

People make choices that violate the principles of decision analysis because they are biased toward a certain consequence. Allais paradox demonstrates this behavior.

Consider the following decisions.

Decision 1: Choose one of the options below;

Option A:
Win a sure $1 million

Option B:
Win $5 million with probability 0.1

Win $1 million with probability 0.89
Win nothing with probability 0.01

In experiments, most subjects selected option A. This choice seems to make sense because most people believe that winning $1 million significantly improves their lives. Therefore, they do not want to trade the certainty of winning $1 million for an opportunity to win more money if this option also entails a small risk of getting nothing.

Decision 2: Choose one of the options below;

Option C:
Win $1 million with probability 0.11
Win nothing with probability 0.89

Option D:
Win $5 million with probability 0.1
Win nothing with probability 0.90

In experiments, most subjects selected option D. This seems reasonable, because it is worth trading 1% chance to win $1 million for the chance to win five times this amount. However, when considered together, these two decisions violate the expected utility axiom. Let $U(0) = 0$ and $U(\$5 \text{ million}) = 1$. Then, because a decision maker prefers option A to B,

$$U(\$1 \text{ million}) > 0.1U(\$5 \text{ million}) + 0.89U(\$1 \text{ million}) \Leftrightarrow$$
$$U(\$1 \text{ million}) > 0.91$$

Because the same decision maker prefers option D to C the

$$0.1U(\$5 \text{ million}) > 0.11U(\$1 \text{ million}) \Leftrightarrow$$
$$0.91 > U(\$1 \text{ million})$$

The two choices violate the maximum expected utility criterion. A rational decision maker would not make these two choices.

One explanation is that the decision maker prefers option A to B in decision 1 because he/she is biased toward the certain amount. Moreover, it is difficult for a decision maker to quantify his/her preferences for very large amounts of money, such as $1 million and $5 million, because the utilities of all these amounts are too close to one.

Aversion to imprecision in probabilistic models
People avoid a venture when they are imprecise about the probabilities of the consequences of the venture. The following example demonstrates this aversion. Consider the following options,

A.   A box contains two marbles that could be black or red. Pick up a ball at random. You win $20 if the ball is red and zero otherwise.
B.   Same option as A but you know that the box contains a red and a black ball.

Most people prefer option B because they know the probability of winning $20, while they do not know this probability in option A.

However, we can think that option A is a compound lottery. A big box contains one million red and one million black marbles. We pick up two marbles at random and put them in a small box. Then we pick up one marble from the small box and win $20 if the ball is red and $0 if it is black.

The probability that the box contains two red marbles is equal to the probability of picking up two red marbles in a row from the big box. This is equal to, $P(\text{red}) = 0.5^2 = 0.25$. The probability of the box containing two black marbles is also 0.25. The probability of a red and a black marble is equal to the sum of the probabilities of picking up a red marble first and then a black one plus the probability of picking up a black marble first and then a red one. Thus the probability of a red and a black marble in the small box is 0.5.

The probability of winning $20 in the second lottery is also 0.5, like the first lottery.

According to the expected utility criterion, options A and B should be equivalent. Yet most people argue that option B is better.

Closing remarks

It is well documented that often people act in a way that is inconsistent with utility theory. However, this fact does not reduce the value of utility theory because this theory is prescriptive, not descriptive. This means that utility theory prescribes how a rational person should make choices; it does not describe how most people actually make choices.

It is important to teach people the rules of clear thinking of decision theory. These rules guide decision makers about how to structure decisions, generate and evaluate alternatives, and perform sensitivity analysis. It is also important to make people aware of biases that often lead to irrational choices.

However, the ultimate test of decision theory is whether people will eventually come to appreciate its value and use it to make informed decisions that are essential to a successful and fulfilling life.

7.6   Conclusion

Decision analysis is a logical approach for making informed decisions. This approach consists of five steps. First, the decision maker defines and frames a decision and decides what she wants to accomplish. Then, she identifies her options. In the third step, the decision maker quantifies her preferences, and estimates the probabilities of the uncertain events that affect the consequences of each course of action. This step involves the development of a predictive model for the consequences of the alternative courses of action. The fourth step is to compare the available options and determine the best one. Finally, the decision maker assesses the robustness of the decision to his/her assumptions about the models for predicting the consequences of alternative options, the probabilities of the outcomes of uncertain events and her preferences toward risk. Decision analysis is iterative; the decision maker appraises the decision quality and repeats one or more of the steps in the decision process as needed.

The authors believe that the first two steps of the decision analysis process are the most critical. One can significantly improve the quality of a decision if one invests time to define correctly the decision problem and his/her objectives.

Using decision analysis one can make good decisions and enjoy better consequences in the long run than by using a heuristic approach and/or intuition. The reason is that decision analysis is based on common sense rules for defining a decision problem, and modeling uncertainty and preferences. This approach enables a decision maker to approach decision problems consistently.

Decision analysis provides the following rules of clear thinking;

1) Know your objectives.
2) Make sure that you work on the correct problem. Ask yourself; am I addressing the right question?
3) Consider many fundamentally different and creative options.
4) Assess carefully the likelihood of uncertain events that affect the consequences of the decision.
5) Compare alternative courses of action on the basis of their consequences.
6) Understand all important issues and include them in the decision model.

Questions and Exercises

Introduction

7.1 Describe an example of a good decision that you made. Did it result in a good or a bad outcome?

7.2 Describe an example in which a bad decision resulted in a good outcome. Explain why the decision was bad in your opinion despite the outcome.

7.3 Decision analysis only aims to facilitate actional thoughts, i.e. thoughts that will result in an action. Why is it necessary to have this condition? Can there be thoughts without action or vice versa?

7.4 A friend tells you that "decision analysis, is a structured approach for making better decisions by using common sense rules. It enables one to select the best course of action in a decision problem." Is this description right or is something critical missing?

7.5 Traditionally, uncertainties are modeled using a worst-case scenario approach. Specifically, we represent the uncertainties with variables and use conservative characteristic or design values to account for the uncertainties. These values represent unfavorable scenarios that are often unlikely. For example in aircraft design, we have two main types of loads: maneuvering and gust loads. Characteristic maneuvering and gust loads are very high and are rarely observed. Similarly, we consider a characteristic value of the ultimate failure stress that is a small percentile of this stress. Then we make design decisions under the requirement that the airplane should be able to sustain these loads.

Do you think that this approach is rational? What are its strengths and weaknesses?

7.6 The manager of an electronics store tells you that he/she orders a new batch of washing machines of a particular brand just when the inventory falls below 20%. The size of the order is always the same. The manager thinks that this simple rule seems to work well.
- What are the pros and cons of this inventory policy in your opinion?
- Under what conditions is this inventory policy close to an optimal?
- When could it fail?

Framing and Structuring Decisions

7.7   Study the drill for oil decision. Suppose that the land owner concluded that all of the options in section 7.1.1 are either too risky or have too low payoff. Think of at least three other options that could be more attractive to the owner.

7.8   Think of a decision where a decision maker did not frame the decision properly because he/she adopted a narrow perspective. Explain how you would have reframed the decision.

7.9   Study the city evacuation, and the grape harvesting decisions, and the decision of the inventor seeking funding from a financier in section 7.1.1. Represent each decision problem by a decision tree. You can draw these trees using a computer program or you can draw them by hand.

7.10   In problem 7.7, consider each possible scenario corresponding to every possible combination of decisions and outcomes of the uncertain events. Explain what will happen in each scenario and describe the consequences of each scenario.

7.11   The vice president of the research and development department of a pharmaceutical company has just finished reviewing a new drug. She will decide whether to provide $10 million funding for further development or stop the project. If the project continues, then the drug will be tested on volunteer patients. If the results are positive the company will market the drug. Otherwise the project will be terminated. If the company decides to market the drug, then the profit will depend on the demand. It is projected that if the demand is high, then the profit will be $250 million but if it is low the profit will be only $50 million. The cost of $10 million for development should be subtracted from the above numbers in order to estimate the net profit. Construct a decision tree for the vice president's decision.

7.12   Construct an influence diagram and a consequence table for the vice president's decision.

7.13   The dynamics of a decision change if multiple decision makers are involved. Think of what additional aspects one might have to consider when the decision makers are: 1) Cooperating and 2) Competing.

7.14   Check if the decision model in problem 7.11 passes the clarity test. Justify your answer.

Solving Decision Problems

7.15   Solve the decision problems below by using backward induction.

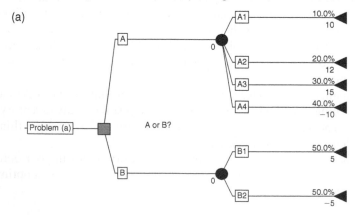

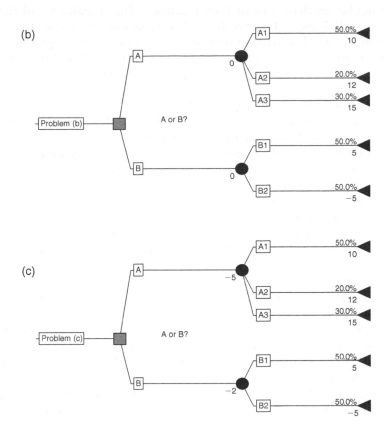

7.16 In which of problems a) to c) the optimum decision can be identified without doing any calculations? Justify your answer.

7.17 Solve the decision problem involving the extended hybrid car–econocar decision in section 7.2.2 by backward induction.

7.18 Construct the risk profiles and the cumulative risk profiles of the decision problems 7.15a to 7.15c.

7.19 Construct the risk profiles and the cumulative risk profile of the decision problems in section 7.2.2. Consider only the optimum strategy.

7.20 Construct the risk profile of the decision problem involving the squeegee development in section 7.3.1. Consider only the optimum strategy.

7.21 A student is considering buying a $50 book to study for a test. His probability of passing the test with only course notes and no book is 0.7. The probability of his failing the test given that he has the book is 0.1. He values passing the test in money amounts at $1,000. How much does he value failing the test if he is indifferent between buying the book and not buying it.

7.22 Two non-repairable testing machines are installed in a factory at a cost of $2,500 each. The machines test slider-crank mechanisms built by the factory which are repaired if the mechanisms' performance characteristics are outside the specifications as found by the testing machines. Each test costs $100. It is possible that before being released, a given mechanism is tested by neither machine, one or both machines. The probabilities of these events are 0.2, 0.7 and 0.1 respectively. The probability that the mechanism performs its intended function successfully after passing through these machines and maintenance, if required, in each case is 0.2, 0.7 and 0.99 respectively. A successful mechanism makes $1,000 profit while a faulty one costs the company $300.

Determine, assuming risk-neutrality:

a. How many mechanisms should a machine be able to test on average before failing, to make sure that the company makes profit?

b. You bought one of the mechanisms from the manufacturer and found it to be working properly. What is the probability that it was: not tested, tested once and tested twice?

c. One of the testing machines broke. Should the company replace it or should it continue with the one remaining? Assume that the testing machines can test 50 mechanisms each before failing. (Hint: Assume that the probability of test occurring is 0.8 and perform a decision tree analysis).

Sensitivity analysis

7.23 Explain two reasons for which a decision maker should perform sensitivity analysis.

7.24 You are planning to perform sensitivity analysis for a decision problem. After listening to your plan, one of your colleagues argues that sensitivity analysis is useless. He/she tells you to consider the problem variables random, estimate their probability distributions and solve the decision problem again so as to maximize the expected profit. What do you think about this argument? How would you try to convince your colleague that sensitivity analysis is useful?

7.25 Mention and explain a main limitation of one-way sensitivity analysis.

7.26 Consider the software development decision in Example 7.12 in section 7.3.1. Derive an equation for the expected profit of the two strategies of a) developing the software and selling it to another company, and b) selling the product directly to consumers.

7.27 In the above problem, assume that the development cost varies from $1.5 to $2.5 million, the revenue from selling the software to another company varies from $4 to $5, and the revenue from selling the software directly to consumers varies from $15 to $25 million. Calculate the expected profit under the best and worst case scenarios and the scenario corresponding to the base values of the variables for both scenarios (a) and (b). Discuss the results.

7.28 Construct a tornado diagram for the risky decision in Example 7.11 in section 7.3.1. Assume that the revenue for a heads up flip of the fair coin varies from $7 to $10. The revenue if the thumbtack lands with its pin up ranges from $6 to $11, and the revenue if the pin lands heads up ranges from $0 to $2. Also assume that the probability that the thumbtack will land with its tip up ranges from 0.5 to 0.8. Discuss the results and explain the observed trends.

7.29 Plot the expected profits of both options to bet on the coin and the thumbtack as a function of the revenues in the above problem.

7.30   Identify the two most important variables in problem 7.28. Explain why these are the most important variables.

7.31   Construct a 2-way strategy region and a 2-way sensitivity graph for the two variables in problem 7.30. Explain if the selection of the optimum decision is robust.

7.32   A risk neutral aircraft manufacturer is deliberating on whether to test the wing of a new design before inspection by the aviation authorities. If the wing deflects more than 1.0 ft at the tip under stated loading conditions, the design is rejected and the manufacturer has to spend $10 million in redesign and certification. The in-house testing and redesign costs only $3 million. The manufacturer estimates that the deflection is uniformly distributed between 0.7 and 1.3 ft. The in-house testing and redesign changes this distribution to a uniform distribution between 0.6 and 1.05 ft. What should the manufacturer do?

7.33   The manufacturer in problem 7.32 now realizes that the redesign costs are not deterministic. Perform a two-way sensitivity analysis over the cost of in-house testing and redesign, and testing and redesign after inspection. Assume ranges of $1 to $5 million and $4 to $15 million respectively.

Utility theory

7.34   Consider the utility function in Equation (7.12). Find the CE of a venture that pays $0 and $20 million each with probability 0.5.

7.35   Find the risk premium of the venture in the above problem.

7.36   People can be risk neutral, risk seeking and risk averse at different points in their utility function. Why do you think this is true? Does this affect the applicability of utility axioms?

7.37   Does a utility function necessarily have to be non-decreasing? (Hint: Imagine cases when the underlying attribute is not money).

7.38   A decision maker's risk tolerance, R, in the equation $U(x) = 1 - e^{-\frac{x}{R}}$ can be elicited from the following lottery.

For what value of x is the decision maker indifferent between $0 and the lottery that pays x with probability 0.5 and charges $0.5x$ with probability 0.5? Then R is equal to x

Verify that this is correct. Is R an approximate or an exact value?

7.39   The sign of the second derivative of the utility function determines if a decision maker is risk seeking, risk neutral or risk averse. Using this information find two example functions for each case.

7.40   Using what you learned in problem 7.39, comment on the utility function of a decision maker, which is the product of the utility functions of two risk seeking decision makers. Assume that the utility functions are always positive and increasing.

7.41   In example 7.18, find the CE of the optimum strategy "Develop program and sell it to a company if successful" using the utility function in Equation (7.12).

7.42   Find the optimum decisions for the three decision trees in problem 7.15 by using the expected utility criterion. Assume an exponential utility function (Equation 7.8) with risk tolerance $R = 5$. Compare the results with those from using the EMV criterion. Discuss if the differences or similarities in the decisions make sense.

7.43 Consider that we scale the utility function in Equation 7.8 by a factor of 2,

$$U(x) = 2(1 - e^{-\frac{x}{R}})$$

Suppose that we use this utility function to find the optimum decisions for the three decision trees in the previous problem. Will this change the optimum decisions compared to the optimum decision in the previous problem? Justify your answers.

7.44 Solve problem 7.42 assuming that risk tolerance $R = 10$. Compare the results with those in problem 7.42 and the results from using the EMV criterion.

7.45 Solve example 7.11 in section 7.3.1 for a risk seeking decision maker whose utility function is shown in the table below:

Profit ($5)	Utility
5	1
4	0.5
−4	0.1
−5	0

Discuss and explain the difference in the results from using the EMV criterion and the expected utility criterion.

Chapter 8

Multiattribute Considerations in Design

Overview of this Chapter

Designers and decision makers in general, seek to maximize their expected utility from a decision as we have discussed already in this book. A utility function maps the attribute level to an abstract notion of satisfaction, thereby facilitating decision making under uncertainty. So far, we have considered only one underlying attribute, which can be a significant simplification. In most decision problems, more than one attribute is involved. As an example, let us consider the design of a car seat. In addition to the cost of the seat, the designer might want to consider the weight, material and physical dimensions of the seat. How does the designer measure the overall worth of the seat when it is clearly measured in terms of multiple attributes and involves tradeoffs? More succinctly how can decision analysis help make decisions under uncertainty when multiple attributes are involved? Table 8.1 shows examples of decision problems that involve multiple attributes.

Thurston (1991) emphasized that evaluation of designs using multiattribute utility theory is essential to selecting the best design that satisfies all the objectives. Over the years, decision-based-design as a field has seen a considerable development, Lewis et al. (2006). Proper modeling of a decision maker's preferences over multiple attributes is complementary to design optimization, i.e. only when we have a mathematical formulation that models tradeoffs between attributes can we maximize a design's overall worth. In this chapter, we present methodologies and formulations that address this important issue.

8.1 Tradeoff Between Attributes

If each of the attributes for a given design problem could be independently optimized, we would not need a multiattribute model of utility. As an example if one could reduce

Table 8.1 Some examples of decision problems and relevant attributes.

Decision problem	Relevant attributes
Aircraft engine design	Dimensions, weight, thrust generated, mean time to failure, cost
Ceiling fan design	Weight, wattage, cost, wind velocity
Electric screwdriver design	Maximum torque, battery life, cost
New diabetes drug	Efficacy, development costs, side effects

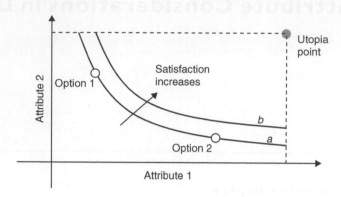

Figure 8.1 Isopreference curves over two attributes.

the weight of an aircraft engine without affecting thrust or cost, it would result in a great design. Unfortunately, this is not possible as most designs involve tradeoffs. The designer has to choose a combination that meets the needs the best. What is that design? How does the designer compare a design with another that sacrifices performance on one attribute to improve performance on another? To answer these questions we need a function that models the tradeoff behavior of the designer in a deterministic way. Given the level of attributes of a design the function should be able to determine its utility or value to the decision maker. As an example, the function should be able to express that a \$10 decrease in price makes up for a 30 minutes decrease in the battery life of an electric screwdriver.

For our discussion in this section, we can assume without any loss of generality that a decision maker seeks to maximize attributes. For attributes that need to be minimized, a suitable transformation of the attributes is enough to establish this. For example, if cost is to be minimized, one can view the decision as one maximizing the negative of cost and thus maximizing the transformed variable '*-cost*'. This simplification helps in the presentation of the material in this section.

Consider Figure 8.1 where a decision maker is trying to maximize attribute 1 and attribute 2 simultaneously. A utopia point, or a point where one has improved the two attributes completely, is shown. Two isopreference curves are also drawn to illustrate tradeoffs between the two attributes. An isopreference curve corresponds to a set of values of the attributes for which the decision maker is indifferent. The decision maker is willing to sacrifice one attribute to improve another and vice versa. In the figure, the decision maker would be indifferent between options 1 and 2, on the isopreference curve *a*. If the decision maker is not indifferent, then the two options are on different isopreference curves. In the figure any option (or design) on isopreference curve *b* is better than any option (or design) on the isopreference curve *a*. This is because for a fixed value of attribute 1 (any vertical line), the point on isopreference curve *b* has a higher value of attribute 2. Similarly for a fixed value of attribute 2 (any horizontal line), the point on isopreference curve *b* has a higher value of attribute 1. In fact, proving that even one point on curve *b* is better than a point on curve *a* is enough

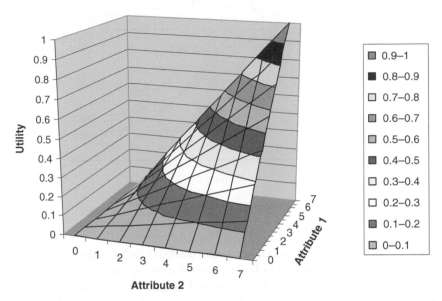

Figure 8.2 A three dimensional graph of a multiattribute utility function showing isopreference curves.

to conclude that any point on curve *b* is preferable to any point on *a*. Isopreference curves, by definition, do not intersect.

Figure 8.2 is a 3-dimensional plot of the utility or value to a decision maker of many options with different combinations of their individual attributes. The underlying function for the plot is given by the function $\frac{xy}{49}$, where x and y are the two attributes of concern to the decision maker. Isopreference curves can be seen as bands on the utility function surface. If viewed from top these bands resemble the isopreference curves shown in Figure 8.1. Utility is shown on z-axis in the figure and is scaled between 0 and 1. We follow this convention from the single attribute case, however a value of 1 in this case means that the decision maker has achieved the utopia point or the best possible level on *all the attributes simultaneously*. On the other extreme, when all the attributes are at their worst level, the utility value is 0. In the given example, only one attribute being at the lowest level is enough to bring the whole multiattribute utility down to zero, but this is because of the type of utility function chosen and it does not always have to be the case.

Pareto Front

While evaluating alternatives, it is also important to account for the constraints imposed by the decision problem itself. For example, in our previous discussion the utopia point is the best possible alternative but it may not be achievable. The obvious question is: how close can we get to it? To answer this question, consider Figure 8.3. Let us say that a certain set of decisions leads to option 1 being realized. For a design problem, this could be the choice of design variables. If there is no other set of decisions

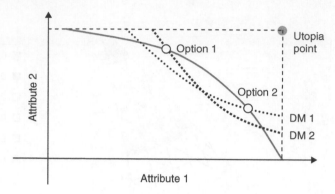

Figure 8.3 Pareto front and isopreference curves.

for which there could be a *simultaneous* improvement in both the attributes we can say that option 1 lies on the Pareto front[1], named after the Italian economist Vilfredo Pareto. This implies that there is no way we could achieve an option that is both above and to the right of option 1 (imagine a rectangle cornered on option 1 above and to the right, any point in this rectangle will dominate option 1). Similarly, for a different set of decisions we could achieve option 2, which is also non-dominated. It is clear from the figure that option 2 sacrifices attribute 2 to gain improvement in attribute 1, when compared to option 1. Similarly, option 1 sacrifices attribute 1 to improve attribute 2, when compared to option 2. We say that neither of these two options *dominates* the other. A set of all such points defines the Pareto front, also sometimes referred to as the non-dominated front. It is clear that no set of decisions (or designs) would give us an option that is simultaneously above and to the right of a point the Pareto front. Similarly any option to the left of and below the Pareto front is inferior to at least one point on the front.

Pareto fronts help us understand a multiattribute decision making process. Consider two decision makers, DM1 and DM2 as represented by their isopreference curves in Figure 8.3. These curves have different slopes, which means that the two decision makers have different tradeoff behaviors. When viewed in conjunction with the Pareto front we can see that decision maker 1 prefers option 1 to option 2 because option 1 is above and to the right of his/her isopreference curve that intersects with option 2. Similarly, decision maker 2 prefers option 2. The optimal option can be found by finding the intersection of the Pareto front with the isopreference curve with the highest value or utility.

Our discussion shows that all that is needed to make a decision that involves multiple attributes is the decision maker's tradeoff behavior, his/her preference order for the attributes and the Pareto front. In practice, this information is not easily obtainable.

---

[1] Pareto fronts are usually presented in the context of attribute minimization problems i.e. the origin is the utopia point, by convention. For consistency between our discussion of multiattribute decision making and Pareto fronts, we chose to present it in the attribute maximization context. This does not affect the treatment of the subject.

Sophisticated optimization methods are required to obtain the Pareto front. While single attribute preference orders are usually available, tradeoff behavior is not and needs to be carefully assessed. Finally a graphical method is not suitable for more than two attributes. The focus of this chapter therefore, is rigorous modeling of tradeoff behavior for multiple attributes, which facilitates decision making.

8.1.1   *Range of negotiability*

For most attributes one can, in theory, define an infinite range, but doing so is counterproductive. As an example, most individuals will only consider a range of $+/-\$100$ million for money because they are unlikely to have a debt of more than $100 million or earn more than $100 million in their lifetimes. This is called the *range of negotiability*. A decision maker only seriously considers alternatives within this range. A designer working on a new wide-body aircraft may consider only between 100 and 1000 passengers because any number outside this range is either highly impractical or infeasible. Similarly, an automotive engineer may believe that putting a less than 100 hp engine in a car will adversely affect its sales while putting more than 1000 hp has limited usefulness. He/she will define his utility function over this range only. Finally, the buyer of the car may consider spending between $10,000 and $30,000 on the car. This is because he/she does not want to spend more than $30,000, while fully understanding that getting a new car for less than $10,000 is highly unlikely.

In multiattribute decision making problems, a complete knowledge of the range of negotiability is critical, even more so than in single attribute problems. Since utility is normalized, one should know a definite point where it has reached a value of one. The decision maker is fully satisfied at this point. A well defined utility function also helps understand the tradeoff behavior of the decision maker. As we shall see later, independence conditions are harder to satisfy if ranges are arbitrarily chosen. The decision maker and/or the facilitator waste a tremendous amount of effort by considering exceedingly wide range of attributes. Determining the range of negotiability is part of the decision framing process as described in chapter 7, and should be followed religiously. As an example, in Figure 8.2, the range of negotiability is between 0 and 7 for both the attributes.

8.1.2   *Value functions vs. Utility functions*

It is important here to make a distinction between value functions and utility functions. Value functions rank alternatives in the order of desirability just like utility functions. Both value functions and utility functions measure only the relative preference and therefore a difference of '*x*' in value or utility between two choices cannot be compared to a difference of '*y*' between any two other choices. Utility functions have the additional property that they also model the risk attitude of the decision maker. Therefore, utility functions can be used to make decisions under risk. Expected utility has the same connotation in multiattribute decision making as it does in single attribute decision making. This means that a decision maker tries to maximize the expected multiattribute utility while making decisions.

In this chapter, we focus mainly on multiattribute utility formulations. Value functions are presented when making assessment of deterministic options, with a utility function defined *over* the value function.

8.2   Different Multiattribute Formulations

It is challenging to assess a multiattribute utility function of a decision maker directly. There are many impediments to accomplishing this task:

1.   Decision makers are usually unable to clearly spell out their tradeoff relationships.
2.   Lotteries to find the indifference points are much harder to construct in the case where multiple attributes are present.
3.   The number of assessments increases exponentially with the attributes. If 10 assessments are required for each attribute then a total of 10^n assessments will be needed for n attributes. For example, for a 5-attribute decision making problem, one would need 100,000 assessments to completely determine the utility function.

There are many approaches to alleviate these problems. One is to find the multiattribute function by utilizing single attribute utility functions as we shall see in section 8.2.1 (Thurston and Carnahan, 1992, Thurston, 2006B). Other approaches model the value function and find the utility over this value function. This approach has the benefit that it does not require us to verify independence conditions, which will also be discussed in the next subsection.

8.2.1   Single-Attribute based formulations

Single attribute based formulations utilize the information from single attribute utility functions to determine the multiattribute utility function. This method has the clear advantage that it requires a small number of assessments. Single attribute utility functions can be assessed as explained in chapter 7.

Attributes however need to satisfy some relatively strong independence conditions for one to be able to use the single attribute utility functions directly. In the following paragraphs we define three different types of independence conditions.

8.2.1.1   Independence conditions[2]

The independence conditions in this section have to do with a decision maker's preferences. These conditions measure how independent the attributes of a problem are when determining the worth of an option or a design to a decision maker. Attributes being independent here has nothing to do with probabilistic dependence and this concept should be carefully understood before proceeding. As an example, the size of a chair may determine its utility independently from the weight of the chair, even though there is almost perfect correlation (probabilistic dependence) between the size of a chair and its weight for a given design and material. In fact, if attributes were not correlated we could optimize them separately and no multiattribute decision theory would be needed.

---

[2] The reader is referred to Clemen (1997) and Keeney and Raiffa (1994) for a complete treatment of the subject.

Preferential Independence

Preferential independence between attributes means that the preference order within an attribute is not affected by the level of other attributes. This is best illustrated with an example. Assume that you are buying an incandescent bulb for an optical imaging experiment and your attributes of concern are the cost, C and the mean time to failure, M (MTTF). You want to minimize cost and maximize MTTF. Let us assume that your ranges of negotiability are [$10, $100] and [100 hours, 1000 hours] for the two attributes respectively. Now arbitrarily fix one of the attributes to a value within its range of negotiability, for example fix the cost at $40. For this value of cost you should be able to say that you prefer higher MTTF to lower. If you fixed cost at $60, then you should still prefer a higher MTTF to a lower one. If this preference order does not change no matter where cost is fixed then we can move to the other attribute. Fixing MTTF at a value within the interval now, you are asked if you prefer less cost to more. MTTF is then changed to a different fixed value and the decision maker is asked the same question again. If the preference order does not change regardless of where MTTF is fixed, then we have established that the attributes are preferentially independent. In general, in the case of two attributes x_1 and x_2, with ranges of negotiability $[x_{1,min}, x_{1,max}]$ and $[x_{2,min}, x_{2,max}]$ we must have for preferential independence:

If

$$x_{11} > x_{12} \Rightarrow U(x_{11}, x_2) > U(x_{12}, x_2) \quad \forall x_2 \in [x_{2,min}, x_{2,max}]$$
or
$$x_{11} > x_{12} \Rightarrow U(x_{11}, x_2) < U(x_{12}, x_2) \quad \forall x_2 \in [x_{2,min}, x_{2,max}]$$

$$(S1)$$

then x_1 is preferentially independent of x_2.

And if,

$$x_{21} > x_{22} \Rightarrow U(x_1, x_{21}) > U(x_1, x_{22}) \quad \forall x_1 \in [x_{1,min}, x_{1,max}]$$
or
$$x_{21} > x_{22} \Rightarrow U(x_1, x_{21}) < U(x_1, x_{22}) \quad \forall x_1 \in [x_{1,min}, x_{1,max}]$$

$$(S2)$$

then x_2 is preferentially independent of x_1.

If both statements S1 and S2 are true then the attributes are mutually preferentially independent. When more than two attributes are involved, preferential independence should exist between every pair of attributes. When this occurs one can say that mutual preferential independence exists for the given set of attributes. We give two examples of cases where preferential independence might not hold:

1. Consider the structural design of an automobile. Increasing the weight initially in an automobile frame is critical to stability and strength, however beyond a certain point one would like to start reducing weight to gain better fuel economy. Therefore weight and fuel economy of an automobile need not be preferentially independent.
2. Consider the choice of wine to accompany a main dish. Most people prefer red wine to white wine with meat, but white to red wine with fish. Thus, attributes "type of food" and "type of wine" are usually not preferentially dependent.

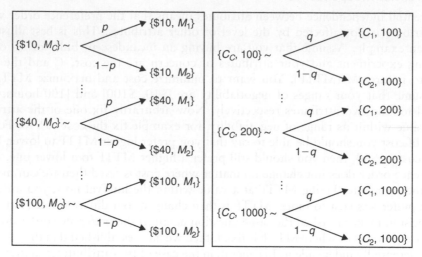

Figure 8.4 Testing for utility independence. The conditions need to hold for all values of $M_1, M_2,$ C_1, C_2, p and q (M_c and C_c will change accordingly for each test).

It should be noted that we are only concerned with independence within the range of negotiability (see statements S1 and S2) and for the decision maker in question. It is possible that two attributes are independent in their ranges of negotiability but not outside it. Additionally, it may be possible that two attributes are preferentially independent for one decision maker and not for another. This holds true for preferential as well as utility and additive independence conditions that we will talk about in the next few paragraphs. Satisfying the independence conditions is a great argument for using proper ranges for attributes.

Preferential independence is generally a weak condition that is easily satisfied and tested for. Even in cases where preferential independence is not satisfied, it is relatively simple to reformulate the objectives so that they do satisfy the conditions. Unfortunately, the multiattribute formulations in this section cannot work just with preferential independence; stricter conditions are needed as we shall see.

Utility independence

Utility independence is a much stronger condition than preferential independence. Utility independence exists when the certainty equivalent between uncertain deals on one attribute are unaffected by the levels of other attributes. In the above example with the light bulb, first fix the cost at a value, say $40, and create a lottery deal between two levels of the MTTF, M_1 and M_2. From the lottery deal find the certainty equivalent MTTF, M_C of the bulb, i.e. the value of MTTF that would make you indifferent between a bulb that costs $40 and has MTTF of M_1 and M_2, with probabilities p and $1 - p$, and a bulb that also costs $40 and but has a known MTTF of M_C. Attributes MTTF and cost are independent, if and only if, probability p does not change with cost.

Figure 8.5 Testing for additive independence. The conditions need to hold for all values of $C_1, C_2,$ M_1 and M_2.

Figure 8.4 demonstrates all the lotteries that need to be assessed to determine utility independence. While the conditions need to hold for all the values of the attributes, testing a few points is generally enough to establish confidence that utility independence exists. The points however need to be chosen carefully so that the whole range of negotiability is well covered.

As with preferential independence, if more than two attributes are present, independence conditions need to be ascertained for all the attributes. If they hold, one can say that the attributes are mutually utility independent. It is a good idea to first check for preferential independence as it is easier. If preferential independence is not found, one can safely assume that single attribute utility functions cannot be directly used.

Additive independence

Additive independence is the strongest independence condition for multiattribute decision making. Consider a set of uncertain deals with equally likely outcomes. This condition implies that the decision maker's valuations of uncertain deals in both the attributes together remains the same, even when values of one of the attributes are switched between the choices of the uncertain deal. An example of the lottery for the bulb purchasing decision should illustrate additive independence. The decision maker should be indifferent between the lotteries given in Figure 8.5, for all values of C_1, C_2, M_1 and M_2. As with utility independence, in practice one does not need to determine additive independence for all the possible points in the ranges of negotiability. Testing for some carefully chosen points is usually enough to ascertain that additive independence holds. It should also be clear from Figure 8.5 that additive independence is symmetric, i.e. if C is additively independent of M, then M is additively independent of C.

As with preferential and utility independence, if more than two attributes are present, additive independence conditions need to be ascertained for all the attributes. If they hold, one can say that the attributes are mutually additively independent. Additive independence implies utility independence.

8.2.1.2   *The additive form of multiattribute utility function*

This is the simplest form of multiattribute formulation and simply assumes that the multiattribute utility is a linear sum of the utilities from the attributes, weighted accordingly. Mathematically:

$$U(x_1, \ldots, x_n) = \sum_{i=1}^{n} k_i U_i(x_i)$$

$$\sum_{i=1}^{n} k_i = 1 \tag{8.1}$$

where U is the multiattribute utility function, x_i's are the attributes, U_i's are the individual attribute utility functions and k_i's are the associated scaling constants.

As is clear from the formulation, the utility function does not allow interaction between the attributes. The contribution to the overall satisfaction level (multiattribute utility) of an attribute is independent of the level of the other attributes. This form of utility function requires additive independence condition as presented before (which implies utility independence). While the conditions are rarely satisfied, this utility function is used extensively because of its ease of use. It is easy to assess the scaling constants, k's associated with the attributes since they are constrained to add up to 1.

8.2.1.3 *The multi-linear form of multiattribute utility function*

If additive independence is not present but utility and preferential independence exist, one can use the multi-linear formulation. This form of multiattribute utility formulation relaxes the constraints that the scaling constants have to add up to one. By allowing them to add up to values different than one they allow for slightly more complex tradeoff behavior. The multiattribute utility formulation is a linear sum of the utilities from the attributes but also includes interaction terms. The interaction term accomplishes two tasks. Firstly, it normalizes the multiattribute utility function so that when individual attribute utilities are equal to 1, so is the multiattribute utility. Secondly, it models how the attributes interact (Keeney and Raiffa, 1994). In the case where the sum of the scaling constants is less than one, the multiattribute utility needs contribution from the interaction terms. In this case the attributes are considered *complements*. In the case where the sum is more than one, the attributes are considered *substitutes*. Examples of attributes that are complements are the availability of a computer and a high speed internet connection. Individually they have value which is significantly enhanced when the two are present together. An example of attributes that are substitutes is the profit of a large company from its subdivisions. Good performance in one division can substitute for below par performance in another.

The multi-linear function for two attributes, x_1 and x_2, is given by:

$$U(x_1, x_2) = k_1 U_1(x_1) + k_2 U_2(x_2) + (1 - k_1 - k_2) U_1(x_1) U_2(x_2) \tag{8.2}$$

where U_1, U_2, k_1 and k_2 are the utility functions and scaling constants associated with the attributes respectively.

The multi-linear form can be extended to multiple attributes using the multiplicative function as shown below. This form allows for interaction between pairs of attributes as well as higher order combinations up to n, where n is the number of attributes.

$$U(x_1, \ldots, x_n) = \frac{1}{K} \left(\prod_{i=1}^{n} (Kk_i U_i(x_i) + 1) - 1 \right) \tag{8.3}$$

where x_i's are the attributes, k's are the scaling constants associated with the attributes and U_i's are the individual attribute utility functions. K is the normalizing parameter required to scale the multiattribute utility between 0 and 1. The value of K is found by using the boundary condition that when the individual attributes are at their best possible level (all U_i's are equal to 1), the multiattribute utility is equal to 1. Mathematically, K is given by the equation:

$$K = \prod_{i=1}^{n} (Kk_i + 1) - 1 \tag{8.4}$$

It is clear that all that is needed to ascertain K is the individual scaling constants. For two attributes the closed form expression is given by $K = \frac{1 - k_1 - k_2}{k_1 k_2}$. For number of attributes equal to n, a polynomial equation in K, of order $n-1$ is formed, which can be solved arithmetically or numerically. While multiple solutions exist for the polynomial equation, we should select the solution where K lies between -1 and 0 when the sum of k_i's is greater than one. On the other hand, a positive value of K should be chosen when k_i's add up to less than one.

When independence conditions do not exist, the facilitator should be prepared to modify the attributes. The facilitator can try to combine two attributes into one. For example, combining the horsepower and weight of an automobile into a single attribute of horsepower per kg, can solve possible lack of independence issue. Another way to alleviate this issue would be to redefine an attribute as deviation from nominal value, for example, blood glucose level. If none of these work the facilitator should use value function based approaches as described in the next section.

8.2.2 Assessing the scaling constants

Here we illustrate with a short example how scaling constants can be assessed for additive and multilinear utility functions.

Example 8.1: Let us consider a small company interested in installing a wind turbine to power its business establishment. The parameters the company is considering are: peak power, P in megawatts, maintenance interval, T_m in years and cost, C in dollars. The ranges of negotiability are [0.5 MW, 3 MW], [1 year, 10 years] and [\$1 million, \$10 million]. The company has linear utility functions for these individual attributes. The company is evaluating two turbines made by two different manufacturers with attributes equal to {2 MW, 3 years, \$2 million} and {3 MW, 4 years and \$4 million}. The company thinks that the three attributes satisfy the additive independence conditions and therefore the additive utility function as shown in equation 8.1, could be used.

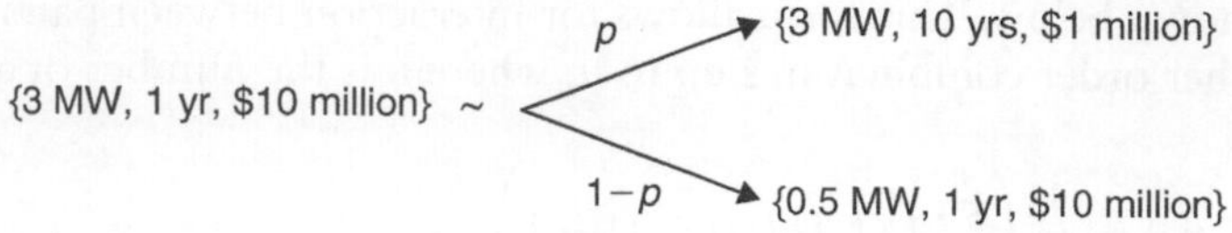

Figure 8.6 Finding the indifference probability and thus the scaling constant associated with peak power for the wind turbine problem.

To assess the scaling constants, the company is asked for its indifference probability for the lottery in Figure 8.6. Note that the company should estimate a probability p, that would make the company indifferent between a certain choice of the best in one attribute and worst in the other two attributes and a lottery that gives the best possible option on all attributes with probability p, and the worst option on all attributes with probability $1-p$. Figure 8.6, shows the lottery to assess the scaling constant for the attribute of peak power.

If the company says that p for the lottery in Figure 8.6 is 0.5, using the additive multiattribute utility formula and the expected utility axiom we get:

$$0.5U(3, 10, 1) + 0.5U(0.5, 1, 10) = k_P U_P(3) + k_{Tm} U_{Tm}(1) + k_C U_C(10)$$

Since the utilities for these extreme cases are known, we have:

$$0.5 \cdot 1 + 0.5 \cdot 0 = k_P \cdot 1 + k_{Tm} \cdot 0 + k_C \cdot 0$$

Therefore $k_P = 0.5$. Similarly other scaling constants are found to be 0.3 and 0.2 for maintenance interval and cost respectively. Note that the scaling constants add up to one because additive independence is satisfied. Using this information and the fact that the individual utility functions are linear we have the following utility function for the company.

$$U(P, T_m, C) = 0.5 \frac{P - 0.5}{3 - 0.5} + 0.3 \frac{T_m - 1}{10 - 1} + 0.2 \frac{10 - C}{10 - 1}$$

Comparing the two turbines, the first turbine has utility 0.544 and the second 0.733. Therefore, the company should purchase the second turbine. Scaling constants for the multilinear and the multiplicative forms shown before can be assessed in a similar way.

8.2.3 *Value function based formulations*

While it is possible to redefine attributes, a value function based approach alleviates lack of independence between attributes. As mentioned earlier, a value function is a deterministic map of the worth of a design or an option (Keeney and Raiffa, 1994). While a value function does not model the risk attitude of a decision maker, it does help identify his/her ranking of different alternatives. We can illustrate this with the help of an example.

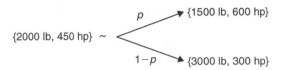

Figure 8.7 Finding the indifference probability to assess Heidi's utility function in the car buying decision.

Example 8.2: Consider Heidi, a racing enthusiast, who wants to buy a high performance car. Money is not an issue for her and the only two attributes that she considers are the weight, w, and the horsepower of the car, h. After spending some time with the facilitator it is found that the two attributes are not even utility independent for Heidi (which precludes use of any of our previously discussed formulations). The facilitator now uses a value function approach to ascertain her utility function.

The facilitator asks Heidi what ranges she is willing to consider alternatives in. The responses are 1500 lb to 3000 lb for weight and 300 hp to 600 hp for horsepower. The facilitator now starts with a candidate car within the feasible range, say 2000 lb and 450 hp. The facilitator now asks Heidi how much the horsepower has to increase if the car weighs 2500 lbs to make her indifferent between the candidate car and this new alternative. Heidi's response is 550 hp. Now the facilitator reduces the weight by 300 lb and asks for a new indifference point to which Heidi says 390 hp. The facilitator uses a value function of the form

$$V(w,h) = \frac{h^q}{w} \tag{8.5}$$

Fitting the values given by Heidi, the facilitator arrives at the value of 1.11 for the parameter q. Now, the facilitator presents Heidi with a simple lottery question using the best and the worst alternatives.

Using her indifference probability, one of the points on the utility function of Heidi can be ascertained. The utility function in this case would be *over the value function* that we established. For $p = 0.7$, we ascertain that $U(2000, 450) = 0.7$. Similarly other points on her utility function can be found. If the exponential form of the utility function[3] is assumed, the utility function can be written as:

$$U(w,h) = 1 - e^{-\frac{V(w,h)}{R}} \tag{8.6}$$

However to normalize the utility between 0 and 1, we need to subtract from the value function the value when the attributes are at the worst level. This number is $V(3000,300) = 0.1873$. Also we need to divide the utility by its value when the

---

[3] The exponential single attribute utility function is widely used for risk averse decision makers. Only one parameter R called the *risk-tolerance* needs to be assessed to fully define the function. The attribute sometimes needs to be redefined so that the utility increases in the right direction, normalization of the utility function may also be required as shown in equation 8.7.

attributes are at the best level i.e. $w = 1500$, $h = 600$. This guarantees that utility will be equal to 1 when the attributes are at the best level. The utility function is given by:

$$U(w, h) = \frac{1 - \exp\left(-\frac{V(w,h) - 0.1873}{R}\right)}{1 - \exp\left(-\frac{V(1500,600) - 0.1873}{R}\right)} \tag{8.7}$$

We calculate R from the indifference probability of 0.7 for the lottery in Figure 8.7 to be 0.2448. Writing the utility function explicitly as a function of w and h, we get:

$$U(w, h) = \frac{1}{0.9209}\left[1 - e^{-\left(\frac{h^{1.11}}{0.2448w} - 0.765\right)}\right]$$

This function can help her compare car options and select the one that maximizes her utility.

It should be kept in mind that although value function based approaches appear simple since they do not require independence conditions, they are limited in that a form of value function needs to assumed. If the form of value function is not chosen appropriately, the tradeoff behavior will not be modeled correctly. Moreover, assessing value functions requires many assessments to model the tradeoffs correctly. The number of assessments required would increase exponentially with the number of attributes, making it tedious for the decision maker as well as the facilitator. Moreover, since tradeoffs need to be made over many attributes, the decision maker may become overwhelmed and not give consistent answers.

8.2.4 Attribute dominance utility and multiattribute utility copulas

A recently proposed approach addresses the issue of modeling the multiattribute utility function using an analogy between probability distributions and utility functions (Abbas and Howard, 2005). The concept of attribute dominance utility is proposed which corresponds to multiattribute utility functions that are equal to 0 when any of the attributes is equal to its lowest acceptable level. This condition is used in addition to the usual ones that utility is increasing in individual attributes and is scaled between 0 and 1. Under these conditions, the utility function resembles a cumulative probability distribution function (except the property of cumulative probability distribution functions that they are n-increasing i.e. have a positive mixed derivative).

The analogy has many intuitive advantages. It brings a lot of results in probability theory into our arsenal. Abbas and Howard show how the attribute dominance formulation allows for notions such as utility inference along the lines of Bayes' rule. As mentioned earlier, in the case of multiple attributes, utility functions do not satisfy the condition of a positive mixed derivative which probability distributions, by definition, do. Abbas (2009) shows how the issue can be resolved using multiattribute utility copulas.

8.3 Solving Decision Problems under Uncertainty using Multiattribute Utility Analysis

Let us revisit the inventor-financier problem (example 7.4) in Chapter 7.

Example 8.3: Let us now look at the problem from a buyer's perspective, i.e. an individual (a decision maker in his own right) who would buy the mechanism from the inventor-financier. He has three attributes to consider: the stroke, the imbalance angle and the cost. He determines that he will be perfectly satisfied with the mechanism if the stroke was equal to 60 mm, imbalance angle was 20 degrees and the cost was $100, i.e. $U(60, 20, 100) = 1$. Similarly he believes that he will be totally dissatisfied if the stroke had an error of 5 mm from nominal, imbalance angle had an error of 2 degrees from nominal and the cost was $1200, i.e. $U(55/65, 18/22, 1200) = 0$.

Since the attributes are mentioned in terms of deviation from the nominal we define three attributes for the buyer: error in stroke, e_s, error in imbalance angle, e_i, and cost, c. The first two errors are the absolute values of the deviations of the stroke and imbalance angle from their nominal values. Next we need to determine single attribute utility functions over these attributes using the methods shown in the previous chapter. The buyer turns out to be risk-averse having exponential single attribute utility functions with risk tolerances given by 2 mm, 2 degrees and $500 for the three attributes.

$$
U_{e_s}(e_s) = \begin{cases} \dfrac{1 - e^{-\frac{5-e_s}{2}}}{1 - e^{-\frac{5}{2}}} & e_s \leq 5 \\[2ex] 0 & e_s > 5 \end{cases}
$$

$$
U_{e_i}(e_i) = \begin{cases} \dfrac{1 - e^{-\frac{2-e_i}{2}}}{1 - e^{-1}} & e_i \leq 2 \\[2ex] 0 & e_i > 2 \end{cases}
$$

$$
U_c(c) = \begin{cases} \dfrac{1 - e^{-\frac{1200-c}{500}}}{1 - e^{-\frac{1100}{500}}} & 100 \leq c \leq 1200 \\[2ex] 1 & c < 100 \\[2ex] 0 & c > 1200 \end{cases}
$$

Next, the buyer tests for independence conditions and realizes that while additive independence conditions are not met, utility and preferential independence conditions are. This allows him to use the multiplicative utility function shown in Equation (8.3). He assesses his scaling constants by comparing an option which is best in one attribute and worst in others, with an uncertain lottery which gives the best option in all attributes with probability p and the worst option with probability $1 - p$. After three such comparisons (one for each attribute as shown in Figure 8.8) he reaches the conclusion that $k_{es} = 0.4$, $k_{ei} = 0.6$ and $k_c = 0.5$.

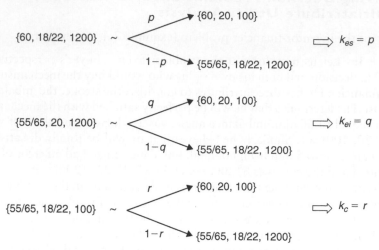

Figure 8.8 Assessing the scaling constants using indifference probabilities for the mechanism purchasing decision. Actual stroke and imbalance angles are shown instead of deviations for clarity.

The normalizing parameter, K for the assessed values of the scaling constants is -0.773, and the multiattribute utility is given by:

$$U(e_s, e_i, c) = \frac{1}{-0.773}[(-0.773 \cdot 0.4 \cdot U_{e_s}(e_s) + 1)(-0.773 \cdot 0.6 \cdot U_{e_i}(e_i) + 1)$$
$$\times(-0.773 \cdot 0.5 \cdot U_c(c) + 1) - 1]$$

Now suppose that the buyer has been offered a mechanism with stroke 58 mm, imbalance angle 21 degrees, at a price of $800. He can choose to accept it or wait for the company to manufacture another one. The parameters in that case could be $\{60, 20, 1000\}$ with probability 0.6 or $\{56, 22, 850\}$ with probability 0.4. To decide, we need to compare the utility of the mechanism he is certain to receive with the expected utility of the mechanism that he would get after the company manufactures it.

The utility from the given mechanism is:

$$U(e_s = 2, e_i = 1, c = 800) = 0.776$$

The expected utility from the mechanism to be manufactured is given by:

$$0.6U(e_s = 0, e_i = 0, c = 1000) + 0.4U(e_s = 4, e_i = 2, c = 850)$$
$$= 0.6 \cdot 0.883 + 0.4 \cdot 0.417 = 0.697$$

Therefore, the buyer should purchase the one that is offered to him now because its multiattribute utility is higher than that of the uncertain one that will be manufactured.

8.4 Conclusions

Most decision problems involve tradeoff between multiple conflicting attributes. In this chapter, we talked about the natural extension of decision analysis methods for one attribute into multiple attributes. Reconciling these attributes with the preferences of the decision maker is a challenging task. Value functions and utility functions allow us to compare different alternatives with different levels of attributes. Utility functions have the added property that they can model the risk attitude of the decision maker. We can use value functions when there is no uncertainty, but we need utility functions when uncertainty is present. We also studied various ways to construct a multiattribute utility function of a decision maker. We demonstrated the direct use of single attribute utility functions when certain independence conditions are met. Finally, the chapter presented some multiattribute utility formulations and illustrated their use with examples.

Multiattribute decision theory is strongly rooted in mathematics, and it provides a normative way to make decisions that are consistent with a set of logical rules (axioms). In engineering problems if the decision maker's preferences are known, then the decision making process can be automated. Numerous additional challenges are encountered when decisions include tradeoff between highly subjective attributes such as health, happiness and comfort. Reconciling normative decision making methods with descriptive decision making is an ongoing field of research.

Questions and Exercises

Tradeoff between attributes

8.1   Think of a decision that you have to make in the near future. Clearly outline the attributes and the sources of uncertainty.

8.2   In problem 8.1, find a candidate solution and try to find isopreference curves between any two attributes of your choice.

8.3   Determine the range of negotiability for your attributes in Problem 8.1.

8.4   Consider the design of a moon-roof in a car. There are many attributes to consider like area, opening and closing mechanism, aspect ratio, and effect on the structural integrity of the car. There is also an issue of leakage during rain if sealing is not proper. Outline how a designer can find his tradeoff preferences and ranges of negotiability for these attributes.

8.5   Write down your attributes of concern for the following decision making problems

   a. Selecting a school to get your master's degree
   b. Buying a car
   c. Picking a destination for a vacation
   d. Designing a new cell phone
   e. Formulating climate policy for the government
   f. Voting in an election
   g. Starting a business

8.6   Determine your ranges of negotiability for the different attributes that you considered in Problem 8.5.

8.7   How can you tell from scaling constants when two attributes are complements, or substitutes?

8.8   Give an example of a decision problem that you could have solved just by using value functions.

8.9   A civil engineer wants a beam to have a certain constant cross sectional area. However, he does not want the beam to be too heavy. He formulates the decision problem using the cross sectional area and weight as two attributes. Now since the beam has a constant cross sectional area, the two attributes are perfectly correlated. Does this violate any of the independence conditions? If yes, which one(s), if no, why not?

Different multiattribute formulations

8.10   Consider an aircraft designer that wants to understand the interaction of system level objectives for his design. He considers three attributes for the design: cost (in millions of dollars), seating capacity and fuel economy in gallons/mile. His ranges of negotiability are [10, 30], [100, 200] and [1, 3]. His scaling constants for the attributes are 0.3, 0.5 and 0.6 respectively. His individual attribute utility functions are linear in the ranges and he prefers: low cost to high, more seating to less and higher fuel economy to lower. Compare the following designs, represented as {cost, seating capacity, fuel economy} using the multiplicative utility function:

a. {15, 150, 3} versus {12, 120, 1.5}
b. {20, 140, 2} versus {27, 110, 2.5}
c. {15, 190, 3} versus {11, 160, 1.6}
d. {30, 110, 1.5} versus {18, 200, 1}

8.11   While performing simulation, an engineer has to decide between accuracy and total duration of simulation. These are conflicting objectives. He has a way of determining the objective value without doing any runs which has an expected error of 5 units from the true value. The simulation brings this value closer at a linear rate of 1 unit per hour. The engineer defines his utility function between 0 and 10 units for the error, and between 0 and 5 hours for the simulation time. It is also known that he is risk averse for both the attributes with risk tolerance equal to 2 units and 2 hours for the two objectives (see exponential utility function in example 8.2). Determine the values of the scaling constants of a simple additive utility function if he is indifferent between running the simulation and simply finding the objective using the alternate method. (Hint: Normalize the utility function within the ranges of negotiability.)

8.12   In the wind turbine selection decision problem in this chapter, the company suddenly decides to add another attribute of distance from the business. What additional aspects does the company have to consider before it can start comparing the options.

Solving decision problems under uncertainty using multiattribute utility analysis

8.13   We can extend the concept of risk attitude to the multiattribute case, based on whether the multiattribute utility function is concave, convex or linear.

Determine, if possible, for each case below the decision maker's risk attitude if their utility function (non-normalized) is given by:

a. $U(f_1, f_2) = f_1 f_2^{1.5}$ $f_1, f_2 > 0$

b. $U(f_1, f_2) = f_1^2 f_2^{0.5}$

c. $U(f_1, f_2, f_3) = f_1 f_2 f_3$ $f_1, f_2, f_3 > 0$

d. $U(f_1, f_2) = a f_1^2 + b f_2^2$ $a, b > 0$

e. $U(f_1, f_2) = a f_1^2 + b f_2^2$ $a, b < 0$

f. $U(f_1, f_2) = a f_1^2 + b f_2^2$ $a < 0, \; b > 0$

g. $U(f_1, f_2, f_3, f_4) = f_1 + f_2 + f_3 + f_4$

Hint: Find the Hessian of the utility function and check to see if it is positive/negative definite for convexity/concavity. This check can be inconclusive for some of the above utility functions.

8.14 Assume that your utility from coffee is dependent on the concentration of caffeine (tablespoons of ground beans to make a cup), concentration of sugar (tablespoons per cup) and concentration of creamer (tablespoons per cup). Your preference is going to be non-monotonic for each of these. Construct a multiattribute utility function. Comment on the tradeoff relationships.

8.15 A company is planning to acquire another. There are two attributes the company is considering, profit in the long run and market share. If it acquires the other company, profit is triangularly distributed between $1 million and $3 million with the most likely value of $1.4 million. Market share is approximately normally distributed with mean of 30% and standard deviation of 6%. If it does not acquire the company, profit and market share are deterministic at $2 million and 20% respectively. Assume that the multiplicative utility function can be used with scaling constants of 0.4 and 0.3 respectively for profit and market share respectively. Also assume that exponential single attribute utility functions as shown in example 8.3 can be used with risk tolerance $500,000 and 10%. Should the company go ahead with the acquisition?

8.16 A designer is considering the design of a turbine blade. He has narrowed down to a few hundred designs based on his simulations. While other performance parameters are all acceptable for the designs, the three attributes he is considering now are weight, nominal length and nominal thickness. Due to complex curvature and width, the three attributes are not perfectly correlated. The designer is indifferent between these three designs: {2 kg, 7 in, 0.34 in}, {3 kg, 6.2 in, 0.34 in}, {2.5 kg, 8 in, 0.50 in}. Find a suitable value function based on this information, which will help him choose the best one among the hundreds, if in the vicinity of these designs he prefers a lighter blade to heavier, shorter blade to longer and thick blade to a thin one.

Glossary

p-value, p. 171   The probability of obtaining a more extreme value of a test statistic than the observed one. This value is also called observed level of significance.

σ-algebra or Borel field, p. 43   A class of subsets of the sample space which is closed under all countable set operations (union, intersection, and complementation).

Additive Independence, p. 492   This term implies that a decision maker's valuations of 50-50 uncertain deals involving multiple attributes remains the same, even when values of one of the attributes is switched between the choices of the uncertain deal.

Aleatory Uncertainty, p. 2   Uncertainty that is due to inherent randomness.

Association or dependence of random variables, p. 356   Two random variables are said to be associated if the values that one assumes are linked to the value of the other. If an expert were to learn the value of one variable then this would affect the expert's beliefs about the value of the other. For example, if an expert learns that the demand of a new car model in the US were higher than what the expert originally expected, then the expert will believe that the demand in Canada would be also higher than expected.

Basic Probability Assignment (BPA) Function, p. 31   A function that assigns a value to a set A that is equal to the part of the probability of A that cannot be assigned to its subsets.

Bayesian estimator of a variable, p. 367   The conditional expectation of the variable given the available evidence.

Bernoulli process, p. 51   A series of Bernoulli trials.

Bernoulli trial, p. 51   An experiment with two possible outcomes.

Body of evidence, p. 31   The collection of all focal elements and the corresponding values of the BPA function.

Calibration of subjective probability assessments, p. 330   The process of comparing these assessments with observed frequencies.

Central credible interval, p. 354   an interval that contains variable X with subjective probability q so that $P(X \leq x_1) = P(X > x_2) = 1 - \frac{q}{2}$.

Central limit theorem, p. 90   In its simplest form, this theorem states the for a population with finite variance, the probability distribution of the standardized sample mean approaches the standard normal distribution with the sample size tending to infinity.

Certainty Equivalent of a gamble, p. 424   The sure amount that is equivalent to the uncertain payoff of the gamble. Different people have different certainty equivalent for the same gamble.

Clarity Check (Test) of a Decision Model, p. 362   The check that addresses the following question: If, in the future, perfect information becomes available regarding all aspects of a decision, will it become possible to determine unequivocally what will happen in every node of an influence diagram or a decision tree representing the decision without using any judgment? No interpretation or judgment must be required to make this determination.

Coherent probabilities, p. 24   A decision maker's probabilities are called coherent if they satisfy the laws of probability theory.

Comprehensive copula, p. 202   A copula family that can model extreme dependence and independence.

Confidence Interval, p. 146   A range that has the following property: If we calculate the same range many times by using a different sample, this range will contain an uncertain parameter, such as a mean value or a long-term relative frequency, with a certain probability (coverage).

Convergence of a sequence of random variables to a value in the mean square sense, p. 120   Sequence $X_1, \ldots, X_n$ converges to a value x in the mean square sense if and only if the mean value of the square of the difference $X_n - x$ converges to zero with n.

Copula, p. 197   A monotonically increasing function that links the marginal CDFs of multiple variables to their joint CDF. For two variables copula $C_\theta(u, v)$ relates the joint CDF of these variables, $F_{XY}(x, y)$, to the marginal CDFs, $F_{XY}(x, y) = C_\theta\{u, v\}$ where, $u = F_X(x)$ and $v = F_Y(y)$.

Coverage, p. 47   Probability that the true value of an unknown parameter lies within a confidence interval.

Credible interval, p. 162 $q\%$-credible interval of variable X is an interval $[x_1, x_2]$ that contains the value of the variable with subjective probability q, $P(x_1 < X \le x_2) = q$.

Critical value, p. 153   A value that separates the rejection and non rejection regions in statistical hypothesis testing.

Cumulative histogram, p. 192   A histogram for which the height of the box in a bin is proportional to the sum of the observed values in this bin and the bins located on the left.

Decision Tree, p. 418   A graphical display of the sequence of choices and uncertain events in a decision.

Decision, p. 1   An irrevocable allocation of resources, or a choice of a course of action among alternatives in order to achieve a desirable payoff.

Decision-Making, p. 7   The process of choosing a course of action among alternatives.

Deductive reasoning, p. 135   The logical process in which a conclusion drawn from a set of premises contains no more information than the premises. For example, consider the premise "probability is a nonnegative number." Using deductive reasoning, we conclude that the probability of failure of a particular pump is nonnegative.

Dependence, p. 74   See association

Descriptive statistics, p. 133   The branch of statistics that focuses on collecting, summarizing and presenting data.

Epistemic uncertainty, p. 2   Uncertainty that is due to lack of knowledge.

Ergodic random process, p. 220   A stationary process is called ergodic if in addition to all ensemble averages being stationary with respect to a change in time, the averages taken along any sample function (time or temporal averages) are the same as the ensemble averages.

Event, p. 2   Set of outcomes of sample space.

Expected value of information in decision making, p. 435   The value gained by using the information to select the best course of action. This value is equal to the expected utility of the decision made by using the information minus the expected utility without the information.

Fault (Failure) tree, p. 341   A graphical representation of a logical model that shows a failure event and possible ways that this event may occur.

Feedback (Decision Analysis), p. 330   The process of checking the adequacy of a distribution that has been fitted to an expert's stated summaries, by telling the expert some implications of the fitted distribution and asking the expert if these implications represent his/her beliefs.

Focal Elements, p. 31   Events whose value of the Basic Probability Assignment function is greater than zero.

Fourier analysis, p. 227   It indicates the transformation of a waveform (signal) from the time domain to the frequency domain and vice versa.

Frame of a Decision, p. 3   A mental boundary that separates those aspects of a decision problem that one considers from those that one neglects.

Framing, p. 3   Selecting certain aspects of a situation or decision while excluding others.

Frequency response function, p. 266   It characterizes the response of a linear, time-invariant system in the frequency domain.

Histogram, p. 139   A graphical representation of observed values of a variable. We construct a histogram by dividing the range of values of a variable into bins or classes and counting the values that lie in each bin. Then we draw a box for each bin with base equal to the bin width, or proportional thereof, and height proportional to the number of observations in that bin.

Hyperparameter, p. 325   A parameter of a probability distribution that is considered as a random variable.

Hypothesis, p. 169   A tentative explanation about a phenomenon.

Imprecision   See Epistemic Uncertainty

Impulse response function, p. 265   It represents the response of a linear, time-invariant system at time t due to a unit impulse that is applied at a different time.

Independent events, p. 49   Two events are independent if knowing whether one event occurred does not change the probability of the other.

Independent experiments, p. 50   Experiments that have the following property; knowing the outcome of one experiment does not change the probability of the other.

Inductive reasoning, p. 135   A logical process in which a conclusion is proposed that contains more information than the observations or experience on which it is based. For example, every swan ever seen is white. Therefore, all swans are white.

Inferential statistics, p. 133   The branch of statistics that analyzes the data and estimates probabilities.

Information economics (in the context of decision making), p. 434   The systematic study of measuring the value of information in support of a decision.

Insufficient reason principle or principle of indifference, p. 47   A decision maker should consider that two or more outcomes are equally likely if there is no evidence that one outcome is more likely than the others (or there is no lack of symmetry).

Isopreference curve, p. 486   A set of points (curves in two dimensions) in the attribute space between which a decision maker is indifferent.

Joint probability distribution, p. 5   The probability distribution of multiple variables considered together.

Level of significance, p. 171   Probability of observing a test value that lies in the rejection region, given that the null hypothesis is true.

Likelihood Function, p. 184   Suppose that the probability density function of random variable X is a function of parameter Θ. We write this function as $f_X(x/\theta)$, where θ is a value of parameter Θ, to show the dependence of this function on the value of θ. If we view $f_X(x/\theta)$ as a function of θ, then the likelihood function will be $L(\theta/x) = f_X(x/\theta)$.

Likelihood of an Event, p. 48   Let A be an event and E the observed evidence, which depends on event A. The likelihood of evidence E is the probability of observing E conditioned upon the occurrence of event A, $P(E/A)$. The term likelihood of an event also refers to its probability.

Long term frequency of an event, p. 18   Frequency with which an event occurs in an infinitely long sequence of experiments. For example, the long term frequency of heads in a coin flip is the relative frequency of heads in an infinitely long sequence of flips of the coin.

Lottery, p. 22   An activity whose payoff depends on the outcomes of uncertain events.

Marginal probability distribution or margin, p. 69   The probability distribution of one random variable.

Maximum likelihood method, p. 184   A method of estimation in which a parameter is estimated by the value that maximizes the likelihood function.

Mean, p. 60   The sum of n numbers divided by n. The mean of a population is called population mean and it is denoted by Greek letter μ. The mean of a sample is called sample mean or average and is denoted by $\hat{\mu}$. The mean of the distribution of a random variable is called its expected value.

Metadecision, p. 408   The process of deciding what to decide.

Multiattribute utility function, p. 485   Any of the numerous formulations that order alternatives involving multiple attributes for a decision maker, taking into account his/her risk preference. Some examples are: additive, multilinear and multiplicative functions.

Net expected value of information in decision making, p. 435   Expected value of the information minus the cost of acquiring it.

Normal or Gaussian random process, p. 227   A random process $x(t)$ is called normal, or Gaussian, if for any set points in time, $t_1, \ldots, t_n$, the joint probability distribution of the corresponding random variables $x(t_1), \ldots, x(t_n)$ is normal.

Null hypothesis, p. 169   A claim about a characteristic of a population. Usually, this claim represents the status quo.

Objective (frequency) probability of an event, p. 16   The long term frequency of the event. According to this definition, the probability of an event is only defined if it is repeatable. We cannot define the objective probabilities of one-time events. We cannot define the probability that a proposition is true, such as "Isaac Newton was born before 1500 AC." This means that objective probability is only usable to quantify random uncertainty, not epistemic uncertainty.

Objective (noun), p. 401   A desired outcome of a course of action.

Observable quantity, p. 362   A quantity whose value can be determined unambiguously from an experiment. No judgments should be required to determine the value of this quantity after the experiment has been completed and the results have become available to the analyst.

Overfitting, p. 354   The practice of eliciting more judgments than needed to fit a parametric probability distribution that the facilitator believes that will represent accurately the subject matter expert's belief.

Parameter, p. 18   There are many definitions of the word parameter. The following two definitions are relevant to probability and statistics:
- Parameter is a quantity that is constant in a particular case, but varies in different cases.
- Parameter is a characteristic of a population, such as the population mean and standard deviation.

Parametric distribution, p. 327   A distribution that is completely defined once we define the values of its parameters.

Parametric family, p. 206   The family of distributions that are obtained by varying the values of the parameters of a distribution.

Pareto front, p. 487   A set of non-dominated solutions for a multiattribute decision problem. None of the solutions is decisively better or inferior to others in the set. This is because if they sacrifice some attributes, they gain improvement in others. Solutions other than those on the Pareto front are dominated by at least one solution of the front.

Population, p. 134   All members in a group about which we want to draw a conclusion. The term finite population refers to a group with finite number of members. The term infinite population refers to the infinite number of values that a continuous random variable can assume.

Posterior Probability Density Function, p. 367   The probability density function updated in view of the available evidence.

Posterior Probability, p. 367   The probability of an event after it is updated on the basis of the available evidence.

Power spectral density, p. 242   A function that provides information about the frequency content of the random process indirectly, indicating the distribution of energy of a random process over the frequency spectrum.

Preferential independence, p. 491   Preferential independence between attributes means that the preference order within an attribute is not affected by the level of other attributes.

Prior Probability Density Function, p. 366   The probability density function before evidence becomes available.

Prior Probability, p. 48   The probability of an event before observing some evidence.

Probability Mass Function, p. 57   A function that shows the probabilities of all possible values of a discrete random variable.

Quantile, p. 95 q-quantile of the probability distribution of variable X is a value x such that $P(X \leq x) = q$.

Quartiles, p. 154   Values that split the range of variable X into intervals such that there is a 0.25 probability that variable X could be less than or equal to the first quartile, a 0.5 probability that variable X could be less than or equal to the second quartile and so on.

Random field, p. 215   It is a generalization of the concept of a random process. A random field is a random function of two or more parameters.

Random process, p. 215   A family or ensemble of time functions. Each time function is called *sample function*, *ensemble member* or *realization* of the random process. It shows the variation of the process with time for a particular realization.

Random sample, p. 85   A set of n observations from a population. The population could be finite or infinite. In the first case, each possible combination of observations is equally likely to be drawn from the population. Mathematically, a random sample is a set of values of independent random variables having the same (population) probability distribution.

Random set, p. 34   The information consisting of focal elements and their probabilities.

Random Variable, p. 42   A function defined over the sample space of an experiment that assigns to every outcome a real value.

Range of negotiability, p. 489   The range of an attribute within which a decision maker is willing to consider alternatives. Outside this range, either the decision maker is totally dissatisfied or believes that it is very unlikely that the attribute will assume a value.

Reliability of a system, p. 6   One minus the probability of failure of the system.

Safety index (or Reliability index) of a system, p. 122   The R-quantile of the standard normal distribution, where R is the reliability of the system.

Sample mean, p. 136   See mean

Sample Space or Universal Set, p. 42   Collection of all possible outcomes of an experiment. The sample space is also called certain event.

Second-order probabilistic model, p. 179   A probability distribution whose parameters are random variables with their own probability distributions.

Sensitivity analysis, p. 15   In a probabilistic study, sensitivity analysis is the process of investigating the effect of an assumption about the input on the final conclusion. For example, a sensitivity analysis involves the study of the effect of changes in the probability distribution of the wave height on the estimated probability of fatigue failure of a joint of an offshore platform.

Singleton, p. 30   A set containing a single element of the sample space.

Spectral analysis, p. 241   It provides information about the frequency content of a random process.

Spectral representation methods, p. 287   They characterize a random process and generate sample functions using a series of harmonic functions or the eigenfunctions of the covariance matrix of the random process.

Stationary random process, p. 218   A process is stationary if all its statistical properties (i.e. all moments) are time invariant, not changing under time shifts.

A random process is wide-sense (weak) stationary if the mean value of the process is constant, and the auto-correlation function does not depend on absolute time.

Statistic, p. 133 A characteristic of a sample or a population.

Statistical hypothesis testing, p. 169 An inferential method for evaluating a claim about a population.

Statistical Summary, p. 168 A characteristic of observed or theoretical probability distributions. These characteristics include a) graphical tools such as histograms, and empirical CDFs, b) numbers such as means, standard deviations, medians and quantiles, and c) ranges such as credible or confidence intervals.

Statistics, p. 41 The totality of methods of recording, organizing, analyzing and reporting quantitative information.

Strong preference, p. 315 The relation between two outcomes for which a decision maker believes that one is strictly more likely than the other.

Structuring a Decision Problem, p. 410 The process of specifying an analytical model representing a vaguely defined decision problem.

Test statistic, p. 170 The statistic used to determine whether a hypothesis is false. This is a value determined from a sample drawn from the population or from the assumed probabilistic model.

Time series, p. 279 It is a sequence of random observations taken over time such as a sequence of the yearly peak temperature. The time series models use past observations to develop statistical models that "closely" represent the phenomenon over time. Time series is used to characterize a random process in the time domain.

Uncertainty, p. 2 The state of being unsure about someone or something.

Utility, p. 7 A measure of the worth of a particular outcome to a decision-maker.

Utility independence, p. 492 Utility independence exists when the certainty equivalent of uncertain deals on one attribute are unaffected by the levels of other attributes.

Value (noun), p. 12 Any object or quality that is desirable to a person as a means or as an end. A person's values determine his/her objectives in a decision.

Value function, p. 489 A deterministic map between a set of attributes and its worth to the decision maker. The value function allows a decision maker to order deterministic alternatives.

Weak preference, p. 315 The relation between two outcomes for which a decision maker believes that one is at least as likely as the other.

References

Abbas, A. E., 2009, Multiattribute Utility Copulas, *Operations Research*, Vol. 57, No. 6

Abbas, A. E., and Howard, R. A., 2005, Attribute Dominance Utility, *Decision Analysis*, Vol. 2, No. 4, December 2005, pp. 185–206

Acar, E., Solanski, K., Rais-Rohani, M., and Horstemeyer, M. F., 2008, "Uncertainty Analysis of Damage Evolution Computed Through Microstructure-Property Relations," *Proceedings of the ASME 2008 International Design Engineering Technical Conferences and Computers and Information in Engineering Conference*, New York City

Agresti, A., and Coull, B. A., 1998, "Approximate is Better than "Exact" for Interval Estimation of Binomial Proportions," *The American Statistician*, 52, pp. 119–126

Alemi, F., 2006, "Probability of Rare Events," *Risk Analysis in Healthcare* (course notes), George Mason University, Fairfax, Virginia http://gunston.gmu.edu/healthscience/riskanalysis/ProbabilityRareEvent.asp

Allwright, D., Hazelwood, V., Maynard, T., 2006, "Estimating copula functions from limited data," Smith Institute, (download from the following link: http://www.smithinst.ac.uk/Events/Copulas/CopulasPaper/copula_paper.pdf)

Almond, R. G., 1995, *Graphical Belief Modeling*, 1st Edition, Chapman & Hall, London

Alpert, M., and Raiffa, H., (1982), "A Progress Report on the Training of Probability Assessors," in Kahneman, D., Slovic, P., and Tversky, A., 1982, *Judgment under Uncertainty, Heuristics and Biases*, Cambridge University Press

Amin, M., 1966, "Nonstationary Stochastic Model for Strong Motion Earthquakes," Ph.D. Thesis, University of Illinois

Ang, A. H-S., and Tang, W. H., 1975, *Probability Concepts in Engineering Planning and Design, Vol. I – Basic Principles*, John Wiley & Sons, New York

Ang, H-S. A., and Tang, W. H., 1984, *Probability Concepts in Engineering Planning and Design*, Volume II-Decision, Risk and Reliability

Aughenbaugh, J. M., and Herrmann, J., 2008, "A Comparison of Statistical Approaches for Assessing Reliability," *International Journal of Reliability and Safety* (under review)

Aughenbaugh, J. M., and Paredis, C. J. J., 2006, "The Value of Using Imprecise Probabilities in Engineering Design," *Journal of Mechanical Design*, Vol. 128, No. 4, pp. 969–979

Aughenbaugh, J. M., and Paredis, C. J. J., 2007, "Probability Bounds Analysis as a General Approach to Sensitivity Analysis in Decision Making Under Uncertainty," *SAE Special Publication*, SP-2119, pp. 397–411

Ben-Haim, Y. and Elishakoff, I., 1990, *Convex Models of Uncertainty in Applied Mechanics*, Elsevier Publishing Company

Ben-Haim, Y., 2001, *Information-Gap Decision Theory: Decisions Under Severe Uncertainty*, Academic Press

Berger, J. O., 1985, *Decision Theory and Bayesian Analysis*, Springer-Verlag, New York, Chapters 3 and 4

Booker, J. M., and McNamara, L. A., 2005, "Expert Knowledge in Reliability Characterization: A Rigorous and Approach to Eliciting, Documenting and Analyzing Expert Knowledge," *Engineering Design Reliability Handbook*, CRC Press, Boca Raton, Chapter 13, pp. 13-1 to 13-31

Borgman, L., 1969, "Ocean Wave Simulation for Engineering Design," *Journal of the Waterways and Harbors Division*, Vol. 95, pp. 557–583

Bossaerts, P. and Plott, C., 2004, "Basic Principles of Asset Pricing Theory: Evidence from Large-Scale Experimental Financial Markets," *Review of Finance*, 8, pp. 135–169

Bossaerts, P., Plott, C., and Zame, W., 2007, "Prices and Portfolio Choices in Financial Markets: Theory, Econometrics, Experiment," *Econometrica*, 75, pp. 993–1038

Brown, D. L., Cai, T. T., and DasGupta, A., 2001, "Interval Estimation for a Binomial Proportion," *Statistical Science*, Vol. 16, No. 2, pp. 101–133

Bury, C., 1999, *Statistical Distributions in Engineering*, Cambridge University Press, Cambridge, UK

Caers, J., and Maes, M., 1998, "Identifying Tails, Bounds, and End-Points of Random Variables," *Structural Safety*, 20, 1–23

Cafeo, J., Donndelinger, J. A., Lust, R. V., and Mourelatos, Z., 2005, "The Need for Non-deterministic Approaches in Automotive Design: A Business Perspective," *Engineering Design Reliability Handbook*, CRC Press, Boca Raton, Chapter 5, pp. 5-1 to 5-18

Castillo, E., 1988, *Extreme Value Theory in Engineering*, Academic Press, San Diego, CA

Caughey, T. K., and Stumpf, H. J., 1961, "Transient Response of Dynamic Systems under Random Excitation," *Journal of Applied Mechanics*, 28(4), p. 563

Charpentier, A., Fermanian, J.-D. and Scaillet, O., 2007, "The Estimation of Copulas: Theory and Practice, in *Copulas: From Theory to Application in Finance*, J. Rank (editor), Section 2, Risk Publications, London, UK

Choi, S.-K., Grandhi, R. V., and Canfield, R. A., 2007, *Reliability-Based Structural Design*, Chapters 6 and 7, Springer-Verlag, London, UK, pp. 203–235

Clemen, R. T., 1997, *Making Hard Decisions*, Second Edition, South-Western College Publications

Coles, S. G., and Tawn, J. A., 1996, "A Bayesian Analysis of Extreme Rainfall Data," Applied Statistics, Vol. 45, No. 4, pp. 463–478

Cook, D., Duckworth, W. M., Kaiser, M. S., Meeker, W. Q., and Stephenson, W. R., 2003, "Principles of Maximum Likelihood Estimation and the Analysis of Censored Data," *Beyond Traditional Statistical Methods*, http://www.public.iastate.edu/~stat415/meeker/ml_estimation_chapter.pdf

Cook, R., 2004, "The Anatomy of the Squizzel The Role of Operational Definitions in Representing Uncertainty," *Reliability Engineering and System Safety*, Vol. 85, Nos. 1–3, July-September, pp. 313–319

Cooke, R. M., 2001, *Experts in Uncertainty: Opinion and Subjective Probability in Science*, Oxford University Press, New York

Cooke, R. M., and Goossens, L. H., 2004, "Expert Judgment Elicitation for Risk Assessments of Critical Infrastructures," *Journal of Risk Research*, 7(6), pp. 643–656

Cooper, G. R., and McGillem, C. D., 1971, *Probabilistic Methods of Signal and System Analysis*, Holt, Rinehart and Winston, New York

Cramer, H., and Leadbetter, M. R., 1967, *Stationary and Related Stochastic Processes*, Wiley, New York, NY

Davenport, W. B., 1970, *Probability and Random Processes*, McGraw-Hill, New York

De Cooman, G., and Troffaes, M. C. M., 2004, "Coherent Lower Previsions in Systems Modelling: Products and Aggregation Rules," *Reliability Engineering and System Safety*, Vol. 85, Numbers 1–3, pp. 113–134

Der Kiureghian, A., and Liu, P-L., 1985, "Structural Reliability under Incomplete Probability Information," *Division of Structural Engineering and Mechanics, University of California at Berkeley*, Report No. CEE-8205049

Der Kiureghian, A., and Liu, P. L., 1986, "Structural Reliability under Incomplete Information," *Journal of Engineering Mechanics*, ASCE, 112(1), pp. 85–104

Donndelinger, J. A., 2006, "A Decision-Based Perspective on the Vehicle Development Process," *Decision Making in Engineering Design*, ASME Press, New York, Chapter 19, pp. 217–225

Ellsberg, D., 1961, "Risk, Ambiguity and the Savage Axioms," *Quarterly Journal of Economics*, Vol. 75, pp. 643–669

Farizal and Nikolaidis, E., 2007, "Assessment of Imprecise Reliability Using Efficient Probabilistic Re-analysis," *SAE 2007 Transactions, Journal of Passenger Cars – Mechanical Systems*, paper number 2007-01-0552

Ferson, S., Nelsen, R. B., Hajagos, J., Berleant, D. J., Zhang, J., Tucker, W. T., Ginzburg, L. R., and Oberkampf, W. L., 2004, *Dependence in Probabilistic Modeling, Dempster-Shafer Theory, and Probability Bounds Analysis*, Sandia Report, SAND2004-3072, Albuquerque, New Mexico

Fonseca, J. R., 2007, "Efficient Robust Design via Monte Carlo Simulation Sample Reweighting," *International Journal for Numerical Methods in Engineering*. Vol. 69, pp. 2279–2301

Fox, E., 2005, "The Role of Statistical Testing in NDA," *Engineering Design Reliability Handbook*, Chapter 26, pp. 26-1 – 26-25

French, S., 1986, "Objective and Subjective Probability," *Decision Theory: An Introduction to the Mathematics of Rationality*, Ellis Horwood Limited, Chichester, West Sussex, England

Fuchs, M, 2008, "Uncertainty Modeling in Higher Dimensions: Towards Robust Design Optimization," Ph. D. Dissertation, University of Vienna, Mathematics Department

Fung, Y. C., 1955, "The Analysis of Dynamic Stresses in Aircraft Structures during Landing as Nonstationary Random Processes," *Journal of Applied Mechanics*, 22, pp. 449–457

Gen, M., and Cheng, R., 2000, *Genetic Algorithms and Engineering Optimization*, Wiley, New York

Genest, C., and Favre A. C., 2007, "Everything You Always Wanted to Know about Copula Modeling but Were Afraid to Ask," *Journal of Hydrologic Engineering*, ASCE, Vol. 12, No. 4, July/August

Genest, C., and Rivest, L. P., 1993, "Statistical Inference Procedures for Bivariate Archimedean Copulas," *Journal of the American Statistical Association*, Vol. 88, No. 423, September, pp. 1034–1043

Genest, C., Quessy, J.-F., and Remillard, B., 2006, "Goodness-of-Fit-Procedures for Copula Models Based on the Probability Integral Transformation," *Scandinavian Journal of Statistics*, 33(2), pp. 337–366

Gersh, W., and Yonemoto, J., 1977, "Synthesis of Multivariate Random Vibration Systems: a Two-stage Least Squares ARMA model approach," *Journal of Sound and Vibrations*, Vol. 52, No. 4, pp. 553–565

Gigerenzer, G., 2002, *Reckoning with Risk*, Allen Lane, The Penguin Press, London

Gosling, J. P., Jeremy E. Oakley, E. J., and O'Hagan, A., 2007, "Nonparametric Elicitation for Heavy-Tailed Prior Distributions," *Bayesian Analysis*, Number 4, pp. 693–718

Gosling, J. P., Oakley, J. E., and O'Hagan A., 2007, "Nonparametric Elicitation for Heavy-Tailed Prior Distributions," *Bayesian Analysis*, Number 4, pp. 693–718

Guedes Soares, C., and Moan, T., 1982, "On the Uncertainties Related to the extreme Hydrodynamic Loading of a Cylindrical Pile," *Reliability Theory and Its Applications in Structural and Soil Mechanics*, Martinus Nijhoff Publications, The Hague, The Netherlands, pp. 575–586

Gumbel, E. J., 1958, *Statistics of Extremes*, Columbia University Press, New York

Gunawan, S., and Papalambros, P. Y., 2006, "A Bayesian Approach to Reliability-Based Optimization With Incomplete Information," Journal of Mechanical Design, Vol. 128, Issue 4, pp. 909–918

Haftka, R. T., Rosca, R. and Nikolaidis, E., 2006, "An Approach for Testing Methods for Modeling Uncertainty," *ASME Journal of Mechanical Design*, pp. 1038–1049, Vol. 128, Issue 5, September

Haldar, A., and Mahadevan, S., 2000, "Commonly Used Probability Distributions," *Probability, Reliability and Statistical Methods in Engineering Design*, John Wiley & Sons, New York, Chapter 4, pp. 63–99

Hasofer, A., 1996, "Non-Parametric Estimation of Failure Probabilities," *Mathematical Models for Structural Reliability*, Eds. F. Casciati, and B. Roberts, CRC Press, Boca Raton, FL, 195–226, 1996

Hazelrigg, G. A., 1996, *Systems Engineering: An Approach to Information-Based Design*, Prentice Hall, Upper Saddle River, Chapter 20, pp. 436–438

Henderson, M., and Meyer, M., 2001, "Exploring the Confidence Interval for a Binomial Parameter in a First Course in Statistical Computing," *The American Statistician*, November, Vol. 55, No. 4, pp. 337–344

Hillier, F. S., and Lieberman, G. L., 2010, *Introduction to Operations Research*, 9th Edition, McGraw Hill Higher Education, New York, Chapter 15

Hora, S. C., and von Winterfeldt, D., 1997, Nuclear Waste and Future Societies: A Look into the Deep Future, *Technological Forecasting and Social Change*, 56, pp. 155–170

Howard, A. R., 1988, "Decision Analysis: Practice and Promise," *Management Science*, Vol. 34, 679–695

Howard, A. R., 2004, "Speaking of Decisions: Precise Decision Language," *Decision Analysis*, June; Vol. 1, No. 2, pp. 71–78

Howard, R. A., *Decision Engineering*, Department of Management Science and Engineering, Stanford University

Howard, R. A., 1989, "Microrisks for Medical Decision Analysis," *International Journal of Technology Assessment in Health Care*, 5, pp. 357–370

Hubbard, D. W., 2007, *How to Measure Anything: Finding the Value of Intangibles in Business*, John Wiley and Sons, Hoboken, New Jersey

Jeffrey, R., 2002, *Subjective Probability*, Notes, Princeton University. Can be downloaded from the following link: http://www.princeton.edu/~bayesway

Jeffrey, R., 2004, *Subjective Probability*, Chapter 1, Cambridge University Press

Johnson, N. L., and Kotz, S., 1972, *Distributions in Statistics: Continuous Multivariate Distributions*, John Wiley & Sons, Inc., New York

Johnson, N. L., Kotz, S., and Balakrishnan, N., 1994, "Continuous Univariate Distributions," Vol. 1, 2nd Edition, John Wiley and Sons, New York

Jordaan, I., 2005, *Decisions Under Uncertainty: Probabilistic Analysis for Engineering Decisions*, Cambridge University Press, Cambridge, United Kingdom

Kahneman, D., Slovic, P., and Tversky, A., 1982, *Judgment under Uncertainty, Heuristics and Biases*, Cambridge University Press

Keeney, R. L., and Raiffa, H., 1994, *Decisions with Multiple Objectives*, Cambridge University Press, Cambridge, United Kingdom

Kelton, W. D., Sadowski, R. P., Swets, N. B., 2010, *Simulation with Arena*, McGraw Hill, New York, 5th Edition

Kennedy, P., 2003, *A Guide to Econometrics, Fifth Edition*, MIT Press

Kim, N.-H., and Ramu, P., "Tail Modeling in Reliability-Based Design Optimization for Highly Safe Structural Systems," *47th AIAA/ASME/ASCE/AHS/ASC Structures, Structural Dynamics, and Materials Conference*, May 1–4, 2006, Newport, RI, AIAA 2006-1825

Knight, F. H., 1921, *Risk, Uncertainty, and Profit*, Houghton Mifflin, Boston

Kotz, S., Balakrishnan, N., and Johnson, N. L., 2000, *Distributions in Statistics: Continuous Multivariate Distributions*, John Wiley & Sons, Inc., 2nd Edition, New York

Kuburg, H. E., "Objective Probabilities," *Reasoning*, pp. 902–904

Law, A. M., 2007, *Simulation Modeling and Analysis*, McGraw Hill, New York, 4th Edition

Lewis, K. E., Chen, W., and Schmidt, L. C., 2006, *Decision Making in Engineering Design*, ASME Press, New York

Li, C., and Kiureghian, A., 1993, "Optimal Discretization of Random Fields," *Journal of Engineering Mechanics*, Vol. 119, No. 6, pp. 1136–1154

Lin, Y. K., 1967, *Probabilistic Theory of Structural Dynamics*, McGraw Hill, New York

Loeve, M., 1963, *Probability Theory*, 3rd Edition, Van Nostrand Company, Inc., Princeton, N.J.

Loève, M., 1978, *Probability Theory, Vol. II, 4th ed.*, Graduate Texts in Mathematics, Vol. 46, Springer-Verlag

Lutes, L. D., and Sarkani, S., 2004, *Random Vibrations*, Elsevier

Madsen, H. O., Krenk, S., and Lind, N. C., 1986, *Methods of Structural Safety*, Prentice Hall, Englewood Cliffs, New Jersey

Maes, M. A., and Breitung, K., 1993, "Reliability-Based Tail Estimation," *Proceedings IUTAM Symposium on Probabilistic Structural Mechanics (Advances in Structural Reliability Methods)*, San Antonio, TX, pp. 335–346

Mark, W. D., 1961, "The Inherent Variation in Fatigue Damage Resulting from Random Vibration," Ph. D. Thesis, Department of Mechanical Engineering, MIT

Matheron, G., 1989, *Estimating and Choosing: An Essay of Probability in Practice*, Spriger-Verlag, Berlin

Melchers, R. E., 1999, *Structural Reliability Analysis and Prediction*, 2nd ed., John Wiley & Sons, Chichester, England

Melsa, J. L., and Sage, A. P., 1973, *An Introduction to Probability and Stochastic Processes*, Prentice-Hall, Englewood Cliffs, New Jersey

Miles, J. W., 1954, "On Structural Fatigue under Random Loading," *Journal of Aeronautical Science*, 21, pp. 753–762

Miner, M. A., 1945, "Cumulative Damage in Fatigue," ASME Transactions, Vol. 67, A159-164

Mitchell, C. R., Paulson, A. S., and Beswick, C. A., 1977, "The Effect of Correlated Exponential Service Times on Single-Server Tandem Queues," *Naval Research Logistics Quarterly*, Vol. 24, pp. 95–112

Moore, R. E., 1979, *Methods and Applications of Interval Analysis*, Society for Industrial & Applied Mathematics

Mourelatos, Z. P., and Liang, J., 2007, "A Methodology for Trading-Off Performance and Robustness under Uncertainty," *ASME Journal of Mechanical Design*, Vol. 128 No. 4, 857–873

Muhanna, R. L., and Mullen, R. L., 2001, "Uncertainty in Mechanics Problems – Interval – Based Approach," *Journal of Engineering Mechanics*, ASCE, Vol. 127, No. 6, pp. 557–566

Mullen, R. L., and Muhanna, R. L., 1999a, "Interval – Based Finite Element Methods," Reliable Computing, 5, pp. 97–100

Mullen, R. L., and Muhanna, R. L., 1999b, "Bounds of Structural Response for All Possible Loadings," *Journal of Structural Engineering*, ASCE, Vol. 125, No. 1, pp. 98–106

Nataf, A., 1962, "Determination des Distribution dont les Marges sont Donnees," *Comptes Rendus de l'Academie des Sciences*, 225, pp. 42–43

Nelsen, R. B., 2006, *An Introduction to Copulas*, Springer Series in Statistics, Springer-Verlag, New York

Neumaier, A., 1990, *Interval Methods for Systems of Equations*, Cambridge University Press

Newland, D. E., 1993, *An Introduction to Random Vibrations, Spectral, and Wavelet Analysis, 3rd Edition*, Logman Scientific & Technical, Essex, England

Nigam, N. C., 1983, *Introduction to Random Vibrations*, MIT Press

Nikolaidis, E., 1995, "Setting Affordable Performance Targets for Consumer Products: A Method Based on Fuzzy Logic," *International Journal of Vehicle Design*, Vol. 17, No. 4/5, pp. 384–395

Nikolaidis, E., 2005, "Types of Uncertainty in Design Decision Making," *Engineering Design Reliability Handbook*, CRC Press, Boca Raton, Chapter 8, pp. 8-1 to 8-20

Nikolaidis, E., 2007, "Decision-Based Approach for Reliability Design," *ASME Journal of Mechanical Design*, Vol. 129, Issue No. 5, May, pp. 466–475

Nikolaidis, E., and Kaplan, P., 1992, "Uncertainties in Stress Analyses on Marine Structures, Parts I and II," *International Shipbuilding Progress*, Vol. 39, No. 417, pp. 19–53; Vol. 39, No. 418, pp. 99–133

Nikolaidis, E., Perakis, A. N., and Parsons, M. G., 1987, "Probabilistic Torsional Vibration Analysis of a Marine Diesel Engine Shafting System: The Input-Output Problem," *Journal of Ship Research*, Vol. 31, No. 1, pp. 41–52

NIST/SEMATECH e-Handbook of Statistical Methods, http://www.itl.nist.gov/div898/handbook/, 2010, "Maximum likelihood estimation," section 8.4.1

Noh, Y., Choi, K. K., Lee, I., Gorsich, D., and Lamb, D., 2009, "Reliability-based Design Optimization with Confidence Level under Input Model Uncertainty," *Proceedings of the ASME 2009 IDETC/CIE, paper DETC 2009-86701*, San Diego, California.

Norton, R. L., 1999, *Design of Machinery*, Second Edition, McGraw-Hill, pp. 7–14

O'Hagan, A., and Oakley, J. E., 2004, "Probability is Perfect, but we Can't Elicit it Perfectly," *Reliability Engineering and System Safety*, Vol. 85, Nos. 1–3, July–September, pp. 239–248

O'Hagan A., Buck, C. E., Daneshkhah, A., Eiser J. R., Garthwaite, P. H., Jenkinson, D. J., Oakley, E. J., Rakow, T., 2006, *Uncertain Judgements: Eliciting Experts' Probabilities*, John Wiley & Sons, Chichester, West Sussex, England

O'Hagan, A., 1988, Probability: *Methods and Measurement*, Chapman and Hall, London

Oakley, J. E. and O'Hagan, A., 2007, "Uncertainty in Prior Elicitations: a Nonparametric Approach," *Biometrika*, Vol. 94, No. 2, pp. 427–441

Oberkampf, W. L., and Helton, J. C., 2005, "Evidence Theory for Engineering Applications," *Engineering Design Reliability Handbook*, CRC Press, Boca Raton, Chapter 10, pp. 10-1 to 10-30

Oberkampf, W. L., DeLand, S. M., Rutherford, B. M., Diegert, K. V., Alvin, K. F., 2000, "Estimation of Total Uncertainty in Modeling and Simulation", Sandia Report SAND2000-0824, Albuquerque, NM, April

Oberkampf, W. L., Helton, J. C., Joslyn, C. A., Wojtkiewicz, S. F., and Ferson, S., 2004, "Challenge Problems: Uncertainty in System Response Given Uncertain Parameters," *Reliability Engineering and System Safety*, Vol. 85, No. 1–3, pp. 11–19

Ochi, M. K., 1990, Applied Probability and Stochastic Processes in Engineering and Physical Sciences, John Wiley & Sons, New York, NY

Otto, K. N., and Antonsson, E. K., 1993, "The Method of Imprecision Compared to Utility Theory for Design Selection Problems," *Proceedings of the ASME Design Theory and Methodology Conference*

Pandey, V., and Nikolaidis, E., 2008, "Using Mechanisms Built in a Design Class to Test Methods for Decision under Uncertainty," *Journal of Structure and Infrastructure Engineering*, Vol. 4, No. 1, February, pp. 1–18

Papoulis, A., 1965, *Probability, Random Variables and Stochastic Processes*, McGraw-Hill, New York

Parzen, E., 1970, "On Models for the Probability of Fatigue Failure of a Structure," Time Series Analysis Paper, Holden Day, San Francisco, Chapter 20, pp. 532–550

Pearson, K., "Contributions to the Mathematical Theory of Evolution. II Skew Variations in Homogeneous Material," *Philosophical Transactions of the Royal Society of London, Series A*, 186, 343–414, 1895

Peebles, P. Z., 1987, *Probability, Random Variables, and Random Signal Principles, 2nd Edition*, McGraw-Hill

Raiffa, H., 1961, "Risk, Ambiguity, and the Savage Axioms: Comment," *Swiss Journal of Economics and Statistics*, LXXV, pp. 690–694

Red-Horse, J. R., and Benjamin, A. S., 2004, "A Probabilistic Approach to Uncertainty Quantification with Limited Information," *Reliability Engineering and System Safety*, 85, pp. 183–190

Reduk, S. J., Aughenbaugh, J. M., Bruns, M., and Paredis, C. J. J., 2006, "Eliminating Design Alternatives Based on Imprecise Information," *Reliability and Robust Design in Automotive Engineering, SAE Special Publication-2032*, pp. 73–85

Rice, S. O., 1944, 1945, "Mathematical Analysis of Random Noise," *Bell Systems Technical Journal*, Vol. 23, pp. 282–332; Vol. 24, pp. 46–156

Rice, S. O., 1954, "Mathematical Analysis of Random Noise," in *Selected Papers on Noise and Stochastic Processes*, N., Wax, editor, Dover Publications, New York

Roberts, J. B., and Spanos, P. D., 1990, *Random Vibration and Statistical Linearization*, Wiley, New York

Ruppert, D., 2004, *Statistics in Finance*, Springer

Russo, J. E. and Schoemaker, P. J. H., 2002, Winning Decisions, Doubleday

Saaty, T. L., 1977, "A Scaling method for Priorities in Hierarchical Structures," *Journal of Mathematical Psychology*, 15, pp. 234–281

Samaras, E., Shinozuka, M., and Tsurui, A., 1985, "ARMA Representations of Random Processes," *ASCE Journal of Engineering Mechanics*, Vol. 111, No. 3, pp. 449–461

SAS Institute, 1993, *SAS/ETS User's Guide, Version 6, 2nd Edition*, SAS Institute, Cary, NC

Savage, L. J., 1972, *The Foundation of Statistics*, Chapters 3 and 4, Dover Publications, New York

Schlosser, J., and Paredis, C. J. J., 2007, "Managing Multiple Sources of Uncertainty in Engineering Decision Making," *SAE Special Publication*, SP-2119, pp. 413–425

Schueller, G. I., 1997, "A State-of-the-art Report on Computational Stochastic Mechanics," *Probabilistic Engineering Mechanics*, Vol. 12, No. 4, pp. 197–321

Scott, M. J., 2005, "Utility Methods in Engineering Design," CRC Press, Boca Raton, Chapter 23, pp. 23-1 to 23-14

Shinozuka, M., 1972, "Monte Carlo Solution of Structural Dynamics," *Computers and Structures*, Vol. 2, pp. 855–874

Shinozuka, M., and Yang, J. N., 1971, Peak Structural Response to Nonstationary Random Excitations, *Journal of Sound and Vibration*, Vol. 14, No. 4, pp. 505–517

Singpurwalla, N. D., 1975, "Time Series Analysis and Forecasting of Failure-Rate Processes," *Society of Industrial and Applied Mathematics*, pp. 483–507

Singpurwalla, N. D., 1978, "Estimating Reliability Growth (or Deterioration) using Time Series Analysis," *Naval Research Logistics Quarterly*, Vol. 25, pp. 1–14

Sinn, H.-W., 1980, "A Rehabilitation of the Principle of Insufficient Reason," *The Quarterly Journal of Economics*, Vol. 94, No. 3, May, pp. 493–506

Sklar, A., (1959), Fonctions de repartition a n dimensions et lers margens, *Publications de L'Institut de Statistiques de Universite de Paris* 8: 229–231

Snedecor, G. W. and Cochran, W. G. (1989), *Statistical Methods*, Eighth Edition, Iowa State University Press

Sobek, D. K., Ward, A. C., and Liker, J., 1999, "Toyota's Principles of Set-based Concurrent Engineering," *Sloan Management Review*, Vol. 40, No. 2, pp. 67–83

Socie, D. F., (2003), Seminar notes, "Probabilistic Aspects of Fatigue," URL: http://www.fatiguecalculator.com

Soll, J. B., and Klayman, J., (2004), Overconfidence in Interval Estimates," *Journal of Experimental Psychology: Learning, Memory and Cognition*, 30, pp. 299–314

Soundappan, P., Nikolaidis, E., Haftka, R. T., Grandhi, R., Canfield, R., 2004, "Comparison of Evidence Theory and Bayesian Theory for Uncertainty Modeling," *Reliability Engineering and System Safety*, Vol. 85, Nos. 1–3, July–September, pp. 295–311

Stavropoulos, C., and Fassois, S., 2000, "Non-stationary Functional Series Modeling and Analysis of Hardware Reliability Series: A Comparative Study using Rail Vehicle Inter-failure Times," *Reliability Engineering and System Safety*, Vol. 68, pp. 169–183

Sudret, B., and Der Kiureghian, A., 2000, "Stochastic Finite Element Methods and Reliability. A State-of-the-art Report," *Report UCB/SEMM-2000/08*, University of California, Berkeley, CA

Thielen, D., 1999, *The 12 Simple Secrets of Microsoft*, McGraw-Hill, New York

Thurston, D. L., 1991, A Formal Method for Subjective Design Evaluation with Multiple Attributes, *Research in Engineering Design*, Vol. 3, No. pp. 105–122

Thurston, D. L., 1999, "Real and Perceived Limitations to Decision-Based Design," *Proceedings of ASME 1999 Design Engineering Technical Conference*, DETC99/DTM-8750

Thurston, D. L., 2001, "Real and Misconceived Limitations to Decision Based Design With Utility Analysis", ASME Journal of Mechanical Design Vol. 123, pp. 176–182

Thurston, D. L., 2006A, "Utility Function Fundamentals," *Decision Making in Engineering Design*, Lewis K. E., Chen, W., and Schmidt, L. C., Chapter 3, pp. 15–20, ASME, New York

Thurston, D. L., 2006B, "Multi-attribute Utility Analysis of Conflicting Preferences," *Decision Making in Engineering Design*, Lewis K. E., Chen, W., and Schmidt, L. C., Chapter 12, pp. 125–134, ASME, New York

Thurston, D. L., and Carnahan, J. V., 1992, "Fuzzy Ratings and Utility Analysis in Preliminary Design Evaluation of Multiple Attributes," *AMSE Journal of Mechanical Design*, Vol. 114, No. 4, pp. 648–658

Todinov, M., 2005, *Reliability and Risk Models, Setting Reliability Requirements*, John Wiley & Sons, Chichester, England, Chapter 2, pp. 19–52

Tversky, A., and Kahneman, D., 1974, "Judgments under Uncertainty, Heuristics and Biases," *Science*, 185, pp. 1124–1131

Tversky, A., and Kahneman, D., 1981, "The Framing of Decisions and the Psychology of Choice," *Science*, 9, pp. 309–339

Vanderplaats, G. N., 2007, *Numerical Optimization Techniques for Engineering Design*, Third Edition, Vanderplaats Research and Development, Inc

Vesely, W. E., Goldberg, F. F., Roberts, N. H., and Haasl, D. F., 1981, *Fault Tree Handbook*, Nuclear regulatory Commission, Washington, D. C.

Von Neumann, J., and Morgenstern, O., 1947, *Theory of Games and Economic Behavior*, Princeton University Press, Second Edition, Princeton

Walker, G., 1931, "On Periodicity in Series of Related Terms," *Proceedings of the Royal Society of London, Series A*, Vol. 131, pp. 518–532

Walley, P., 1991, *Statistical Reasoning with Imprecise Probabilities*, Chapman and Hall

Wang, W. and Wells, M. T., 2000, "Model Selection and Semiparametric Inference for Bivariate Failure-Time Data," *Journal of American Statistical Association*, 95(1) pp. 62–76

Welch, P. D., 1967, "The Use of Fast Fourier Transform for the Estimation of Power Spectra: A Method Based on Time Averaging over Short, Modified Periodograms," *IEEE Trans. Audio Electroacoustics, Vol. AU-15*, pp. 70–73

Wilson, E. B., (1927), "Probable Inference, and the law of Succession and Statistical Inference," *Journal of the American Statistical Association*, Vol. 22, pp. 209–212

Winkler, R. L., 1967, The Assessment of Prior Distributions in Bayesian Analysis, *Journal of American Statistical Association*, 62, pp. 776–880

Winman, A., Hansson, P. and Juslin, P. 2004, "Subjective Probability Intervals: How to Reduce Overconfidence by Interval Evaluation," *Journal of Experimental Psychology: Learning, Memory and Cognition*, 30, pp. 1167–1175

Wirsching, P. H., and Chen, Y.-N., 1988, "Considerations of Probability-Based Fatigue Design for Marine Structures," *Marine Structures*, 1, pp. 23–45

Wirsching, P. H., and Light, M. C., 1980, "Fatigue under Wide Band Random Stresses," *Journal of Structural Design*, ASCE, Vol. 106, No. ST7

Wirsching, P. H., Paez, T. L., and Ortiz, H., 1995, *Random Vibrations: Theory and Practice*, John Wiley and Sons, Inc., New York, NY

Wolfers J., and Zitzewitz, E., 2004, "Prediction Markets," *Journal of Economic Perspectives*, Volume 18, Number 2, pp. 107–126

Woloshin, S., Schwartz, L. M., Byram, S., Fischoff, B. and Welch, H. G., 2000, "A New Scale for Assessing Perceptions of Chance: A Validation Study," *Medical Decision Making*, 20, pp. 298–307

Wood, K. L., Antonsson, E. K. and Beck, J. L., 1990, "Representing Imprecision in Engineering Design: Comparing Fuzzy and Probability Calculus," *Research in Engineering Design*, 1, pp. 187–203

Wood, K. L., Antonsson, E. K., 1990, "Modeling Imprecision and Uncertainty in Preliminary Engineering Design," *Mechanisms and Machine Theory*, Vol. 25, No. 3, pp. 305–324

Wu, J. S., Apostolakis, G. E., and Okrent, D., 1990, "Uncertainties in System Analysis: Probabilistic Versus Nonprobabilistic Theories," *Reliability Engineering and System Safety*, 30, pp. 163–181

Yan, Y., 2006, Enjoy the Joy of Copulas, (download from the following link: www.stat.uiowa.edu/techrep/tr365.pdf)

Yang, J. N., and Liu, S. C., 1980, "Statistical Interpretation and Application of Response Spectra," *Proceedings 7th WCEE*, Vol. 6, pp. 657–664

Yang, J. N., and Shinozuka, M., 1971, "On the First Excursion Failure Probability in Stationary Narrow-Band Random Vibration," *ASME Journal of Applied Mechanics*, Vol. 38, No. 4, pp. 1017–1022

Yang, J. N., and Shinozuka, M., 1972, "On the First Excursion Failure Probability in Stationary Narrow-Band Random Vibration-II," *ASME Journal of Applied Mechanics*, Vol. 39, No. 4, pp. 733–738

Yule, G. U., 1927, "On a Method of Investigating Periodicities in Disturbed Series, with Special Reference to Wolfer's Sunspot Numbers" *Philosophical Transactions of the Royal Society of London, Series A*, Vol. 226, pp. 267–298

Zhang, G., Nikolaidis, E., and Mourelatos, Z. P., 2009, "An Efficient Re-Analysis Methodology for Probabilistic Vibration of Large-Scale Structures," *ASME Journal of Mechanical Design*, May, pp. 051007-1-051007-1

Zhou, J., Mourelatos, Z. P., and Ellis, C., 2008, "Design Under Uncertainty Using a Combination of Evidence Theory and a Bayesian Approach," *SAE Special Publication*, SP-2170, pp. 157–170

Structures and Infrastructures Series

Book Series Editor: Dan M. Frangopol

ISSN: 1747–7735

Publisher: CRC/Balkema, Taylor & Francis Group

1. Structural Design Optimization Considering Uncertainties
    Editors: Yiannis Tsompanakis, Nikos D. Lagaros & Manolis Papadrakakis
    2008
    ISBN: 978-0-415-45260-1 (Hb)

2. Computational Structural Dynamics and Earthquake Engineering
    Editors: Manolis Papadrakakis, Dimos C. Charmpis,
    Nikos D. Lagaros & Yiannis Tsompanakis
    2008
    ISBN: 978-0-415-45261-8 (Hb)

3. Computational Analysis of Randomness in Structural Mechanics
    Christian Bucher
    2009
    ISBN: 978-0-415-40354-2 (Hb)

4. Frontier Technologies for Infrastructures Engineering
    Editors: Shi-Shuenn Chen & Alfredo H-S. Ang
    2009
    ISBN: 978-0-415-49875-3 (Hb)

5. Damage Models and Algorithms for Assessment of Structures
    under Operating Conditions
    Siu-Seong Law & Xin-Qun Zhu
    ISBN: 978-0-415-42195-9 (Hb)

6. Structural Identification and Damage Detection using Genetic Algorithms
    Chan Ghee Koh & Michael John Perry
    ISBN: 978-0-415-46102-3 (Hb)

7. Design Decisions under Uncertainty with Limited Information
    Efstratios Nikolaidis, Zissimos P. Mourelatos & Vijitashwa Pandey
    ISBN: 978-0-415-49247-8 (Hb)